INDUSTRY 4.0 – SHAPING THE FUTURE OF THE DIGITAL WORLD

PROCEEDINGS OF THE 2ND INTERNATIONAL CONFERENCE ON SUSTAINABLE SMART MANUFACTURING (S2M 2019), 9–11 APRIL 2019, MANCHESTER, UK

Industry 4.0 – Shaping The Future of The Digital World

Edited by

Paulo Jorge da Silva Bartolo

The University of Manchester, UK

Fernando Moreira da Silva

CIAUD, Faculty of Architecture, University of Lisbon, Portugal

Shaden Jaradat

The University of Manchester, UK

Helena Bartolo

School of Technology and Management, Polytechnic Institute of Leiria, Portugal

CRC Press
Taylor & Francis Group
Boca Raton London New York

CRC Press is an imprint of the
Taylor & Francis Group, an **informa** business
A BALKEMA BOOK

CRC Press/Balkema is an imprint of the Taylor & Francis Group, an informa business

Typeset by Integra Software Services Pvt. Ltd., Pondicherry, India

Library of Congress Cataloging-in-Publication Data

Applied for

Published by: CRC Press/Balkema
Schipholweg 107C, 2316XC Leiden, The Netherlands
e-mail: Pub.NL@taylorandfrancis.com
www.routledge.com – www.taylorandfrancis.com

ISBN: 978-0-367-42272-1 (Hbk)
ISBN: 978-0-367-82308-5 (eBook)
DOI: 10.1201/9780367823085
https://doi.org/10.1201/9780367823085

Industry 4.0 – Shaping The Future of The Digital World – da Silva Bartolo et al. (eds)
© 2021 Taylor & Francis Group, London, ISBN 978-0-367-42272-1

Table of contents

Design education

Urban spaces

Industry 4.0 – Shaping The Future of The Digital World – da Silva Bartolo et al. (eds)
© 2021 Taylor & Francis Group, London, ISBN 978-0-367-42272-1

Preface

"Industry 4.0 – Shaping the future of the digital world" contains papers presented at the 2nd International Conference on Sustainable Smart Manufacturing (S2M) held in Manchester, UK, between 9 – 11 April 2019. The conference was organised by The University of Manchester (UK) in collaboration with the University of Lisbon and the Polytechnic Institute of Leiria (Portugal).

This Conference included the themes of: Smart Manufacturing (*e.g.* cyber-physical systems, machine learning, reconfigurable assembly systems); Additive Manufacturing and Virtual Prototyping; Materials for Healthcare Applications and Circular Economy; Design Education; and Urban Spaces. It was part of the "Industry 4.0 Summit & Expo" which also included policy forums, strategy-focussed workshops and an exhibition for key industrial players, celebrating and demonstrating Manchester's manufacturing heritage and innovation.

Manchester was the birthplace of the first industrial revolution, and home to the fathers of the nuclear age (Ernest Rutherford) and modern computer (Alan Turing). It was where Andre Geim and Konstantin Novoselov discovered graphene. It is also where world-leading and pioneering work on industrial digitization is conducted.

Designed to be a major forum for stimulating, cross-disciplinary discussions on the most recent innovations, trends, and challenges on a variety of topics, S2M brought together more than 120 researchers from over 20 countries.

I am deeply grateful to authors, participants, reviewers, Scientific Committee, session Chairs, student helpers and administrative assistants, for contributing to the success of this conference.

Paulo Bartolo
Conference Chair
Professor of Advanced Manufacturing
The University of Manchester

Industry 4.0 – Shaping The Future of The Digital World – da Silva Bartolo et al. (eds)
© 2021 Taylor & Francis Group, London, ISBN 978-0-367-42272-1

Committee members

CONFERENCE CHAIRS

Paulo Jorge da Silva Bartolo
The University of Manchester, UK

Fernando Moreira da Silva
CIAUD, Faculty of Architecture, University of Lisbon, Portugal

Helena Bartolo
School of Technology and Management, Polytechnic Institute of Leiria, Portugal

CONFERENCE MANAGER

Shaden Jaradat
The University of Manchester, UK

ORGANISING COMMITTEE

Shaden Jaradat
The University of Manchester, UK

Boyang Huang
The University of Manchester, UK

Henrique Almeida
Polytechnic Institute of Leiria, Portugal

Ana Lemos
Polytechnic Institute of Leiria, Portugal

Flávio Craveiro
University of Lisbon, Portugal

Industry 4.0 – Shaping The Future of The Digital World – da Silva Bartolo et al. (eds)
© 2021 Taylor & Francis Group, London, ISBN 978-0-367-42272-1

Keynote speaker

Gideon Levy
Technology Turn Around, Switzerland

Robert Kirkbride
Parsons' School of Constructed Environments, USA

José Rui De Carvalho Mendes Marcelino
Almadesign, Portugal

Ajay Malshe
University of Arkansas, USA

Axel Demmer
Fraunhofer Institute for Production Technology IPT, Germany

Andres Harris
Foster & Partners, UK

Richard Bibb
Loughborough Design School, Loughborough University, UK

James Evans
Manchester Urban Institute, The University of Manchester, UK

Tan Ming Jen
Nanyang Technological University, Singapore

Ian Gibson
University of Twente, The Netherlands

Industry 4.0 – Shaping The Future of The Digital World – da Silva Bartolo et al. (eds)
© 2021 Taylor & Francis Group, London, ISBN 978-0-367-42272-1

International scientific committee

Konrad Wegener, *ETH Zurich, Switzerland*
Luca Iuliano, *Politecnico di Torino, Italy*
Luigi Galantucci, *Politecnico di Bari, Italy*
Luis Paschoarelli, *UNESP, São Paulo, Brazil*
Marco Casini, *University of Roma 1, Italy*
Maria da Graça Carvalho, *University of Lisbon, Portugal*
Mario Monzón, *Las Palmas University, Spain*
Mattheos Santamouris, *University of Athens, Greece*
Markus Keane, *National University of Ireland, Galway, Ireland*
Miroslaw Skibniewski, *University of Maryland, USA*
Neri Oxman, *MIT, USA*
Olaf Diegel, *Lund University, Sweden*
Paula Trigueiros, *Universidade of Minho, Portugal*
Paula Vilarinho, *University of Aveiro, Portugal*
Paulo Ferreira, *Iberian Institute of Nanotechnology, Portugal*
Paulo Lisboa, *Liverpool John Moores University, UK*
Paulo Lourenço, *University of Minho, Portugal*
Paulo Vila Real, *University of Aveiro, Portugal*
Ricardo Gonçalves, *New University, Portugal*
Rita Almendra, *University of Lisbon, Portugal*
Rita Moura, *PTPC, Portugal*
Rita Newton, *University of Salford, UK*
Sandra Lucas, *Eindhoven University of Technology, Netherlands*
Steve Evans, *University of Cambridge, UK*
Terry Wohlers, *Wohlers Associates, USA*
Theo Salet, *Eindhoven University of Technology, Netherlands*
Tugrul Ozel, *Rutgers University, USA*
Victor Ferreira, *Habitat Cluster, Portugal*
Wilfried Sihn, *TU Wien, Austria*

Smart manufacturing

Industry 4.0 – Shaping The Future of The Digital World – da Silva Bartolo et al. (eds)
© 2021 Taylor & Francis Group, London, ISBN 978-0-367-42272-1

Towards a digital revolution in the UK apparel manufacturing: An industry 4.0 perspective

I.W.R. Taifa, S.G. Hayes & I. Duncan Stalker
Department of Materials, Faculty of Science and Engineering, The University of Manchester, Manchester, UK

ABSTRACT: The study explores the potential for developing a virtual distributed manufacturing network for UK apparel manufacturing using a simulation approach for small and medium-sized enterprises (SMEs). The digitalisation is on enabling an equitable order allocation system (sharing or dividing) amongst the SMEs to ensure the apparel manufacturers' survival. The solution provides a potential model for proper production planning and scheduling service in developing strong networks. For the scheduling service, the SMEs are required to upload available capacity to the private cloud server, retailers load orders to cloud services, and the virtual factory scheduler service (VFSS) allocates and optimises work across factories. This study gives an overview of using mixed methods. Arena® software, together with the synthetic data, were utilised to simulate a model for a single retailer to multiple manufacturers. Thus, through the simulation approach as a component of Industry 4.0, it is possible to perform order allocation equitably.

1 INTRODUCTION

1.1 *Research background*

The UK textile industry was the leading industry for the first industrial revolution (late 18th century). In 1784, the first mechanical weaving loom was invented. Since then, much has been discovered. People are now applying the fourth industrial revolution concepts - Industry 4.0 – which began in 2011. Since the 1970s to the mid-2000s, the UK textiles and apparel (T&A) experienced a difficult period where many job opportunities and the domestic production vanished, and the country witnessed high importation of apparel products (Jones & Hayes 2004), mainly from China, Turkey, Bangladesh (Turker & Altuntas 2014), etc. The T&A sector needs revitalisation to benefit much from the digitalisation, which has both tangible and intangible benefits to manufacturers and retailers (Küsters, Praß, & Gloy 2017). In the business context, digitalisation is the use of information technology and digital media to develop a business process. Such a transitioning process in this sector can lead to Textile 4.0 (Chen & Xing 2015). The transformation can be either for all phases, i.e. design, manufacturing, distribution and sales, or to the ordering process. Thus, manufacturers need orders from retailers, both domestically and globally, to increase profits.

The UK T&A sector has both large manufacturers and SMEs. For large manufacturers, there is more possibility of securing orders easily from British retailers on their own, and possibly, they can run production without collaborating with others. Yet, for the SMEs, many of them are of low capacity that cannot enable them to secure orders from retailers on their own. This requires SMEs to work as an extended enterprise (EE) to obtain orders. To do so, SMEs need a virtual distributed manufacturing network that can assist in competing profitably with the foreign manufacturers who have advantages of highly containerised distribution networks, cheap land or renting cost, cheap raw materials, and cheap labours compared to the UK. Thus, the management for the UK SMEs needs to offer a quick response to highly volatile customers' apparel orders in season.

1.2 *Research motivation*

The UK T&A sector requires a quick decision for it to remain competitive in the market. The nature of its business outlook needs that both manufacturers and retailers must collaborate to tackle the routine nature of intense competition and continuously evolving scenario (Lectra 2018). Lectra (2018) further states that this sector needs a fully digitalised system. The digitalisation is needed for the conceiving process, designing and transfer process before enforcing for more technical development. Lectra (2018) states further that there is a need for having well-interlinked supply chain partners. It is thus correct that "fashion is now rapidly becoming a predominantly digital industry – one where huge volumes of data, digital collaboration, online social interaction, digital marketing, and e-commerce come together to create and sell a physical product to a digital-native demographic" (Lectra 2018, p.3). Still, the sustainability of the T&A sector will depend on the members' willingness and commitment as digitalisation needs collaboration in an extended enterprise. So, there is a need for having a bigger picture of effecting digitalisation in apparel manufacturing industries as a means of responding to the fourth industrial revolution concepts.

1.3 *Research main objective*

This study explores the potential for developing a virtual distributed manufacturing network (VDMN) for UK apparel manufacturing using a simulation approach. The intention is to enable SMEs to secure orders from British retailers that on their own would not be able to secure enough orders. The digitalisation is on enabling an equitable order allocation (sharing) amongst the SMEs to ensure the apparel manufacturers' survival. Figure 1 is a schematic diagram of the scheduling service. Factories are required to upload available capacities to the private cloud server, and retailers load orders to cloud service while the VFSS allocates and optimises the work across factories. Note that LSP is a logistics service provider that transports partial garments between factories, and MRP is the manufacturing resource planning.

2 BACKGROUND

2.1 *Related studies*

Order allocation is not a new research area. Hardly any study accomplished order allocation in T&A industries regarding Industry 4.0. In other industries, including automobile, there is much information about Industry 4.0 on transforming their systems. Some T&A firms use information technologies to transform their systems, but hardly any study explored the potential for a VDMN for order allocation equitably among different manufacturers working as extended enterprises. The differences between this research and the previous studies are based on the used methods, considered factors in allocating orders and an Industry 4.0 viewpoint. For example, Renna & Perrone (2015) studied the order allocation within an industry cluster consisting of numerous suppliers which give original equipment manufacturers and used a JAVA package to simulate. Their study was for the suppliers within a cluster grounded on the utilisation evaluation.

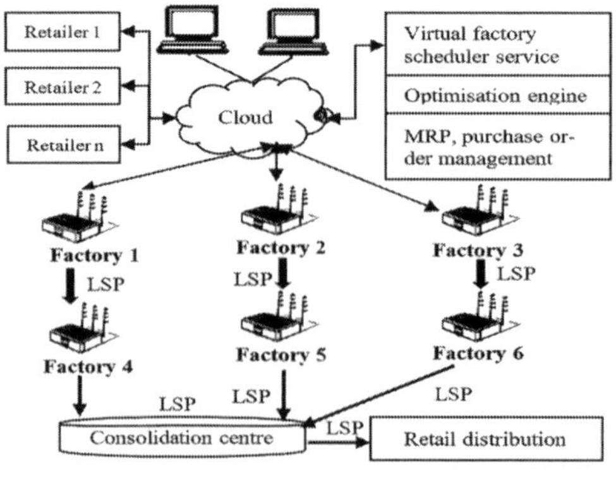

Figure 1. A schematic diagram of the scheduling service.

2.2 *Industrial 4.0 overview*

Digitalisation is the act of deploying digital technologies to transform the available business models and come up with new revenue and value-producing opportunities (Gartner 2018). Digitalisation is also called Industry 4.0 (Hounshell 2018; Taifa & Vhora 2019). There are nine technologies for Industry 4.0 that are vital in transforming industrial production (Gerbert et al. 2015). Generally, in Germany, such technologies form Industry 4.0, and they include simulation, cloud, additive manufacturing, augmented reality, cybersecurity, industrial internet of things, autonomous robots, vertical and horizontal system integration, and big data and analytics (Gerbert et al. 2015). As per this study, the key technologies considered include simulation, which is integrated with the vertical and horizontal system. This is because the study involves manufacturers and retailers as part of the vertical and horizontal systems integration, then uses simulation to drive decision making. Hence, this works squarely in Industry 4.0. Industry 4.0 has an objective of transforming "production by fully integrating the production flow, and consequently, leading to a change in the relationship between suppliers, producers and customers, improving the efficiency and the product customisation" (Oliveira & Sommer 2017, p.3). Unlike the earlier three industrial revolutions (18th to 20th centuries), Industry 4.0 is more automated, decentralised, and controlled interdependently (Umachandran et al. 2018). Hounshell's (2018) view is that digitalisation must change the present and future business viewpoints by enabling technology that achieves four aspects, namely: (a) creation of the decentralised decision making support; (b) to create automated assistance and support; (c) to create information transparency, and (d) interoperability - an ability of a computer systems to use and exchange information acceptably.

3 RESEARCH METHODOLOGY

3.1 *Methodology*

Research approaches follow under either qualitative, quantitative or mixed approach. This research deals with a mixed method. The qualitative part comprises of interviews, questionnaires, documentary review and observations while the quantitative part can be sub-divided into the experimental, inferential, and simulation approach. Thus, this research deployed the computer simulation approach.

3.1.1 *Simulation approach*

This study considered modelling and simulation phases, as shown in Figure 2 and Table 1.

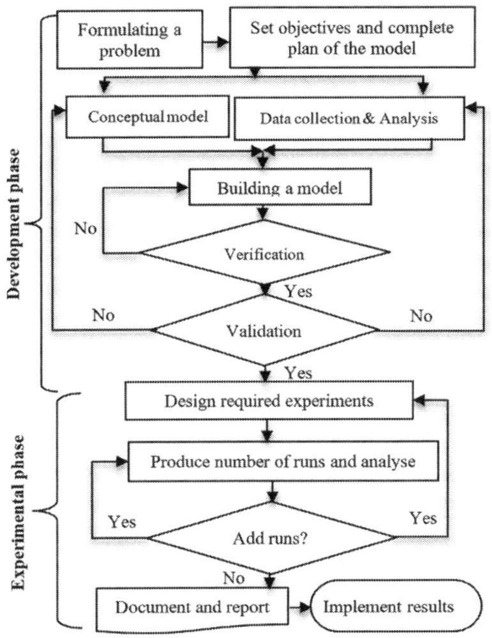

Figure 2. Phases for modelling and simulation (Banks 1998).

Table 1. The fundamental phases for modelling and simulation.

Phase	The explanation for the phases
Problem formulation	Define the order allocation problem together with all the required objectives.
Model conceptualisation	Abstracting a model that comprises elements which help to understand the model to be developed.
Data collection	Determine, stipulate, and gather data to support the model development process.
Building a model	The conceptualised model can be captured using relevant simulation software.
Verification	Evaluate the model's transformational accuracy and check whether a model is created in the right way.
Validation	Compare the developed model with reality. Perform an evaluation of the model representational or behavioural accuracy and checks whether the right model is developed. Validation is for a model.
Analysis	Scrutinise the simulation outputs to make interpretations and recommendations.
Documentation	Make supportive information and then implement the obtained optimal or potential solution.

3.2 *Problem formulation*

Manufacturers receive apparel orders from their customers (retailers). The frequency of order arrival varies from one factory to the other. Several decision criteria are required to rank each manufacturer. This study initially considered five criteria - quality, delivery, cost or price, capacity, capability - in allocating orders equitably to the SMEs. Equitable order allocation amongst the SMEs was categorised into four major aspects. Table 2 summarises the four scenarios.

a) When a single manufacturer (SM) receives bulk orders (BO) from a single retailer (SR); here, a traditional approach can be used to process the received apparel orders.
b) When SM receives BO from multiple retailers (MRs) - such a scenario is considered as a multi-sourcing process.
c) When multiple manufacturers (MMs) (factories) receive BO from SR, in this case, manufacturers need to prioritise their capacity.
d) When MMs receive BO from MRs, the Industry 4.0 (digitalisation) concept can be used.

3.3 *Conceptual model*

Figure 3 depicts a conceptual model for scenario 'D' in Table 2. The inputs (see Figure 3) include logistics services, raw materials, capital, tools and spare parts, workforce, among others. The ordering process can be executed either in an extended enterprise way using EDI or through a traditional approach where a single retailer order directly from the manufacturer.

Table 2. Possible ordering categories for the UK textiles and apparel sector when placing orders in bulk.

Scenario	Apparel retailer	Apparel manufacturer	Decision criteria (points)	Code
A	Single	Single	Quality, delivery, cost or price, capacity, capability	SRSM
B	Single	Multiple	Quality, delivery, cost or price, capacity, capability	SRMM
C	Multiple	Single	Quality, delivery, cost or price, capacity, capability	MRSM
D	Multiple	Multiple	Quality, delivery, cost or price, capacity, capability	MRMM

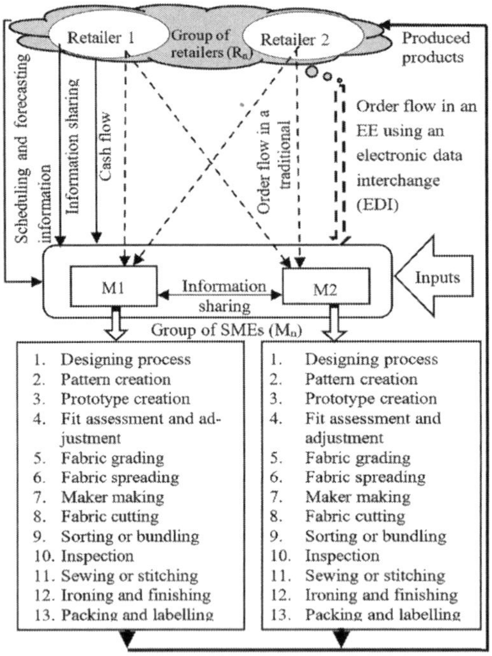

Figure 3. Conceptual model: multiple retailers to multiple manufacturers.

3.4 *Simulation software and collected data*

Arena® software simulated the system with data such as the interarrival time of orders, order quantities, operation time, and probability distribution function for each operation, operation (processing) time, number of machines, retailers and manufacturers.

4 HYPOTHETICAL SCENARIO

The SRMM model was developed. The manufacturers - Mx_1 and Mx_2 - receive orders from a retailer (Rb_1). The interarrival time of Rb_1 follows a Poisson distribution while the retailers' demands follow POIS (120) on the presumed basis. Retailer's arrival rate is POIS (30). The 'create' module named 'order arrival_Rb_1' was used to receive orders. Rb_1 uses EDI to place orders. The first order creation was set to 0 (i.e. the production begins without any order)., and the units are days. The maximum arrivals were set to 360. The received orders undergo initial screening at the initial 'decide' module named 'orders to be accepted.' At this stage, 97% of the received orders are assumed to be accepted, and 3% are rejected. The next stage is an order distribution (allocation) process amongst Mx_1 and Mx_2. Order allocation was executed in the second 'decide' module named 'order distribution' using the decision criteria in Table 2 (Taifa et al. 2020). Mx_1 and Mx_2 secured 48.81% and 51.19% of the total orders, respectively. Mx_1 starts to process the received apparel orders. Of the received orders, there are two product categories - Polo shirt and trousers - with the distribution of 65% and 35%, respectively.

Similarly, Mx_2 distributes orders using the 'decide' module named 'apparel-types-distribution at Mx_2'. Of the received orders, there are also two different product categories - Polo shirt and trousers - with the distribution of 55% and 45%, respectively. This distribution was executed using the 'decide' module named 'apparel-types-distribution at Mx_2.' Then, Mx_1 and Mx_2, and Rb_1 discuss the received orders to finalise all the formalities. Such processes were set using the 'enter' module named 'enter manufacturers.' Under this module, the order undergoes a 'delay' of NORM (5, 2) (in days) to allow completion of the designing process, pattern creation, prototype creation, assessment and adjustment, and the fabric grading. The manufacturer set up production processes using the 'enter' module named 'production planning' with a delay of a NORM (1, 1) (in days).

Subsequently, the accepted orders wait in a queue so as an operator can assign the sequence of each product type. The processing time for this stage follows NORM (2, 0.5) and NORM (1, 0.5) (in days) for Mx_1 and Mx_2, respectively. Furthermore, the finalised orders are then processed at four different workstations under the 'process' module, namely "cutting process workstation Mx_1", "sewing process workstation Mx_1", "inspection process workstation Mx_1", and "finishing process workstation Mx_1." Mx_2 also has the same types of workstations. Each product category was assigned its production time in Table 3, whereby O&M is an operator and machines and RS represent the needed resources. Under the inspection workstations, it was assumed that 5% and 10% of the products require a repair (rework) at Mx_1 and Mx_2, respectively. For the 'assign' module, the apparel products were assigned the picture type. For Mx_1, the shirt and the trouser were assigned the blue and the green balls, respectively, while the red and the yellow balls were assigned to the shirt and the trouser for Mx_2, respectively. Other multiple assignments were made about the sequencing processes.

For the four workstations, it was assumed to be having two different machines - new and old machines. The old machine is 10% slower than the new one. Also, the transfer times between all the four workstations follow an exponential distribution with EXPO (10) minutes. The batch module was used to group 15 completed products to enable the packaging process using the 'process' module - EXPO (10) minutes. The batched products are then split using the 'separate' module to allow proper counting of the outputs at the 'dispose module' named 'order to leave

Table 3. Manufacturing data for the SRMM Model.

Operations	Product type for Mx_1	Delay type (in minutes) for Mx_1	RS	Operations	Product type for Mx_2	Delay type (in minutes) for Mx_2	RS
Cutting Mx_1	Polo shirt	Norm (35,3)	O&M	Cutting	Polo shirt	Norm (30,3)	O&M
	Trouser	Norm (15,3)	O&M	Mx_2	Trouser	Norm (17,3)	O&M
Assembly	Polo shirt	Norm (20,3)	O&M	Assembly	Polo shirt	Norm (18,3)	O&M
Mx_1	Trouser	Norm (35,3)	O&M	Mx_2	Trouser	Norm (32,3)	O&M
Inspection	Polo shirt	Norm (4,1)	O&M	Inspection	Polo shirt	Norm (3,1)	O&M
Mx_1	Trouser	Norm (4,1)	O&M	Mx_2	Trouser	Norm (3,1)	O&M
Finishing	Polo shirt	Norm (11,3)	O&M	Finishing	Polo shirt	Norm (9,3)	O&M
Mx_1	Trouser	Norm (11,3)	O&M	Mx_2	Trouser	Norm (8,3)	O&M

the factory.' All arriving products at the batch module were positioned in a queue while waiting for the necessary batch size of 15 products. Once all the tasks are completed, Mx_1 and Mx_2 collect the manufactured products at the distribution centre (DC) using logistics services for an average of 3 days. From DC to the retailer, the 'route module' named 'route for transporting products' was used and followed NORM (7, 2) (in days). The 'hold' module used to accumulate enough products before transporting them to Rb_1.

Some of the considered assumptions in developing the SRMM model (Figure 5) include: the raw materials are plentily available; accepted orders cannot leave the system before completing all processes; each machine has one operator; there is no disruption factor to interfere the production processes.

4.1 *Run length and number of replications*

Manufacturers are assumed to operate for two shifts of 8 hours each, making a total of 16 hours a day. In a week, they work for six days: thus, making 4992 hours a year. The simulation model starts without any received order - all machines and operators are idle. The actual simulation results were gathered after the warm-up period (WUP) to allow the model to reach a steady-state condition. The model was run for 60 days (2 months) to determine the reasonable WUP, which was found to be 960 hours. So, the total simulation runtime was 5952 hours per year. Equation 1 (half-width method) computed the number of replications (N_m) (Kamrani et al. 2014).

$$N_m = \left(\frac{STD(m) \times t_{m-1,1-\frac{\alpha}{2}}}{\bar{X}_m \times \in}\right)^2 \quad (1)$$

where m is an initial number of replications which was assumed to be 5 and its waiting times in hours; the allowable percentage error (ϵ); the sample mean ($\bar{X}_m$); standard deviation (STD_m); and $t_{m-1,1-\frac{\alpha}{2}}$ = student's t-distribution with m - 1 degree of freedom. A confidence interval (significance level) of 90% (α = 0.1) was considered. ϵ of 10% with $t_{4,\ 0.95}$ gives a p-value of 2.132. N_m was found to be 22 replications. Model verification was performed while the validation processes were not performed.

5 RESULTS

The total apparel orders processed are 963. Mx_1 and Mx_2 secured and completely processed an average of 644 and 319 orders, respectively (Figure 4). Still, this shows the simple way of allocating orders equitably. Other results are not reported in this paper. The model can also generate results on the key performance indicators for the four categories (Table 2), such as order distribution, manufacturing (production) throughput time, resources utilisation, production scheduling, lead times, and lateness, earliness and tardiness.

6 CONCLUSION

The developed SRMM model, together with other models to be developed, will enable smooth retailing between the retailers and the SMEs. Such models are also expected to be vital support in creating an alignment of the multi-sites production processes to enable a virtual factory. The virtual factory will be established with enough capacity to service the retail demands in an agile manner. This research will

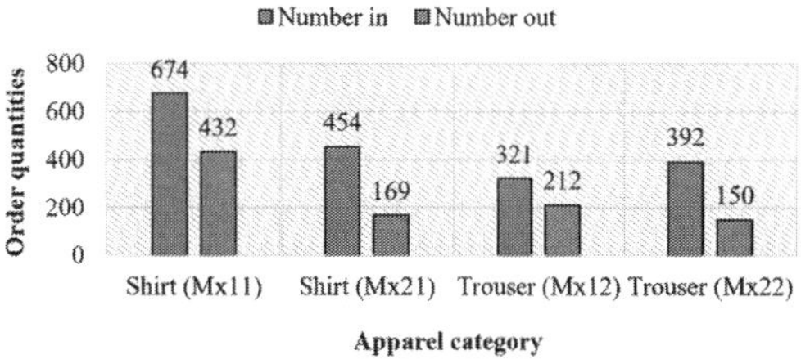

Figure 4. Order entities processed from the Arena® software.

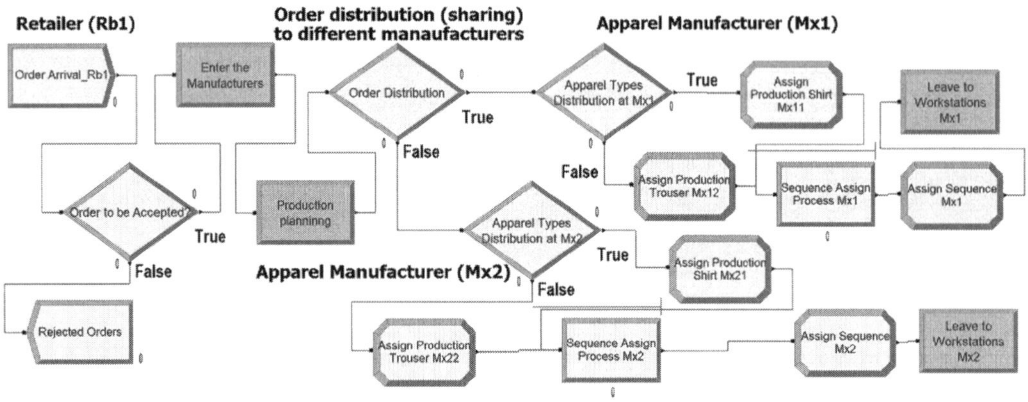

Figure 5. Part of the SRMM model from the Arena® software.

further be integrated with other systems for capturing vital information for allocating orders from retailers to SMEs working as an EE. Thus, the study has explored the potential for a VDMN in the UK apparel manufacturing using a simulation approach.

REFERENCES

Banks, J. 1998. *Handbook of Simulation: Principles, Methodology, Advances, Applications, and Practice*. New York: John Wiley & Sons, Inc.

Chen, Z. & Xing, M. 2015. Upgrading of textile manufacturing based on Industry 4.0. *Advanced Design and Manufacturing Engineering; Proc. of the 5th intern. conf.*, Shenzhen, China, 19-20 *September* 2015. Atlantis Press.

Gartner. 2018. Digitalization. *Gartner*. Available at: http://www.gartner.com/it-glossary/business-intelligence-bi/ [accessed 31 Oct 2018].

Gerbert, P., Lorenz, M., Rüßmann, M., Waldner, M., Justus, J., Engel, P. & Harnisch, M. 2015. Industry 4.0: The future of productivity and growth in manufacturing industries. *The Boston Consulting Group*. Available at: https://www.bcg.com/publications/2015/engineered_products_project_business_industry_4_future_productivity_growth_manufacturing_industries.aspx [accessed 30 Oct 2018].

Hounshell, L. 2018. The industrial internet of things, digitalization and the future of business. *Forbes Technology Council*. Available at: https://www.forbes.com/sites/forbestechcouncil/2018/07/30/the-industrial-internet-of-things-digitalization-and-the-future-of-business/ [accessed 31 Oct 2018].

Jones, R.M. & Hayes, S.G. 2004. The UK clothing industry: Extinction or evolution? *Journal of Fashion Marketing and Management: An International Journal* 8(3): 262–278.

Kamrani, M., Abadi, S. & Golroudbary, S. 2014. Traffic simulation of two adjacent unsignalized T-junctions during rush hours using Arena software. *Simulation modelling practice and theory* 49: 167–179.

Küsters, D., Praß, N. & Gloy, Y.S. 2017. Textile Learning Factory 4.0 – Preparing Germany's Textile Industry for the Digital Future. *Procedia Manufacturing* 9: 214–221.

Lectra, 2018. *The Digitalization of Fashion*. Available at: https://lectrafashionplm.lectra.com/sites/lectrafashionplm.lectra.com/files/lectra_white_paper_digitalize_en_pap.pdf [accessed 2 November 2018].

Oliveira, P. & Sommer, L. 2017. Globalization and digitalization as challenges for a professional career in manufacturing industries - differences in awareness and knowledge of students from Brazil and Germany. *Education Sciences* 7(55): 1–13.

Renna, P. & Perrone, G. 2015. Order allocation in a multiple suppliers-manufacturers environment within a dynamic cluster. *International Journal of Advanced Manufacturing Technology* 80(1-4): 171–182.

Taifa, I.W.R., Hayes, S.G. & Stalker, I.D. (2020). Development of the critical success decision criteria for an equitable order sharing in an extended enterprise. *The TQM Journal*. Available at:https://doi.org/10.1108/TQM-05-2019-0138. [accessed 11 March 2020].

Taifa, I.W.R. & Vhora, T.N. (2019). Cycle time reduction for productivity improvement in the manufacturing industry. *Journal of Industrial Engineering and Management Studies* 6(2): 147–164.

Turker, D. & Altuntas, C. 2014. Sustainable supply chain management in the fast fashion industry: An analysis of corporate reports. *European Management Journal* 32(5): 837–849.

Umachandran, K., Jurčić, I., Corte, V. Della & Ferdinand-James, D.S. 2018. Industry 4.0: The new industrial revolution. *Big data analytics for smart and connected cities*: 138–156. USA: IGI Global.

Industry 4.0 – Shaping The Future of The Digital World – da Silva Bartolo et al. (eds)
 ISBN 978-0-367-42272-1

Towards fault detection and self-healing of chemical processes over wireless sensor networks

B. Dorneanu & H. Arellano-Garcia
LS Prozess- und Anlagentechnik, Brandenburgische Technische Universität Cottbus-Senftenberg, Cottbus, Germany
Department of Chemical and Process Engineering, University of Surrey, Guildford, UK

H. Ruan, A. Mohamed & P. Xiao
Institute for Communication Systems, Home of 5GIC, University of Surrey, Guildford, UK

M. Heshmat
Surrey Space Center, University of Surrey, Guildford, UK
Department of Mathematics, Faculty of Science, Sohag University, Sohag, Egypt

Y. Gao
Surrey Space Center, University of Surrey, Guildford, UK

ABSTRACT: This contribution introduces a framework for the fault detection and healing of chemical processes over wireless sensor networks. The approach considers the development of a hybrid system which consists of a fault detection method based on machine learning, a wireless communication model and an ontology-based multi-agent system with a cooperative control for the process monitoring.

1 INTRODUCTION

Modern engineering systems and manufacturing processes are becoming increasingly complex, and operating in highly dynamic environments. In the age of high competition and stringent environmental and safety regulations, the role of maintenance as an effective tool to increase profit margin, improve plant reliability and reduce safety and environmental hazards has become increasingly important (Nguyen & Bagajewicz 2010). Fault detection and diagnosis are important issues in chemical engineering applications. The early detection of process faults can help avoid or reduce productivity loss and its associated costs. The petrochemical industry, for example, loses an estimated of 20 billion dollars every year and rated abnormal event management (AEM) as their number one problem that needs to be solved (Venkatasubramanian et al. 2003). Research conducted by Oneserve in partnership with British manufacturers found that 3% of all working days are lost annually in manufacturing due to faulty machinery, amounting to a cost of £180 billion a year.

In industrial processes faults happen due to design error, implementation errors, human operator errors, wear, aging, or environmental aggressions. A fault can be defined as a deviation of at least one characteristic property or parameter of the system from the standard condition. Fault diagnosis is a subfield of control engineering which includes fault detection, fault isolation, and fault identification. Fault detection techniques work on capturing the fault and estimating the time of fault occurrence. Then it comes to the fault isolation to determine the location of the fault in the system and the fault identification to estimate the size and the type of fault (Severson et al. 2015). Fault detection and isolation (FDI) methods can be classified into (Zhong et al. 2018): a) model-based methods, which provide a description of the dynamic behaviour and a physical understanding of the system, but are difficult to account for modelling errors and uncertainties; b) knowledge-based methods, which combine rules derived from first principles, rules with no underlying first principles from physics and simple limit checks; and c) data-driven/signal-based methods, often used in monitoring of large-scale industrial applications as they do not require a lot of computation and are compatible with real-time constraints of dynamic complex systems, but require pre-processing steps for extracting information on the system. In the following, a novel framework for fault detection in chemical processes will be presented, based on a combination of ontology-based multi-agent systems, cooperative MPC, communications over wireless sensor networks and fault detection using machine learning algorithms.

The reminder of this paper is structured as follows: Section 2 introduces the communication network and data transfer scheme. Section 3 provides an overview of multi-agent systems (MAS) and ontology for fault detection, followed by machine learning techniques for fault detection in Section 4. Section 5 introduces the overall system architecture combining the latest wireless communication techniques with fault detection. Finally, conclusions are drawn in Section 6.

2 COMMUNICATION NETWORKS AND DATA TRANSFER

Wireless sensors are considered as one of the key enables in industry and process automation, more specifically in intelligent chemical/process plants. Unlike conventional wired sensors, wireless sensors can be deployed in hazardous areas of the plant in a plug-and-play mode and they reduce deployment and maintenance overhead. It is envisioned that many sensors will be deployed in large-scale plants to pro-vide reliable, timely and up-to-date measurements, thus enabling fast and accurate decisions by the control system. Some of these sensors may provide periodic measurements with different inter-measurement times, i.e. traffic inter-arrival times, while others will be configured to operate on an event-basis. For in-stance, a temperature sensor can be configured to transmit measurement when a certain condition is met and suspends transmission otherwise. In such scenarios, reliable network access, massive connectivity support and timely scheduling will become critical considerations.

The legacy wireless systems have been designed primarily for human initiated mobile broadband communications. They are highly suboptimal for latency-critical, narrow band, short-bust, sporadic traffic (e.g. measurement data, such as temperature, pressure, humidity, etc.) generated by sensors. Consequently, a new design paradigm is needed to support large numbers of heterogeneous sensing devices with diverse requirements and unique traffic characteristics. Compared to the sensors in traditional Internet-of-Things (IoT) networks, those deployed in extreme environments need to operate in harsh (sometimes hazardous) conditions, thus are prone to wear and tear, and cannot be easily replaced, posing major challenges in designing resilient wireless networks for reliable communications.

Recently, the first release of the fifth-generation (5G) cellular system has been standardised, and key use cases for narrow-band communications have been identified for ultra-reliable and low latency communications (URLLC) and massive machine-type communications (mMTC). The former is characterised by extremely low end-to-end transmission latencies (1 ms user plane latency and 20 ms control plane latency) and high reliability figures (e.g. 99.999% success rate within 1 ms with 0 ms mobility interruption time), while the latter supports extremely high connection densities up to 1,000,000 devices per km2 with typically low data rates (ITU 2017, Nokia 2018). Figure 1 maps the main parameters of each 5G use case.

It can be noticed that low latency and high connection density are the main requirements for the scenarios under consideration for plant maintenance. For

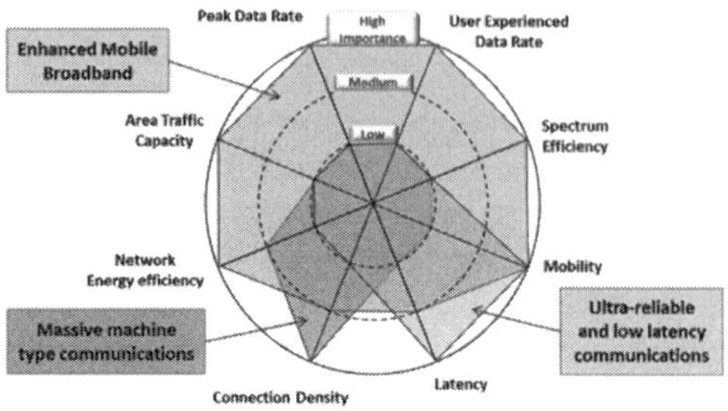

Figure 1. 5G use cases (3GPP, 2015).

instance, the fault detection and prediction algorithm will require fast transmission of measurement data to provide timely decisions. In addition, it is envisioned that large number of sensors will be deployed within a plant, and the same network can support multiple plants/factories. In other words, both the mMTC and the URLLC system characteristics should be taken into account for the scenarios under investigation.

3 MULTI-AGENT SYSTSEMS AND ONTOLOGY FOR FAULT DETECTION

Ontologies and multi-agent systems (MAS) can be used in the development of knowledge-based FDI techniques. An ontology is a formal representation of a set of concepts within a domain and the relation-ships between those concepts. One of the ad-vantages of ontologies is that they can be processed by logic reasoners or inference engines so that hid-den relationships between elements of a system can be discovered. The ontology serves as a library of knowledge components to efficiently build intelligent systems and as a shared vocabulary for communication between interacting human and/or software agents (Batres 2017).

Furthermore, an agent can be defined as an entity which is placed in an environment and senses different parameters that are used to make a decision based on the goal of the entity. The MAS is a computerised system composed of multiple interacting agents exploited to solve a problem. The salient features of a MAS, which include efficiency, low cost, flexibility, and reliability, make it an effective solution to solve complex tasks (Dorri et al. 2018).

Ontology and MAS have been often used in development of FDI methodologies. One of these approaches includes information contained in the operating, safety and control procedures for diagnosis of complex process plants (Hangos et al. 2008). A MAS technology-based chemical plant supervisory system has been developed that connects the chemical equipment and is able to monitor the entire enterprise (Wang & Zhang 2008). The proposed system can al-so be integrated with the current systems through and interface.

Other approaches of ontology and MAS-based process supervision have been developed using a conceptualisation of equipment, control systems and hazards, or to demonstrate how description logic (DL) reasoning could be used to detect and diagnose faults, without the help of external agents (Musulin et al. 2013).

For the management of abnormal situations, an agent-based approach called ENCORE, using Onto-Safe, explicitly captures the hierarchy of the offshore platform, comprising the entire process at the highest level and the individual instruments and equipment at the lowest (Natarajan & Srinivasan 2014). Each FDI method in ENCORE is modelled as an agent specialised in monitoring different aspects at varying levels and scope.

A robust fault detection technique based on consensus-based multi-agent approach for sensors networks makes use of the information interaction and coordination among the neighbouring networks for the fault detection (Jiang et al. 2014).

More recently, a framework for automatic generation of a flexible and modular system has been proposed for fault detection and diagnosis (Steinegger et al. 2017). These methods gather the information from various engineering artefacts using ontology and generates fault detection and diagnosis functions based on structural and procedural generation rules.

4 MACHINE LEARNING FOR FAULT DETECTION

There are two types of machine learning (ML) methods towards fault detection. a) Supervised learning (SL) approaches (Wang et al. 2015), where the fault detection ML model needs to be trained with some expert knowledge. This means that there should be some previous information on how faulty the data is. Most real scenarios, these prerequisites are not available, as the faults are not known until they are detected, recorded and analysed for the first time. b) Un-supervised learning (USL) approaches (Costa et al. 2014), which do not depend on expert knowledge of faults. USL includes options such as clustering (Jyoti & Singh 2011) or self-organising map (SOM) methods (Chen & Yan 2012). However, very limited USL methods deliver effective performance due to the high complexity cost, especially for those hierarchical approaches, which force the algorithms to run over and over again on pairwise samples of large datasets. In most scenarios, due to limitation of hard-ware, the hierarchical approaches are infeasible as the hardware cost would be much higher and the performance are still unpredictable.

5 FAULT DETECTION VIA WIRELESS SENSOR NETWORKS

As no one fault detector can work well for complex processes (Venkatasubramanian et al. 2003, Miljkoviç 2011), a hybrid algorithm is included in the

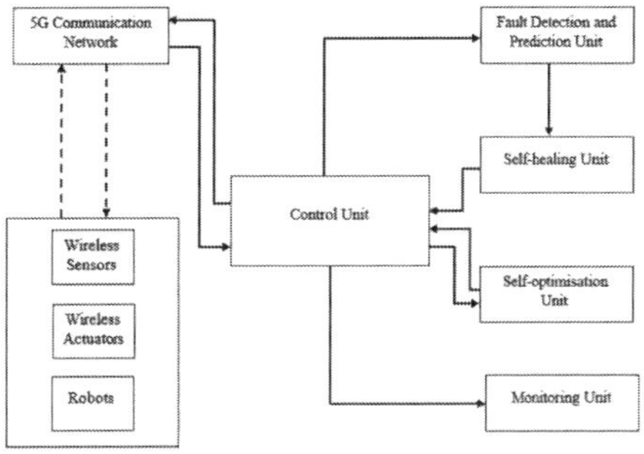

Figure 2. High level overview of the overall system model.

framework, illustrated in Figure 2, for robust fault detection capability.

The framework consists of a network of four main components. The first one is a wireless sensor network (WSN) that facilitates data management for better data-based fault detection. The WSN transmits over a 5G communication network. The second component is an efficient fault detection algorithm that could analyse the data and classify it in faulty or normal. The third component is a knowledge-based and a model-based fault detection monitoring system, based on ontology and MAS. Finally, the fourth component is a cooperative MPC system that takes the required measures to ensure stable process operation.

Using information on the process' structure and behaviour, equipment information, and expert knowledge encoded into the ontology, the system is able to detect any faults. The integration with the monitoring system will facilitate the detection and optimise the controller's actions.

A low data rate is considered for all sensors. In addition, most sensors are assumed to be static, i.e. deployed in fixed locations. Since humanoid robots and autonomous robots can be considered as an element of the control and measurement system, it is assumed that a small number of sensors are moving with a low speed (e.g. ≤ 3 km/h).

A centralised control mechanism is considered where the sensors are connected to a fusion node via wireless links.

The latter can also be used to send commands to actuators within the plant. Consequently, the measurements are transmitted in the uplink to the control unit with the commands being transmitted in the downlink. The network consists of a heterogeneous set of periodic and event-triggered sensors with mixed requirements, characteristics and traffic models.

Considering heterogeneity of the plant and the associated sensors, a statistical model rather than a deterministic model is chosen for the sensor transmission events. The number of incoming packets (or events when each event generates a single packet) per unit of time follows the Poisson distribution while the packet interval is modelled as an exponential distribution. This results in probability-based transmissions that can be controlled by the arrival rate and the inter-arrival time. Figure 3 shows an example of the probability-based transmission.

Given this traffic model coupled with the network model illustrated in Figure 2, the fault detection framework will propose a low latency network access scheme for wireless sensors networks in intelligent chemical/process plants. The proposed scheme considers the 5G frame structure and allows massive numbers of sensors to access the network simultaneously without collisions to request resources for measurement transmission.

A low overhead and low latency approach is being developed, and simulation along with theoretical evaluations are used to show gains of the proposed over conventional access schemes.

Once the measurement data is received at the network side, it will be routed to the control unit for further processing. This data can be used for fault detection and prediction, as well as to optimise the plant operating parameters.

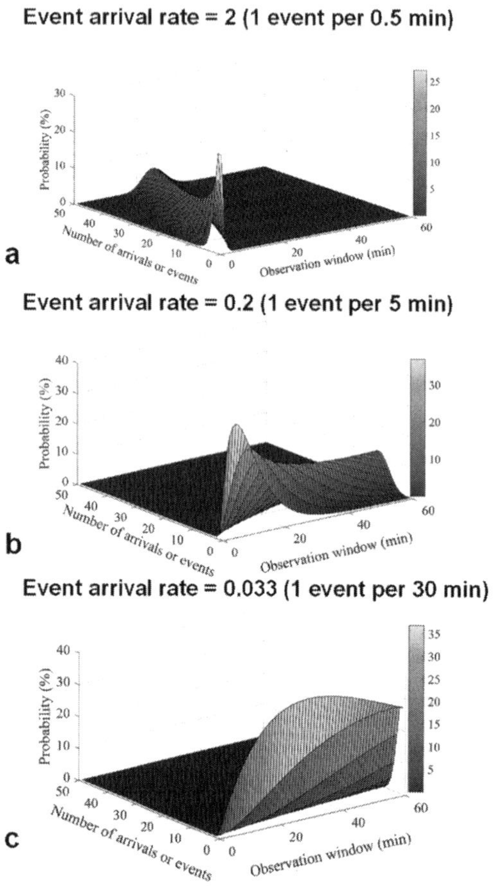

Figure 3. Transmission probability for: a) high periodicity sensor/probability event; b) medium periodicity sensor/probability event; c) low periodicity sensor/probability event.

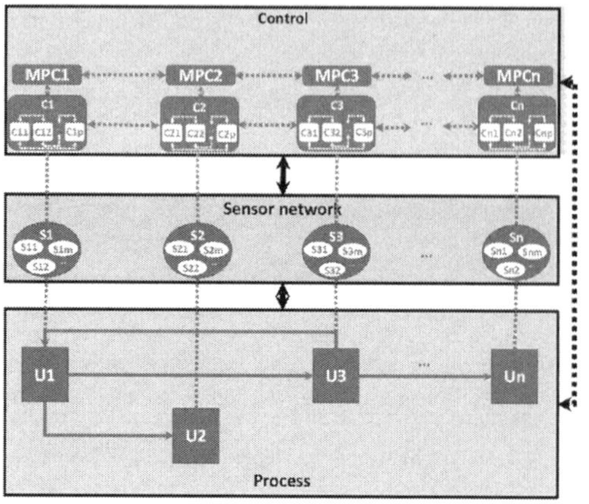

Figure 4. Cooperative control framework.

Decisions of the control unit will be routed to the 5G base station (known as g-NodeB), from where the commands will be sent to the wireless actuators.

The control unit is developed based on cooperative control algorithm, based on distributed MPC approaches (Rocha & Oliveira 2016).

Under the cooperative framework, the distributed controllers share information about their state with other controllers, as shown in Figure 4. Thus, the interactions between the processing units will not be lost, as the local controller will optimise a local objective function, with input from other sections of the process.

For the fault detection, a two-stage method based on a hybrid learning approach is applied, which utilises both SL and USL. In the first stage, an unsupervised learning algorithm (a K-aware K-mean – KKM - clustering) is applied to separate the whole dataset into clusters based on their similarities and select the optimal K value (within a given range) for the best clustering performance. The cluster with the smallest amount of data which is also much smaller than the penultimate cluster, is perceived as the faulty data cluster. The proposed KKM-clustering method is summarised in Figure 5.

Initialize: *range of* $K: 2 \leq K \leq 8$; *Repitition* $= 1000$; *iteration* $N = 100$; 135000 *samples* $= d(1), d(2), \ldots, d(135000)$.

For $K = 2, \ldots, 8$:

–*For iteration* $= 1, 2, \ldots,$ *Repetition*:

– – *randomly choose centroids* $c(1), \ldots, c(K)$ *from the samples*

–*For* $n = 1, 2, \ldots, N$:

– – *For* $k = 1, 2, \ldots, K$:

– – –*For* $i = 1, 2, \ldots, 135000$:

– – – – $cluster\_index(i,k) = argmin_{c(1),\ldots,c(k)} \|d(i) - c(k)\|_2^2$

– – –*update* $c(k) = \frac{1}{length(cluster_{index(i,k)})} \sum_{cluster\_index(i,k)} \|d(i) - c(k)\|_2^2$

– – $c(k,K) = \frac{1}{Repetition} \sum c(N,k)$

–$C(K) = min_k c(k,K)$

$C_{faulty} = min_K C(K)$

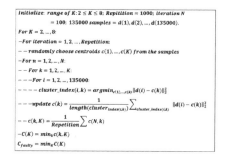

Figure 5. KKM clustering algorithm.

The value of K varies only between 2 and 8. For each value of K, the algorithm is repeated 1,000 times, for reproducible results. In each repetition, 100 iterations are conducted to iteratively compute and update the centroids and cluster indices of the data clusters. Then, at each K-mean algorithm for each K value, the smallest data cluster is selected. Finally, the selected cluster with the smallest size is perceived as the faulty data cluster, while the corresponding value of K is selected as the best K. Please not that ‖·‖_2 denotes the Euclidean distance, cluster_index(i,k) the distance between sample *i* and cluster *k*, used to index both data points and the clusters they are located in, *c(k)* the index of cluster *k*, c*(N,k)* the cluster *k* after N iterations, *c(k,K)* the cluster *k* with a certain value of K, *C(K)* the smallest cluster with a certain value of K, and C_faulty is the perceived faulty data cluster.

In the second stage, a single SL method is used for training on the pre-processed data obtained from stage 1, and then used to classify faults from the unknown dataset after training. Various classical SL algorithms will be applied, to compare their accuracy.

Preliminary results were obtained for a set of data collected from a pilot plant that produces sodium ion solution as sodium chloride for sale to fine chemical, pharmaceutical, and food industry, which is illustrated as the Process in Figure 4. The data was collected from March 21st – 24th, 2017, which aggregated 30MB in total. The data was collected by 43 fixed sensors set up on the units of the pilot plant with a frequency of 1 Hz (i.e. one sample per second).

A total collection time of 8 h was considered for each day, which delivered 135k data samples. Every sample measured 43 variables, consistent with the number of installed sensors. Data points include both floating point and Boolean data. For stage 2, the whole dataset with labels is split into training and testing subsets of 100k and 35k, respectively for SL so that a specific fault detection ML model can be established.

For the pre-processing of the raw data, measurements are standardised using the standard scalarization method to remove the means. In order to have an intuitive visualisation of the data, the principal component analysis (PCA) is used to project the original unvisualisable data from higher-order feature space onto a 2D plane based on its first two principal eigen components, illustrated in Figure 6.

Most of the data is distributed with the second principal component value less than 2, except for some outliers on the top right corner which have values larger than 30. The proposed KKM algorithm is then used to detect the outliers and categorise them as faults. A mini-batch data packaging method is used to reduce the computational complexity in a moderate level, where the samples are packaged into small batches. Thus, the algorithm has to be run on different subsets of the original data instead of all of them. The mini-batch size is set to 1,000, which only takes 1,000 samples per iteration instead of the 135k. Figure 2 shows the results after applying the KKM algorithm for the data in stage 1. As can be seen, the outliers (in crosses) are correctly detected. The result also indicates the optimal K value is 4, where the outlier data is represented by the smallest cluster.

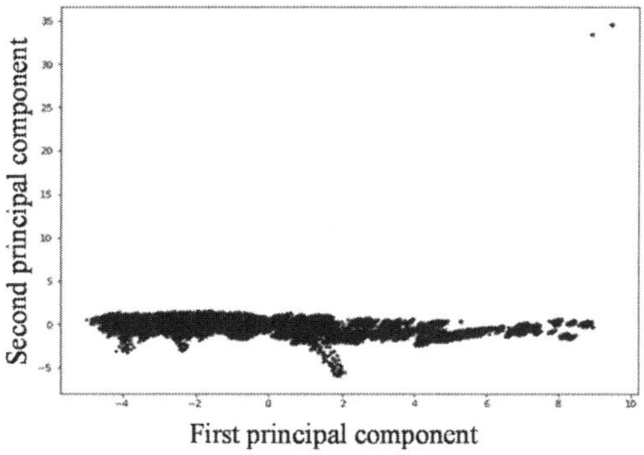

Figure 6. Data visualization after applying PCA with 2 principal components.

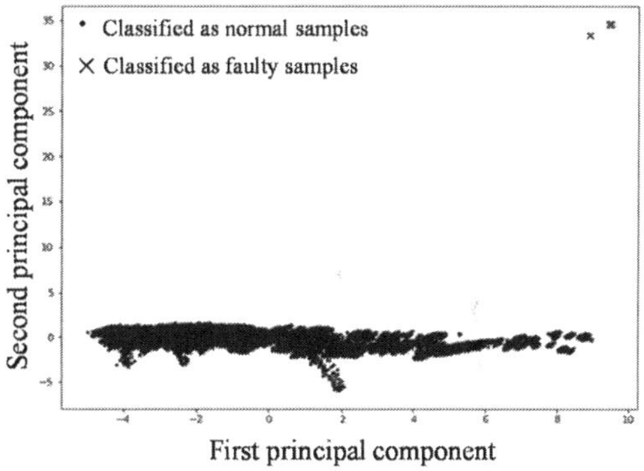

Figure 7. Data visualization after applying the KKM algorithm.

In the 2nd stage, several SL classification algorithms are compared regarding the detection accuracy, based on the expert knowledge obtained from stage 1. The results are summarised in Table 1. For the multilayer perceptron (MLP) algorithm, only one hidden layer is tested, which proves adequate for the task, while the support vector machines (SVM) only linear kernels are considered for low complexity purposes.

The hyper parameters are properly tuned to produce the best accuracy performance. As shown in Table 1, all algorithms are able to achieve 95% accuracy, where K-nearest neighbours (KNN), decision tree (DT), MLP, and SVM give the best performance, with accuracy >99%.

Table 1. Detection accuracy for various classifiers.

Classifier	Detection accuracy
Logistic regression	97.5%
KNN	100.0%
DT	99.4%
Linear discriminant analysis	95.4%
MLP with one hidden layer (-10-)	99.0%
Linear SVM	99.5%

6 CONCLUSIONS

A framework for the detection of faults and healing in chemical plants is proposed. The framework is developed on an ontology-based MAS that considers models for the physical process, the cooperative MPC system used for controlling the plant, a model for the wireless communication networks between the WSN implemented inside the plant, and a two-stage fault detection approach based on USL and SL. Preliminary results have been obtained for the fault detection within a mini plant operated at the University of Surrey. For the future work, the operation of the plant will be analysed over a longer period of time, under controlled disturbances to assess the healing capabilities of the novel approach. Fault prediction algorithms will be implemented as well within the framework.

ACKNOWLEDGEMENTS

This project is carried out within the framework of the EP/R001588/1 Stepping towards the industrial 6th Sense project. The financial support of the EPSRC is gratefully acknowledged.

REFERENCES

3GPP, 2015. Industry vision and schedule for the new radio part of the next generation radio technol-ogy, 3GPP RAN workshop on 5G, Phoenix, USA.

Batres, R. 2017. Ontologies in process systems engi-neering, Chemie Ingenieur Technik 89 (11), 1421–1431.

Chen, X. & Yan, X. 2012. Using improved self-organising map for fault diagnosis in chemical indus-try process, Chemical Engineering Research and De-sign 90 (12): 2262–2277.

Costa, B.S.J. et al. 2014. A new unsupervised ap-proach to fault detection and identification, Pro-ceedings of the 2014 International Joint Conference on Neural Net-works: 1557-1564.

Dorri et al. 2018. Multi-agent systems: A survey, IEEE Access 6: 28573–28593.

Hangos, K.M. et al. 2008. A procedure ontology for advanced diagnosis of process systems, Knowledge-Based Intelligent Information and Engineer-ing Sys-tems: 501-508.

ITU – International Telecommunication Union, 2017, Report ITU-R M.2410.0: Minimum require-ments related to technical performance for IMT-2020 radio interface(s).

Jiang, Y. et al. 2014. A consensus-based multi-agent approach for estimation in robust fault detection, ISA Transactions 53 (5): 1562–1568.

Jyoti, K. & Singh, S. 2011. Data clustering approach to industrial process monitoring, fault detection and isolation, International Journal of Computer Applica-tions 17 (2): 41–45.

Miljkoviç, D. 2011. Fault detection methods: A lit-erature survey, Proceedings of the 34th International MIPRO Convention: 110-115.

Musulin, E. et al. 2013. A knowledge-driven ap-proach for process supervision in chemical plants, Computers & Chemical Engineering 59: 164–177.

Natarajan, S. & Srinivasan, R. 2014. Implementation of multi agents based system for process supervision in large-scale chemical plants, Computers & Chemi-cal Engineering 60: 182–196.

Nguyen, D. & Bagajewicz, M. 2010. Optimisation of pre-ventive maintenance in chemical process plants, Indus-trial & Engineering Chemistry Research 49 (9): 4329–4339.

Nokia, 2018, Self-Evaluation: URLLC and mMTC evalu-ation results, 3GPP RAN Workshop on 3GPP submis-sion towards IMT-2020, Brussels, Belgium.

Rocha, R.R. & Oliveira-Lopes, L.C. 2016. A coop-erative distributed model predictive control for non-linear sys-tems with automatic partitioning, Computer Aided Chemical Engineering 38: 2205–2210.

Severson et al. 2015. Perspective on process monitor-ing of industrial systems, IFAC-PapersOnLine 48 (21): 931–939.

Steinegger, M. et al. 2017. A framework for auto-matic knowledge-based fault detection in industrial conveyor systems, Proceedings of the 22nd IEEE In-ternational Conference on Emerging Technologies and Factory Automation: 1–6.

Venkatasubramanian, V. et al. 2003. A review of process fault detection and diagnosis: Part I: Quanti-tative model-based methods, Computers & Chemical Engin-eering 27 (3): 293–311.

Wang, X. et al. 2015. Fault detection and classifica-tion for complex processes using semi-supervised learning algorithm, Chemometrics and Intelligent La-boratory Systems 149 (B): 24–32.

Wang, Y. & Zhang, Y. 2008. Multi-agent based chemical plant process monitoring and management system, Pro-ceedings of the 4th International Confer-ence on Wire-less Communications, Networking and Mobile Computing: 1–4.

Zhong, J.H. et al. 2018. Fault diagnosis of rotating machin-ery based on multiple probabilistic classifiers, Mechan-ical System and Signal Processing 108: 99–114.

Industry 4.0 – Shaping The Future of The Digital World – da Silva Bartolo et al. (eds)
© 2021 Taylor & Francis Group, London, ISBN 978-0-367-42272-1

Pattern recognition based on neural network for sign language interpretation system using flexes sensor

D.A. Kadhim & O. Obaid
Alkafeel University, Alnajaf, Iraq

ABSTRACT: Sign Language Recognition has emerged as one of the important areas of research in Computer Vision. The difficulty faced by the researchers is that the instances of signs vary with both motion and appearance. In this paper, we introduce a sign language interpreter based on hardware and software .We suggest using gloves builder with hardware to recognize hand gestures. In this work we suggest to use a gyroscope to solve many problems accompanied the previous works (to our knowledge, our work is the first to have used gyroscope for this purpose). In addition, we suggest including neural network in the software to enhance the recognition and make the gloves suitable for any language. The proposal recovers many problems and enhances recognitions. The performance of recognition, continuous words reaches up to 83.5.

1 INTRODUCTION

The understanding of Deaf culture as well as the recognition of Sign Language is of major importance of the inclusion of the Deaf community in everyday life. In particular, it is important that Sign Language is given a greater standing among language learning and acquisition agendas should be recognized as an important minority language for deaf people and their communication. Traditionally the First and Secondary education of deaf individuals becomes either in special schools, where teaching of courses becomes from professionals. Who knows the sign language and the main method for educating deaf individuals is the optic – acoustic one. One of the most important roles in the deaf community is the interpreter of Sign Language. The role of the interpreter is to solve the communication problems.

2 RELATED WORK

Kawale et al. (2018) introduced basics lessens to the corresponding hole between goofy and standard individuals and propose to encourage idiotic individual's way of life. Arya et al. (2017) presented, a low cost system for interpreting speech, by utilizing the help of sensors (contact, accelerometer, flex sensors, etc.) they try to collect information through hand signals and lead theses gesture by actuators (display screen, speaker, etc.). They depend upon the hand signs and the finger pattern sensors fixed on gloves will provide a range of data by detecting varying resistance of flex sensors. Ansari et al. (2016) designed a glove to help deaf person's communication with others. By means of a glove based on deaf communication interpreting system. The gloves are supplied with five flex sensors, accelerometer and tactile sensors. For each particular signs shaped by the fingers, the flex sensor provides a rational change in resistance and accelerometer measures the position of the hand. All movement of hand processed by Microprocessor. The system contains a text to speech conversion (TTS) block, which interprets the matched signs. Aarthi et al. (2016) Suggested use talking glove, which attached to the sensor (flex sensor) that are changing its resistance according to the bending. The output voltage of the sensor, converted into digital form which are processed using ARM7 processor

3 SYSTEM DESCRIPTIONS

The system focuses on finding a simple way to interpret a sign language based on hardware and software. The hardware was designed and built by using ATM processor; flex sensor, small LCD screen and gyroscope, to detect movement of the hand, which is transmitted to processor for processing, and interpret hand movements into words displayed on LCD screen (see Figure1.).

Sign Language depends on series of sign formed by fingers, which represent the type of words or descriptions of some specific objects. The goal of this research is to transmit the hand gestures to the processor, which is interpreted to English characters and words. The software part based on BP NN (Back Propagation Neural Network). The stages of this proposal are outlined (see Figure 2).

The whole process passes through two phases. Each phase consists of three stages. First, building hardware, which consists of flex sensors, generates a signal according to the movements of the fingers. For each finger, there is a sensor in response to the any

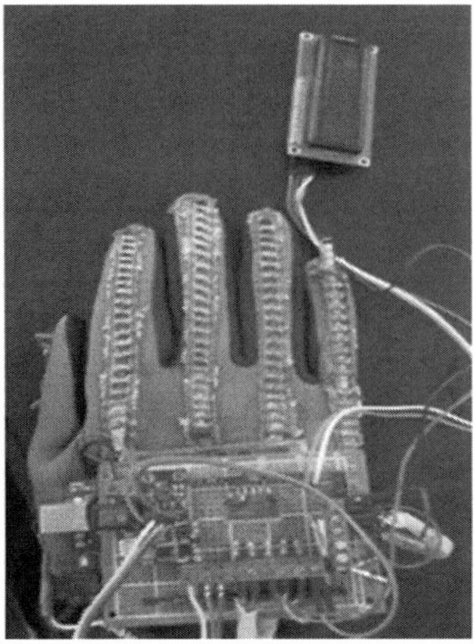

Figure 1. Glove hardware.

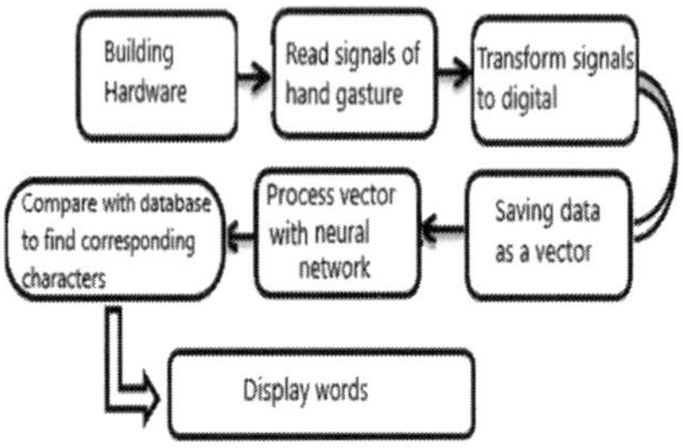

Figure 2. Proposed system block diagram.

movement of the finger (there are five sensors for each hand). When the fingers bend, then the flex sensors are activated and send signals with specific resistance related to the amount of finger bending (according to the angle created due to the finger movements) as shown in Figure 3.

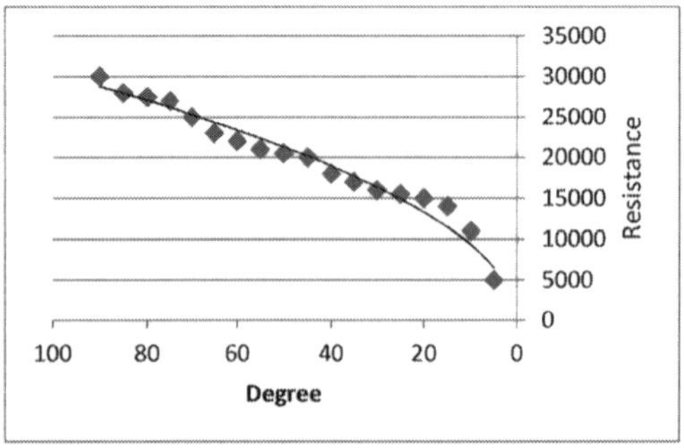

Figure 3. The relation between the angles of flex sensor and resistance produced.

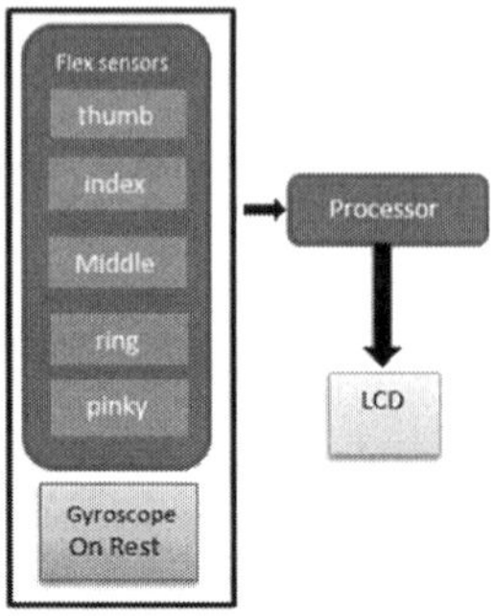

Figure 4. Hardware block diagram.

The second part or item in hardware is the gyroscope which response to rotation of the hand rest in three dimensions (x, y, and z). The signals were produced from sensors and gyroscope processed at ATM processor. While the last part of the hardware is the LCD screen to display the results. (see Figure 4.) .

The second stage of this proposal is software part. At this stage, we propose to use neural network (NN) to train the system on the signals produced from the sensors and gyroscope. The NN consist of one input layer and 10 hidden layers, each have 12 nodes. The NN is back propagation system. The system train on all possible vectors. Vector is the group of signals produced from sensors and gyroscopes due to the different movements of the fingers.Vector represents us:

$$\mathbf{V} = [F1\ F2\ F3\ F4\ F5\ F6\ F7\ F8\ F9\ F10\ G1\ G2]$$

Where F: represent Flex signal from fingers. In addition, G: represent the Gyroscope signal for both hands (XYZ).

4 RESULTS

We test the gloves with all characters and different words; the performance of the system was 83.5%. The drawback in the performance due to

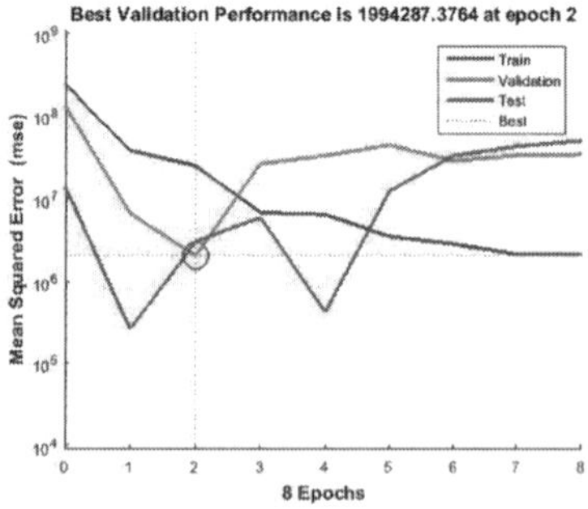

Figure 5. Relation between mean square error and number of epochs.

some problems in the reading of gyroscopes movements.

We determine the Number of epochs according to the mean square error, we found that the (MSE) reduced with increasing number of epochs at the training stage to be at steady state at epoch 8. Epochs more than epoch 8 has no effect on the performance of recognition for reducing the error (see Figure 5).

5 CONCLUSIONS

In this paper, we propose a glove with building hardware to recognize sign language and interpret to English words. To our knowledge this first work use gyroscope to enhance recognition and solve the problems with other similar works. Gyroscope solves problems with age peoples and persons with Parkinson's disease, which make hands unstable and that has an effect on reading hand movements. In addition, it enhances the recognition by using the neural network, which highly enhances recognitions, this useful for training the system for any language.

The performance of recognition input words through the gloves was 83.5%. The main problem happens to sign produced from handholding.

REFERENCES

Aarthi M & Vijayalakshmi P. 2016 .Sign Language to Speech Conversion. *fifth international conference on Recent Trends in Information Technology*. IEEE. India.

Anetha. K, Rejina Parvin J. 2014.Talk-A Sign Language Recognition Based On Accelerometer and SEMG Data . *International Journal of Innovative Research in Computer and Communication Engineering*. Volume 2 (Special Issue 3) : 206–215.

Ansar, A.H. i, Sanjay Jori, Shesherao, Ashwini & chandrakant Prajakta.2016. Givening Voice To Mute People Using Flex Sensor. *IJARIIE* Volume 2 (Issue-3) :3020–3027.

Arya Meenakshi, Patil Romali, Karande Pooja, & Jadhav Divya. 2017. An automated Sign to Speech Portable Translator for Vocally Challenged using Android and Touch Sensors. *International Journal of Engineering Technology Science and Research* Volume 4 (Issue II) : 379–383.

Bhat Sachin, Amruthesh M, Ashik, Chidanand Das & Sujith. 2015 .Translating Indian Sign Language to text and voice messages using flex sensors. *International Journal of Advanced Research in Computer and Communication Engineering* Volume 4 (Issue 5):316–321.

Jadhav Shubham, Shah Pratik, Bagul Avinash Hoshing Paag& Wadhvekar Ashutosh. 2016. Review on Hand Gesture Recognition using Sensor Glove. *International Journal of Advanced Research in Computer and Communication Engineering* Volume 5 (Issue II):201–215.

Kawale Nisha, Kaspate Pradnya, Vanjari Hruchika & Sarode Prachi.2018. Sign Language Using Flex Sensor. *International Journal of Innovative Research in Computer and Communication Engineering* Volume 6 (Issue 3): 635–640.

Khan Fauzan, Surve Akshata, Shaikh Faiz, & Ali Amjed. 2017 .Sign Language Glove using Arduino . International *Journal for Research in Applied Science & engineering Technology (IJRASET)* Volume 5 (Issue III):

Kushwah Mukul, Sharma Manish, Jain Kunal & Chopra Anish. 2017 .Sign Language Interpretation Using Pseudo Glove. *Advances in Intelligent Systems and Computing*. Dehradun : springer.

Lokhande Priyanka, Prajapati Riya & Sandeep Pansare. 2015. Data Gloves for Sign Language Recognition System. *International Journal of Computer Applications* (0975 – 8887):11–14.

Potdar R. & Dr. Yadav D. M. 2014. Innovative Approach for Gesture to Voice Conversion: Review. *International journal of innovative& Development*. Volume. 3 (Issue 6): 459–462.

Industry 4.0 – Shaping The Future of The Digital World – da Silva Bartolo et al. (eds)
© 2021 Taylor & Francis Group, London, ISBN 978-0-367-42272-1

A framework for intelligent monitoring and control of chemical processes with multi-agent systems

B. Dorneanu & H. Arellano-Garcia
LS Prozess- und Anlagentechnik, Brandenburgische Technische Universität Cottbus-Senftenberg, Cottbus, Germany
Department of Chemical and Process Engineering, University of Surrey, Guildford, UK

M. Heshmat
Surrey Space Center, University of Surrey, Guildford, UK
Department of Mathematics, Faculty of Science, Sohag University, Sohag, Egypt

Y. Gao
Surrey Space Center, University of Surrey, Guildford, UK

ABSTRACT: Industry 4.0 is transforming chemical processes into complex, smart cyber-physical systems that require intelligent methods to support the operators in taking decisions for better and safer operation. In this paper, a multi-agent cooperative-based model predictive system for monitoring and control of a chemical process is proposed. This system uses ontology to formally represent the system knowledge. By integrating the cooperative-based model predictive controller with the multi-agent system, the control can be improved, and the process can be converted into a self-adaptive system.

1 INTRODUCTION

Modern chemical plants have evolved into extremely large and elaborate industrial systems, exhibiting scale, structure and behaviour complexity. Under these conditions, the plant operators find it extremely difficult to manage all the information available, infer the desired conditions of the plant and take timely decisions to handle abnormal operation (Natarajan & Srinivasan 2014). The plant control is usually implemented via a computerised Distributed Control System (DCS) with a large number of control loops, in which autonomous controllers are distributed throughout the system, but there is a central operator control.

In the context of Industry 4.0, the chemical plant is upgraded and transformed into a cyber-physical system (CPS) by the addition of elements such as the smart sensors, Internet of Things (IoT), big data analytics or cloud computing. Although this increases the system's complexity, in the same time it offers the opportunity for enabling the plant make smart decisions through real-time communication and cooperation between humans, machines, sensors and all the other components of the CPS (Zhong et al. 2017).

There is a need for the deployment of intelligent systems for process engineering applications that should automatically adapt efficiently and accurately to the continuously new requirements of Industry 4.0. Modelling and computation tasks become much more complex as the size of the system continues to increase. Considering the large number of devices existent in a CPS, distributed methods are needed to transfer the computational load from centralised to local (decentralised) controllers.

This has led to the motivation of applying multi-agent systems (MAS) methodologies as a solution to distributed control as a computational paradigm (Xie & Liu 2017). MAS consist of autonomous entities known as agents that collaboratively solve tasks, while having the flexibility of learning and making autonomous decisions (Dorri et al. 2018).

The increase in the CPS complexity requires the introduction of conceptualised, structured representations of the system and its components in order to enable simplification and automation. One of these is represented by semantic web technologies, in particular ontology-based, which can tackle the integration of multiple knowledge representations (Batres 2017, Ekaputra et al. 2017).

In this paper, a framework for the development of intelligent monitoring and control of a chemical process is described. The framework integrates the advantages of the various technologies available within Industry 4.0, the MAS of the resulting CPS and the cooperative control, which are discussed in Section 2. Section 3 describes the aims and outlines of this framework. Conclusions and future work are presented in Section 4.

2 ONTOLOGIES AND MULTI-AGENT SYSTEMS FOR CONTROL OF CHEMICAL PLANTS

2.1 *Ontologies*

An ontology is a formal representation of a set of concepts within a domain and the relationships between those concepts, with the goal to limit complexity and organise information into data and knowledge. The ontology uses a formal language to encode this representation. Ontologies are useful for generating new conclusions from existing data and providing the structure and the semantics needed for validating information. The development of ontologies was motivated by the need to enable knowledge sharing between software agents in order to exchange queries and assertions to their knowledge bases (Gruber 1993). According to Pohjola (2003) processes can be described in terms of objects having four attributes: purpose, structure, state, and performance. The purpose indicates the rationale of human-made artefacts; the structure is about the connectivity and part information of an object; the state is a set of properties of an object and their values; while performance refers to the evaluation of the state in comparison with the purpose.

A similar effort is made in (Batres et al. 1999), where a multidimensional framework was proposed which represents the artefacts in three dimensions: behavioural, physical and operational. The behavioural dimension refers to physicochemical phenomena, the physical dimension is related to the plant and equipment, and the operational dimension deals with management and operational aspects. The three dimensions of this framework served as the basis for ontologies aimed at supporting the design and operation of process plants.

Within the context of Industry 4.0, a substantial amount of supporting data and information related to various aspects of the potential network elements of the resulting CPS need to be shared and communicated (Gilchrist 2016, Lu 2017).

Several ontologies have been proposed in the field of process systems engineering. OntoCAPE (Marquardt et al. 2010) is perhaps one of the most widely used. It is organised into layers that separate general classes and relations from those specific to particular domains and applications. The uppermost layer is the meta-layer which describes the design principles and provides guidance for extending the ontology. The next layers are the upper layer which defines the principles of general systems theory, the conceptual layer which covers chemical engineering classes and relations for entities such as unit operations, equipment, materials, physical properties, and mathematical models. Finally, on the bottom lies the application-oriented layer, which extends the ontology to specific classes such as those representing specific process units (e.g. chemical reactor) and provides classes and relations for particular software applications.

Another example of a widely used ontology is the ISO 15926 Ontology (Batres et al. 2007), which is developed for long-term data integration, access, and exchange and to support the evolution of data through time.

2.2 *Multi-agent systems*

MAS has received tremendous attention from scholars in different disciplines (Dorri et al. 2018, Kravari & Bassiliades 2015, Xie & Liu 2017), making agent-based technologies a powerful tool for engineering applications. The agent is an entity placed in an environment that senses different parameters used to make a decision based on a goal of the respective entity. The goal of its agent is to solve an allocated task with some additional constraints. Knowledge of its neighbours may be used as well. Based on this decision, the entity performs the necessary actions on the environment. The knowledge together with the history of previous actions taken and the goal are fed into an inference engine which decides on the appropriate actions to be taken by the agent.

The MAS is a computerized system composed of multiple interacting agents exploited to solve problems that are difficult or impossible for an individual agent to solve. The salient features of MAS, including efficiency, low cost, flexibility, and reliability, make it an effective solution to solve complex tasks. The efficiency of a MAS originates from the fact that a complex task is divided into multiple smaller tasks, each of them assigned to a distinct agent. Each individual agent will decide on the proper action to be taken to ensure the task is solved, using multiple inputs (e.g. history of actions, interaction with neighbouring agents, agents' goals, etc.).

There are several properties that agents can have (McArthur et al. 2007, Dorri et al 2018): Sociability - agents can share their knowledge as well as request information from other agents to improve performance in reaching their goals; Autonomy – agents execute independently the decision-making process and take appropriate actions; Proactivity – agents use history, sensed parameters, and information of other agents to predict possible future actions; Connectivity – agents' performance and functionality rely on the communication layer, especially the connection topology and associated protocols; Mobility – agents can be static or mobile.

Intelligent agents can be classified into several types with respect to the decision-making mechanisms (Liu 2017): Purely reactive agents – make decisions using only present information, without referring to historical data; Belief-desire-intention (BDI) agents – built using symbolic representations of the intentions, beliefs, and desires of agents, and with several software layers that can be incorporated in their architecture.

A comparative review of the existing agent platforms that can be used is presented in Kravari &

Bassiliades (2015), which proposes a classification to help users understand which platforms exhibit broadly similar properties, and what choices can be made in various situations.

2.3 *Ontologies and multi-agent systems for industrial applications*

Ontologies and MAS have been used in the past decades to solve problems or improve/add new capabilities for industrial processes.

An ontology-based scheme has been used for describing sensors and their features for sensor net-works (Xue et al. 2015), where ontology help in providing an effective management and data sharing system for the sensor networks. Another example is the development of manufacturing process ontologies that combine formal concept analysis with a set of criteria for characterisation of classes of processes (Akmal & Batres 2013). Other applications consider the description of physical entities such as production processes, equipment and products, and the relationships of operation logic and operation sequences (Shi et al. 2017). With the support of the ontology description, the manufacturing system can make automatic adjustments for ensuring the completion of the tasks when there are changes in the internal requirements or the external environment. An agent-based method has been used for the coordination of tasks in chemical plants (Nikraz & Bahri 2005).

Ontologies have been utilised as a communication language between the interacting software agents and human users for retrieval of desirable process modelling components libraries (Yang et al. 2008) and for the development of a multi-agent technology-based chemical plant supervisory system that realises the connection between the chemical equipment, and monitors the entire enterprise, while also being able to integrate with current system through an interface (Hairui & Yong 2008).

Capability to manage abnormal situations have been added using the OntoSafe framework, which provides semantics for abnormal situation management (Natarajan et al. 2012), for the development of a multi-agent based distributed intelligence system – ENCORE (Natarajan & Srinivasan 2014), which contains three types of agents that can cooperate with each other: the plant information manager agent, the process supervision agent and the user interface agent. An offshore oil and gas production process is used to illustrate the effectiveness of the approach.

In a different ontology-based framework for process supervision in chemical plants, a conceptualisation of equipment, control systems and hazards has been developed which includes the semantics of each modelled term in order to obtain a heavyweight ontology (Musulin et al. 2013). The framework has been applied on a knowledge-driven approach to demonstrate how description logic (DL) reasoning could be used to support process supervision, detecting and diagnosing faults, without the help of external faults.

Other applications of ontology-based methods focus on the development of a decision-support system (J-Park Simulator) for the design and operation of eco-industrial parks (Zhou et al. 2017). Based on OntoCAPE, J-Park Simulator is used to perform plant-wide process simulation and optimisation.

Furthermore, ontology-based methods have been used to enhance maintenance decisions based on the knowledge gathered through the process of monitoring (Elhdad et al. 2013). The monitoring process is based on signals triggered during the plant safety shutdown process.

Agent-based models have been used to evaluate the dynamic behaviour of a global enterprise, considering both system-level performance and the components' behaviour, as to be used for predicting the effects of local and operational activities on plant performance and improving the tactical and strategic decision-making at the enterprise level (Behdani et al. 2009).

One of the strengths of ontology-based methods is that they can integrate heterogeneous systems (Ekaputra et al. 2017). In the context of Industry 4.0, personalised customisation requires more agile and flexible processes, which makes system reconfigurability a crucial feature for the enterprise to remain competitive. To this end, MAS was introduced to intelligently trigger a system reconfiguration in order to restore its performance to the original levels (Farid 2015, Wan et al. 2017). However, most of the existing systems must be suspended when reconfiguration takes place, as this procedure may lead to disorder and uncertainty. The reconfiguration approaches should minimise the leading time while ensuring the system's stability.

2.4 *Intelligent systems for monitoring and control*

An intelligent system has the capability to apply knowledge in an intelligent manner through its properties of perception, reasoning, and decision-making from incomplete information. Such systems have be-come a practical way for a variety of engineering applications in multiple domains including mechanical engineering, automotive engineering, electrical engineering, control engineering, civil engineering, bio-medical engineering, and micro/nanoengineering (Wong et al. 2017). MAS has been used for monitoring and assessing air-quality attributes, using data coming from a meteorological station (Athanasiadis & Mitkas 2004), for the development of intelligent based technology for efficient data acquisition and expert fault-tolerant systems in the smelting process (Siti et al. 2005) or the design of a MAS for integrated management of greenhouse production (Kasaei et al. 2011).

Furthermore, MAS has been used for the development of an intelligent multi-agent quality control sys-tem (IMAQCS) for controlling the quality of cement production processes (Mahdavi et al. 2013). IMA-QCS uses a rule-based quality control mechanism that serves as an online quality control method acting after the faults occur.

Classical process control systems, such as proportional-integrative-derivative (PID) control utilise measurements of a single process output variable (e.g. temperature, pressure, level, etc.) to compute control actions needed to be implemented by a control actuator so that this output variable can be regulated at a desired set-point value (Christofides et al. 2013).

Model predictive control (MPC) addresses the question of (practical) optimal control of dynamical systems under process constraints and economic incentives (Saltik et al. 2018). MPC simultaneously adjusts all inputs to control all outputs while accounting for all process interactions (Forbes et al. 2015). Centralised control, in which all subsystems are con-trolled via a single agent, can account for the plant-wide interactions, which often makes the system difficult to coordinate and maintain for large-scale plants (Stewart et al. 2010). Typically, large-scale plants are controlled in a decentralised fashion: each subsystem is controlled independently, without interchange of information between different subsystems, which often leads to a big loss of information when the interactions are strong (Ferramosca et al. 2013). In these situations, decentralised control may not provide good performance and may not even stabilise the system (Stewart et al. 2011). A structure that preserves the topology and flexibility of decentralised control and at the same time offers great control properties is the distributed control approach (Rocha & Oliveira-Lopes 2016). This approach can be divided into two types: non-cooperative, where each local controller optimises a local objective function, and cooperative, where each local controller optimises a global objective function. However, most of the distributed MPC algorithms only consider the stabilization of a priori has known set-point, with very few exceptions considering other cooperative tasks (Müller et al. 2011). As stabilisation of some set-point is often not the primary control objective, but the optimal operation with respect to a performance criterion (e.g. cost, profit, etc.) more generic cooperative MPC frameworks have been proposed (Köhler et al. 2018). In a cooperative distributed form of MPC agents optimise plans locally and exchange information.

The attractive features of MPC have been exploited in a distributed implementation combining machine learning techniques to perform negotiation of multiple dependencies between sub-systems in a cooperative multi-agent environment, and over a multi-agent platform (Javalera et al. 2010). This approach is based on negotiation, cooperation and learning and provides speed, scalability and computational effort reduction.

In the following section, the development of a framework for applying an ontology-based MAS for the cooperative control of a chemical plant will be discussed.

3 INTELLIGENT SYSTEM FOR MONITORING AND CONTROL OF PROCESS SYSTEMS

In order to achieve highly flexible dynamic optimal control in the context of Industry 4.0, a cooperative distributed MPC approach will be considered. To this end, the cooperative control problem will be investigated. The cooperative control problem has received attention in distributed coordination and control of dynamic agents in areas such as automotive (Rochefort et al. 2014) or wind farms (Badihi et al. 2016) where the focus is typically on networks of heterogeneous agents. For chemical engineering applications, the states of the agents have different dimensions and the processes have non-identical dynamics, which will make impossible to achieve an internal state agreement (Dorneanu et al. 2018).

In the following, the proposed framework for a co-operative-based MPC approach is discussed for a chemical process. The chemical process is a mini-plant that produces sodium ion solution as sodium chloride for sale to fine chemical, pharmaceutical, and food industry.

A cooperative MPC protocol is implemented to achieve the control of the plant. The protocol is defined using a simple algorithm to reach an agreement regarding the state of N agents (Olfati-Saber et al. 2007). The MPC controller (Figure 1) receives information from the process unit, in the form of the output variable, y, as well as from the neighbouring units in the form of a cooperation variable, v. Based on these exchanges of information, the controller will correct the input variables, u, to the system.

For adding a monitoring feature to this system, as well as to improve its control capabilities, a MAS system is defined that could be integrated with the MPC. The MAS has two main tasks: a) to decide the optimal connectivity between the distributed MPCs for a safer and better operation; b) to monitor the system and detect any deviation in the behaviour. The proposed MAS consists of follower agents (FAs), a coordinator agent (CA) and a monitor agent (MoA).

The FA's main task to follow its control unit operation and keep track of the equipment behaviour and its relationship with the other control units. The FA could stop the operation of the equipment should

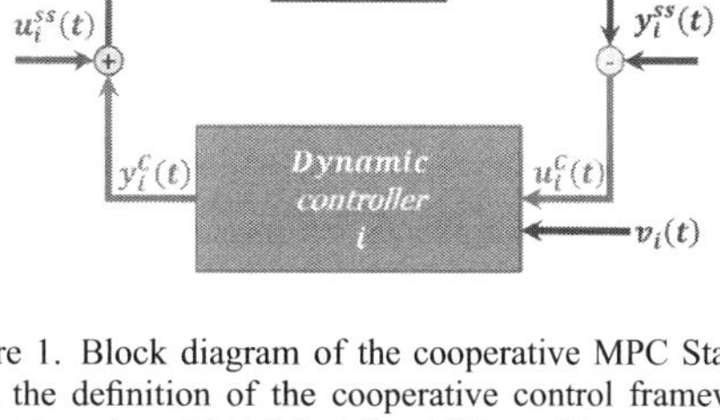

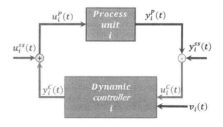

Figure 1. Block diagram of the cooperative MPC Starting from the definition of the cooperative control framework, an ontology-based MAS is defined (Figure 2).

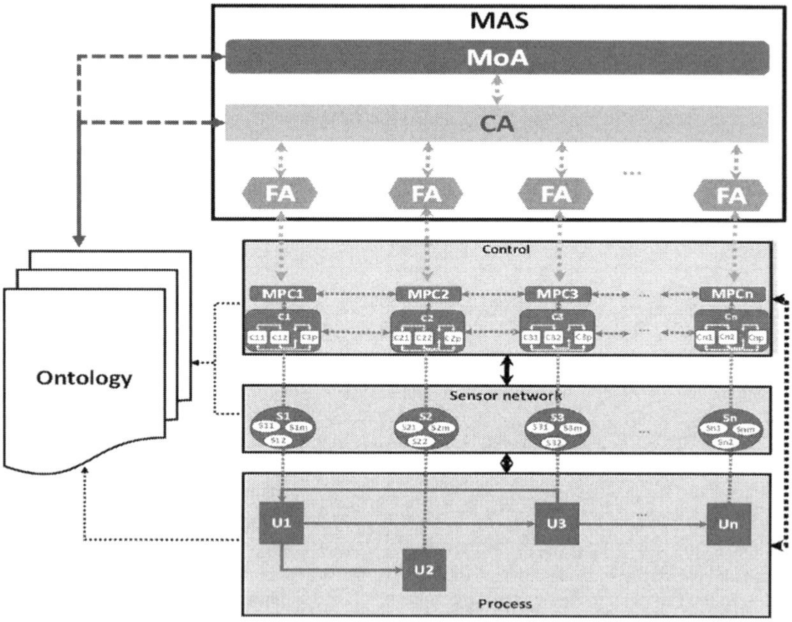

Figure 2. The MPC-MAS-based intelligent monitoring and control framework of chemical processes.

there be an indication from the coordinator agent towards this.

The CA decides the optimal connectivity between the controllers based on the information from the ontology resources and the FAs. It enables better and safer system operation.

The MoA is an upper layer of system monitoring and management. The agent analyses the information gathered from the CA and the ontology to predict abnormalities and system behaviour deviations and report the expected failures and recommendations to deal with them to the operator.

The ontology is a formal representation of the system that includes the system's models, flowcharts, as well as the description behaviour models.

4 CONCLUSIONS

To tackle the chemical plant complexity problem, an ontology-based MAS cooperative MPC framework is introduced. The MAS makes the cooperative MPC controller more efficient by establishing the optimal connected communities. It also takes advantage of the communication between the various elements of the plants which can be enhanced through technologies such as wireless sensor networks, Internet-of-Things, machine-to-machine communications.

Another capability of the proposed system is the monitoring capability. Based on the knowledge from the ontology and the agents' sharing capabilities, the system can detect faster any deviation compared to standard operation. This framework can easily be adapted for other control approaches by very simple modifications in the structure and the objectives. A practical demonstration in a pilot plant environment is envisaged for the future.

ACKNOWLEDGEMENTS

This project is carried out within the framework of the EP/R001588/1 Stepping towards the industrial 6th Sense project. The financial support of the EPSRC is gratefully acknowledged.

REFERENCES

Akmal, S. & Batres, R. 2013. A methodology for developing manufacturing process ontologies, J Jpn Ind Manage Assoc 64: 303–316.

Athanasiadis, I.N. & Mitkas, P.A. 2004. An agent-based intelligent environmental monitoring system, Management of Environmental Quality 15 (3): 238–249.

Badihi, H. et al. 2016. Model-based active fault-tolerant cooperative control in an offshore wind farm, Energy Procedia 103: 46–51.

Batres, R. et al. 1999. A multidimensional design framework and its implementation in an engineering design environment, Concurrent Engineering 7 (1): 43–54.

Batres et al. 2007. An upper ontology based on ISO 15926, Computers & Chemical Engineering 31 (5-6): 519–534.

Batres, R. 2017. Ontologies in process systems engineering, Chemie Ingenieur Technik 89 (11): 1421–1431.

Behdani, B. et al. 2009. Agent-based modelling to support operations management in a multi-plant enterprise,

Proceedings of the 2009 IEEE International Conference on Networking, Sensing and Control: 323–328.
Christofides, P.D. et al. 2013. Distributed model predictive control: A tutorial review and future re-search directions, Computers & Chemical Engineering 51: 21–41.
Dorneanu, B. et al. 2018. Towards the cooperative-based control of chemical plants, Computer Aided Process Engineering 43: 1087–1092.
Dorri et al. 2018. Multi-agent systems: A survey, IEEE Access 6: 28573–28593.
Elhdad, R. et al. 2013. An ontology-based frame-work for process monitoring and maintenance in petroleum plant, Journal of Loss Prevention in the Process Industries 26 (1): 104–116.
Ekaputra et al. 2017. Ontology-based data integration in multi-disciplinary engineering environments: A review, Open Journal of Information Systems 4 (1): 1–26.
Farid, A.M. 2015. Designing multi-agents systems for resilient engineering systems, Proceedings of the 7th International Conference on Industrial Applications of Holonic and Multi-Agent Systems: 1–6.
Ferramosca A. et al. 2013. Cooperative distributed MPC for tracking, Automatica 49 (4): 906–914.
Forbes, M.G. et al. 2015. Model predictive control in industry: Challenges and opportunities, IFAC-PapersOnLine 48-8: 531–538.
Gilchrist, A. 2016. Industry 4.0: The Industrial Internet of Things, Apress, Berkley.
Gruber, T.R. 1993. A translation approach to portable ontology specifications, Knowledge Acquisition 5 (2): 199–200.
Hairui, W. & Yong, Z. 2008. Multi-agent based chemical plant process monitoring and management system, Proceedings of the 4th International Conference on Wireless Communications, Networking and Mobile Computing: 1–4.
Javalera, V. et al. 2010. Distributed MPC for large scale systems using agent-based reinforcement learning, IFAC Proceedings Volumes 43 (8): 597–602.
Kasaei, S.H. et al. 2011. Design and development a control and monitoring system for greenhouse conditions based-on multi agent system, BRAIN 2 (4): 28–35.
Köhler, P.N. et al. 2018. A distributed economic MPC framework for cooperative control under conflicting objectives, Automatica 96: 368–379.
Kravari, K. & Bassiliades, N. 2015. A survey of agent platforms, Journal of Artificial Societies and Social Simulation 18 (1): 11–29.
Lu, Y. 2017. Industry 4.0: A survey on technologies, applications and open research issues, Journal of Industrial Information Integration 6: 1–10.
Mahdavi, I. et al. 2013. IMAQCS: Design and implementation of an intelligent multi-agent system for monitoring and controlling quality of cement production processes, Computers in Industry 64: 290–298.
Marquardt et al. 2010. OntoCAPE: A re-usable ontology for chemical process engineering, Springer-Verlag Berlin Heidelberg.
McArthur, S. et al. 2007. Multi-agent systems for power engineering applications – Part I: Concepts, approaches, and technical challenges, IEEE Transactions on Power Systems 22 (4): 1743–1752.
Müller, M.A. 2011. A general distributed MPC framework for cooperative control, IFAC Proceedings Volumes 44 (1): 7987–7992.
Musulin, E. et al. 2013. A knowledge-driven approach for process supervision in chemical plants, Computers & Chemical Engineering 59: 164–177.
Natarajan et al. 2012. An ontology for distributed process supervision of large-scale chemical plants, Computers & Chemical Engineering 46: 124–160.
Natarajan, S. & Srinivasan, R. 2014. Implementation of multi agents based system for process supervision in large-scale chemical plants, Computers & Chemical Engineering 60: 182–196.
Nikraz, M. & Bahri, P.A. 2005. An agent-oriented approach to integrated process operations in chemical plants, Computer Aided Process Engineering 20: 1585–1590.
Olfati-Saber et al. 2007. Consensus and cooperation in networked multi-agent systems, Proceedings of the IEEE 95 (1): 215–233.
Pohjola, V.J. 2003. Fundamentals of safety conscious process design, Safety Science 41 (2-3): 181–218.
Rocha, R.R. & Oliveira-Lopes, L.C. 2016. A cooperative distributed model predictive control for non-linear systems with automatic partitioning, Computer Aided Chemical Engineering 38: 2205–2210.
Rochefort, Y. et al. 2014. Model predictive control of cooperative vehicles using systematic search approach, Control Engineering practice 32: 204–217.
Saltik, M.B. et al. 2018. An outlook on robust model predictive control algorithms: Reflections on performance and computational aspects, Journal of Process Control 61: 77–102.
Shi, Z. et al. 2017. An ontology-based manufacturing description for flexible production, Proceedings of the 2nd International Conference on Advanced Robotics and Mechatronics: 362–367.
Siti, M. et al. 2005. The use of multi agent system for monitoring and control of the smelting process in the mining metallurgical sector, Proceedings of the IEEE Power Engineering Society Inaugural Conference and Exposition in Africa: 326–331.
Stewart, B.T. et al. 2010. Cooperative distributed model predictive control, Systems & Control Letters 59: 460–469.
Stewart, B.T. et al. 2011. Cooperative distributed model predictive control for nonlinear systems, Journal of Process Control 21: 698–704.
Wan et al. 2015. Online reconfiguration of automatic production line using IEC 61499 FBs combines with MAS and ontology, Proceedings of the 43rd Annual Conference of the IEEE Industrial Electronics Society: 6683–6688.
Wong, P.K. et al. 2017. Engineering applications of intelligent monitoring and control 2016, Mathematical Problems in Engineering 2017 (2945861): 1–2.
Xie, J. & Liu, C.C. 2017. Multi-agent systems and their applications, Journal of International Council on Electrical Engineering 7 (1): 188–197.
Xue, L. et al. 2015. An ontology based scheme for sensor description in context awareness system, Proceedings of the 2015 International Conference on In-formation and Automation: 817–820.
Yang, A. et al. 2008. A multi-agent system to facilitate component-based process modelling and design, Computers & Chemical Engineering 2 (10): 2290–2305.
Zhong et al. 2017. Intelligent manufacturing in the context of Industry 4.0: A review, Engineering 3: 616–630.
Zhou, L. et al. 2017. Towards an ontological infra-structure for chemical process simulation and optimisation in the context of eco-industrial parks, Applied Energy 204: 1284–1298.

Industry 4.0 – Shaping The Future of The Digital World – da Silva Bartolo et al. (eds)
© 2021 Taylor & Francis Group, London, ISBN 978-0-367-42272-1

The transport of domestic animals and the Internet of Things: Guidelines for the development of a smart carrier

C.V.C. Albuquerque & R.A. Almendra
CIAUD, Lisbon School of Architecture, Universidade de Lisboa, Portugal

ABSTRACT: The transport of pets is an ever-expanding market. Although transport companies give some attention to this specific audience, the service they offer is not always as expected and the outcomes can be disastrous. Through exploratory research, we found that these journeys can be stressful and cause great anxiety to those involved, especially when the visual contact between the guardians and their animals is lost. Despite the existence of products that could enable this interaction, when pets and guardians are physically distant, the carriers used in these journeys have not evolved accordingly during the last decades. The objective of this paper is therefore to propose guidelines for the implementation of smart technology in this niche market. Through the Internet of Things, that is, the exchange of data between real-time monitoring systems and autonomous devices it is possible to develop a product capable of providing welfare and safety to these animals.

1 INTRODUCTION

The transport of pets is a field that has not yet been explored to its fullest and lacks documental records. The current legislation varies not only across countries but also across companies within the same country. This variation not only makes it difficult to identify objectively the weaknesses of this service but it also makes it difficult to develop guidelines that would solve the problem at a more comprehensive and international level.

The US Department of Transportation (DOT) is one of the few organizations concerned with collecting statistical information on the incidents involving this type of transport by requiring the US airlines to submit a monthly report (Johnson, 2013). Outside the North American context, this subject is still obscure and the only source of information on what happens is: newspapers, articles, digital magazines or websites that are concerned with this issue (mostly non-governmental organizations and related associations).

The majority of international means of transport, whether by air, land or water, consensually accepted the Live Animal Regulations by the International Air Transport Association (IATA). This document, developed in the late 1970s, which is now in its 45th edition, contains a chapter detailing how containers for the transport of domestic animals should be. Curiously, since then, the design of these containers has barely evolved. However, the bond between guardians and their domestic animals, which for a long time were considered as mere properties, have strengthened considerably. The old and traditional conventions of what is considered a family nucleus are no longer the only acceptable ones, at least in many contemporary societies that have made room for animals to be seen as effective members of a family (Bland, 2013). Due to this, it is not surprising that many establishments, such as restaurants, shopping malls, and hotels, motivated by these cultural changes, have a more pet-friendly attitude. This new posture not only paved the way for a different niche market but also invited people to travel more frequently with their animals.

While the agents responsible for the transport of domestic animals show some concern for their well-being and safety it is clearly not widespread, most companies see and treat them as baggage (Bland, 2013). Unfortunately, because there are no viable alternatives, guardians end up undergoing this standard treatment, even if this causes afflictions and doubts, especially on long-haul flights when contact with animals is impossible.

2 THE PET TRANSPORT CONTEXT

According to the Intergroup president Janusz Woiciechowski (2015), even though the current legislation on the transport of animals exists for more than a decade, there are still reports of animals that are subjected, without any respect for their welfare, to extremely long journeys. According to Woiciechowski, the European Commission has the duty to prosecute those Member States that are continually non-compliant with this legislation and, as a consequence, cause immense suffering. Although this legislation is more suitable for livestock animals the same issues occur in the pet transport context.

"More and more now, families consider their pets to be members of the family and want to include them on trips. Unfortunately, airlines don't consider animals a member of your family. They consider them cargo" (Bland, 2013).

Statistics show that the risk of fatal incidents when travelling with pets in the aircraft cargo compartment is low (Coren, 2012). However, once the animals are shipped as baggage, they are also treated as so (Bland, 2013), in other words, in the same way that baggage can get damaged, so can the animals.

The data presented by airlines in reports about incidents involving domestic animals is, for the most part, far from reliable (Coren, 2012). Access to this type of information is very diffuse and inconclusive in most transport facilities throughout the world. DOT is one of the few accessible sources to promote this type of statistical data. However, even though DOT has established a standard measure that is supposed to force American airlines to submit monthly reports (Johnson, 2013) on possible animal incidents (such as disappearances, injuries or deaths), this information is in fact unreliable. The reason for this is very straightforward: companies do not meet the standards and DOT does not push them to do so (Coren, 2012).

Faced with the lack of concise and reliable information about what in fact happens behind the scenes of domestic animal transport, we chose to investigate isolated cases to try to understand what risks and circumstances might have led to an incident. Fortunately, an increasing number of people who have gone through this experience, many of who suffered grave consequences, are disseminating them through the media and exposing what the airlines try to hide.

"I'm not sure if airlines imagined five years ago that [pet travel] would become the business that it has. We are a mobile world. They are moving and want to bring pets. I hope airlines can continue to carry these pets and continue to focus on these safety standards" (Johnson, 2013).

US airlines are seeking to meet the guardians' needs and are earning around U$50 billion a year. Yet, accommodating pets requires a much more complex system than simply securing a space in the aircraft basement. As an example, to illustrate possible initiatives that could be adopted by transport companies, the United Airlines created the SafePet program which includes training staff on how to handle containers carefully, offers assisted tracking for when pets travel separately and provide air-conditioned vehicles to make the animals more comfortable. (Johnson, 2013).

As mentioned previously there is no comprehensive international standard, each country, region or transportation company develops its own internal rules based on their experience, making this a very broad topic. Despite that, most companies require the animals to be transported inside an appropriate container. Due to this lack of regulation covering all means of transport, most companies eventually adopt the IATA specific requirements for this kind of containers.

There is a subtopic in the IATA's *Live Animal Regulations* (LAR) named "Container Requirement 1" (under the topic "8.3. Container Requirements") that describes the necessary features that this type of container should have. The subtopic specifies the design, materials and necessary ventilation openings for the construction of carriers for these animals, more specifically for dogs and cats. The standards set out therein serve to ensure basic safety and welfare conditions for the animals, as well as for the people involved in these operations. We could assume that, because of these restrictions and the lack of guidelines that relate this niche market to the current technological context, the containers have hardly evolved in the last decades.

3 THE PET PRODUCTS TRENDS

Unlike the transport context, it is notorious that home products for pets have benefit most from the technological advancement and the ever-growing number of smart devices that use the Internet of Things (IoT). Probably because guardians spend many hours away from home, due to longer work shifts, options that make life more practical or that enable communication between them and their animals, even when they are distant, have become more prominent.

We were able to identify three recurring trends that have innovative features: activity monitoring; autonomous functions; and virtual interaction. The first one is a small smart device that can be placed on the leash, such as *Voyce*, *Whistle* or *Fitbark*. Through micro-sensors, these products monitor a number of activities, including heart rate, exercise, distances traveled, calories lost or gained, respiratory and sleep quality. The data collected is transmitted to a mobile app and uploaded to the pet's profile and can be then sent to veterinarians so they can follow more accurately their patients' health conditions.

The second trend we have identified involves products that execute autonomously day-to-day tasks, such as *PetNet* and *Catspad* smart feeders. These products act using sensors to detect the need to replenish the food or water in the containers or work according to a pre-programmed schedule determined by the guardians.

The third one is related to products such as *PetCube* and *Petzy* that allow real-time interactions with animals when the guardians are absent. Through cameras and laser projections, the guardians can record videos, take pictures and also play with their animals and encourage them to exercise.

These trends show the increasing demand from owners to be constantly present in the life of their pets. Even when they are away they want to know what is happening with their four-legged family members. They want to monitor them and be able to act to secure their well being.

4 THE INTERNET OF THINGS AND THE MOBILITY CHALLENGE

When the objects that surround us contain a microcomputer that is connected to the Internet, they can be called "smart", that is, they act more intelligently than those that do not carry similar systems (Fleisch, 2010). These things, or objects — such as radio frequency identification (RFID) tags, sensors, actuators, or mobile phones — are capable of interacting and cooperating with each other through a single addressing scheme to achieve common goals (Atzori et al., 2010). This is the basic concept of the Internet of Things – IoT.

The vast majority of Internet-based services have humans as their main users, but IoT attributes almost completely exclude humans from having to intervene directly. In most IoT applications, smart things communicate with each other or with other computers connected to the network in a machine-to-machine mode (Fleisch, 2010). When humans need to be involved, for example to make decisions, they often act using computers or smartphones in a human-to-machine mode (Fleisch, 2010).

For an IoT system to work, it is essential to have an uninterrupted connection between all devices in the network, so that the data exchange is accurate and complete. When this exchange occurs in an environment where the devices (sensors, robots or autonomous vehicles) are predominantly mobile, the Internet of Mobile Things – IoMT – happens. Although the mobility factor brings with it an even greater openness to new fields of action, it also introduces a challenging issue that is disconnection. The disconnection of nodes and loss of uninterrupted connection has a negative impact on the quality of a network performance, i.e., possible data loss, delay in transfers and possible failures of autonomous functions (Bouaziz et al., 2017).

Regarding the network, classical IoT applications can be categorised into two dimensions: reach and mobility. Reach refers to the geographical distribution of devices. It describes whether the devices are deployed in a small area, i.e., within a few kilometres of each other, or are scattered over a wider area. Mobility refers to the perception of device movements and whether they need to communicate while moving (Dhillon et al., 2017). There is no knowledge of a comprehensive Internet-focused approach to Mobile Things (IoMT) in which connectable devices can be moved or can move independently and yet remain accessible and controllable from anywhere at any time (Talavera et al., 2015).

There are still some issues regarding connectivity and the IoT, which affect the uninterrupted and good quality of its operation and may jeopardise the preventive actions to which the system would be proposed. However, improvement towards the emergence (or the establishment) of new and better communication networks, such as the 5G, a new standard of cellular networks (Dhillon et al., 2017) and the personal area Bluetooth 4.0 (Talavera et al., 2015) are still in development. Therefore, it is expected that in the near future the existing problems encountered to ensure the good performance of an IoT or IoMT system will be reduced, or perhaps completely solved.

5 CROSSING INFORMATION TO UNDERSTAND THE DEMAND

To fully understand the needs of this niche market we decided to combine information from the different parts involved: the transport companies, the target audience and the pet carrier market. By combining these sources of information, we were able to identify the real demands of this service to develop guidelines for its improvement

It is worth mentioning that this paper is a synthesis of a Master's dissertation. To be concise we chose to describe the main highlights of the methodologies used and the most relevant results.

5.1 *The transportation companies*

To better comprehend the fragilities of this market we adopted two different methodologies: a case study and an exploratory research. We first carried out a case study by selecting real reports about incidents that ended tragically. We elected five stories that happened in different parts of the world (China, Germany, United States, Brazil and Portugal) and two different kinds of transport (airplane and bus). By analysing these cases we could identify two main problems: death caused by a long exposure to extreme weather conditions and the impossibility to help the animals because they were inaccessible.

For the second methodology we collected as much information as possible on the internal rules from a series of means of transport. For that, we selected a sample of transport companies, such as: *TAP Air Portugal, Delta Airlines, and National Rail*, among others. Additionally, a sample of specialised websites was also selected, such as: *PetTravel, Petmate, Petplan, etc.* Our goal was to develop an accessibility scale to indicate whether or not each means of transport allow guardians to rescue their animals in case of danger. Finally, we created a pyramid graph where at its base is the most accessible means of transport and at the top the most restrictive one (Figure 1). By analysing the rules applied by the companies and the government we concluded that airplanes are least accessible means of transport, followed by buses, the ferry boats or cruise ships, trains and, at the bottom, cars as the most accessible ones.

These two methodologies make it clear the need for guardians to have the perception of being

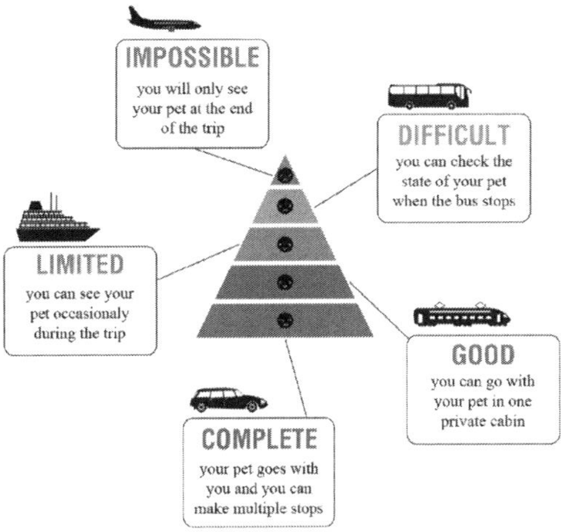

Figure 1. Level of contact tutors can have in transport with their respective pets on a scale from the impossible to the complete (Albuquerque, 2018).

able to help their pets even though they are not physically in contact with them. The need to be ubiquitous points to a solution that allows a virtual interaction between guardians and their animals. This feature, as mentioned previously, is already found in pet products but has not yet been used in pet carriers.

5.2 *The target audience*

We conducted an online questionnaire about the transport of pets to be able to better understand the target audience. The questionnaire was created using the *Google Forms* digital tool, and it was accessible from October 31st to November 2nd, 2016. We chose to conduct it online so that we could reach as many people as possible, since sharing over the Internet enables a wider dissemination.

In this way, the objective of the online questionnaire was to trace the profile of the people who usually travel with animals; to outline their experience, collect their opinion about the existing pet carriers and, finally, to offer the opportunity to make suggestions and to share their experience.

The online questionnaire obtained 98 responses. Among the respondents: 43.9% were between 26 to 35 years old; 27.6% between 18 and 25 years old; 13.3% from 36 to 45 years old; 10.7% from 46 to 55 years old; and 4.5% were 56 years old or above. Regarding gender, 88.8% were female and 12.2% male.

We discovered that 91.8% of the respondents preferred to travel by car, 84.4% stated it was very important to maintain physical and/or visual contact with the animals, 81.3% considered it very important to allow the animal to get familiarised with the carrier before the trip and 79.6% stated that they would feel more secure if they could monitor the animal during the journey. These were some of the highlights we obtained using this approach.

In conclusion, using all the information gathered we were able to create a group of personas that represents the main public we should address our research to. In short lines: an audience that is mostly female, who have small to medium-sized dogs, are experienced in travelling with animals and are unsatisfied with the options available in the market.

5.3 *The pet carrier market*

Since changing the whole system of transport is a long-term goal, we opted to investigate short-term alternatives. For this reason, we decided to make a comparative analysis of the carriers that stand out the most in the international market. We selected six options that fit into a list of characteristics considered pertinent to the investigation. Of the six products we analysed, two were not containers, but presented good solutions related to this market, so we decided to add them to the analysis. The selected products were: *Gunner Kennel's G1 Series; Skudo's Travel Carrier; Curver's Bunkbed 3 in 1; Aridus Den; Sleepypod Air and Raden Smart Luggage.*

Once chosen, we listed the positive and negative elements of each of the products and then compared them and obtained a group of 15 characteristics. In the end, we were able to do a SWOT analysis (Figure 2) of this market that allowed us to identify which paths to follow and which ones to avoid.

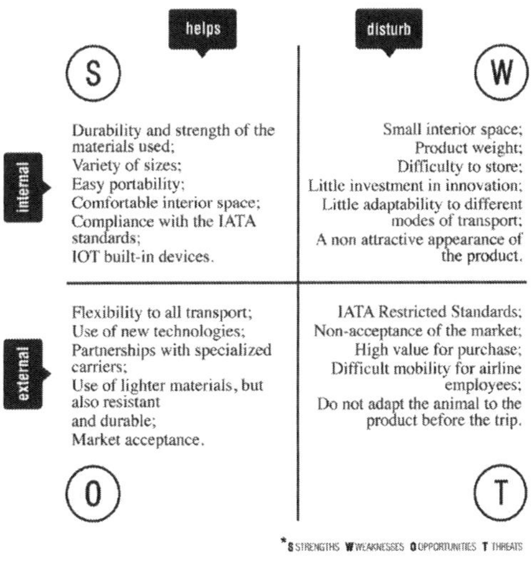

Figure 2. SWOT analysis of the pet carrier market (Albuquerque, 2018).

6 RESULTS

The data obtained during the investigation allowed us to create a set of guidelines that will help to create a smart carrier especially designed to transport pets. These guidelines aim to improve the quality of this type of service, considering the well being and safety of these animals, as well as reducing the stress and anxiety the guardians may experience during a journey.

Why a smart carrier? By crossing the information from the different factors that influence pet transport we decided that it would be more effective to change the common denominator to all means of transport and laws, i.e. the use of a carrier. For this new carrier to be able to meet all the needs that were identified, including: temperature control, activity monitoring, remote aid, previous familiarisation, it must have a built-in IoT system that allows a machine-to-machine interaction (to control all devices autonomously) and a machine-to-human interaction (so that guardians can monitor and/or provide aid remotely).

6.1 *Guidelines for the development of a smart carrier*

- The carrier should have an embedded IoT system to allow an autonomous performance when the data collection sensor detects a risky situation.
- The carrier should have embedded sensors to constantly assess the condition of the internal environment.
- The carrier should have an autonomous temperature control system to act in case of exposure to severe weather conditions.
- The carrier should have a real-time monitoring system so that guardians can watch over their animals even though they are physically distant.
- The carrier should have a double rigid plastic wall to ensure the safety of the animals in the event of accidents (a car crash, for instance).
- The carrier should have an attractive appearance to encourage its use prior to travel in the home environment.

7 DISCUSSION AND FUTURE WORK

The purpose of this article was to promote the improvement of a market that is little discussed and hardly changed in recent decades. Additionally, it is the first step towards developing a smart carrier for pets. However, even if we have formulated a set of guidelines for the development of this type of product, we do not have empirical information as to whether a product endowed with these characteristics will be able to improve the welfare and safety of animals when travelling.

As airlines do not divulge what actually happens when animals are harmed (Coren, 2012), we imagine that the existence of a real-time monitoring system would generate evidence against the company itself. That said, we believe that the first obstacle to the implementation of these guidelines would be its acceptance by the transport companies.

The second obstacle could be the IoT system because there are still issues to maintain an uninterrupted connection and if the devices are mobile (IoMT) the problem could be substantially more difficult to resolve (Bouaziz et al., 2017). Probably the biggest challenge would be the human-to-machine interaction because of the distance between the user and the product and because its mobility. Since the machine-to-machine interaction takes place within shorter distances and demands a less complex communication network, it should not present connectivity problems.

To know if a smart carrier will solve or at least mitigate the problems involving this type of transport a prototype should be developed and submitted to usability and safety tests. Nevertheless, even if this prototype achieves the expected objectives in a simulated testing environment, its capabilities will only be truly proven when it is subjected to actual journeys using different modes of transport.

Despite the difficulties that may be encountered, we believe this is the correct path to take. Airlines may be pressured to adopt more transparent policies to meet their customers' requirements. Hopefully, in the near future, new and more powerful communication networks will be able to maintain an uninterrupted connection. New generations of cellular network (Dhillon et al., 2017) or personal area networks (Talavera et al., 2015) may solve the problem of mobile devices.

REFERENCES

Albuquerque, C. V. C. 2018. *Casota Smart: Contentor Híbrido para o Transporte de Animais Domésticos em Viagens de Longa Duração*. Lisbon: Unpublished Master's Dissertation from the Faculty of Architecture of the University of Lisbon.

Atzori, L, Iera, A., Morabito, G. 2010. *The Internet of Things: A survey*. Article in Elsevier webpage. Accessed November 21, 2017. www.elsevier.com/locate/comnet.

Bland, A. 2013. *Is Taking Your Pet on an Airplane Worth the Risk?* Article in Smith Sonian Mag webpage. Accessed November 11, 2016. http://www.smithsonianmag.com/travel/is-taking-your-pet-on-an-airplane-worth-the-risk-6241533/?no-ist.

Bouaziz, M., Rachedi, A., Belghith, A. 2017. *EC-MRL: An energy-efficient and mobility support routing protocol for Internet of Mobile Things*. 14th IEEE Annual Consumer Communications & Networking Conference (CCNC).

Coren, S. 2012. *Is It Safe to Ship Dogs or Cats by Air?* Article in Psychology Today webpage. Accessed November 4, 2016. https://www.psychologytoday.com/blog/canine-corner/201209/is-it-safe-ship-dogs-or-cats-air.

Fleisch, E. 2010. *What is the Internet of Things? An Economic Perspective*. Zurique: Auto-ID Labs White Paper.

Johnson, M. 2013. As Pet Deaths Continue, Airlines Pressured to Change Their Ways. Article in CNN webpage. Accessed November 7, 2016. http://edition.cnn.com/2013/01/11/living/mnn-pet-airline-deaths/.

Talavera, L. E., Endler, M., Vasconcelos, R., Cunha, M., Silva, F. J. S. 2015. *The Mobile Hub Concept: Enabling applications the Internet of Mobile Things*. The 12th IEEE International Workshop on Managing Ubiquitous Communications and Service.

Industry 4.0 – Shaping The Future of The Digital World – da Silva Bartolo et al. (eds)
© 2021 Taylor & Francis Group, London, ISBN 978-0-367-42272-1

Development of production monitoring systems on labor-intensive manufacturing industries

R. Hartono
Universitas Pasundan, Bandung, Indonesia

S. Raharno, M.Yanti Pane, M. Zulfahmi, M. Yusuf & Y. Yuwana
Institut Teknologi Bandung, Bandung, Indonesia

H. Budi Harja
Politeknik Manufaktur Bandung, Bandung, Indonesia

ABSTRACT: This paper discusses how to monitor production activities in the workshop where most of its production activities are carried out manually. The product is made through five stages of the manufacturing process on five workstations. Currently, all production activities are done almost without planning so that production targets are difficult to evaluate. Therefore, a production management system is needed to enable an easy evaluation of the production target. The first step is to monitor all production activities that occur inside the workstation. Each workstation is added with a monitoring device that will record the time of login/logout of materials, operators, and tools to/from workstations. In addition, the monitoring device also records the real start and finish operation time. This monitoring system, material stock, operator working hours, tool time usages, production targets, material needs, and several other issues related to production activities can be easily handled in a timely manner.

1 INTRODUCTION

The Technology of Information and communication have been experiencing a rapid development while, on the other side, it influences a new lifestyle, a new form of society which could not have been prediction before. Society nowadays can fulfil their needs easily through support by robust internet facility in wide range. The internet has an important role of helping to form this current society, digital society. Consumption lifestyle of goods and services also easily change when new information or new trend going viral. Lifestyle change puts pressure on suppliers of goods and services, they also must change the way they do business. The good providers are the ones who fulfil society needs more effectively than competitors.

The manufacturing industry as a supplier of goods is also part of the business that needs to be changed to keep pace with today's digital society. The Manufacturing industry needs to respond fast to consumer needs with smart and appropriate way without additional costs. They need to act smart by implement the new concept of technology which is Industry 4.0.

The concept of industry 4.0 has been developed by many researchers as an effort to create new form of manufacturing industry which can have quick response and adaptive to the change of rapid consumer needed. The concept of industry 4.0 contains two main factors, which are interoperability and consciousness. Interoperability consists of digitalization, communication, standardization, flexibility, real-time responsibility, and customizability. Predictive maintenance, decision making, intelligent presentation, self awareness, self optimization and self configuration consisting of awareness. In other words, this concept means to form a collaboration of information and communication technology to be applied in manufacturing industry (Qin et al. 2016).

This cooperation requires several hardware supports such as good sensor, central database, communication device, and so on to be applied to manufacturing industry. All this items is integrated to be one system which called Cyber-physical System or CPS. CPS is a cooperation of computational entity which connect the physical item and surrounding intensively which could capture all on going event real time also provide and use access data service to process data with internet connection in the same time (Liu & Jiang 2016).

The first concept of industry 4.0 was introduced in 2011 in Germany. During the launch it was said that this concept will be easily implement to automated manufacturing industry which use robot, automatic transportation, automatic machine and other automatic device in their production unit. For countries with populations that are lacking in productive age can also be a reason to apply this concept. However, this concept does not stop there; many frameworks can be developed as long as they still have two main factors, namely interoperability and awareness.

Could the concept of industry 4.0 implement in manned manufacturing industry? Is this concept implementation require automatic device? Is the industry 4.0 implementation target means reduce in workers?

In the developing country, such as Indonesia, the most industry is manned manufacturing industry. Production process involved non-automated machine and a lot of workers. Industry problems in developing countries are the lack of effectiveness and efficiency caused by delays in raw materials from suppliers, excess stock which means cost burdens, lack of optimization of assets such as machinery and workers, unbalanced workloads, inaccurate production costs, delays in fulfilling orders to customers, and other factors related to the management of the production process on the shop floor. Most of the problem cause are happened due to unsmooth information flow between several production unit. This accurate information flow is the main factor of good production plan which will reduce the problem cause above.

In Indonesia, the concept of industry 4.0 appears to solve problems in the manufacturing industry whose main target is to increase the effectiveness and efficiency of production, less production costs, and fulfilment of orders on time. This concept should develop without the consequence to change the existing line production, change the non-automatic machine to be automatic or any un-needed additional cost. The implementation also should have less effect to society because of employee rationalisation.

The next chapter will discuss about the research of one of manned manufacturing industry in Indonesia from un-planned manufacturing to be planned manufacturing. Second chapter will tell about the current workshop condition. Third chapter will tell about the implementation of the concept of industry 4.0 which will start by first phase which is monitoring. Monitoring means knowing the real time activities of many events in the workshop which will help improve the effectiveness and efficiency of the production process by implementing planned production. Fourth chapter will explain about the simulation and and last chapter 5 will tell about conclusion.

2 CURRENT WORKSHOP CONDITION

2.1 *Product*

The product of this workshop is an air conditioning for train carriage. This product is put above the passenger train carriage to control the air conditioning inside the carriage. This product consists of four air conditioners which are placed in the main casing. Figure 1 is shown the product from above.

2.2 *Workshop lay out*

This workshop is divided into several areas. All these areas are then called work station (WS). Each

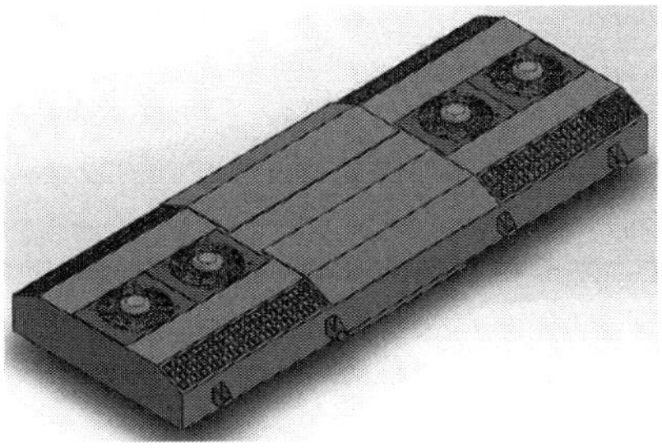

Figure 1. AC for passenger train carriage.

work station also been group by their activity type. First group is storing. Work station which include in this group is warehouse. This workstation is used to store material which received from out of workshop and to be sent to out of the workshop.

The second group is called main production line. This production line is consisting of two production line which consist of five workstation each. So, in total ten workstation is part of this group. Activities in this main production line are product assembly and inspection. Duration of process for each workstation assume similar.

The third group is called minor production line. Activity in this production line is preparing materials for assembly process in main production line. This group is consisting of four work station, which are cutting insulation WS, wiring assembly WS, piping assembly WS, and Sealant and Insulation WS. The workshop layout can be seen in Figure 2.

3 OPERATION MONITORING IN WORKSHOP

3.1 *Modelling the workshop*

To monitor the real condition of operation at the workshop, the physical systems need to be configured into virtual system model. Main physical systems that need to be configured into virtual system model are consisting of product type model, line production model, order model, product model and operation model. Product type model is the representation of the product design that need to be made. Product type model is combination of hierarchy model, structure model, and process model which related each other. The product type model can be seen in Figure 3.

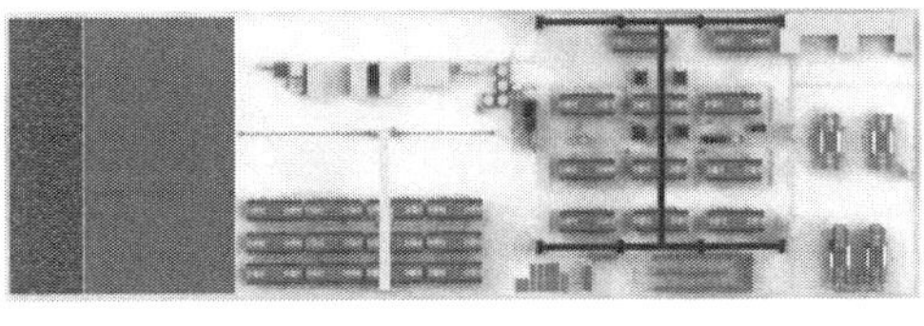

Figure 2. Lay out workshop (top view).

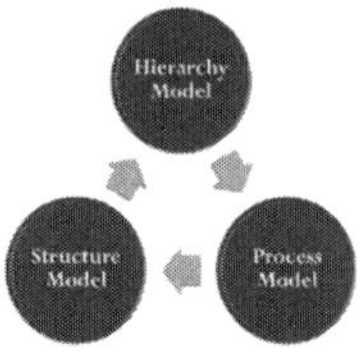

Figure 3. Product type model.

Hierarchy model explain about the levelling of a product from its brand, model, and Variant. Structure model explain about components that compile the product. Process model explain the processes of each component from raw material into become a final product. The hierarchical model, structural model, and process model can be seen in Figures 4 and 5.

Production line model describe about the arrangement of the workstation which will be used in process production. Processes for every component is mapping to the planned workstation, because each workstation is dedicated to certain process. This model can be seen in Figure 6.

Order model describe about the order that made by customer. Order model has attribute of product type order, quantity, and due date of when final product needs to be ready. From this order then product model can be made. Product model is the representation of real product in the workshop. Ordering models and product models can be seen in Figure 7.

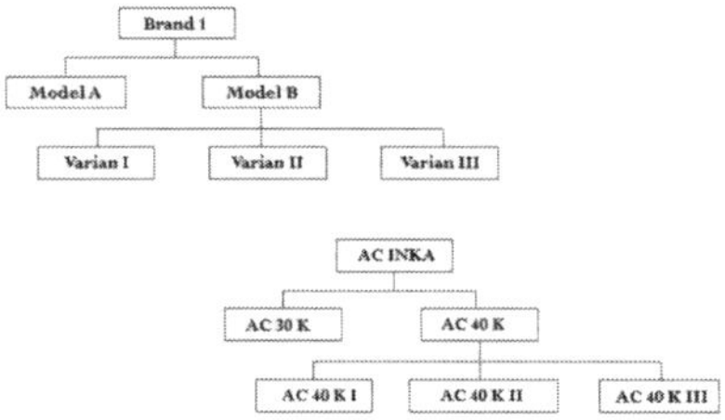

Figure 4. Hierarchical model.

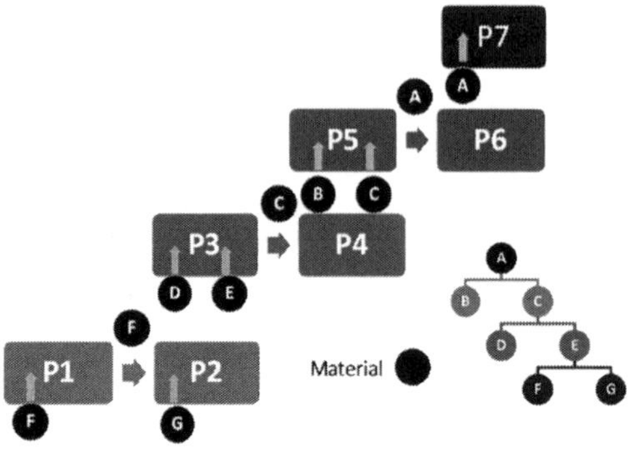

Figure 5. Process model and structure model.

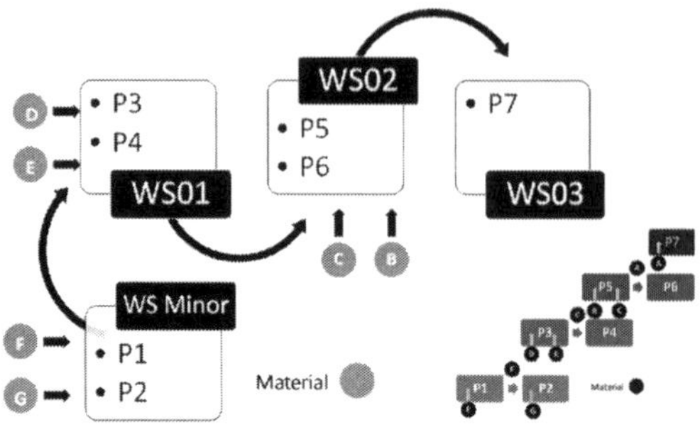

Figure 6. Production line model.

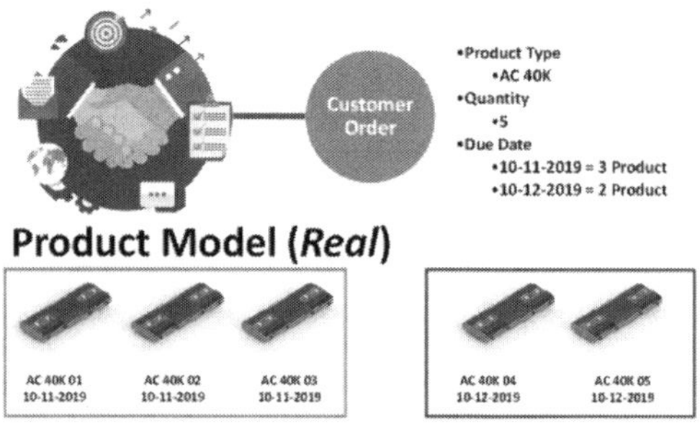

Figure 7. Order model and product model.

From the product model then operation model can be made. Operation model is the transformation of process model into actual model. The relationship between the operating model and the process model can be seen in Figure 8.

Operation model has several attributes which are materials, operator, tool, machine, workstation, scheduled start operation, scheduled finish operation, actual start operation, and actual finish operation. The operation attribute is then arranged by the schedule start operation and workstation. This will form matrices of operation schedule in a certain workstation. These operations need to be monitored. The operation matrix can be seen in Figure 9.

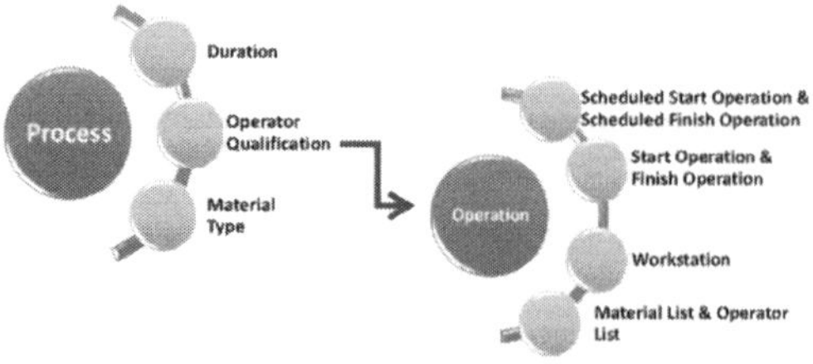

Figure 8. The relation between operation model and process model.

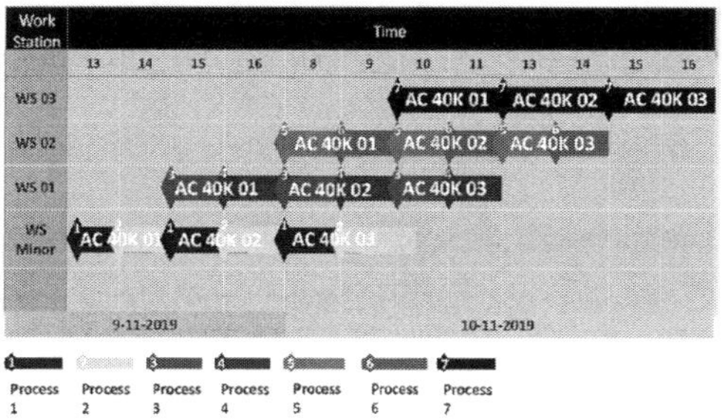

Figure 9. Operation matrices.

3.2 *Monitoring method in workshop*

Workshop is divided into several workstation which coincide each other thus no empty space between the workshops. Each workstation is support by operation monitoring device. Type and quantity for the device is manage as needed. The concept of monitoring method is shown in Figure 10.

The monitoring device that been installed at the workstation is used to record the material and tool event when enter and exit the workstation, also the operator when login and logout of the workstation. The device also records the actual start and finish operation. When any event is happened in a certain workstation, the device will renew the status of each model in virtual system according to physical system that represented.

3.3 *Operation monitoring device*

An operation monitoring device is consisted of minicomputer, barcode scanner, RFID reader, and a monitor. Minicomputer is used to receive signal from barcode scanner and RFID reader and process the signal which then send the signal to server. Barcode scanner is used to scan the material which enter/exit from/to workstation. RFID reader is used to record the tome of operator login/logout from/to workstation. Printer is used to print the serial number of materials which been an output of each operation. Monitor is used to show operation status in each workstation.

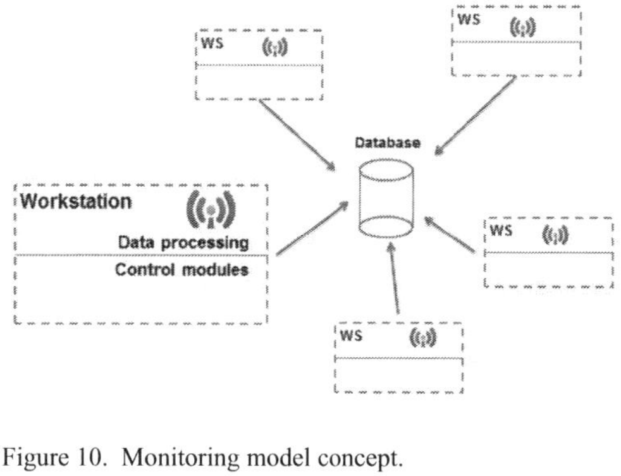

Figure 10. Monitoring model concept.

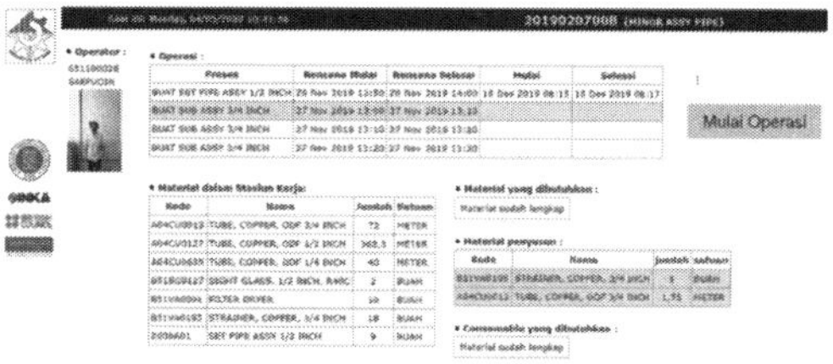

Figure 11. Interface of monitoring device.

3.4 *Interface of monitoring device*

Monitor in each workstation is show the operation status which happens at the dedicated workstation. Status which been shown is including the operator in the workstation, material in the workstation, material that need to be provided for the current operation but not available yet, the latest operation that had been done, current operation, and next operation that need to do. An example of the interface is shown in Figure 11.

4 SIMULATION

Monitoring devices that had been made need to be simulated before executing in the actual operation. Simulation is done by scan the barcode which represent actual material that enter the workstation. When a material had been scanned in a certain workstation, the material location then moves to new location in the database.

Operator login/logout from/to workstation also simulates by tap the RFID card into RFID reader. At the first tap, the operator is considered login to the workstation. When the same operator makes a second tap, it then considers that operator had been logout from the workstation.

The record of actual start operation is begun when the operator touches the start operation button at the screen monitor. All material that been used for certain operation also recorded. The actual finish operation is recorded when the operator touches the finish operation button at the screen monitor. When the finish operation button been touched, the system also will create serial number of materials to dedicated material in certain workstation.

The workstation will also have a central monitor that will display the activity of the workshop which consist of the several workstations. This will help the management to know what happened in each workstation without visiting there anytime. If the workstation shows no operation for a long time, it could be mean a problem and need to be checked.

Several data could be got from the simulation. This data also could be manipulated to give benefit information to the company.

From the simulation, event of material enter/exit from/to workstation can give information about material stock, quantity, and material position. It could be monitored anytime.

Material data combined with purchase data can provide information when purchasing material experiences a delay in delivery. This can be valuable information to evaluate the supplier.

From login/logout operator data and the operating time can give actual record of operator attendance time and their working time. The management also could be informed when several schedules is delay from target.

5 CONCLUSIONS

This system is developed from the concept of industry 4.0 in manufacturing industry, which could be generally used in developing countries. The early phase of the concept used the definition of CPS and applied it into monitoring systems. This research work is mostly about how to build and define this monitoring system from actual condition into a virtual model, which can be recorded and controlled.

From this system, many data can be captured and used as needed. The data gathered from system database could give valuable information to the management for decision making in production planning.

REFERENCES

Liu, C. & P. Jiang (2016). A cyber-physical System Architecture in Shop Floor for Intelligent Manufacturing. *9th International Conference on Digital Enterprise Technology. Procedia CIRP 56*, 372–377.

Qin, J., Liu, Y. & R. Grosvenor (2016). A Categorical Framework of Manufacturing for Industry 4.0 and Beyond. *Changeable, Agile, Reconfigurable & Virtual Production Conference. Procedia CIRP 52*, 173–178.

Industry 4.0 – Shaping The Future of The Digital World – da Silva Bartolo et al. (eds)
© 2021 Taylor & Francis Group, London, ISBN 978-0-367-42272-1

Importance of maturity assessments in promoting adoption of i4.0 concepts in SMEs

H.F. Castro, A.R.F. Carvalho & F. Leal
Intelligent & Digital Systems, R&Di, Instituto de Soldadura e Qualidade, Grijó, Portugal

H. Gouveia
R&Di Programmes, R&Di, Instituto de Soldadura e Qualidade, Taguspark, Oeiras, Portugal

ABSTRACT: Most European countries have developed oriented initiatives to widespread the Industry 4.0 (i4.0) paradigm and boost its implementation within their economies to take full advantage of its potential value creation. However, the reality shows that this effort doesn't reach most companies, particularly SMEs. Previous studies indicated that Portuguese SMEs need different approaches to increase their knowledge about i4.0 concepts. An effective way of assisting SMEs to grasp its actual condition in terms of preparedness to face i4.0 challenges consists in running an on-line diagnosis. This paper contributes with a methodology developed by ISQ in Portugal, with the aim of stimulating the awareness and adoption of i4.0 concepts by SME. In this context, we present a platform, named SHIFTo4.0, which assesses the level of i4.0 maturity and proposes multiple guidelines considering 6 diferent dimensions. As result, we present a case study which demonstrates the importance of the SHIFTo4.0 on-line platform.

1 INTRODUCTION

Industry 4.0 (i4.0) is undoubtedly one of the most challenging issues of today's developed economies. The i4.0 concept encompasses: (*i*) digitalization of horizontal and vertical value chains; (*ii*) new enabling technologies; (*iii*) innovation in products and services; and (*iv*) the creation of new business models. Due to the its high impact, it is the central focus of the 4th industrial revolution.

Technologies have a key role in the evolution of conventional operations, methods, and businesses, leading organizations to rapidly transform and adapt for the new emerging economic scenarios. Therefore, the adoption of i4.0 principles becomes a necessity which can be more easily satisfied by large organizations/companies. On the other hand, small and medium enterprises (SMEs) need higher support to implement the majority of the i4.0 concepts. To take advantage of the i4.0 benefits, SMEs need to get on board in a challenging journey that includes the modernization of the legacy systems and processes. Towards this scenario, it is important to show the benefits of i4.0 adoption to SMEs through pilot demonstrations of specific business cases in order to catch their attention.

Aware of this reality, ISQ has launched in Portugal a project, named SIM4.0 – Intelligent Monitoring Systems 4.0. This project aims to increase the SMEs awareness on i4.0, promoting the adoption of its concepts. The goal is to develop pilot demonstrations of i4.0 solutions for the shop floor. In addition, the SIM4.0 project contemplates awareness i4.0 activities such as open workshops and seminars. These activities showed that SMEs need a different approach to get involved in the i4.0 paradigm and its challenges.

The main SMEs concerns are: (*i*) understanding the implications and impacts of i4.0 in their environments; (*ii*) knowing their i4.0 maturity level; and (*iii*) how to start a i4.0 project (*e.g.*, a pilot) in their companies. The standard workshops on i4.0 are too abstract, providing scant clues concerning how to approach a i4.0 transformation. It indicates that the majority of SMEs need a guidance and even a roadmap to develop and implement i4.0 concepts as well as to define priorities.

Based on these findings, ISQ developed an online self-assessment platform[1] that allow the company to get a picture of its maturity level regarding i4.0. It is based on multiple questions which consider 6 dimensions with 18 different fields encompassing: strategy and organization, smart factory, smart operations, smart products, data-driven services, and human resources. This paper presents a description of the platform as well as the maturity model to assess the readiness of the Portuguese SMEs for i4.0. Additionally, the platform provides a report describing the maturity level of the company.

Moreover, it is provided a set of recommendations to implement in order to achieve a higher maturity level considering a time frame of 5 years. The report aims to be a discussion document, assisting the company to reflect on: (*i*) its current capabilities; (*ii*) strategies; and (*iii*) action plans. As result, we present a case study which demonstrates the importance of the SHIFTo4.0 platform.

This paper is organised as follows: Section 2 presents a brief review of i4.0 maturity models, Section 3 describes the base model and methodology used in this study, whereas Section 4 describes some examples of its application. Finally, Section 5 summarises some of the outcomes of this work.

2 BRIEF REVIEW OF I4.0 MATURITY MODELS

Several models for assessing the readiness or maturity have been published and reviewed concerning i4.0 domain (Gökalp et al. 2017; Mittal et al. 2018). There is an extensive number of approaches to assess the maturity level of companies. According to ISO/IEC (2015), the maturity models should be objective, impartial, consistent, repeatable, comparable, and representative of the assessed organizational units.

Schumacher et al. (2016) suggest a maturity index that can be used to calculate the maturity level of an SME concerning i4.0 technologies as well as digital and smart practices. The maturity index is based on various organizational aspects. To validate this index empirically, they deployed qualitative and quantitative methods. The index involves nine dimensions: (*i*) strategy; (*ii*) leadership; (*iii*) customers; (*iv*) products; (*v*) operations; (*vi*) culture; (*vii*) people; (*viii*) governance; (*ix*) technology. This model contemplates a set of relevant questions about i4.0. However, it is imperative that the company managers need to have a good understanding about i4.0 topics in order to explore further the concepts in their companies.

Schuh et al. (2017) developed a maturity index with the help of a four-stage methodology, known as Acatech report. The maturity index considered six development stages: (*i*) computerization; (*ii*) connectivity; (*iii*) visibility; (*iv*) transparency; (*v*) predictability; and (*vi*) adaptability. The Acatech report's objective was to define a maturity index that can evaluate the present i4.0 stage of the organization and find the recommendations, which can lead them to a higher maturity stage. The model finds the present maturity level of the organization guiding towards the next level of maturity. However, this is not a self-assessment instrument, being necessary to have a considerable level of understanding to answer the survey questions. As a result, from the SMEs viewpoint, this approach might be expensive and time-consuming.

Geissbauer et al. (2016) assessment model is based on four stages: (*i*) digital novice; (*ii*) vertical integrator; (*iii*) horizontal collaborator; and (*iv*) digital champion. In addition, it incorporates seven dimensions: *(i)* digital business models and customer; (*ii*) digitization of product and service offerings; (*iii*) digitization and integration of vertical and horizontal value chains; (*iv*) data analytics as core capability; (*v*) agile IT architecture; (*vi*) compliance security, legal and tax; and *(vii)* organization, employees and digital culture. Furthermore, Geissbauer et al. (2016) recommend the following steps for the digital success: (*i*) map the i4.0 strategy; (*ii*) create initial pilot projects; (*iii*) define needed capabilities; (*iv*) become a 'virtuoso' in data analytics; (*v*) transform into a digital enterprise; and (*vi*) actively plan an ecosystem approach. Geissbauer et al. (2016) assume the presence of digitally integrated supply chains that require a real-time update of the product with the collaborators. This scenario may not be possible for the SMEs because of financial constraints. Finding an economical and high tech supplier might be time-consuming for the SMEs which already have a high dependence concerning customers/suppliers.

"IMPULS – Industrie 4.0 Readiness" is another important readiness (Lichtblau et al. 2015). It is based on a comprehensive dataset which includes details about dimensions, items, and the approach for assessment. The IMPULS model is scientifically well grounded where its structure and results are explained transparently. This maturity model is well fitted to SMEs, due to its comprehensive and simple approach. This model is based on 5 levels of readiness, which clearly defines boundaries between each level. This allows to establish a set of rules that can act as advices to increase the level of readiness. "Level 0" indicates that the company does not meet any requirements of i4.0. The transition from "level 0" to "level 1" may include a drastic change in the organizational culture of the company once it involves the adoption of new technologies, learning skills, etc. Therefore, this shift may probably take longer, requiring more resources, and involving more unforeseen problems in comparison of the transition from "level 1" to "level 2". It is important to consider both the present level of its organization and the move towards the i4.0 vision using a self-assessment tool.

Realizing the benefit of such approach, ISQ developed a model and tool that would assist Portuguese companies, particularly SMEs, to prepare themselves for future challenges derived from the new industrial paradigms. In this work, we present SHIFTo4.0 on-line self-assessment tool, adapted from VDMA to the Portuguese reality, which presents 3 essential features recommended by the literature (Bruin et al., 2005) (*i*) descriptive, *i.e.*, characterizes the i4.0 maturity automatically using a self-assessment tool; (*ii*) prescriptive, *i.e.*, presents an automatic guideline/roadmap to achieve a desired level in a defined time frame; and (*iii*) comparative, *i.e.*, performs a benchmark within the Portuguese reality.

3 BASE MODEL & METHODOLOGY

3.1 *Selection of the base model*

After an exhaustive review on self-assessment and maturity models for i4.0 readiness, ISQ decided to take as main reference, the IMPULS for Industrie 4.0 from VDMA. The six key dimensions of this model are: strategy and organization, smart factory, smart operations, smart products, data-driven services and human resources. It allows to build both a structured and focused insight about SMEs i4.0 maturity and a strategic action plan to improve their performance (i.e., maturity level). Additionally, the model contains attributes, such as, close interaction and involvement with German leading companies and experts as well as a validation from the mechanical engineering industry. As a result, we obtained an on-line self-assessment tool named SHIFTo4.0.

3.2 *SHIFTo4.0 methodology*

SHIFTo4.0 explores the structural attributes, motivators and hurdles of the Portuguese companies. It is built upon the requirements of the IMPULS model.

Our model is based on 6-dimensional assessment that is the core of i4.0: strategy and organization, smart factory, smart operations, smart products, data-based services, and human resources. The core of the survey is the definition and the measurements of the 18 indicators used to describe in detail the 6 dimensions.

SHIFTo4.0 survey contains several conditional questions that depends of previous answers in order to obtain a more fluid and logical completion of the survey.

To acquire enough information from the company regarding i4.0, we formulate 40 questions. The number of questions used is a result of a trade-off between information needed for the model and the respondents answering time. The SHIFTo4.0 survey has 2 knockout dimensions: smart factory and smart products. It allows to include companies from any economic domain, namely SMEs that provide services instead of goods.

In the first part of the survey, the respondents are asked to provide information about the structure of their companies. This information is used primarily to ensure that the survey is representative. The second part of the survey contains general questions about i4.0 to identify the level of implementation regarding i4.0 concepts. Companies are required to provide feedback on their i4.0 strategy implementation, their equipment infrastructure functionalities, the collected data, autonomous production, data-driven services, and employees. Additionally, companies are asked about the main drivers in implementation of i4.0 solutions and the main obstacles.

Considering the answers, the companies are classified between 0 to 5 level according to their i4.0 maturity. Each level is named as follows: (*i*) 0 - Outsider, (*ii*) 1 - Beginner, (*iii*) 2 - Intermediate, (*iv*) 3 - Experienced, (*v*) 4 - Expert, and (*vi*) 5 - Top performer.

The final classification of maturity is obtained using the weighted average, as follows (Lichtblau et al., 2015): (*i*) strategy and organization - 25 %; (*ii*) smart factory - 14 %; (*iii*) smart operations - 10 %; (*iv*) smart products - 19 %; (*v*) data-based services - 14 %; and (*vi*) resources human - 18 %. The maturity level of each dimension is based on the lowest classification obtained by one or more fields within that dimension. Each question is classified according to some pre-set requirements, and the field is classified based on the lowest classification obtained by one or more questions within that field.

The survey sampling pool was obtained by an empirical data collection gathered by the SHIFTo4.0 online survey, with the collaboration of IAPMEI (Portuguese Agency for Competitiveness and Innovation of SMEs), and some Portuguese economic clusters.

The SHIFTo4.0 tool provides a final report with a description actual condition of the company regarding i4.0. The report provides a graphical representation of the level of maturity of each dimension as represented by Figure 1. Moreover, the tool recommends a set of guidelines based on the desired maturity level, within a time frame of 5 years. This report constitutes a reflection tool to deploy and improve i4.0 readiness and maturity of company, in compliance with the ISO/IEC (2015).

4 EXAMPLE OF APPLICATION

SHIFTo4.0 is ready for providing i4.0 maturity evaluations. In this paper, we present an example of application considering a foundry industry. Firstly, the company has obtained a maturity level of 1.14, as shown in Figure 1. Considering this assessment, the SHIFTo4.0 provided an evaluation report with a roadmap which incorporates an action plan for a timeframe of 5 years.

This roadmap identifies several actions to take, such as the need a conducting analysis of the production chain, and an autonomous control introduction. Furthermore, the report suggests getting academic collaboration and/or knowledge from research institutions.

Following the report recommendations, the company applied humidity sensors in the circuit line which, in conjunction with the process information already available, increase the quality control of the green sand anywhere in the circuit. It allows the prediction of defects related to bad preparation of the

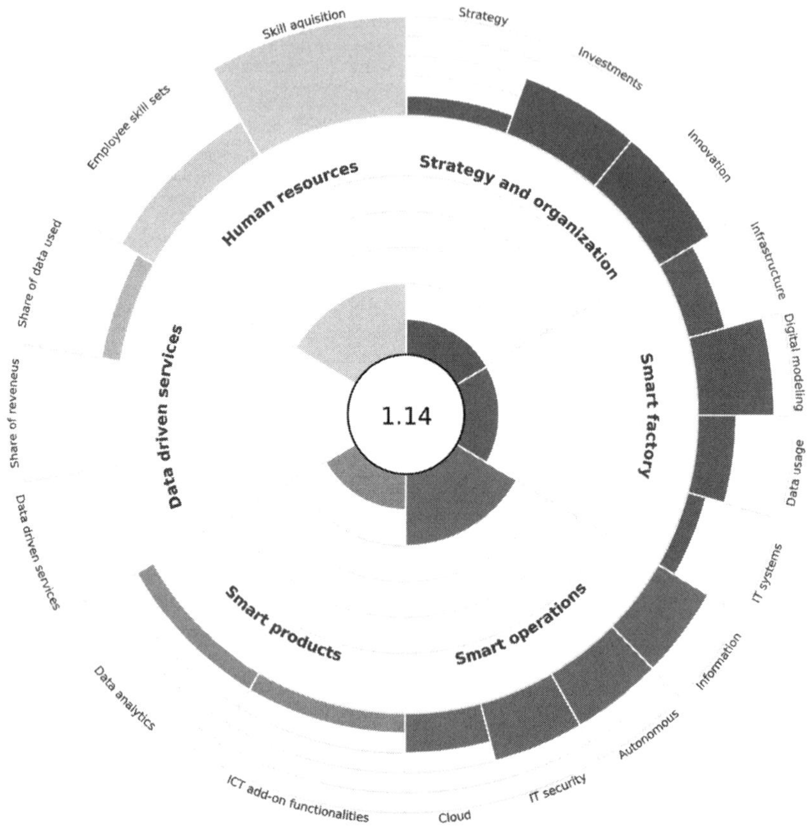

Figure 1. Graphical representation of the latest self-assessment result. The achieved level (0 to 5) for each of the 18 fields is represented, allowing an interpretation of the result for each Dimension. The global level is represented in the middle.

green sand. Additionally, the company employed RFID technology to identify both the moulds and the number of times that a concrete mould was used. It helps to ensure the correct moulds assembly in the moulds-plate, eliminating human errors.

Data analytics is one of the most important concepts that drives i4.0. The majority of industries ignore the data generated throughout their productive processes. According to the recommendation from the report, the company implemented data analytics methods to predict and minimize the product defects derived from the sand parameters. The predictive methods were based in the data provided by sensors put into use.

The maturity level obtained after the implementation of recommendations increased the global result to 1.78 as shown depicted in Figure 2. This improvement was achieved in less than one year showing the effectiveness of the SHIFTo4.0 tool.

Figure 2 demonstrates that strategy and organization and smart factory dimensions experienced substantial evolution, after applying the SHIFTo4.0 recommendations. The strategy field was the most affected increasing 2 levels. A putative reason for this maturity improvement is the analysis and reflection about their i4.0 strategy positioning after the recommendations given by SHIFTo4.0. It leads to new implementations in the fields of innovation, as well as, in the IT systems, with the involvement of new data acquisition through RFID technology and data analysis.

The introduction of new sensors and the RFID technology took into consideration: overall equipment; IT infrastructure; its integration; and that it complies with the company's i4.0 strategy.

5 CONCLUSIONS

The development and application of SHIFTo4.0 tool (which is one outstanding by-product of SIM4.0 project) in the Portuguese industry has already shown to be more effective in increasing the i4.0 awareness of SMEs than most conventional dissemination activities, as workshops, seminars, publications, etc. The report of the SHIFTo4.0 survey, is a key element to reflect the future regarding strategic organization of the company. The self-assessments survey allows the company to acknowledge its actual position towards i4.0. Additionally, it reports the main gaps and

Figure 2. Graphical representation of the latest self-assessment result after implementing the recommendations given by the SHIFTo4.0 tool. The achieved level (0 to 5) for each of the 18 fields is represented, allowing an interpretation of the result for each Dimension. The global level is represented in the middle.

existing challenges and provides a roadmap which help the company to increase i4.0 maturity level, with a coordinated and integrated effort, in a foreseeable future.

The application case presented above is a demonstration of SHIFT04.0 methodology effectiveness. Based on the self-assessment output, a strategic plan for i4.0 implementation was designed taking into account the recommendations of the assessment. Action plans in different areas of the organization were prioritized considering the company strategy. The improvement of the company i4.0 maturity level is a clear evidence that SHIFT04.0 has the potential to promote the implementation of i4.0 concepts in SME.

ACKNOWLEDGENTS

This work is financed by FEDER funds through COMPETE 2020 in the context of the SIAC system (Collective Actions Support System): 03/SIAC/2016; investment project nº 026752.

REFERENCES

Bruin, T. De, Freeze, R., Kaulkarni, U., & Rosemann, M. (2005). Understanding the main phases of developing a maturity assessment model. In (Eds.) *Australasian Conference on Information Systems*.

Geissbauer, R., Vedso, J., & Schrauf, S. (2016). 2016 Global Industry 4.0 Survey Industry 4.0 : Building the digital enterprise. *PWC* Available at: https://www.pwc.com/gx/en/industries/industries-4.0/landing-page/industry-4.0-building-your-digital-enterprise-april-2016.pdf [Accessed 1 February 2019]

Gökalp, E., Şener, U., & Eren, P. E. (2017). Development of an assessment model for industry 4.0: Industry 4.0-MM. In *Communications in Computer and Information Science*, 128–142.

ISO/IEC. (2015). INTERNATIONAL STANDARD ISO/IEC Information technology — Process assessment — Requirements for performing process assessment, *2015*.

Lichtblau, K., Stich, V., Bertenrath, R., Blum, M., Bleider, M., Millack, A., Schmitt, K., Schmitz, E., & Schröter, M. (2015). Industrie 4.0 Readiness. *VDMA Impuls-Stiftung*. Available at: https://industrie40.vdma.org/documents/4214230/26342484/Industrie_40_Readiness_Study_1529498007918.pdf/0b5fd521-9ee2-2de0-f377-93bdd01ed1c8 [Accessed 1 February 2019]

Mittal, S., Khan, M. A., Romero, D., & Wuest, T. (2018). A critical review of smart manufacturing & Industry 4.0 maturity models: Implications for small and medium-sized enterprises (SMEs). *Journal of Manufacturing Systems*, 49, 194–214.

Schuh, G., Anderl, R., Gausemeier, J., ten Hompel, M., & Wahlster, W. (2017). *Industrie 4.0 Maturity Index. Acatech Study.* Available at: http://www.acatech.de/fileadmin/user_upload/Baumstruktur_nach_Website/Acatech/root/de/Publikationen/Proj ektberichte/acatech_STUDIE_Maturity_Index_eng_WEB.pdf [Accessed 1 February 2019]

Schumacher, A., Erol, S., & Sihn, W. (2016). A Maturity Model for Assessing Industry 4.0 Readiness and Maturity of Manufacturing Enterprises. In *Procedia CIRP*, 52, 161–166.

SHIFTo4.0. [ONLINE] Available at: https://shift4.isq.pt [Accessed 1 February 2019].

SIM4.0. [ONLINE] Available at: https://sim4.isq.pt/ [Accessed 1 October 2018].

Industry 4.0 – Shaping The Future of The Digital World – da Silva Bartolo et al. (eds)
© 2021 Taylor & Francis Group, London, ISBN 978-0-367-42272-1

Fashion segment and Industry 4.0: Brazilian textile and clothing industry – perspective into the new industrial paradigms

E. Pereira das Neves & D. Nogueira da Silva
São Paulo State University, Bauru, Brazil
CIAUD, Lisbon School of Architecture, Universidade de Lisboa, Lisbon, Portugal

L.C. Paschoarelli
CIAUD, School of Architecture, Arts and Communication, São State University, Bauru, Brazil

F. Moreira da Silva
CIAUD, Lisbon School of Architecture, Universidade de Lisboa, Lisbon, Portugal

ABSTRACT: The Brazilian textile and clothing industry have featured in global scenario, not only for being amongst the five biggest industrial complexes of the segment, but also because of their professionalism and creativity. The sector in the last years, although slow and timid, reveals a set of issues that challenges the adequacy of the industrial segment within the Industry 4.0 reality. This paper aims to analyze and summarize the economic reality and the challenges faced by the textile and clothing industry, and the perspectives of investments regarding their adaptation to the new industrial paradigms. It was observed that the Brazilian fashion industry is late regarding the adoption of new technologies, mainly because of financial difficulties and the lack of government support. However, this industry needs to promote itself by adapting to alternative models of smart manufacturing, bridging the gap between supply chain and consumers.

1 INTRODUCTION

The historical context of Brazil's textile and clothing production dates back over 200 years ago and revels the sector as responsible to promote several other industrial sectors. This phenomenon made this industry the greatest driving force of the national industrial revolution (Abit, 2015), similarly to what happened in other parts of the world. Sector's plurality makes the textile and clothing chain a multipurpose and multi-disciplinary scenario, which involves various aspects such as business, sales, arts, crafts, technology, cultural and social paradigms, physics, chemistry, biology, among other (Abit, 2018a).

The sector is qualified as one of the most complex industrial sectors in the world, standing out for its productive potential as well for developing activities ranging from fibers and filaments' production until confection (CNI, 2017). The dimension of its industrial park makes it the fifth largest textile industry in the world and fourth largest in clothing production (CNI, 2017; Abit, 2018b).

However, despite maintaining a prominent position, the Brazilian industry has suffered from the economic policies and economic instabilities over the years, which has been affecting the technological advance of the sector. This reality confronted with the new economic and industrial paradigms supported by the recent industrial phenomenon named Industry 4.0.

First proposed in Germany in 2011, the new economic concepts of Industry 4.0 are based on the articulation of technological and human resources to reach highest level of operational and productive efficiency through automation and digitization (Lu, 2017).

This new reality requires a comprehensive and global view of how technology can affect lives, reshaping the economic and the social, cultural and human environment. Emerging challenges require substantial transformations that encompass new business strategies and greater investment in technology. Initiatives that can be a major challenge for the Brazilian industrial reality, since it is still behind in relation developed countries, which appear as strong competition for the country. According to a survey conducted by Fiesp[1] in partnership with Senai[2]-SP, of the 227 companies addressed, 32% did not hear about the new industrial concept and only 30% already had started some alterations aligned with the new technological demand. Fiesp Only 18% of this group considered themselves adapted to the 4.0 background regarding information technology infrastructure. Regarding investments, only 38% had invested up to 5% of the earnings in new technologies (Fiesp, 2018).

The late arrival of Brazil in the discussion about the new industrial model is due to the difficulties of management and development resulting from bureaucracies and also because of the political and economic instability and the national industrial profile, which is composed of micro and small business. Due a lack of resources, the smaller companies have restricted access (or even no access) to the technologies that are relevant to the solution of exiting problems, which consequently restricts the growth of the company.

The sector has emerged as a pioneer regarding new plans of development, strategies and action that contributes to the awareness of companies about the need of their adequacy to the new industrial intelligence. This movement is because some institutions such as the CNI[3], Abdi[4], Abit[5], Sebrae[6] and Senai-Cetiqt[7] that, along the last decade, have been promoting the discussion about the new industrial reality and have organized the new paradigms implantation.

This paper, based on a literature review, aims to analyze and provide a summarize of the economic reality and the challenges of textile and clothing industry, and the perspectives of investments regarding their adaptation to the new industrial paradigms.

2 BRAZILIAN CONTEXT: INDUSTRY 4.0 AND THE TEXTILE AND CLOTHING SECTOR

2.1 *Industry 4.0: Through a historic technological evolution*

As other periods of industrial evolution, technology has determined the rupture with obsolete systems and paradigms, promoting revolutions within the industry as much as in the consumer habits and the whole social, economic and consumer environment. At the end of the nineteenth century, the electricity advent and the Ford's assembly lines standards the Second Industrial Revolution was established. In the1960s, the digital revolution, computer advent and the democratization of the internet overlapped the industrial and social standards of that moment and contributed to the Third Industrial Revolution.

Throughout this evolutionary process, the industrial production structure faced different phases until confronts the current dynamic markets. The industrial production changed from artisanal to mass, being characterized by the segmentation of different industrial stages and by the standardization and specialization of the activity within the production line. The saturation of the market turned this process into a failed strategy, driving the industrial race for differentiated products as well as for leaner production that could reduce the waste. (Lasi et al., 2014; Brettel et al., 2014).

In the current scenario, the continuous advances in digital technologies have immersed the society into a global integration process, propelling the industrial sector to open up to new tools and systems involving Cyber-Physical Systems (CPS), Internet of Thing (IoT), Internet of Service (IoS), Robotics, Big Data, Augmented reality, Artificial Intelligence and Cloud Computing. These technologies adoption has become central smarter industrial processes development. In this industry standard devices, machines, productive modules end products work together through the exchange of information, creating, thus, an intelligent industrial environment (Pereira e Romero, 2017).

This integrated structure allows the constant acquisition of information regarding stock, problems and repairs requirement, production demand order, production alterations, and other, contributing to machines and products operating without the manual lab force intervention (Nadais, 2017). With this, equipment, human factors, productive process and products are factors that must be reviewed to provide more productiveness, revenue growth, employment and investments.

In the apparel context, the textile and clothing segments consist in a dynamic industrial phenomenon that propose frequent changes into the cultural, socioeconomic and aesthetic environment. It has intensified this reality along the last years because of the ease and agility in the exchange of information among individuals approaching different worlds. Reflecting this, there is a new consumer eager for products that can promote individualization and can satisfy expectations and emotional needs. The consumer is anxious for news and impatient to experience them, which implies in product release in shorter time intervals and fast delivery. Within this perspective, incorporating new technologies is substantial for industries framework into new productive demands. In addition, alternative business models are emerging, creating a decentralized and horizontal production process where all elements of the manufacturing chain are important to deal with the information data as well with the technology capacity in order to determinate the productive demands and products.

2.2 *The evolution of the textile and clothing sector: The impact of new economics scenario into the industrial strategies.*

In the 1970s, incentives policies and fiscal protections imposed by Brazilian Government created a protected environment for the textile and clothing industry, creating a scenario of import restrictions in favor of the national product. In this context, the industry had a glimpse of the profits maximization from the minimization of technological investments, because the importation of machinery and equipment was prevented because of the import restrictions. This trend differed from the industrial and economical strategies applied on foreign countries, which promoted the mechanization and digitization of fashion industrial chain in response to the paradigms of globalization.

With the national economy opening to the foreign market (beginning of 1990s), the Brazilian textile and clothing industry was not prepared, suffering enormous impact caused by imported products from China and USA (mainly).

The impact of the reduction of import restrictions and the changes in the capital flow on the industrial segments of textile and clothing sector was not homogenous, although it was negative in almost all of them.

The industry was exposed to a strong international competition and, therefore, longing for quick strategies that could promote domestic product both nationally and globally. The companies that already were working with the external market, previously perceived the need to keep up to date with new technologies, being benefited by foreign financial plans and secured in the long term. (Gorini and Martins, 1998)

In the last decade, the global financial crisis has substantially altered the investments and market conditions, leading major international players to change their strategies to face the new scenario of global competition. In Brazil, the saturation and questioning of certain business models occurred, such as the indiscriminate use of low-skilled labor from China, for example, as a resource to get products with lower price; or/and the direct purchase of Chinese low-cost (and low-quality) end products (Bruno, 2016).

2.3 *The textile and clothing industry in Brazil: Numbers of the sector*

According to "The State of Fashion 2019" study, designed in partnership between MCKinsey & Company and The Business of Fashion (BOF), the world textile and clothing had positive performed during 2018, although it still suffers from the final crisis of recent years. However, for 2019, the report foresees a smaller growth if compared with the previous year due the economic previsions that show slower global growth and ruptures in some commercial relations (MCKinsey&Company and BOF, 2019)

The *"Brasil Têxtil 2018"* report, edited and published by IEMI, shows that between 2013 and 2017, there was a reduction in the number of industrial units operating in Brazil. Regarding textile industry, the reduction reached 17%, while the clothing sector was 17,8%. The reduction progression was only interrupted in 2017, when both sectors had a slight recovery. Despite the numbers, the Brazilian production is still not significant in world trade. It dominates only 0,3% of exports and holds the 40º position in the production ranking of the main textile and clothing exporting countries. Most of the volume of production is from Asia, principally China. In whole, the Asian production represents over than 50% of the world's textile and clothing trade (Abit, 2017).

The CNI's report (2018), in 2017, set the resumption of the investments within the industries after three years decline. It is expected that investments to improve the productive process and to increase the capacity of the manufacturing line become the fundamental goal of the industries, due to the prevision of greater demand and a tendency of investments oriented to the domestic market. However, the Confederation explains that due the high dependence on own financial resources the industries have a difficult to invest, which remains an obstacle to investment.

Besides that, Brazil faces the dynamics of its verticalized productive chain, where the production is focused on domestic supply, corroborating with a deficit trade balance (CNI, 2017). This scenario intensifies discussions about the sector maturity towards the global competition, even when it is possible to find a positive scenario still to be explored.

2.4 *Industry 4.0: The challenges*

In 2017, according to the latest survey of FIESP and CIESP (2019) in Brazil, there were over 49,000 establishments destined to making clothing products and accessories, plus another 10,000 related to textile production. Together, both sectors represent around 18% of the establishments of the manufacturing industry.

CNI divides the industries of this sector in three distinct segments: (1) Fibers and filaments; (2) Textile; and (3) Apparel (clothing). Using published data by IEMI, the confederation shows which among the total of the companies, 71,7% can be classified as micro enterprise, 25,1% as small enterprise, 2,8% as middle size enterprise and, 0,3% as big enterprise (CNI, 2017). Micro and small enterprises represent 96,8% of the manufacturing units, being clothing manufacture the majority profile. In addition, the confederation shows that the clothing sector concentrates around 75% of the workforce of the entire Brazilian sector.

Over the last 20 years, the government has launched a set of strategies with the principal goal to guarantee the success and the safety of micro and small companies. In 2009, the Individual Microentrepreneur favored the formalization of millions of companies were under illegal situation. The environment is still hostile to entrepreneurs, because of obstacles such as: labor taxation, that inflates the employers' costs approximately 100% on wages; and, the lack of credit (lending) with acceptable rates (taxes) and longer terms, which suffocates the companies' cash flow (Dana, 2016).

As the economy slows and the cost of credit rises, small and medium companies face greater challenges regarding to keep the production and invest in technology, because much of the capital comes from own company. Also, the familiar and manual production character of most national industries do not facilitate the Industry 4.0 paradigms implantation.

According to a CNI survey published in 2017, the industrial segment of clothing and fashion accessories is one industry that uses the least technology in the manufacturing process: only 29% of national companies adopt new technologies, which are, generally, used for the manufacturing chain development of (Audaces, 2017).

In addition, the companies readaptation process to a new structural and governance profile must be adapted to the new internal socio-economic changes. These changes resulted from emphases of development polices, which evidence new types of consumption and to interfere with the industrial jobs offer and demand and with the institutional capacities. The lack of manual labor contributes to the process of industrial automation and robotization, since such strategy seems to be an appropriate solution.

With this industrial reality there is a set of changes regarding the consumption that encompass issues such as individualization and the "physical translation" of the individual needs into the product as a solution to supply the emotional and biophysical needs and expectations of the user.

In fashion value chain, fashion consumption is demanding products that have style and symbolic elements with short life cycles. A demand that requires more frequent launching of small collections at shorter intervals, which are characterized by products of physical and aesthetic quality and cheap price.

This scenario implies the reduction of any activity or investment not adding value to the final product, which has promoted the paradigms of lean manufacturing and smart factories. The 4.0 model of manufacturing presents a decentralized production, it is no longer the managers of companies who dictate what to be used and/or produced, but the consumer, who holds and determines the demand of products that must be produced to satisfy personal needs and expectations.

For this purpose, companies must invest in fast and versatile manufacturing process and in suppliers that also fit in the same managerial and productive ways. The information flow becomes essential for the production operation, which highlight the importance of promoting investments in broadband and mobile network, revising old models of telecommunications.

However, all this technology reflects on the quality of the professional skills that involved in the manufacturing processes that must be capable of comprehend and use new technologies into problems solution and, consequently, into the ultimately products compatible with emerging demands.

2.5 *Expectations in the sector*

Audaces, a world leading company in technological innovation of the fashion manufacturing, believes that the 4.0 industry implantation in the textile and apparel must begin with investments that benefit some production stages, such as cutting and sewing sectors. The company mentions four examples in which factories could benefit (Audaces, 2017):

(1) Appropriation of technologies that enable the machinery itself understand the lack of a certain fabric and notify suppliers through a request; this tool contributes for decreases the time between stages and avoids manufacturing gaps, which optimizes and make the process of production faster; (2) The understanding of the system can improve the distribution logistics of the end product itself once it can calculate the demand being produced and the destination. Through information flow the systems themselves can match schedules and routes, decreasing the time spent in translation and reducing costs; (3) Machinery that, besides create digital clothing pattern, can also transfer it automatically to the fabric that are going to the cutting stage. In this process, the used technology can recognize that the remains of this fabric are compatible with other pattern, which minimizes the waste and guarantee a more sustainable production; (4) The machines could be able to monitor, diagnose and warn when an equipment within the manufacturing process needs revisions. This would give the opportunity to reorganize the production schedule so that this temporally interruption doesn't disturb the final delivery.

By Audaces vision, digital integration along the productive chain is essential to the efficiency of productivity, and to the relationship between industries and supplies: it is all about the compatible technologies employment and the exchange of information. It is a reflex of the industry framework into the models of smart manufacturing, in which machinery and inputs "converse" throughout industrial productive stages, occurring in an autonomous and integrated way (CNI, 2016).

In the concept, smart manufacturing has the purpose of optimizing production and product transactions by full using advanced information and manufacturing technologies (Kusiak, 1990). Soon, the concept of the new industry goes beyond the production integration and distribution process, involving all stages of the value chain.

The technology adoption within the manufacturing environment does not involve only operating process, machinery and exchange of information. It is also applied in the new materials development for garments and sewing, patterning and construction techniques. In the Textile and Apparel segment, mainly, a set of new technologies can be applied in the development of technological, fabrics providing fabrics with new properties and functions, such as fire resistant and antibactericides fabrics.

Senai\Cetiqt (2018) has been essential to promote the adequacy of the segment into the new 4.0 Industry. The confederation declares that soon it will be possible to order an exclusive garment from the factories and have it less than one hour later. In 2018 the organization presented an organizational plan of the first "Factory Model 4.0" that has the aim to explain what Industry 4.0 is and how and where the technology can be used in favor of an integrated and productive manufacturing which is characterized by technological investments and by promoting the approximation of the consumer with the company.

Besides being innovative, the "Factory Model 4.0" also provides a set of information throughout the processes using technologies such as IoT and Big Data, which are stored in the cloud. These data analyze with detail the process of production, which contributes to guide strategic decisions (Senai-Cetiqt, 2018).

To prepare the entrepreneurs and professionals involved with the Fashion segment, Senai-Cetiqt has promoted a postgraduate course about the application of the reality 4.0. It aims to capacitate individuals to be prepared and, consequently, be capable to apply and use the technology.

3 CONCLUSION

Developing countries, such as Brazil, has to struggle with new technologies implantation, intensifying the discussion about the country be late and unprepared in some industrial segments in face the new industrial context, and its path to be adequate and competitive.

In apparel industrial context, although it stands out in the world scenario due the creativity and professionalism, this reality has been observed, because of the size of the industries that characterize the sector: micro and small enterprise. Despite that, along the recent years, reflecting also the timid resumption of the sector after a period of economic crisis, some institutions and organization, such as Abit, Senai, CNI, Sebrae (and other) have been promoting studies that aim to identify and analyze the current industrial profile of the sector. Some strategies have been developing to make the industries aware about the Industry 4.0.

The Industry 4.0 is intensifying several discussions about the future of Brazilian Textile and Apparel Industry. This paper highlights the importance of further studies to identify what is happening within the segment companies, and how the entrepreneurs are dealing with the new reality. Are they prepared? Are they aware about the Industry 4.0?

All these issues should be studied in more detail to produce knowledge regarding the new industrial and consumer reality. Such issues must be treated considering the perspective of sustainability, which is no longer a differential but a mandatory strategy.

And, finally, how should design act with such paradigms? How can the strategies and tools of design be applied to contribute to the industrial advance and to help the industry become competitive?

ACKNOWLEDGMENT

This study was conducted with the support of: CIAUD, Universidade de Lisboa; CAPES (Process 88881.188811/2018-01), and, CNPq (Process 207371/2017-2).

REFERENCES

ABIT, 2015. Associação Brasileira da Indústria Têxtil e de Confecção. Agenda de Prioridades: Têxtil e Confecção – 2015 a 2018. Abit (online source). Available at: [http://www.abit.org.br/conteudo/links/publicacoes/agenda_site.pdf], 24.10.19.

ABIT, 2018a. Associação Brasileira da Indústria Têxtil e de Confecção. Pessoas: transformando e sendo transformadas na era digital. Abit (online source). Available at: [http://www.abit.org.br/noticias/artigo-tempo-de-transformar-e-ser-transformado], 24.01.19.

ABIT, 2018b. Associação Brasileira da Indústria Têxtil e de Confecção. Perfil do setor. Dados Gerais do setor referentes a 2017 (updated 2018). Available at: [http://www.abit.org.br/cont/perfil-do-setor], 24.10.19.

AUDACES, 2017. AUDACES. Indústria 4.0 na confecção. Portal Audaces. Available at: [Available at: [http://g1.globo.com/economia/blog/samy-dana/post/pequenas-empresas-crescem-no-brasil-mas-os-desafios-ainda-sao-grandes.html/], 23 jan 2019.

Brettel, M. et al., 2014. How Virtualization, Decentralization and Network Building Change the Manufacturing Landscape: An Industry 4.0 Perspective. World Academy of Science, Engineering and Technology International Journal of Information and Communication Engineering Vol:8, No:1.

Bruno, F.S., 2016. A quarta revolução Industrial: do setor e de confecção: A visão de Futuro para 2030. São Paulo: Estação das Letras e Cores.

CNI. 2016/2017/2018. Confederação Nacional da Indústria. Desafios para a indústria 4.0 no Brasil/Confederação Nacional da Indústria. – Brasília: CNI. at:[http://www.portaldaindustria.com.br/publcacoes/2016/8/desafios-para-industria-40-no-brasil/], 20.01.19.

DANA, S. 2016. As pequenas emoresas crescem no Brasil, mas os desfios ainda são grandes. Blog do Samy Dana, Portal G1. Available at: [http://g1.globo.com/economia/blog/samy-dana/post/pequenas-empresas-crescem-no-brasil-mas-os-desafios-ainda-sao-grandes.html/], 20.01.19.

FIESP, 2018. Federação das Indústrias do Estado de São Paulo. Portal Fiesp – Agência Indusnet Fiesp. Fiesp identifica desafios da indústria 4.0 no Brasil e apresenta proposta. Available at: [https://www.fiesp.com.br/noticias/fiesp-identifica-desafios-da-industria-4-0-no-brasil-e-apresenta-propostas/] 20.01.19.

FIESP\CIESP, 2019. Depto. De Economia, Competitividade e Tecnologia. Panorama da Indústria de Transformação Brasileira. 17ª. Edição.

Gorini, A.P.S., MARTINS, R.F. 1998. Novas tecnologias e organização do trabalho no setor têxtil: uma avaliação do programa de financiamento do BNDES. Revista do BNDES, Rio de Janeiro, v. 5, 10.

Kusiak A., 1990. Intelligent Manufacturing Systems, Prentice Hall, Englewood Cliffs, New Jersey, p. 532

Lasi H. et al., 2014. Industrie 4.0. IRTSCHAFTSINFORMATIK. doi: 10.1007/s11576-014-0424-4.

Lu, Y., 2017. Industry 4.0: A survey on technologies, applications and open research issues. Journal of Industrial Information Integration, 6, June 2017, pp. 1–10.

MCKinsey & Company and The Business of Fashion. The State of Fashion 2018.

Nadais, J., 2017. Artigo de Opinião-Industria 4.0. Fibrenamics, Volume 14.

Pereira, A.C., Romero, F. 2017. A review f the meanings and the implications of the Industry 4.0 concept. Procedia Manufacturing, vol. 13, pp.1206–1214.

Senai-Cetiqt. 2018. Com Indústria 4.0, fábricas inteligentes vão melhorar a competitividade do setor da moda. Portal Fiems (online source). Available at: [http://www.portaldaindustria.com.br/publicacoes/2016/8/desafios-para-industria-40-no-brasil/], 20.01.19.

GLOSSARY OF ORGANISATIONS

Abdi – Brazilian Agency for Industrial Development is an institution associated to the Ministry, Foreign Trade and Services (MDIC) that operates in strategic areas of Industry through incentive programs, investments, training and action.

Abit -Brazilian Textile and Apparel Industry Association is one of the most important entities among the economic sectors of the Country.

CNI - Brazilian National Confederation of Industry is the official and highest-level organization representing Brazilian Industry.

Fiesp - The Federation of Industries of the State of São Paulo is a legal Brazilian industry entity that is associated with the Brazilian National Confederation of Industry (CNI) It represents about 130 million of a diverse set of industries.

IEMI – Research Market Intelligence is responsible for attend the industry and entities demand for numerical and behavior information about market.

Sebrae – Brazilian Micro and Small Business Support Service is a non-profit private entity with the mission of promoting the sustainable and competitive development of small businesses.

Senai- National Service for Industrial Training is the largest institution of professional and technological education in Latin America that was established and maintained by the CNI.

Senai-Cetiqt – Professional Senai's school unit that also has a center for professional development to the tectile industry in Brazil. The unit is located in Riachuelo, Rio de Janeiro (Cetiqt: Technology Center of Chemical and Textile industry).

Industry 4.0 – Shaping The Future of The Digital World – da Silva Bartolo et al. (eds)
© 2021 Taylor & Francis Group, London, ISBN 978-0-367-42272-1

Jumping to Industry 4.0 through process design and managing information for smart manufacturing: Configurable virtual workstation

S. Raharno
Insititut Teknologi Bandung, Bandung, Indonesia

G. Cooper
University of Manchester, Manchester, UK

ABSTRACT: This paper presents a conceptual framework for first step of Industry 4.0 implementation in Indonesia's traditional manufacturing industry. The main objective of this research was formulating a suitable framework for Indonesia's traditional manufacturing industry in order to jump to Industry 4.0 principles without automation of the main manufacturing process. The work focuses on a configurable virtual workstation as a framework to achieve information transparency that is one of design principles in industry 4.0 concept. The configurable virtual workstation was designed to link the physical production process with the virtual world by enriching digital plant information with sensor data. A web-based laboratory scale software for implementing the framework has been developed. The software handles interactions between the users and the system.

1 INTRODUCTION

The 4th Industrial Revolution is sweeping the globe and many countries have ambitious plans to adopt new digital technologies and principles to make this a reality (European Commission, 2018; Federal Ministry for Economic Affairs and Energy & Federal Ministry of Education and Research, 2019; Ministry of Industry, Indonesia, 2018). Manufacturing in the US and Europe is largely automated and moving to Industry 4.0 involves only improving connectivity but for many countries in the world manufacturing still involves a large amount of manpower due to the low cost of wages in comparison to the global economy it is not cost effective to automate processes. Effectively many of the industries in these countries is still operating at industry 2.0 (without automation) but how can they jump to Industry 4.0 to remain competitive in the global market? In the fast-changing scenery of manufacturing industry, the manufacturer is required to be flexible and highly efficient to gain competitive advantage. One way is by implementing or adopting the Industry 4.0 concepts in the design of processes and information management to enable smart manufacturing. In general, the implementation of the Industry 4.0 in the manufacturing industry may be achieved by developing intelligent production elements that are capable to exchange information for autonomous decision making to control production. To enable information exchange and decision making, each production element needs to be equipped with sensors to collect data coupled with a well-designed manufacturing process. In addition, each production element needs to be able to process data obtained from the sensors to analytically extract useful and meaningful information.

At present, most manufacturing industry in Indonesia, particularly small and medium enterprises, are still categorized as traditional manufacturing industries with little automation. Although they are using modern machines such as CNC machines, they are still in the stage of using manpower as the basis of the production system and relatively few that utilize automation-based production systems, that is the signature of the Third Industrial Revolution. To enable these industries to survive and even increase its competitiveness, the Industry 4.0 concept needs to be considered for future development, without having to go through the full automation phase first.

The implementation of Industry 4.0 concepts is not simple and straightforward, but it would involve large numbers of production elements, sensors and data, development of intelligent production system, processing of large volume of data to extract useful information, secure transmission of data. All of these would require additional redesign of processes and job changes that may hinder the implementation due to tendencies of the workers to resist new technologies. To alleviate this potential problem, a strategy of implementation needs to be developed in a conceptual framework, which avoids the need for expensive automation, utilizes smart process design and information exchange and considers social factors to improve workforce satisfaction.

2 LITERATURE REVIEW

Atluru et al. (2012) dealt with developed a numerical control (NC) machine supervision system. The system is able to determine the condition of the cutting tool, inspect the product, make the optimum cutting tool trajectory, determine the position of the cutting tool in real time, and determine the condition of the machine tool. This system is only implemented in NC machine tools. NC machine tools can be upgraded to be a smart machine. Velandia et al. (2016) discuss the location monitoring of products (crankshaft) in real time by utilizing RFID tags embedded in the product. Any equipment used to make the product comes with an RFID tag reader. This system is only able to determine the time interval of the manufacturing process on each piece of production equipment. Monostori et al. (2017) discuss the linkages between the development of computer science, communication and information technology, and technology and manufacturing science. This has an effect on the current production system leads to Cyber Physical Production System (CPPS). CPPS is considered very promising but the realization requires a very significant effort and economic investment. Konstantinov et al. (2017) discuss the monitoring of an electric motor assembly unit. The electric motor assembly unit is replicated in the virtual system so that the activity that occurs in the real assembly unit can be simulated and monitored through a virtual system. The system designed is only for the electric motor assembly unit. Their paper focuses on getting and applying the real-time location information as positioning, tracking and monitoring work in progress (WIP) in the manufacturing workshop. Huang et al. (2017) propose a framework of a RTLS based on RFID and UWB technology to sense the location and the status of manufacture resources such as material, tools, equipment and workers. Isaksson et al. (2017) dealt with the impact of digitalization manufacturing on the future of control and operation in manufacturing industry according the Internet of Things (IoT), which is related cyber-physical systems, industry 4.0 and smart manufacturing. Mrugalska and Wyrwicka (2017) discuss the application of some parts of industry 4.0 which support a lean production system. It allows creating a smart network of machines, products, components, properties, individuals dan ICT systems in the entire value chain to have an intelligent factory. Tamás (2016) states to implement industry 4.0 in manufacturing, which in mass production, productivity and specified cost requires the IOT and cyber physics system as well as big data created a significant research potential regarding the more efficient actuation and continuous improvement of the logistic system. This is because of the reduction of lead time of the tasks led to the increase of the optimized collaboration between systems objects (source and drain, material handling equipment, staff). Qin & Sheng (2017) describe aspects that relate to future digital design and manufacturing and its future challenge. It explains the social aspects on Industry 4.0 in social, economy, and environmental challenges, but not specify the technology in Industry 4.0. Ramadan et al. (2017) proposes a lean module for dynamic value stream (DVSM), namely the real-time manufacturing cost tracking system (RT-MCT) which is able to track the development or accumulation of actual product costs during the flow of products along a value stream. Liu & Xu (2017) propose a new generation of machine tools, i.e. Machine Tool 4.0, as a future development trend of machine tools. Machine Tool 4.0, otherwise known as Cyber-Physical Machine Tool (CPMT), is the integration of machine tool, machining processes, computation and networking, where embedded computers and networks can monitor and control the machining processes, with feedback loops in which machining processes can affect computations and vice versa. Sang & Xu (2017) present a Control System as a Service (CSaaS) as a scenario whereby the control system is decoupled from the machine tool and exists in the cloud, so that control of a machine tool becomes a cloud-based service. Machine tools connected to the cloud are treated as the local resource. The machining jobs are scheduled and distributed among the connected machine tools considering their capability and availability. Basl (2017) shows the Czech company's attitude about awareness of the existence of a trend known as Industry 4.0, higher penetration of the principles of Industry 4.0, and improvement in terms of delivery of available information on Industry 4.0 to the employees. Qin et al. (2016) propose combining the intelligence within the production system of automation levels which are machine, production process, and factory system to made hierarchical framework.

All of these propose adding connectivity to automated production systems. This automation is not available in all industries around the world (Indonesia, India, China etc.). But how can industry jump to Industry 4.0 without large investment in automation? This paper proposes using information management to enable smart manufacturing systems. The main objective of this research was formulating a suitable framework for Indonesia's traditional manufacturing industry in order to jump to Industry 4.0 principles without automation of the main manufacturing process. The work focuses on a configurable virtual workstation as a framework to achieve information transparency that is one of design principles in industry 4.0 concept.

3 CONFIGURABLE VIRTUAL WORKSTATION

There are several relatively well-known Industry 4.0 design principles. One of them is information transparency which means every activity should be monitored and known. Furthermore, the information about the activities can be used to assess production efficiency and enable smart redesign of manufacturing processes if necessary. This concept could avoid the need to automate all processes, thus jumping the problem of automating all production steps but gaining the advantages of smart information management and utilization.

To achieve this all production elements needed to be equipped with a controller and a sensor to obtain data. The main problem with this is that not all production elements have controllers and are able to send data as many of the processes are manual. Even though some production equipment has been equipped with controllers and sensors like CNC Machines, that system only uses these to ensure the function of the machine tools. The next problem is that it is impossible to plug in the sensor and controller due to technical aspects. Also, operators usually will refuse to use the sensor and controllers if it is installed on them personally or if it interferes with their daily tasks. But in several cases, they may accept limited sensors such RFID cards or wrist band tags. The installation of sensors and controllers in all production elements will also increase cost for some industries, especially small and medium industries – making these technologies a barrier to adoption.

This study uses the concept of a low-cost virtual workstation. This could be in the form of adapted mobile phone technology or low-cost computing such as an Arduino. Figure 1 shows all the area of the shop floor is defined as the workstation in the virtual environment which is not only the area that has production equipment but also the working area of the product and the operator, or area which doesn't have production equipment. Every virtual workstation has the ability to manage their members like machines, products, and operators. Some of the members in a virtual workstation are permanent like machines and some of them are temporary like product and operators. A virtual workstation can have more than one member of each membership and it is not necessary to have all members.

Generally, the usage of this concept will cause every production element like the operator, the machine, and the product always in one of visual workstations that is already defined as long as they are still in the production area. Physically, every virtual workstation will be equipped with a controller and sensors depending on their needs. Installation of authentication sensors can be used to detect the existence of the product and operator.

Each workstation will be equipped with a controller and sensors (see Figure 2). The function of the controller is to bridge the real workstation with the virtual workstation. This controller is connected with a server database where the data collected is saved and processed. Each controller is also equipped with several modules which depends on the type and function of the workstation area defined. For example, module authentication. This module is a must for all virtual workstations. This module is used to detect login or logout object from related virtual workstation. Another example is cutting condition recording module. This module will only fit to a virtual workstation which has machine tools as part of its member. The module is used to monitor and record each machining process condition. With this concept, the virtual workstation can be implemented to all

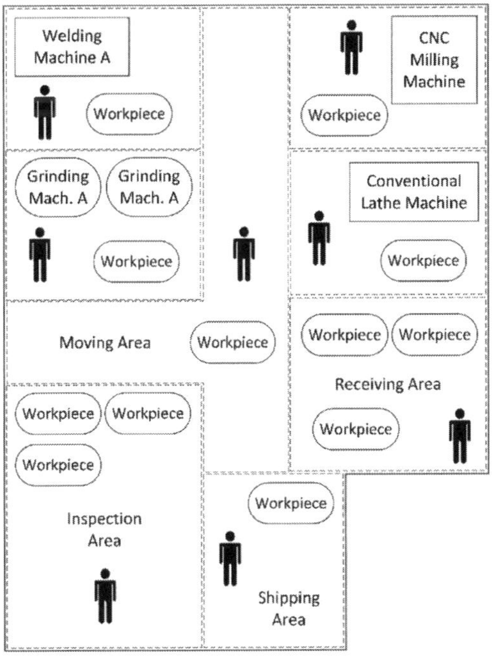

Figure 1. Concept of virtual workstation.

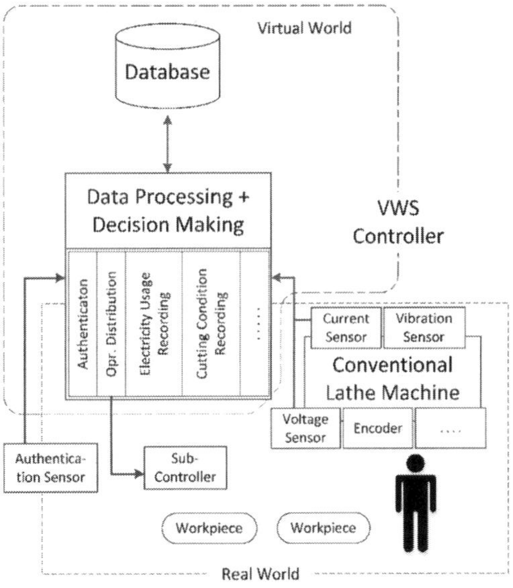

Figure 2. Virtual workstation's architecture.

types of equipment, condition, and each factory area by configuring every required module.

4 A CASE STUDY

This section presents a case study implemented in the air-condition unit manufacturing plant at Indonesia's national train manufacturer, PT INKA. The experiment used several machine tools, like 3-axis CNC milling machine and convention lathe machine. The objective is to detect the existence of the operator and work base and also operating time of the production equipment.

The virtual workstation (VWS) is controlled using a Raspberry Pi 3 model B and Wemos Mini D1. The system is programmed using Microphyton language. Modules which implement to main controller is written with phyton language. The main controller is communicated with controlling sensor wirelessly via MQTT protocol. Main controller also connects with data base MS SQL Server.

For this case study, the virtual workstation is configured to have two function/modules which are operator and work base detection and also access right to the equipment. In this case a web-based laboratory scale software has been developed. By using a web browser, the status of this virtual workstation can be monitored online (see Figure 3). According to Figure 3, it is shown that active virtual workstations are VWS01 and VWS04. It also shown there are one working operator and one workpiece at

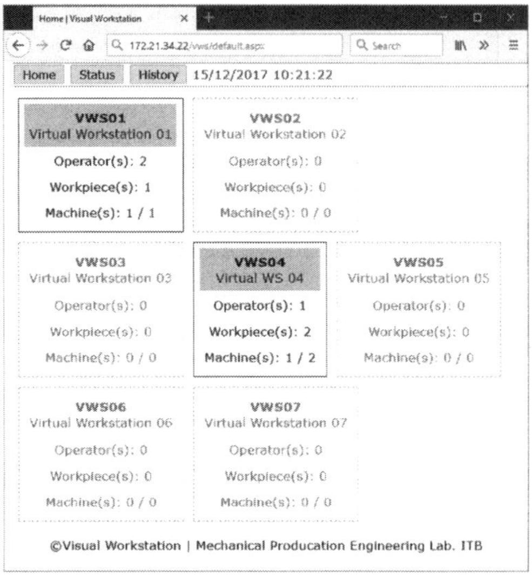

Figure 3. Monitoring status of visual workstation using web browser.

VWS01, currently. At VWS01, there is one running machine from one registered machine of the workstation. While at VWS04 there is one running machine from two registered machine of the workstation.

The virtual workstation used in this case study is equipped with RFID sensor which is used to read

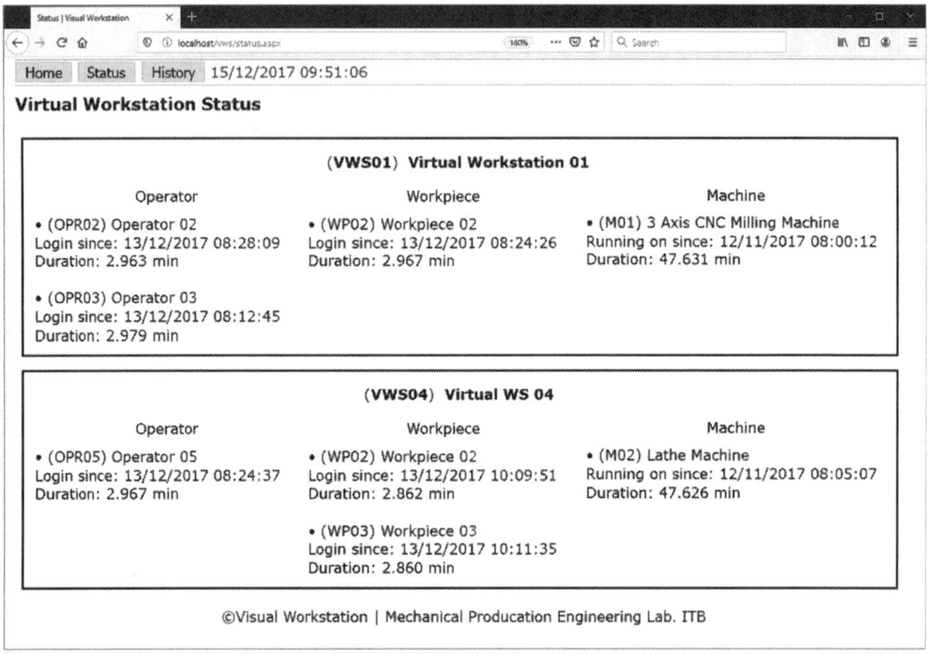

Figure 4. Detail of visual workstation's status.

operator ID card or workbase ID. When the operator is taps their ID, the controller at the virtual workstation will check whether the operator has right to enter the workstation or not. If the operator has the right to access the workstation, the controller will update the database. Next, the controller will check whether the operator has the right to operate the machine or not. If the operator has right, then the controller will order the machine to activate the machine tool.

Detail of the status of virtual workstation also can be monitored by using web browser (see Figure 4). At the time, there are two operators at VWS01 namely OPR02 and OPR03 which had been login for 2.963 minutes and 2.979 minutes, respectively. The current process at VWS01 is processing of workpiece with ID WP02 which had been in this workstation about 2.967 minutes.

5 CONCLUSIONS

The proposed conceptual framework has ability to link between real world and virtual world on the shop floor even without automation. Operators, workpieces and machines' status could be monitored using this method. This virtual workstation could be configured depending on the type of workstations and needed functions. This concept could be implemented as first step of Industry 4.0 implementation. A web-based laboratory scale software for implementing the framework also has been developed. The software handles interactions between the users and the system. Although it needs further work to fully explore the success of implementation, the principle of using information management to jump to industry 4.0 in Indonesia to create smart manufacturing systems seams very feasible.

REFERENCES

Atluru, S., Huang S.H., Snyder J.P. 2012. A smart machine supervisory system framework, in Int. J. Adv. Manuf. Technology, 2012, vol. 58, pp. 563–572, DOI 10.1007/s00170-011-3405-4.

Basl J. 2017. Pilot study of readiness of Czech companies to implement the principles of Industry 4.0, Management and Production Engineering Review, June 2017, vol. 8, no. 2, pp. 3–8.

European Commission. 2018. Digitizing European Industry. https://ec.europa.eu/digital-single-market/en/policies/digitising-european-industry. Accessed Feb 2019

Federal Ministry for Economic Affairs and Energy & Federal Ministry of Education and Research, Germany. 2019. Platform Industrie 4.0. https://www.plattform-i40.de/I40/Navigation/EN/Home/home.html. Accessed Feb 2019

Great Britain. Dept. for Business, Energy and Industrial Strategy. 2017. Made Smarter Review 2017

Huang S., Guo Y., Zha S., Wang F., Fang W. 2017. A real-time location system based on RFID and UWB for digital manufacturing workshop, The 50th CIRP Conference on Manufacturing Systems, Procedia CIRP 63, 2017, pp. 132–137.

Isaksson A. J., Harjunkoski I., Sand G. 2017. The impact of digitalization on the future of control and operations

Konstantinov S., Ahmad M., Ananthanarayan K., Harrison R. 2017.The Cyber-Physical e-machine Manufacturing System: Virtual Engineering for Complete Lifecycle Support. The 50th CIRP Conference on Manufacturing Systems, Procedia CIRP 63, 2017, pp. 119–124.

Liu C., Xu X. 2017. Cyber-Physical Machine Tool – the Era of Machine Tool 4.0, The 50th CIRP Conference on Manufacturing Systems, 2017, pp. 70–75.

Ministry of Industry, Indonesia. 2018. Making Indonesia 4.0. 2018

Monostori L., Kadar B., Bauernhansl T., Kondoh S., Kumara S., Reinhart G., Sauer O., Schuh G., Sihn W., Ueda K. 2017. Cyber-physical systems in manufacturing. The 50th CIRP Conference on Manufacturing Systems, Procedia CIRP 63, 2017, pp. 631–638.

Mrugalska, B., Wyrwicka, M.K. 2017. Towards Lean Production in Industry 4.0, 7th International Conference on Engineering, Project, and Production Management, Procedia Engineering 182, 2017, pp. 466–473.

Qin J., Liua Y., Grosvenor R. 2016. A Categorical Framework of Manufacturing for Industry 4.0 and Beyond, Changeable, Agile, Reconfigurable & Virtual Production Conference, 2016, pp. 173–178.

Qin S. F., Sheng K. C. 2017. Future Digital Design and Manufacturing: Embracing Industry 4.0 and Beyond, Springer, 2017, pp. 1047–1049.

Ramadan M., Al-Maimani H., Noche B. 2017, RFID-enabled smart real-time manufacturing cost tracking system, Int. J. Adv. Manuf. Technol., 2017, pp. 89:969–985, DOI 10.1007/s00170-016-9131-1

Sang Z., Xu X., 2017. The framework of a cloud-based CNC system, The 50th CIRP Conference on Manufacturing Systems, 2017, pp. 82–88.

Tamás B. I. 2016. Waste reduction possibilities for manufacturing systems in the industry 4.0, 2016, IOP, pp. 1–8.

Velandia D. M. S., Kaur N., Whittow W. G., Conway P. P., West A.A. 2016. Towards industrial internet of things: Crankshaft monitoring, traceability and tracking using RFID, Robotics and Computer-Integrated Manufacturing 41, 2016, pp. 66–77.

Industry 4.0 – Shaping The Future of The Digital World – da Silva Bartolo et al. (eds)
© 2021 Taylor & Francis Group, London, ISBN 978-0-367-42272-1

A framework for Industry 4.0 implementation: A case study at King Saud University

H. Alkhalefah
Advanced Manufacturing Institute, King Saud University, Riyadh, Saudi Arabia

ABSTRACT: Industry 4.0 is the amalgamation of many terms and concepts which were introduced in the last decade including factory of future, smart manufacturing, intelligent manufacturing, and collaborative manufacturing to name a few. Industry 4.0 is a stupendous concept and owing to its numerous benefits, it is considered to be a pioneering idea for the future. However, still most of the research work conducted in this field discusses about theoretical models. Therefore, there is a need for the real implementation of this trail-blazing concept. Hence, in this research work a framework is developed and presented based on the case study that deals with implementation of Industry 4.0 at King Saud University. In conclusions, the paper highlights the challenges, issues and need for future work required.

1 INTRODUCTION

Steam power and mechanization kicked of the first industrial revolution, followed by Henry Ford's production optimization through assembly lines in the second, and the third arrived with the computer and automation revolution that drove the latter 20th century up through the recent past. Industry 4.0 is here and its cyber-physical systems will provide the perfect environment for additive manufacturing to flourish.

Industry 4.0 is a revolution in manufacturing, and it brings a whole new perception to the industry on how manufacturing can team up with new technologies to get maximum output with minimum resource utilization. It is a concept in which manufacturing is unified with information technology (Adolph et al. 2016). The result of this alliance is the development of factories that are "smart," i.e., they are highly effective in resource utilization, and they adapt very rapidly to meet management objectives and present industry circumstances. The information technology part of Industry 4.0 includes a cyber-physical system (CPS), cloud computing, and the Internet of Things (IoT). Since the information technology part is progressing day by day, therefore, it is required from the manufacturing side also to be updated. Therefore, in Industry 4.0, a paradigm shift has been occurred from conventional manufacturing processes to additive manufacturing, robotics, and use of virtual reality (VR)/augmented reality (AR) techniques in manufacturing. Its fundamental concepts associated with virtual environment comprises Internet of Things (IoT), Big Data, Cloud Computing etc., whereas its physical realm includes Autonomous Robots and Additive Manufacturing Figure 1 shows a schematic of Industry 4.0 based factory.

The extracted information from the products, machines, or production lines includes a considerable amount of statistical data to be exchanged and analyzed. Further sources of data are design records, customers' order, suppliers' delivery, stock and logistics related information. Altogether, this large quantity of data is defined as Big Data, which is another major concept in Industry 4.0. Furthermore, cloud computing, which is related to the processing of all the available information, can also be considered as one of the most important terms in the virtual industrial world. All of these cyber technologies help to guarantee the efficient use of existing information for smart manufacturing of future (Wang & Wang 2016). Table 1 shows the technology realm of Industry 4.0.

A plethora of research is going on around the world in the field of Industry 4.0. The invention of the term Industry 4.0 does not automatically mean that the subject can be proficiently mastered and implemented. It is therefore necessary to show that not only a vision been created, but also that it can be competently implemented. Therefore, under the vision of 2030, Saudi Arabia is also willing to improve its manufacturing industry and keen to invest in futuristic technologies for the sustainable growth of the country. As a leading university in the region, King Saud University at Advanced Manufacturing Institute (AMI), started a research project to implement the pioneering Industry 4.0 concept. The objective is to build the local know-how of this innovative concept and moreover, to train and prepare the young workforce of the nation so that they will apply this knowledge when they will join the workforce in the industry.

In this paper, the opportunities, challenges, capabilities, and key players of Industry 4.0 in Saudi

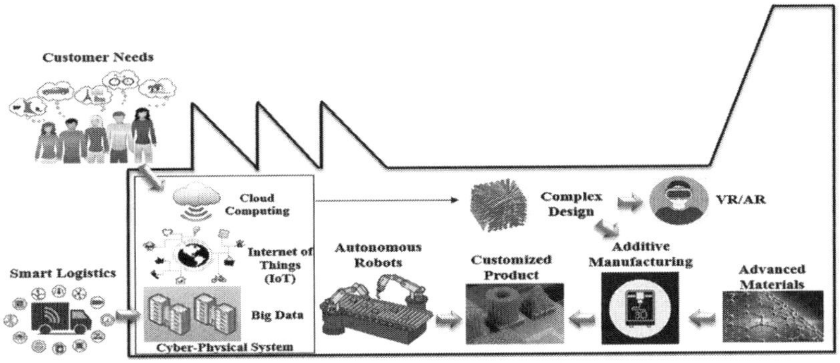

Figure 1. Schematic of Industry 4.0 based factory.

Table 1. Technology Realm of Industry 4.0.

Realm	Technology
Physical → Digital	• Sensors and Controls • Augmented Reality/VR • Wearable devices
Digital	• Cloud Computing • Optimization & Prediction • Visualization • Cognitive & high-performance computing • Digital Design and Simulation
Digital → Physical	• Additive Manufacturing • Advanced Materials • Autonomous Robotics

Arabia are explained briefly. Then, the implementation of this concept at the Advanced Manufacturing Institute, King Saud University is explained in detail. In the conclusion section, challenges, issues faced, and future research directions are presented.

2 GETTING STARTED

Fourth Industrial Revolution can be an implausible accelerator to realize Vision 2030 by enabling greater trade and integration, localizing supply chains, rapidly upskilling a young population and driving innovation (Zugravu 2018). Governments, together with businesses and members of civil society, have five cross-industry and cross-technology areas of action to drive inclusive adoption of technologies and foster a healthy Industry 4.0 adoption: up-skilling the workforce, investing in innovation, integrating the country into the global value chain, as well as strengthening the institutional framework and ensuring sustainable resource management.

Industry 4.0 would enable Saudi producers to successfully compete locally and internationally. The technologies enable small-scale, on-location production, and would make the production process more flexible, involving fewer steps, shorter lead times, lower capital requirements, and even batch sizes of one (Zugravu 2018).

Saudi has to work hard to prepare its future workforce by refining basic education, equipping workers with the correct skills, and incentivizing them on joining the private sector. The technologies of the Industry 4.0 hold a greater promise: that of swift upskilling. Saudi companies and governments can apply augmented reality and virtual reality training to guarantee quick development of market-ready skills.

Recently, the Business Incubators and Accelerators Company (BIAC), a subsidiary of Saudi Technology Development and Investment Company (TAQNIA) has announced the formation of a new joint venture company to support and develop Industry 4.0 technology applications in the Kingdom (Saudigazette 2019). Nawaf Alsahaf, Chief Executive Officer of BIAC, said: "The joint venture will train and employ numerous male and female Saudi creators and engineers, specializing in Industry 4.0 related technologies. The purpose of the new company will be to provide advanced problem-solving solutions to local companies, develop new intellectual property and establish specialist production capacity within the Kingdom."

Saudi Aramco's Uthmaniyah Gas Plant (UGP) has been renowned by the World Economic Forum (WEF) as a "Lighthouse" manufacturing facility, a leader in technology applications of the Fourth Industrial Revolution. Saudi Aramco is the first energy company globally to be included in this select group of manufacturing sites (Eye of Riyadh 2019). To attain the vision 2030 of the Saudi Arabia in addition to the National transformation program 2020, KACST (King Abdulaziz City for Science and Technology) established the innovation center for industry 4.0. Moreover, KACST contributed the implementation of the fourth industrial revolution techniques in the region to build up the importance of the techniques and the concepts of this revolution through localization and adoption (KACST 2019). Saudi Industrial Development Fund (SIDF) mentioned in one of their reports that the industry currently accounts for approximately 12% of GDP (Gross Domestic Product), and the Vision 2030 targets to increase its contribution to higher levels with an emphasis on the shift to high-tech industries towards the 4th Industrial Revolution (SIDF 2019).

Therefore, the government, as well as private organizations are paying a lot of attention to implement Industry 4.0 in Saudi Arabia. To achieve this objective, existing and new projects will have to use modern machinery and technologies, which in turn will lead to the development of the industrial base in Saudi Arabia and help industry to play its key role in economic growth and job creation.

3 IMPLEMENTATION FRAMEWORK OF INDUSTRY 4.0 AT ADVANCED MANUFACTURING INSTITUTE

As mentioned above, there is a need of implementation of Industry 4.0 concept. Kamble et al. (2018) mentioned that the case study approach is the most popular approach used by researchers to demonstrate this technology. Therefore, in this paper, a case study of Industry 4.0 implementation is presented. At the AMI, various pillars or components of Industry 4.0 that are related with manufacturing sectors are already presented such as AR/VR, additive manufacturing (AM) unit, reverse engineering setup, dimensional quality inspection, and autonomous robots. Therefore, the main goal is to integrate all these systems with the help of cyber-physical systems. The focus of this work was to explore the technical issues faced during implementation rather than from business or commercial side. Figure 2 shows the various facilities available at the AMI.

As shown in Figure 2, AMI is equipped with state of the art technology in the fields required for Industry 4.0 in the manufacturing side. Now, the major task is to make a CPS system that will develop a machine-to-machine communication network between. The developed model will be used for research purpose, therefore commercial aspects are not integrated in the developed model.

Now let us consider a scenario that will show how this framework will work. The various steps are as follows:

- A customer sends his needs to the system with the help of cloud computing, and IoT.
- The customers' demands then will be transformed into engineering specification by the professionals initially, then at later stages, the decision making programs and with the help of deep learning

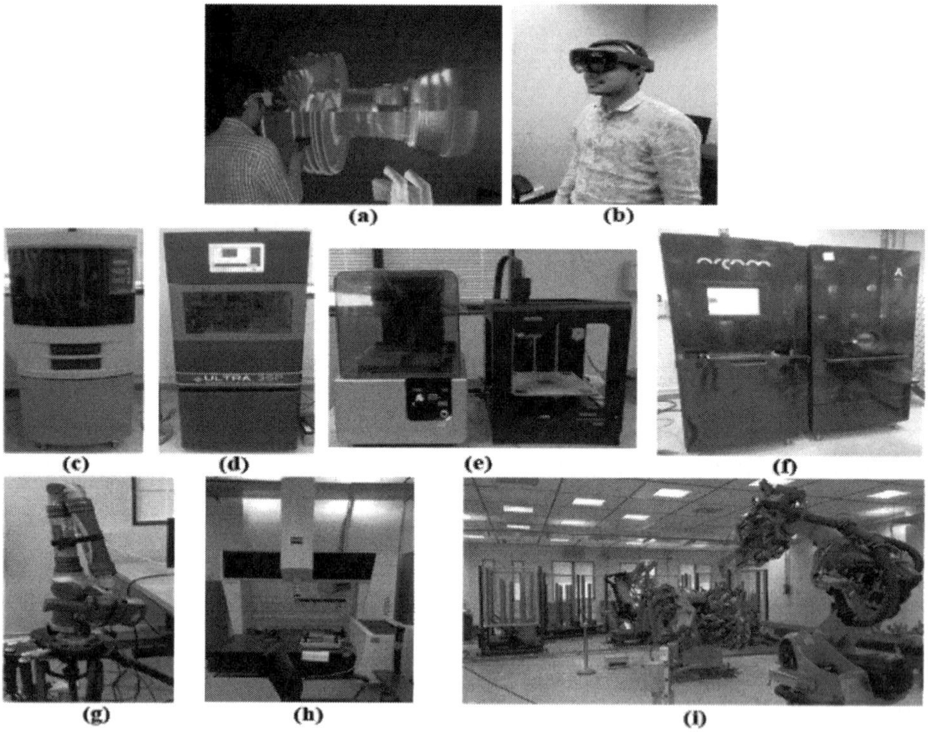

Figure 2. Different facilities available at AMI for Industry 4.0 a) VR system; b) AR system; c) FDM 3D printer; d) SLA 3D Printer; e) Table top 3D printers; f) Metal 3d Printer (EBM); g) Faro Scanner (for RE); h) CMM (for inspection); i) Industrial Robots.

algorithms these needs will be transformed into technical specifications automatically

- Then a digital model will be sent back to the customer through the same channel.
- The customer reviews the design and can make suggestions for improvement according to his needs. After the final design, an approval will be taken.
- Before making the physical part with the help of VR/AR, the virtual model of the product can be shared with the customer.
- All this data handling will be managed by Big Data analytics.
- The design will be analyzed, based on that a manufacturing process will be suggested, either it will be through AM, through hybrid (additive-subtractive) or through subtractive only.
- The final design model then will be sent to the appropriate machine based on the type of technology required.
- Then the quality assurance team will inspect the part and will send it to the customer.
- Meanwhile, the products guarantee, recycling information, and sustainable life cycle information will be stored and shared with the customer who can view it anytime through the internet.
- RFID tags will be used for smart logistics and to reduce supply chain hitches.
- Other than that, the machine will be equipped with sensors that will be used for predictive/preventive maintenance
- Automatic storage/retrieval system will be used to store the raw material and it will keep updating the inventory levels.

Figure 3 shows a framework for Industry 4.0 at AMI.

4 CONCLUSIONS AND DISCUSSIONS

Industry 4.0 is a huge and phenomenon with continuous improvement. Most of its pillars have many advantages, and needs to be explored. The implementation of Industry 4.0 at the AMI is going on successfully until now. The complete system once will be integrated, will fulfill all the capabilities as mentioned in the scenario.

There are still many issues and challenges need to be tackled both technology-wise as well as business model wise. Based on experience and literature the following open issues are there for Industry 4.0 implementation:

- The need of highly-skilled manpower
- High cost for implementation
- Legal aspects concern data possession and data privacy aspects
- Data analytics and secure data transmission
- Standardization of data transmission
- Machine retrofitting with sensors or alike
- High internet speed is required

There is a fear that machines will replace the humans and the job market will be down, however,

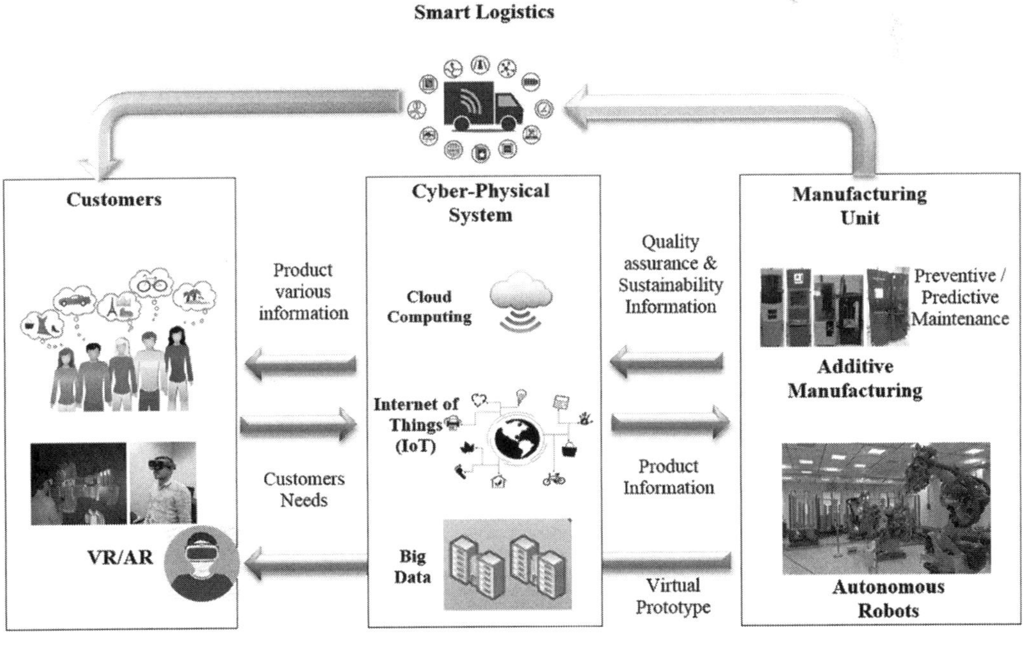

Figure 3. A framework for AM in Industry 4.0.

in actual, there will be more need of manpower but the skilled one. Therefore, in a developing country like Saudi Arabia, there is a lot of potentials to implement this pioneering technology, since the manufacturing sector is growing here. Government is also making efforts and big investment to make manufacturing sector flourish in Saudi Arabia. Since manufacturing industries are at very low stage here, therefore it is easier to implement the new technology, as no resistance will come from the organizations' workforce like in the case of established industries.

Future research studies may examine the effect of various Industry 4.0 technologies on the level of process integration that is realized by the organizations. This will assist to recognize the important technologies that help to accomplish process integration in the different industrial setups.

REFERENCES

Adolph, Lars, et al., 2016. German Standardization Roadmap: Industry 4.0. Version2. DIN, Berlin.

Eye of Riyadh, 2019. 'Saudi Aramco Recognized as a Leader in the Fourth Industrial Revolution', *Eye of Riyadh*, dated: 2019-01-22, available online https://www.eyeofriyadh.com/news/details/saudi-aramco-recognized-as-a-leader-in-the-fourth-industrial-revolution, accessed on 2nd March, 2019.

KACST, 2019. Innovation Center for Industry 4.0, *King Abdulaziz City for Science and Technology*, available online: http://ici4.sa/en/establish-innovation-center-for-industry-4-0/, accessed on 3rd March, 2019.

Kamble, S.S., Gunasekaran, A., & Gawankar, S.A. 2018. Sustainable Industry 4.0 framework: A systematic literature review identifying the current trends and future perspectives. *Process Safety and Environmental Protection* 117: 408–425.

Saudigazette, 2019. BIAC, Simera Innovate launch JV to accelerate Industry 4.0 in Saudi Arabia, *Saudi Gazette*, dated: 2018-02-13, available online: http://saudigazette.com.sa/article/559059/BUSINESS/BIAC-Simera-Innovate-launch-JV-to-accelerate-Industry-40-in-Saudi-Arabia, accessed on 2nd March, 2019.

SIDF, 2019. Industrial Development in Saudi Arabia, *Saudi Industrial Development Fund*, available online: http://www.sidf.gov.sa/en/IndustryinSaudiArabia/Pages/IndustrialDevelopmentinSaudiArabia.aspx, accessed on 3rd March, 2019.

Wang, L. & Wang, G. 2016. Big Data in Cyber-Physical Systems, Digital Manufacturing and Industry 4.0. *International Journal of Engineering and Manufacturing* 6 (4): 1–8.

Zugrav A., 2018. 'Fourth Industrial Revolution 'an incredible accelerator for Saudi Arabia to achieve Vision 2030', *Saudi Gazette*, dated: 2018-08-11, available online: http://saudigazette.com.sa/article/540938/BUSINESS/Fourth-Industrial-Revolution-an-incredible-accelerator-for-Saudi-Arabia-to-achieve-Vision-2030, accessed on 2nd March, 2019.

Industry 4.0 – Shaping The Future of The Digital World – da Silva Bartolo et al. (eds)
© 2021 Taylor & Francis Group, London, ISBN 978-0-367-42272-1

Health and safety in smart industry: State-of-the-art and future trends

O.J. Bakker, T. Kendall & P. Bartolo
School of Mechanical, Aerospace and Civil Engineering, Faculty of Engineering, The University of Manchester, UK

ABSTRACT: Industry 4.0, also known as smart manufacturing, is the name of what is considered as the fourth industrial revolution. The drivers behind are globalization, mass customization, and the rapid increase in computational power which makes it now possible to process big data and perform advanced analytics. In response, manufacturers require production systems that have more resilient, predictive and adaptable capabilities to achieve a truly agile production that is capable of lean, competitive production of mass customized products. The digitization of manufacturing has many implications not only on the shop floor. The biggest potential improvements identified in the survey is that Industry 4.0 brings to health and safety in the form of advanced monitoring of the worker and their environment through wearables, vision, and other sensor systems. The implementation of these technologies in turn also challenges that require further addressing, these challenges are not only technical but also organizational in nature.

1 INTRODUCTION

This paper is motivated two trends. The first is the megatrend of digitization. On a wider scale, the digitization of society results in the development of smart products, smart homes through the Internet of things (IoT). Application of cyber-physical systems and advanced robotics enable driverless vehicles. Traditional business models also change: many physical shops have lost customers due to the rise of online shops and sale platforms. It also enabled the sharing economy where corporations provide platforms such "peer-to-peer car sharing" that proves a significant competitor to traditional taxi services, "peer-to-peer house sharing", which has changed the traditional hospitality industry, and "peer-to-peer money transfers" that provide alternatives for international bank transfers. These developments have provided many people with often relatively cheap goods and services, ordered by simply pushing a button on an app or website from the comfort of one's own home, delivered to the door. Meanwhile reports appeared in the media raising concerns about the working conditions of many "gig workers" and drivers who deliver all the parcels. Other news reports also talk about the latter category of workers being replaced by driverless vans and trucks. Advanced automation and digitalization of manufacturing mean a radical change in how advanced and emerging economies produce. New technologies on the shopfloor introduce new hazards, but also provide technologies that keep humans out of harm's way in dangerous areas in production sites etc. The second trend is that Lloyd's Registry Foundation and HSE observed that the decrease in H&S incidents observed since the implementation of the Health and Safety at Work etc. Act in 1974 is plateauing. From these two trends, a key research question arises: Can the technology underpinning Smart Manufacturing be used to reduce H&S incidents? A special focus of this this scoping study: to link Industry 4.0 readiness level to H&S management. The rest of the paper is organized as follows. Firstly, the search method to find literature is discussed. After this, the underpinning technologies and the transformative nature of Industry 4.0 are charted and reviewed. From here the emerging risks and opportunities are identified and discussed. Finally, before drawing conclusions, Industry 4.0 maturity and implementation are considered.

2 SEARCH METHOD

Health and safety, more formally called Occupational Health and Safety (OHS) is a multidisciplinary subject area concerned with the safety, health, and welfare of people in the workplace. It this therefore not surprising to see publications in wide ranging disciplines, such as law, ergonomics, production systems etc. More intuitively than scientifically back up, many authors distinguish four different fields within OHS (Badri et al., 2018): (1) organization of work, (2) OHS management systems, (3) management of occupational risks, OHS legislative and regulatory Framework.

In this publication, the first three areas are of concern. The fourth area, studying how health and safety are dealt with in the law and regulations flowing from these laws are of less interest at this moment, as firstly the authors are looking for technical solutions, the legislation still has to catch up with digitization (Badri et al., 2018), and thirdly on the longer term this work seeks to inform future legislation.

As the focus of this literature survey is on the application and implementation of Industry 4.0 technology to improve health and safety on the shopfloor, both scientific publications and 'grey literature' (foresight publications and business news articles) are analyzed. Initial searches in Scopus using a more restrictive search resulted in a very small number of publications. To cast the net wider, these restrictions were lifted, and "Google Scholar" was used to find scientific publications and 'normal' "Google" for grey literature. The key words used to search were: Keywords Health and safety + each of the following names of Industry 4.0: smart factory, smart Industry, smart manufacturing, factory of the future, Industry 4.0, digital manufacturing. This resulted in more sifting work, as in many papers there are just the statements that Industry 4.0 has a large potential to be beneficial for health and safety, which apart from a wide shown consensus does not add further information. After identification of key papers, double search was carried out, going through the references in those papers and download those. A few of the key literature review papers are:

The review paper by Badri et al. (2018) is the closest to where this survey is focussed on, it does address the area of legislation briefly. Our unique focus on impact and solutions distinguishes this work from theirs.

Bragança et al. (2019), are concerned with the health and safety aspects of collaborative robots.

Villani et al (2018) review safety aspects of human robot collaboration for both collaborative robots and restrained industrial robots.

The book chapter by Brocal et al (2018), is concerned with novel manufacturing technologies and emerging risks

Pavón et al. (2018), Ranavolo et al. (2018) and the methodology Podgórski et al. (2018) discuss health and safety frameworks based on smart PPE and other wearables. A notable omission in the field are the remote controlled and autonomous robots that can enter places that are dangerous for humans. Which is probably explicable since there is little direct engagement of these robots with the field of health and safety.

3 DIGITAL TRANSFORMATION

The name Industry 4.0 stands for the fourth industrial revolution. The first industrial revolution started in England in the second half of the 18th century with the water and steam powered mechanization of the textile industry. This revolution ended mid 19th century. Late 19th century and early 20th the second industrial revolution or wave of mechanization arrived in the form of electrical power enabled mass manufacturing. The third industrial revolution is characterized by computer programmable automation and happened in the last decades of the 20th century. The term fourth industrial revolution is then coined by the German government, who had commissioned a strategic study. Some initial results were revealed at the Hannover Messe in 2011, and the final report in 2013 (Kagerman et al. 2013). The fourth industrial revolution, enabled by a larger computational power and wireless networks, adds cyber physical systems (CPS), connectivity and decentralized decision making.

3.1 *Underpinning emerging technologies for the digital manufacturing revolution*

Lasi et al. (2014) described the pull for a rapid, flexible and agile production, which in turn is then enabled by advanced manufacturing technologies. Today, most researchers and practitioners follow Boston Consulting Group's (BCG's) 9 pillars of Industry 4.0 group and examine emerging technologies underpinning the fourth industrial revolution (Gerbert et al., 2015): (1) Big data and analytics. (2) Autonomous robots. (3) Simulation. (4) Horizontal and vertical system integration. (5) The industrial internet of things (IIoT). (6) Cybersecurity. (7) The cloud. (8) Additive manufacturing. (9) Augmented reality. In their report, the OECD lists a number of key emerging technologies (OECD, 2017), partially overlapping, partially complementary to BCG's 9 Pillars of Industry 4.0: digital technologies, industrial robotics, industrial biotechnology, nanotechnology, additive manufacturing, advanced materials.

3.2 *Industry 4.0's 8 smart transformations*

Not only technologies themselves can bring threats and opportunities, also the impacts, or transformations aimed for, can change the working environment and impact the occupational health and safety of workers. With industry looking for impact, going to an industry association helping its members to reap the benefits of manufacturing can give some insight into the real changes aimed for, with the obvious limitation that the downsides might not be discussed. SmartIndustry.nl is a platform set up by the Dutch Government, trade associations and the Netherlands Organization for Applied Scientific Research TNO. SmartIndustry (2018) mentions 8 smart transformations: (1) Advanced Manufacturing. (2) Flexible Manufacturing. (3) Smart Products. (4) Servitization. (5) Digital Factory. (6) Connected Factories. (7) A Sustainable Factory. (8) Smart Working.

4 EMERGING RISKS

Many scientific publications on risk in a broad context can be found that can be distinguished in terms of organizational, functional or economic risk and how to govern these. Examples are Florin (2013) discussing 23 risk governance deficits and 10 contributing factors. This paper has been written in the framework of the EU FP7 iNTeg-Risk project, which was "aimed at improving the management of emerging risks in the innovative industry". (iNTeg-Risk website, 2013). Herrmann (2018) identifies the following major technological risks or challenges: the lack of standardization, information security, availability of appropriate IT infrastructure, availability of fast internet, inordinate system complexity, unadapted organization of management. From the scientific literature, to the best knowledge of the authors, the publications by Brocal and co-workers (Brocal and Sebastián, 2014; Brocal et al., 2018) form the most comprehensive peer reviewed contributions in the field. Brocal and Sebastián (2014) set up a risk assessment model based on the existent risk assessment framework such as ISO31010 (risk assessment techniques). For their list of new and emerging risks, Brocal and Sebastián (2014) and Brocal et al. (2018) rely on the forecast studies from EU-OSHA. Most up-to-date references from EU-OSHA are: Bradbrook et al. (2013) Stacey et al. (2018). Independently from EU-OSHA organized workshops, Hauke et al. (2018) carried out an online survey among industry experts that showed a similar outcome. The British Safety Council (2018) also published a foresight study. Summarising their most important findings:

Exposure to new and dangerous materials on the shopfloor. These materials can be in form of liquids, powders and fumes, and are linked to nanotechnology, additive manufacturing and bioproduction.

Cybersecurity: a hack or a data breach can not only reveal intellectual property, trade secrets and other commercial sensitive data, but also expose data of employees, such as identity, contact details and also medical and occupational health (and safety data). On the other hand, a hack into the productions system can also give hackers access to the SCADA and PLC level of the shopfloor control, potentially them to override controls and turn processes and equipment into a hazard by abruptly stopping the process or steering it outside the process window.

An interruption in electricity supply will make the process stop abruptly, which can lead to dangerous situations. Note that this does not hold for some of the equipment that is fail-safe designed, or a company that uses emergency power supply or completely uninterruptable power supply to continue working or allow for a controlled shutdown, potentially after finishing the current work cycle. In addition, electrical power shutdown will also affect the network and the controls of the manufacturing system and the data flow for decision making and transmission of real-time instructions to a worker.

The worker's wellbeing can be deteriorated by stress from continuous change and ever increasing performance targets. In addition, stress can lead to bad decision making and imprecise working, which can lead to hazardous situations.

Human-robot interaction is actually increasing. Not only are robots added to the human work environment to alleviate some of the human worker's tasks, the British Safety Council (2018) found that humans are replacing robots in the automotive sector to increase the process flexibility for achieving higher levels or product customization.

The increased chance of exposure to laser light several forms of additive manufacture.

Health and safety assessments can be incomplete. Due to the increased flexibility, especially with the application of robots and co-bots it is necessary that health and safety assessments cover all the possible scenarios.

Increased system complexity also increases the worker's cognitive stress levels, which can lead to fatigue, poor judgement etc. resulting in potentially dangerous situations.

Gig workers are workers without permanent contract, they are often less trained with the equipment they use when working, resulting in a higher level of incidents.

It should be concluded here that the flexibility of and labour that comes with the digitization of work and has a large influence on the wellbeing but also on the health and safety of a work. Similar for less or lower skilled and trained workers, there is a skill gap (SmartIndustry.nl, 2018) and the increased load of cognitive stress, due to the nature of work changing from physical to more cognitive having a larger impact on lower skilled workers (Work4.0, 2017). The Worldbank actually calls for a new social contract, where governments and employers commit to more training to improve the social inclusion (Worldbank, 2019).

5 OPPORTUNITIES

Drones and other automated or autonomous systems work or do inspection in dangerous or hard to reach areas. This type of technology significantly improves the occupational health and safety conditions of workers. However, this is prevention and elimination through avoidance and is not measure that improves occupational health and safety of the human worker's workspace. For this reason, similar to other publications in the field, the authors decided to not further discuss this technology.

Exoskeletons are applied in the medical field for the rehabilitation of patients affected by muscle deceases or muscle shrinkage due to not being able to use muscles in certain body parts due long-term forced rest. In industry, exoskeletons can be utilized to prevent musculoskeletal disorder, reduce fatigue and assist people lifting heavier loads.

The last and most studied area of opportunity is the area of "wearables". Some applications are on the market already, whereas others, more complex, are still in the research phase to prove their technology.

When starting with the state-of-the-art industry, the health and safety apps are commonly referred to as "the connected worker. Also, consultancy companies, notably LNS Research (Bussey, 2018) and Deloitte (Perkins and Thomson, 2017; Tan, 2018). Wearables can be used as a sensor, as tool to convey information to the worker and to communicate with remote help. This will off-load the cognitive load from the human worker.

From a more academic perspective, Podgórski et al. (2018) define a paradigm shift for OHS risk assessment in flexible manufacturing environments. Firstly, they claim, it will have to happen on an individual basis. The OHS system has to be context aware, such that it know has the system has changed, this happens through the wearables worn by the worker. For most results, a mix of probabilistic and deterministic risk assessment will be carried out. An open question is whether this will be acceptable by future regulations. Adjiski et al. (2019) designed an alternative architecture for a similar application, without calling it risk assessment, to achieve real-time safety in mines. Gregori et al. (2017) designed a system that can measure several aspects of worker well-being with sensors and the worker and distribute on the shopfloor. The system was partially evaluated with a prototype system implemented in a woodworking work cell. Kaare and Otto (2015) suggested monitoring using smart healthcare technology, such as haemoglobin value, blood sugar etc. and fuse this data with the context and shopfloor environmental data to measure the workers' well-being. Only two papers found presented a whole architecture to define real-time safety or online risk assessment. It is probably the right moment observe the shift of EHS software from a focus on compliance with occupational health and safety regulations to active, real-time safety measurement (Bussey, 2018).

6 INDUSTRY 4.0 READINESS AND IMPLEMENTATION

Basl (2018) conducted a literature review on Industry 4.0 readiness. A total of 18 models are compared. For Basl's intentions this was sufficient, as his main objective was establish the dimensions of the assessments, to derive a meta-model on Industry 4.0 maturity in a further study (Basl and Doucek, 2019). Basel found that the models in circulation are actually quite different in assessment of Industry 4.0 dimensions, depending on interest they can be more focused on: manufacturing capabilities, management/strategy/sociological aspects, digital capabilities, supply chain.

Of special interest for this literature survey is that only one of the assessed models in Basl's study considers health and: Roadmap Industry 4.0 by Pessl et al. (2017). This is very much in line with the general view of health and safety, as a sort of afterthought, a matter of compliance, which is more a cost (Romero, 2018). The Industry 4.0 maturity index by Pessl et al. (2017) is strongly focussed on (change) management and sociological aspects. Not all the models reviewed in Basl's study consider the dimension of change management and other sociological aspects. However, for a long-term, structured roadmap for implementation, it is strongly advisable to follow a more structured approach to innovation, such as the open innovation model, which is very appropriate for digitally networked innovation (Veile, 2018).

7 CONCLUSIONS AND OUTLOOK

The main question asked in the beginning of this survey, can the technology advancement that comes with Industry 4.0 reduce the number of incidents can be partially answered. As it is still too early to draw any statistical conclusions. However, on the one hand, it has been observed that the new manufacturing technologies bring exposure to new emerging risks; autonomous systems can be used to eliminate the need for human presence in dangerous situations; exoskeletons can be applied to alleviate physical exertion of the human workers to prevent work-related musculoskeletal diseases; and wearables integrated in the shopfloor's data and decision can be used for proactive prevention.

Other observations from the literature are that, firstly the transformations caused by digital revolution and emerging risks and opportunities have been reviewed. Secondly, the Industry 4.0 Readiness for occupational health and safety measures has been discussed. It was found that based on more commonly applied maturity indices, also the Industry 4.0 maturity levels for occupational health and safety management can be established. Using the open innovation philosophy for the design and implementation of new technology, a basic roadmap to integrate H&S in smart manufacturing can be composed.

As main future trend, the development of monitoring and EHS software can be identified. As there will be a shift towards using wearable devices to actively reduce health and safety incidents and integrate it into the manufacturing system's network, data will be used to both monitor for hazards, but also to improve the human worker's efficiency and quality of production. Therefore, for monitoring and EHS Software, there will be a shift from compliance-based health and safety administration towards integration with productivity, quality and risk (PQR) programs in integrated management systems. Herein also lies the largest challenge to overcome: making sense of all the data, not only the sheer amounts of it, but also from an organizational point of view. It should be noted that this data challenge also needs to be overcome in the broader context of Industry 4.0. Solving the data challenges will undoubtedly see significant effort from research and practitioners in the future.

For the optimal integration of occupational health and safety with the digital work, there are still lacunas in knowledge to model the human worker in the manufacturing system. For this to happen, there is a need of systematic research in various scientific disciplines on the study of human behaviour: human interaction with technology, organization of work and related stress levels. Additionally, virtual task analyses will form an integral part of this approach to cover as many scenarios as possible before the actual work is carried out on the shopfloor.

Apart from the main future trend to apply wearables to monitor and inform workers for hazards, whilst simultaneously boost their productivity, the human centred-manufacturing and -design approach will have the following implications for improving occupational health and safety. Firstly, design effort will be focussed on ergonomics and human comfort. System design will show a trend towards the implementation of an integrated approach for safety by design (prevention through design), although even that domain is undergoing rapid changes, due to Industry 4.0 (Clegg, 2018).

ACKNOWLEDGEMENTS

The reported research is part of the Discovering Safety programme, funded by the Lloyd's Register Foundation, supported by the Thomas Ashton Institute, the support of which is greatly appreciated.

Furthermore, the authors wish to express their gratitude to S. Naylor and J.P. Gorce from HSE for their insightful discussions.

REFERENCES

Adjiski, V., Despodov, Z., Mirakovski, D., & Serafimovski, D. 2019. System architecture to bring smart personal protective equipment wearables and sensors to transform safety at work in the underground mining industry. *Mining-Geological-Petroleum Engineering Bulletin*, 34(1): 37–44.

Badri, A., Boudreau-Trudel, B., & Souissi, A.S. 2018. Occupational health and safety in the industry 4.0 era: A cause for major concern? *Safety Science*, 109: 403–411.

Basl, J. 2018. Analysis of industry 4.0 readiness indexes and maturity models and proposal of the dimension for enterprise information systems. In A. M. Tjoa, M. Raffai, P. Doucek, and N. M. Novak (eds). *Research and Practical Issues of Enterprise Information Systems*. Cham. Springer International Publishing.

Basl, J. & Doucek, P. 2019. A metamodel for evaluating enterprise readiness in the context of industry 4.0. *Information*, 10(3): 89.

Bradbrook, S. et al. 2013. *Green jobs and occupational safety and health: Foresight on new and emerging risks associated with new technologies by 2020*. European risk observatory report, European Agency for Safety and Health at Work (EU-OSHA), Luxembourg.

Bragança S., Costa E., Castellucci I., & Arezes P.M. 2019. A Brief Overview of the Use of Collaborative Robots in Industry 4.0: Human Role and Safety. In: Arezes P. et al. (eds) *Occupational and Environmental Safety and Health. Studies in Systems, Decision and Control*, Vol 202. Springer, Cham

British Safety Council 2018. *Future risk: Impact of work on health, safety and wellbeing, a literature review*. Technical report, British Safety Council, London, UK.

Brocal, F., Sebastián, M.A. & González, C. 2019. Management of Emerging Public Health Issues and Risks: Multidisciplinary Approaches to the Changing Environment. In B. Roig K. Weiss & V. Thireau (eds), *Management of Emerging Public Health Issues and Risks: Multidisciplinary Approaches to the Changing Environment*, Chapter 2. Academic Press.

Brocal Fernández, F. & Sebastián Pérez, M.A. 2015. Analysis and modeling of new and emerging occupational risks in the context of advanced manufacturing processes, In Procedia Engineering 100, 1150–1159. *25th DAAAM International Symposium on Intelligent Manufacturing and Automation, 2014*.

Bussey, P. *The Connected Worker: Modernize Safety and Risk Management with Digital Innovation*. LNS Research. website, accessed on 15-03-2019, URL: https://blog.lnsresearch.com/the-connected-worker-modernize-safety-and-risk-management-with-digital-innovation.

Clegg, R., Cooper, R. & Ross, C. 2018. *Foresight review on design for safety: Protecting lives from the start*. Technical Report No.2018.2, Lloyd's Register Foundation, London, UK.

Florin, M.V. 2013. IRGCS approach to emerging risks. *Journal of Risk Research*, 16(3-4): 315–322.

Gerbert, P. et al. 2015. *Industry 4.0: The future of productivity and growth in manufacturing industries*, Technical report, Boston Consulting Group.

Gregori, F. et al. 2017. Digital manufacturing systems: A framework to improve social sustainability of a production site. Procedia CIRP 63, 436–442. *Manufacturing Systems 4.0 – Proceedings of the 50th CIRP Conference on Manufacturing Systems*.

Hauke, A., Flaspöler, E. & Reinert, D. 2018. Proactive prevention in occupational safety and health: How to identify tomorrow's prevention priorities and preventive measures. *International Journal of Occupational Safety and Ergonomics*, 26(1): 181–193.

Herrmann, F. 2018. The smart factory and its risks. *Systems*, 6 (4).

iNTeg-Risk 2013. iNTeg-Risk: *Early recognition, monitoring and integrated management of emerging, new technology related risks*. iNTeg-Risk website, accessed 15-03-2019.URL: http://www.integrisk.eu-vri.eu/.

Kaare, K.K. & Otto, T. 2015. Smart health care monitoring technologies to improve employee performance in manu-facturing. Procedia Engineering, 100: 826–833. *25th DAAAM International Symposium on Intelligent Manufac-turing and Automation, 2014*.

Kagermann, H., Wahlster, W. & Helbig, J. 2013. *Securing the future of german manufacturing: Recommendations for implementing the strategic initiative INDUSTRIE 4.0 – final report of the industrie 4.0 working group*, Technical report, Plattform INDUSTRIE 4.0.

Lasi, H., Fettke, P., Kemper, H.-G., Feld, T. & Hoffmann, M. 2014. Industry 4.0. *Business & Information Systems Engineering*, 6(4): 239–242.

OECD, 2017. *The next production revolution: Implications for governments and business*. Technical report, OECD, Paris.
Pavón, I. et al. 2019. Wearable Technology for Occupational Risk Assessment: Potential Avenues for Applications. In P.M. Azares et al. (eds) *Occupational Safety and Hygiene VI Book chapters from the 6th International Symposium on Occupation Safety and Hygiene (SHO 2018), March 26-27, 2018, Guimarães, Portugal*, Chapter 79: 447–452. CRC Press.
Perkins, B. & Thomson, R. 2017. *The connected worker: Clocking in to the digital age*. Technical report, Deloitte.
Pessl, E., Sorko, S.R. & Mayer, B. 2017. Roadmap industry 4.0 implementation guideline for enterprises. *International Journal of Science, Technology and Society*, 5(6): 193–202.
Podgórski, D. et al. 2017. Towards a conceptual framework of OSH risk management in smart working environments based on smart PPE, ambient intelligence and the internet of things technologies. *International Journal of Occupational Safety and Ergonomics*, 23(1): 1–20.
Ranavolo, A. et al. 2018. Wearable monitoring devices for biomechanical risk assessment at work: Current status and future challenges a systematic review. *International Journal of Environmental Research and Public Health*, 15(9).
Romero, D. et al. 2018, Digitalizing occupational health, safety and productivity for the operator 4.0. In I. Moon, G. M. Lee, J. Park, D. Kiritsis & G. von Cieminski (eds). *Advances in Production Management Systems. Smart Manufacturing for Industry 4.0*, Springer International Publishing, Cham, pp. 473–481.
Smartindustry.nl 2018. *Slimme transformaties {Smart Transitions}*. smartindustry.nl website, in Dutch, accessed 15- 03-2019.URL: https://smartindustry.nl/wiki-smart-industry/transformaties/.
Stacey, N. et al. 2018. *Foresight on new and emerging occupational safety and health risks associated with digitalisation by 2025*. European risk observatory report, European Agency for Safety and Health at Work (EU-OSHA), Luxembourg.
Tan, A. 2018. *The connected worker: Charging up the business services workforce*. Technical report, Deloitte.
Veile, J. W., Kiel, D., Müller, J. M. & Voigt, K.-I. 2018, How to implement industry 4.0? an empirical analysis of lessons learned from best practices. In B. Nunes, A. Emrouznejad, D. Bennett & L. Pretorius, (eds), *IAMOT 2018 Proceedings', International Association for Management of Technology (IAMOT), Aston Business School, Birmingham, UK.*
Villani, V., Pini, F. Leali, F. & Secchi, C. 2018. Survey on humanrobot collaboration in industrial settings: Safety, intuitive interfaces and applications. Mechatronics, 55: 248–266.
Work 4.0 2017. *Re-imagining work – white paper*, White paper, Federal Ministry for Labour and Social Affairs.
World Bank 2019. *World development report 2019: The changing nature of work*, Technical report, World Bank Group.
Zhou, C., Damiano, N., Whisner, B., & Reyes, M. 2017. Industrial internet of things: (iiot) applications in underground coal mines. *Mining Engineering*, 69(12): 5056.

Industry 4.0 – Shaping The Future of The Digital World – da Silva Bartolo et al. (eds)
© 2021 Taylor & Francis Group, London, ISBN 978-0-367-42272-1

BIM - based machine learning engine for smart real estate appraisal

T. Su & L.H. Li
School of Engineering, Cardiff University, Cardiff, UK

ABSTRACT: Various machine learning algorithms such as Artificial neural networks (ANNs), Support vector machine (SVM) and Bayesian neural network have been used to improve the accuracy performance of real estate price forecasting. But little research and practice has focused on estimating the price of housing from the construction perspective. Building information modeling (BIM), as a new technology for project information exchange and information management, has been developed for many different industry-specific applications such as automated code checking, energy performance analysis, collaborative design, lifecycle management in the Architecture, Engineering, Construction and Facility Management domain. By integrating BIM and machine learning technologies, this paper proposes a smart comprehensive model which can be used to forecast the price of a new building at the design stage. Furthermore, the smart price estimation engine could be integrated in the whole lifecycle of the building industry.

1 INTRODUCTION

Real estate appraisal is central to various stakeholders for different purposes: government taxation, bank mortgage, house transactions, and property investment (Ahn, 2012). There are several real estate appraisal issues: (1) significant changes of the real estate market value periodically; (2) the huge size of property inventories with different types such as houses, factories, schools; (3) uncertain factors such as low transparency of the market data; (4) subjective factors in determining the impact of weight (Kettani, 2015). Various machine learning algorithms such as ANNs, SVM and Bayesian neural network have been used to automate the traditional real estate appraisal process. Although the application of machine learning algorithms on real estate appraisal has achieved certain improvement in the efficiency and accuracy of house price valuation, little research and practice has focused on the price forecasting of a building in construction industry. According to Rafiei (2015), to decide build or not build during an uncertain economic climate, construction companies have to search the business, economic and real estate journals for the price of housing.

Building information modeling (BIM), as a new technology for project information sharing and as a digital representation of physical and functional characteristics, is believed to facilitate the information interoperability and integration of the construction industry (Volk, 2014). The application of BIM models has the capability to create, collect, store and manage the project information with high quality based on the needs of different stakeholders. To ensure information exchange and interoperability in the AEC industry, an open and neutral exchange data standard named Industry Foundation Classes (IFC) is developed by buildingSMART. Each IFC file contains various geometric information such as curves and surfaces and a wide range of semantic information such as manufacturer contacts, transportation cost, procurement schedule and weather condition (Bloch, 2018). The adoption of BIM models and IFC standard can contribute to real estate appraisal both as a database and a platform to support the automation process and incorporate price forecasting application in the whole lifecycle of a building. Therefore, a BIM-based smart real estate appraisal method incorporating machine learning is proposed to forecast the price of a new building at the design stage.

The proposed smart real estate appraisal model contains a database, an appraisal model and a BIM-Machine learning interactive system. The database has three types of data: the real estate trade data, economic data, and the attribute data. The appraisal model is based on one of the machine learning algorithms namely ensemble learning. The BIM-Machine learning application programming interface (API) ensures the application of price forecasting of a new building at the design stage.

The following contents are organized as follows. Section 2 introduces the real estate appraisal knowledge related to this study. Section 3 presents the proposed smart real estate appraisal model. Section 4 draws the conclusions of this paper.

2 REAL ESTATE APPRASIAL THEORY

Real estate appraisal is required by different stakeholders for different objectives: real estate developers for investment, financial institutions for mortgage, local authorities for property taxation, real estate agencies for transactions. Land and building are two inseparable factors during the property valuation process, due to the sensitive relationship between geographical locations and economic and social changes.

For property valuation, an accurate analysis and estimation of the market price of the property and recent property transactions is essential. The appraisal model should therefore be a representation of the attributes of properties, the underlying fundamentals of the market culture and geographical location.

Different cities or countries have different social and economic culture, which means different standards and real estate appraisal methods. There are two types of property valuation methods: traditional methods and advanced methods.

2.1 *Traditional appraisal methods*

According to Pagourtzi (2003), traditional property valuation methods contain:

- Sales Comparison method
- Investment method
- Profit method
- Development method
- Multiple regression method

Sales comparison method is currently the most widely used method, which heavily depends on the accuracy, dimension and quality of sale transaction data. The appraisal process involves firstly analyzing the differences between the subject property and similar sale properties in the same market area and then adjusting the selling price of the transaction houses to the subject house. The characteristics considered between the subject property and comparison property normally are the selling date and built date, livable spaces, number of rooms, garage types and property sizes, cooling and heating conditions. According to Pagourtzi (2003), "distance" is used to measure the differences between the subject property and the sale property. The "distance", D is calculated as follows:

$$D = \sqrt[\lambda]{\sum_i [A_i(X_{i-}X_{si})]^\lambda + \sum_j \left[A_{j\delta}^{-}\left(X_{j,}X_{sj}\right)\right]^\lambda} \quad (1)$$

where λ = Minkowski exponent lambda; A_i = weight associated with the ith continuous characteristic; A_j = weight associated with the jth categorical characteristic; X_i = value of the ith characteristic in the sale property; X_j = value of the jth characteristic in the sale property; X_{si} = value of the ith characteristic in subject property; X_{si} = value of the jth characteristic in subject property; $\sum_i$ = summation of terms of ith characteristics; $\sum_j$ = summation of terms of j characteristics; $\delta(\mathrm{a,b})$ = inverse delta function. Adjusted price (AP) is calculated as follows:

$$AP = \text{Sales price} - \sum\nolimits_1^i Comparable_i * Weight_i \quad (2)$$

2.2 *Advanced appraisal methods*

According to Pagourtzi (2003), advanced property valuation methods contain:

- Artificial neural networks (ANNs)
- Hedonic pricing method
- Spatial analysis method
- Autoregressive integrated moving average (ARIMA)

Artificial neural networks (ANNs) models are used to mimic human brain's learning process, which mainly contain an input layer, several hidden layers and an output layer (Mitchell, 1997). The model must be firstly trained from a set of training data and be evaluated about the outcome accuracy from another set of test data. With accepted training set mean absolute error and test set mean absolute error, the trained model can be used for price prediction. Most hidden layers have two functions: the weighted summation functions and the transformation functions. Both functions deal with the data from input layer (the sale date and built date, livable spaces, number of rooms, garage types and house sizes, cooling and heating conditions) to the output layer (the sale price). According to Pagourtzi (2003), the weighted summation function is calculated as follows:

$$Y_j = \sum\nolimits_j^n X_i W_{ij} \quad (3)$$

where Xi = the input values; Wij = the weights of input values for each the jth hidden layer nodes. The transformed function for a regular sigmoid transformation is calculated as follows:

$$Y_T = \frac{1}{1 + e^{-y}} \quad (4)$$

2.3 *Current Gaps*

Although there are still reservations about the method's reliability, the comparable method is still recognized as an accurate and reliable appraisal method by the UK and American literature. For the advanced estimate methods, ANNs has been used to improve the accuracy and efficiency of real estate appraisal forecasting, Bayesian neural networks has been used to solve the uncertainty and subjectivity of property valuation, the ARIMA method has tried to do property evaluation based on time-series data (Prado, 2018). Although the application of those machine learning algorithms has achieved certain improvement in the efficiency and accuracy of real estate appraisal, a plethora of research shows unstable performance of machine learning algorithms with different sources of house sale data. For the building industry, little research and practice has focused on the price forecasting of a building at the design stage.

3 SMART REAL ESTATE APPRASIAL (SREA) MODEL

The proposed smart real estate appraisal model contains a database, an appraisal model and a BIM-Machine learning interactive system. Firstly, to reduce the number of attributes and calculate the best weight of each attribute, a gradient descent optimization algorithm is applied to preprocess the real time house trade data. After that, 17 attributes are selected to build the smart real estate engine and to define the related IFC datatypes. The selected 17 attributes contain two types of data: the attribute data and geographical data. Based on related real estate appraisal theory, new economic data will be added to the IFC datafile. Finally, a BIM-Machine learning interactive system is used to connect the smart real estate engine to the BIM platform.

3.1 *The feature selection of real estate data*

17 attributes of properties for real estate appraisal are selected based on the property transaction cases from 47 different cities in America. The feature selection uses an optimization algorithm named gradient descent to reduce the number of attributes for the model building. In the neural network theory, although the increasing number of variables makes machine learning model more accurate with the prediction, the large number of input variables also needs a huge amount of training data which make the training process impractical for the intensive computation. This is known as dimensionality curse (Rafiei, 2015).

Gradient descent is an iterative optimization algorithm that can be used to minimize the cost function and find the best weights of attributes to build the training model. A cost function is used for as a measurement of how wrong the prediction outcome from the sale price, which is calculated as follows:

$$\text{Cost Function} = \frac{\sum_{i=1}^{m} (ModelGuess(i) - Saleprice)^2}{m} \tag{5}$$

The Gradient descent optimization algorithm is performed using the software called PyCharm which is a python based integrated development environment for computer programming. Two powerful python packages named Numpy and Scikit-learn are used on the Pycharm platform: Numpy's argsort function providing the list of array indexes connecting to each element in the array in order and Scikit-learn gives an array of the feature importance for each feature. 17 attributes out of 63 are selected to do the smart real estate engine training and to define the related IFC datatypes. Table 1 shows the 17 attributes and their weights on feature selection. The most important attributes determining the house price are the property size, locations of the city, the size of garage, the year when it was built, the number of bathrooms and bedroom, and the heating and cooling conditions.

3.2 *Define related IFC datatype*

Industry Foundation Classes (IFC) is based on a standard object-oriented programming language named Express which contains three kinds of IFC classes: object, relation and resource. The object classes define the geometrical information, the identity of the object and its functionalities, the relation classes define various relations between object classes and their functions, the resource classes contain the attributes in the functionalities of objects (Vanlande, 2008). The IFC4 Add2, the newest version which published in 2016 by buildingSMART organization, was aiming for overall project procurement and to provide a more efficient way to define the semantic relations between objects (Liebich, 2013).

Table 1. Weight of attributes on feature selection.

Attribute	Weight Percentage
Total floor area	22.42%
City locations	20.64%
Livable area	12.67%
Without garages	8.81%
Garage area	7.16%
Garage attach	0.65%
Garage detach	0.10%
Have fireplace	5.86%
Year of built	5.44%
Full bathrooms	4.37%
Bedroom number	4.19%
Have a pool	3.53%
Stories	1.12%
Central cooling	0.31%
Central heating	0.23%
Carport area	1.89%
Half bathroom	0.61%

The 16 attributes (except for the price data) selected from the real estate sale data are grouped as three types: geometric data (livable area, stories, size of house, size of bathroom and bedroom), geographical data (city location), resource data (garage type, year built, number of rooms, cooling and heating condition). On the other hand, IFC schema is divided into five groups:

1) Geometric data
2) Semantic data
3) Objects representing building components and space
4) Attributes of objects
5) Relationship between objects

Figure 1 shows the general framework of the data mapping hierarchy between BIM data and real estate trade data. The geometric data from IFC schema is directly mapped into the geometric data of real estate sale data. Information such as price data and weight of attribute are not included in IFC schema, which are defined by mapping relationships between objects and attributes of objects respectively. The objects representing building components and space data are mapped into geographical data. Semantic information from BIM is mapped into resource data in real estate. After data mapping, the IFC data can be extracted from BIM models and be used for real estate appraisal.

3.3 *SREA engine training*

The real estate sale data from 47 different cities in America is divided into two groups: 70% in the first group for training dataset and 30% in the second group for test dataset. The training of the smart real estate appraisal model uses an ensemble learning method, which bases on multiple machine learning algorithms together to achieve better accuracy performance of the prediction. The basic data structure of the ensemble learning algorithm is a decision tree. Figure 2 shows the general workflow of the ensemble learning model. To ensure the complexity of a machine learning model, there are basically two ways: one is increasing layers and branches to a decision tree as many as possible, the other one uses a number of separable decision trees of which each decision tree contributes to the model a little by dealing with one small part of the problems. The ensemble learning method belongs to the latter one. The trees are built on each other. Each new tree fixes a small part of the errors where the previous trees have, until the learning speed and the prediction accuracy are reasonable.

After splitting the training data and test data, the model hyperparameters are tested iteratively on PyCharm to find the best fit of the learning speed and the complexity of the patterns to be discovered behind the training data. The hyperparameters contain the number of decision trees, the learning speed, the maximum depth of each tree and the minimum numbers of a feature required to be at a leaf node. Model hyperparameters are set for training the SREA engine as follows:

- Decision tree numbers: 1000
- Learning rate: 0.1
- Maximum depth: 6
- Minimum sample leaf: 9

Before the SREA model being used for prediction of house price, it is tested by the test dataset with accepted test set mean absolute error. The python library Scikit-learn provides a mean absolute error function which checks every prediction our model makes and gives an average of the error. Figure 3 shows the python code of the SERA model.

3.4 *BIM supported SREA model*

The smart real estate appraisal model contains a database, a SREA model and a BIM-Machine learning interactive system. Figure 4 presents the framework of the BIM supported SERA model workflow. Firstly, related attributes for real estate appraisal are selected based on gradient descent optimization which recognized as feature selection. Secondly, the selected attributes are further used for defining related IFC

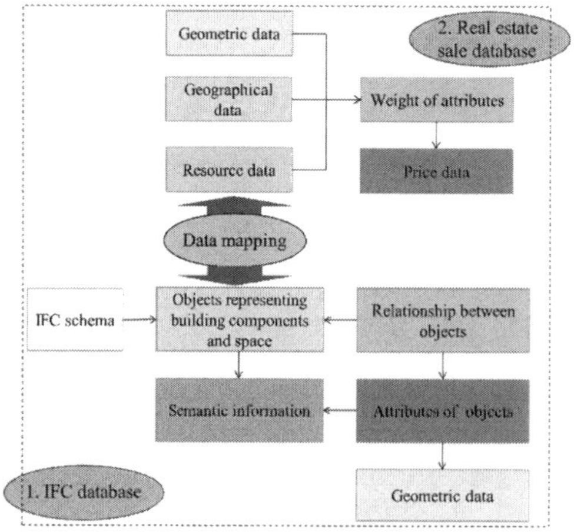

Figure 1. Data mapping between BIM and real estate trade datatype.

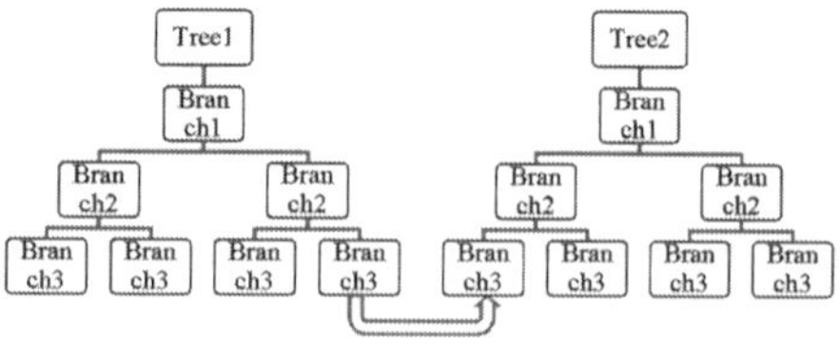

Figure 2. General workflow of the ensemble learning model.

Figure 3. The python code of the SREA engine.

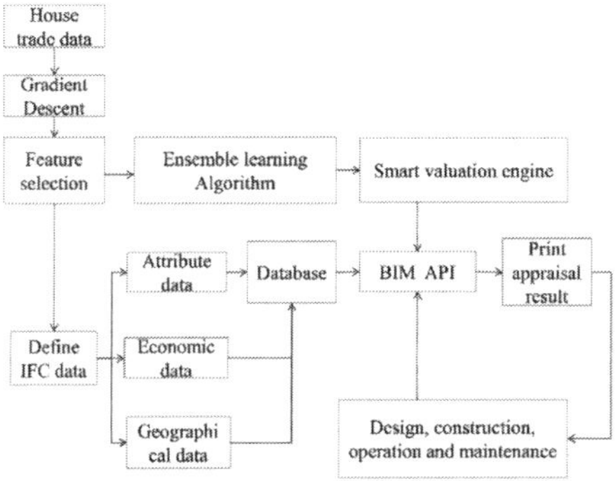

Figure 4. BIM supported SREA model workflow.

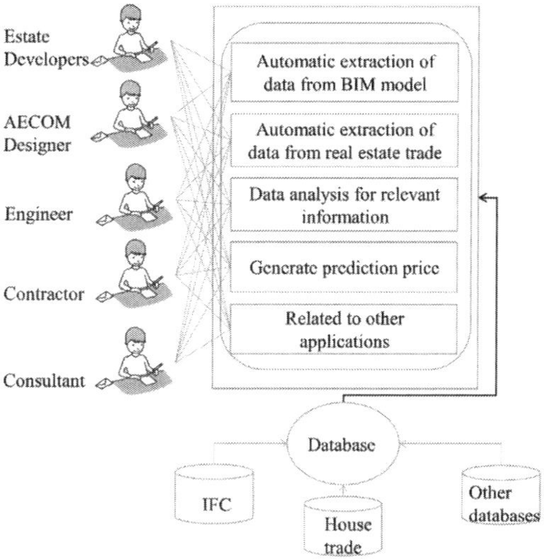

Figure 5. The core concept of the Revit API connecting BIM and Machine learning platform.

datatypes and training the SREA model. Finally, a Revit API connecting the smart real estate engine to the BIM platform is proposed to be constructed for the BIM-Machine learning interactive system.

Figure 5 shows the core concept of the Revit API connecting BIM and machine learning platform. With the database consists of IFC, house trade data and other social and economic data, the API is able to do automatic extraction of data from BIM models and real estate trade database and automatically generate prediction price after data analysis.

4 DISCUSSION AND CONCLUSION

In this paper, a comprehensive smart real estate appraisal model integrating BIM and machine learning algorithms was presented for house price estimation at the design stage. The proposed smart real estate appraisal model contains a database, an advanced appraisal model and a BIM-Machine learning interactive system. The first contribution of this paper is using a gradient descent optimization algorithm for minimizing the cost function, reducing the data dimensionality and finding the best weight of attributes automatically. The selected attributes and weights data are further used for the definition of new IFC datatypes and for training the SREA engine. Secondly, a data mapping framework is presented for the integration of BIM data and real estate trade data, with two new kinds of datatype created namely the Price data and the Weight of attributes. Thirdly, the smart real estate appraisal engine is trained by an ensemble learning algorithm using python language. The trained model is efficient and accurate for house price prediction. Finally, to incorporate BIM and the SREA engine, a Revit API is proposed to enable house price prediction at the design stage. Such a model could be of great value for the smart decision making of construction industry, when construction companies facing an unstable economic climate regarding construction. Further, the model can be applied to other construction activities in the lifecycle of a building such as procurement and 4D schedules.

REFERENCES

Ahn, J.J. and Byun, H.W. 2012. Using ridge regression with genetic algorithm to enhance real estate appraisal forecasting, *Expert systems with application* 39(9): 8369–8379.

Bloch, T. and Sacks, R. 2018. Comparing machine learning and rule-based inferencing for semantic enrichment of BIM models, *Automation in Construction* 91: 256–272.

Kettani, O. and Oral, M. 2015. Designing and implementing a real estate appraisal system: The case of Quebec province, Canada, *Social-Economic Planning Sciences* 49: 1–9.

Liebich, T. 2013. IFC4-the new buildingSMART Standard.

Mitchell, T. 1997. *Machine Learning*. New York: MaGraw-Hill.

Pagourtzi, E. 2003. Real estate appraisal: a review of valuation methods, *Journal of Property Investment & Finance* 21(4): 383–401.

Prado, M. 2018. *Advances in Financial Machine Learning*. New Jersey: John Wiley & Sons, Inc.

Rafiei, M.H. and Adeli, H. 2015. A novel machine learning model for estimation of sale prices of real estate unites, *Journal of Construction Engineering and Management* 142(2): 1–10.

Vanlande, R. 2008. IFC and building lifecycle management, *Automation in Construction* 18(1): 70–78.

Volk, R. Stengel, J. and Schultmann, F. 2014. Building information modeling (BIM) for existing buildings-literature review and future needs, *Automation in Construction* 38: 109–127.

Industry 4.0 – Shaping The Future of The Digital World – da Silva Bartolo et al. (eds)

Data analytics applied to sand casting foundry industries

F. Leal, H.F. Castro & A.R.F. Carvalho
Intelligent & Digital Systems, R&Di, Instituto de Soldadura e Qualidade, Grijó, Portugal

H. Gouveia
R&Di Programmes, R&Di, Instituto de Soldadura e Qualidade, Taguspark, Oeiras, Portugal

ABSTRACT: Data analytics is one of the most important concepts that drives industry 4.0. Most industries give a limited use to the process data. Specifically, foundry industries that involve complex processes that can be enhanced by data analytics. In this context, this paper explores data analytics methods for foundry industries with sand casting to minimise the percentage of product defects by controlling sand parameters, such as compactability and compressive strength. The present work contributes with two models to predict: (i) sand parameters; and (ii) casting rejections. First, we apply a Multiple Linear Regression to predict the sand parameters using the initial moisture sand, the amount of water introduced, and the bentonite. Then, a rule-based model is applied to predict the percentage of casting defects based on sand parameters data. Experiments with an industrial dataset show that the proposed methods improve the accuracy of the predictions when compared to a baseline approach.

1 INTRODUCTION

Industry 4.0 (i4.0) is a term coined in this century which can be adumbrated as the application of novel information and computational technologies to strategically automate the industrial processes using an intelligent and integrated approach, or smart manufacturing. Smart manufacturing aims to optimise the industrial processes taking advantage of Internet of Things, Cyber-Physical Systems, Cloud Computing and, Data Analytics (Zhong et al. 2017). This paper applies data analytics techniques to foundry industry in order to maximise the sustainability of the process as well as the profits reducing the casting rejections. This work is part of a demonstration action developed in the scope of project SIM4.0, Intelligent Monitoring Systems 4.0.

The current scenario of foundry industry requires optimization of operational costs which can be achieved by i4.0 implementation (Kozłowski et al. 2019; Lewis 2016) namely with predictive analytics. Foundry is one of the oldest manufacturing methods encompassing multiple processes - moulding, melting, pouring, solidification, shot blasting, and fettling. There are distinct casting methods such as permanent mould casting, centrifugal casting, die-casting, investment casting, shell casting, and sand casting (Aramide et al. 2009). On one hand, sand casting (sand moulding) is widely used once it is one of the cheapest methods. On the other hand, sand casting is a complex method generating multiple casting rejections and, consequently, large amount of offscourings (Saikaew & Wiengwiset 2012). Regarding this scenario, we present a methodology to ensure the quality and efficiency of the sand circuit for iron casting industries. The methodology relies on multiple sensors as well as data analytics methods integrating multiple i4.0 concepts. In this case study, the sand circuit incorporates three steps: (*i*) pre-mixer; (*ii*) muller; and (*iii*) GSM sand laboratory. The pre-mixer has the role of cooling the sand, pre-preparing it for another moulding. Then, the muller prepares the sand, adding an appropriated amount of water. Finally, the GSM laboratory as a control system, measures the green sand features (humidity, compactability, and compression strength).

Each step of the circuit involves sand-related data which influences the quality of the products. Using this data, the study aims to apply data science techniques to sand-related processes information, mainly, for preparing foundry industries to predictive analysis which is one of the main concepts of i4.0. In this context, we address multiple statistical analysis to correlate the casting rejections with the sand features. Therefore, we propose a linear and rule-based model to predict the compactability, compressive strength and casting rejections. By applying these models, it is possible a real-time quality control of the green sand anywhere in the circuit.

This paper is organised as follows. Section 2 presents a review on data analytics. Section 3 describes the proposed method. The experiments and results are reported in Section 4. Finally, Section 5 summarises and discusses the outcomes of this work.

2 RELATED WORK

Technology plays an important role in industries. Increasingly, the industries need to automate their processes to cope with the increasing market demand. It raises the i4.0 concept or the fourth industrial revolution which embraces a number of contemporary automation, data exchange, and manufacturing technologies (Mittal et al. 2018; Zhong et al. 2017).

In casting industry, the data-driven combination of different process parameters is a key issue for foundry i4.0 implementation. Applying data analytics to prevent cast rejections is a challenging task. The casting process involves large amounts of data that is difficult for humans to analyse and discover important information that, ultimately, will make the processes even more effective. Artificial Intelligence (AI) methodologies are able to explore large amounts of data to model the process using multiple variables. It relies on, essentially, data mining and machine learning algorithms for learning the context, and, then, to provide useful information.

An adequate preparation of the green sand is fundamental to ensure the quality of the final product (Saikaew & Wiengwiset 2012). The defects are mainly caused by incorrect values of the process parameters. Therefore, we survey multiple research works which apply data analytics methods to optimize the sand casting or predict the casting rejections caused by the poor quality of the sand.

Multiple studies have explored the research problem of optimizing the sand parameters employing: (*i*) Taguchi method (Chandran R.M. 2013; Guharaja et al. 2006; Kumar et al. 2011; Upadhye & Keswani 2012); (*ii*) Artificial Neural Network (ANN) (Karunakar & Datta 2007; Singha & Singh 2015); and (*iii*) a mixture experimental design, response surface methodology, and propagation of error (Saikaew & Wiengwiset 2012). However, very few researches have explored the casting rejection prediction. Patil & Inamdar (2014) approach this topic applying ANN to predict casting defects. Ihom et al. (2014) applies a Multiple Linear Regression to predict the permeability of the green sand.

2.1 *Contributions*

This paper explores data analytics methods for sand casting data to prevent casting rejections. The main goal is to help the sand casting foundry industries to introduce AI methods in the productive process. The related work shows that predictive methods concerning casting rejections have been unexplored. Therefore, this paper contributes with predictive models which enables both to calculate the sand casting parameters (compactability and compressive strength) and casting rejections. Similarly, to Ihom et al. (2014) the current work proposes a MLR model to predict sand parameters (compactability and compressive strength). However, while previous works use ANN to make predictions, we use a rule-based method once it is easily interpretable. In this context, this paper helps the sand casting foundry industries to employ predictive manufacturing which is a key concept of i4.0.

3 PROPOSED METHOD

This paper uses data analytics mechanisms for foundry industries to prepare them for the i4.0 reality. In this context, we explore the sand casting processes and their corresponding information. Figure 1 describes the proposed method which involves: (*i*) green sand system components; (*ii*) predictive algorithms; and (*iii*) the evaluation of the results.

The green sand system involves a muller and a GSM laboratory. At the entrance of the muller, sand moisture and temperature are measured to calculate the amount of water needed to obtain the appropriated sand parameters. Then, the GSM measures the compactability and the compressive strength to control the quality of the mixture. As predictive models, we employ a Multiple Linear Regression (MLR) and a rule-based model (M5-rules). While MLR aims to predict the compactability and compressive strength using the muller-related information, the M5-rules aims to obtain a prediction concerning the number of casting rejections. Finally, we experiment and assess our method with sand casting-related data using Root Mean Squared Error (RMSE) and Mean Absolute Error (MAE).

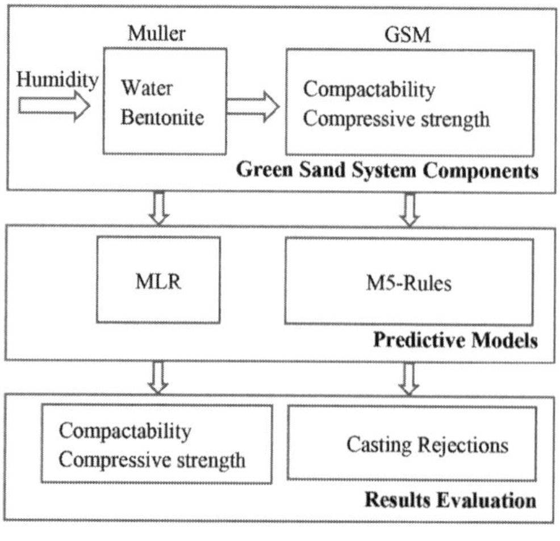

Figure 1. Proposed method.

3.1 *Green sand parameters*

The correct composition of the green sand is essential to avoid casting rejections. In this context, it is fundamental to control distinct parameters to enhance the product quality such as: (*i*) moisture content; (*ii*) compactability; (*iii*) compressive strength; and (*iv*) bentonite. The amount of water in the mixture activates the sand clay causing the development of plasticity and resistance. Furthermore, the sand moisture also influences other sand properties.

To obtain a good moulding, *i.e.*, with correct and unchanging dimensions till the casting solidification, the sand should have a certain mechanical resistance (compressive resistance) which is guaranteed by both the amount of water as well as the amount and quality of the bentonite. In addition, the compactability indicates the amount of water needed in order to maintain the moulding features, being one of the most important indicators of the quality of the mixture. Finally, the bentonite is the main source of plasticity along with water. It must have sufficient fire resistance to not become inert during the leakage process.

Table 1 contains the reference values of green sand parameters for iron casting.

It is common to find flaws in sand casting mainly due to the poor quality of the green sand. Therefore, the sand parameters must be controlled to avoid casting rejections. Table 2 enumerates multiple casting defects originated by the weak preparation of the green sand (J G Professor, Rajkolhe & Khan 2014; Sai et al. 2017).

Table 1. Reference values of the green sand parameters.

Setting	Reference Value
Moisture	3.3 – 4.3 %
Compactibility	40 – 44 %
Compressive Strength	170 – 210 kN/m^2
Bentonite	8 – 10 %

Table 2. Casting defects originated by green sand.

Defect	Cause
Blow holes	Excessive moisture
Shot metal	High moisture
Sand Inclusion	Low compactability
	Low bentonite content
Shrinkage	Low compressive strength
Penetration	High compactability
Roughness	Coarse sand
Erosion	Low bentonite content
Swelling	High compactability
	Low bentonite content

The main goal of this paper is to apply data analytics to control efficiently the sand parameters and minimize casting rejections associated to the sand quality. In this context, we present prediction models to control compactability and compressive strength. Additionally, considering the sand-related data, we apply a rule-based model to predict the amount of cast rejections considering different green sand samples.

3.2 *Predictive models*

The prediction models aim to predict different variables of the green sand circuit. As compactability and compressive strength are crucial to ensure the quality of the final product, we employ an MLR to predict the corresponding values in order to control them anywhere in the sand circuit. In addition, we use a rule-based model to predict the amount of casting rejections considering different samples of sand.

3.2.1 *Multiple linear regression*

MLR is typically applied to multivariate scenarios in order to predict one or more continuous variables based on other data set attributes, *i.e.*, by identifying existing dependencies among variables (Ober 2013).

Equation 1 displays the model of the MLR with k regression variables where $\in_i$ is the disturbance, β_0 is the intercept, and β_i (i = 1 to k) are the partial regression coefficients, representing the rate of change of Y as a function of the changes of $X = \{x_1, x_2, \ldots, x_k\}$ (Fletcher, 2009).

$$Y_i = \beta_0 + \beta_1 x_1 + \beta_2 x_2 + \ldots + \beta_k x_k + \in_i \quad (1)$$

We use Ordinary Least Squares (OLS) to estimate the unknown parameters (β_i) of this linear regression model. OLS minimises the distance between the observed responses and the responses predicted by the linear approximation (Kiers & Smilde 2007). Equation 2 represents the OLS method where x_i and y_i are the observations, while $\hat{x}$ and $\hat{y}$ are the predictions.

$$\hat{\beta} = \frac{\sum (x_i - \hat{x})(y_i - \hat{y})}{\sum (x_i - \hat{x})^2} \quad (2)$$

3.2.2 *M5-rules model*

Predicting values is a hard task involving complex mathematical models. However, mainly in the industrial world, it is important to present the results in a comprehensive way. The rule-based models are one the most popular methods used in machine learning once it can be easily interpreted. The artificial intelligence is a new challenge in the industrial domain whereby it is relevant to start with the most understandable methods.

Rule-based models exhibit some if-then-reading to express statements regarding a domain, *e.g.*, IF condition THEN action. Each rule defines a small and independent piece of knowledge. A rule set represents the knowledge structure of a domain and the respective relations. The rules are composed by a condition which satisfies a certain situation. Therefore, the essence of a rule-based representation is to use a series of cycles to cover all knowledge.

In this paper, we apply the M5-rules model to predict casting rejections. The M5-rules generates rules from model trees which are built repeatedly. At each iteration, it is selected a rule considering the best leaf of the tree. According Holmes et al. (1999), this method produces rule sets that are as accurate but smaller than the model tree constructed from the entire dataset.

3.2.3 *Evaluation metrics*

The predictive accuracy metrics measure the error between the predicted value and the real value. It is the case of the Mean Absolute Error (MAE), which measures the average absolute deviation among the predicted value and the real value, or the Root Mean Square Error (RMSE), which highlights the largest errors (Willmott et al. 1985).

Equation 3 and Equation 4 represent both error functions where $\hat{r}_m$ represents the value predicted for the moulding m, r_m the real value of the moulding m in the testing partition, and M the total number of mouldings.

$$MAE = \frac{\sum_{m=1}^{M} \left| \hat{r}_m - r_m \right|}{M} \tag{3}$$

$$RMSE = \sqrt{\frac{\sum_{m=1}^{M} \left(\hat{r}_m - r_m \right)^2}{M}} \tag{4}$$

These metrics were determined using the off-line protocol which knows the full data set in advance. The first 80 % of the data set (train partition) is used to build the model and the remaining 20 % (test partition) is for evaluation. This protocol uses the test partition to determine the RMSE and MAE.

4 EXPERIMENTS AND RESULTS

We have conducted several experiments with a sand casting data set to evaluate the proposed method. The data processing was implemented in JAVA.

The system is running on an Intel Core i5-8250U which holds 8 GB RAM, and 256 GB hard-disk. The experiments involved predictive manufacturing models to obtain sand casting parameters (compactability and compressive strength) and sand casting rejections.

4.1 *Dataset*

The dataset was prepared with sand casting data containing information about 2985 mouldings.

The dataset encompasses the following data: (*i*) mould identification; (*ii*) mixture time; (*iii*) data at muller (humidity, amount of water introduced, temperature, bentonite, sand used, and new sand); (*iv*) GSM data (temperature, compactability, and compressive strength); (*v*) product ID; (*vi*) amount of products moulded; (*vii*) product weight; and (*viii*) casting rejections.

While the MLR uses the data at muller entrance to predict the sand parameters, the M5-rules combines both data, *i.e.*, at muller and GSM, to predict casting rejections.

4.2 *Predictive models results*

To perform predictive manufacturing, we have conducted two different experiments: (*i*) prediction of compactability and compressive strength using a MLR; and (*ii*) prediction of casting rejections using a rule-based model.

4.2.1 *Compactability prediction*

Compactability plays an important role in sand casting foundries. It is related with the reduction of the volume of sand mixed with clay and water after undergoing compression applied by compaction.

Towards this scenario, we apply the MLR to predict the compactability value considering as independent variables the sand moisture (h) at muller entrance as well as the quantity of water introduced (w) for a concrete amount of sand, and the bentonite (b). The result of the model is depicted in Equation

$$Compat = 3.8h + 0.2w - 0.7b + 42.1 \tag{5}$$

Figure 2 Compares the error between the real and the predicted values. The model presents a RMSE of 14 % and MAE of 10 %. The results show that the compactability can be predicted with a relevant accuracy using a linear model.

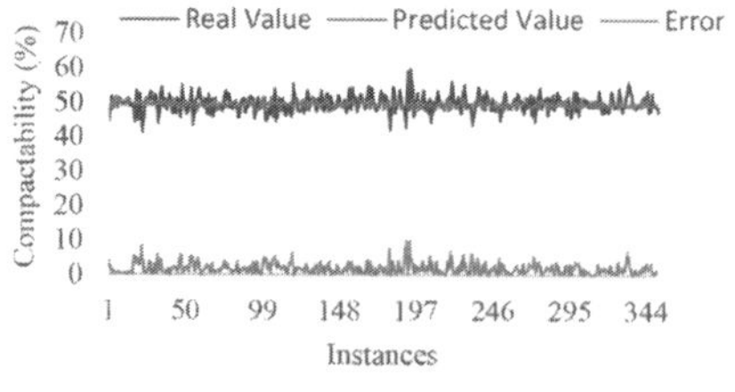

Figure 2. Comparison of the predicted and the real values.

To detect statistical differences between the baseline (constant value of compactability) and the proposed approach for compactability prediction, we compare both results in terms of error. As baseline, we consider a regressor that makes predictions using a constant prediction, the average of the compactability values of the dataset. The results show that the proposed approach presents an improvement of 5 % concerning the baseline version.

4.2.2 *Compressive strength prediction*

Compressive strength depends on grain size, shape and distribution of sand grains, amount of clay, and moisture content.

Towards this scenario, we apply the MLR in order to predict the compressive strength value considering as independent variables the sand moisture (h) at muller entrance as well as the quantity of water introduced (w) for a concrete amount of sand, and the bentonite (b). The result of the model is depicted in Equation 6.

$$Compress = 50.8h + 4.3w + 67.5b + 617.9 \quad (6)$$

Figure 3 Compares the error between the real and the predicted values. The model presents a RMSE of 8 % and MAE of 6 %. The results show that the compressive strength can be predicted with a relevant accuracy using a linear model.

As used in compactability parameter prediction, detection of statistical differences between the baseline (constant value of compressive strength) and the proposed approach for compressive strength prediction is applied and compared both results in terms of error. As baseline, we consider a regressor that makes predictions using a constant prediction, the average of the compressive strength values of the dataset. The results show that the proposed approach present an improvement of 20 % concerning the baseline version.

4.2.3 *Casting rejections prediction*

The aim of foundry industries is to decrease the casting rejections. In this context, we propose a rule-based model to predict the percentage of defects that occur due to the sand quality considering multiple attributes: (*i*) data at muller (water (w), used sand (*uSand*), new sand (*nSand*), and bentonite (b)); (*ii*) data at GSM (compactability (*compact*), compressive strength (*compress*), and temperature (T)); and (*iii*) product weight (*Pweight*). Applying the M5-rules model, it is generated a set of rules. These rules indicate the conditions to occur casting rejections (Table 3). Figure 4 compares the error between the real and the predicted values. The model presents a RMSE of 24 % and MAE of 17 %. The results show that the rule-based models can provide a prediction of the casting rejections with a relevant accuracy.

To detect statistical differences between the baseline (constant value of cast rejections percentage) and the proposed approach for cast rejections prediction, we compare both results in terms of error. As baseline, we consider a model that makes predictions using a constant prediction, the average of the cast rejections values of the dataset. The results show that the proposed approach present an improvement of 10 % concerning the baseline version.

Table 3. Rule based model to predict percentage of sand casting defects.

Rule 1

IF
$weight\ P \leq 0.825$

compact

THEN
$defects = 0.28 \times w + 0.01 \times uSand - 0.08 \times compact$

$+0.018 \times b - 0.005 \times weightP - 0.15$

Rule 2

IF
$b \leq 10.9$

$compress 1758$

THEN
$defects = -0.014 \times w + 0.06 \times uSand - 0.02 \times compact$

$+0.018 \times compress + 5.17$

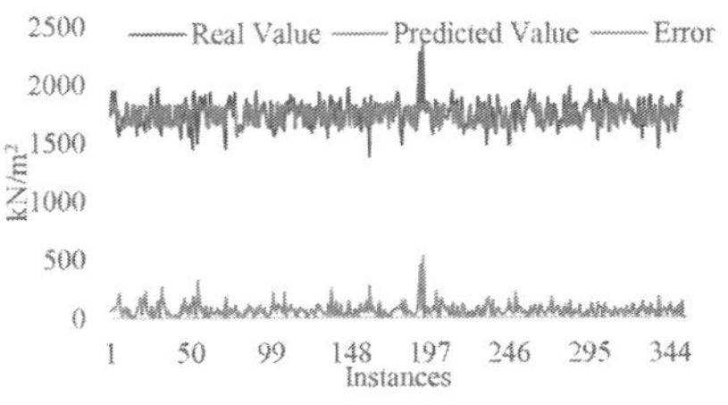

Figure 3. Comparison of the predicted and the real values.

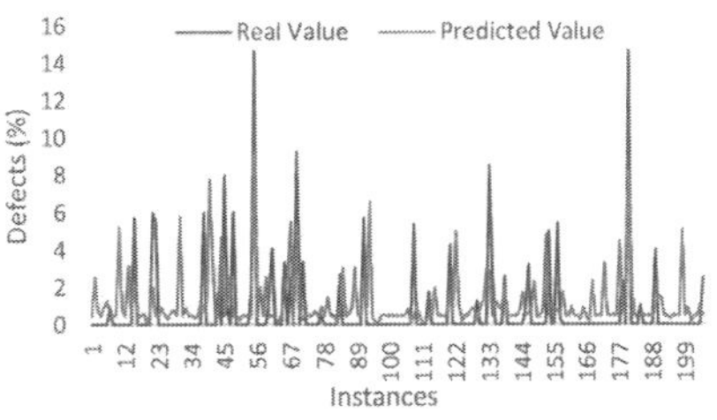

Figure 4. Comparison of the predicted and the real values.

5 CONCLUSIONS

Under i4.0 context, data analytics allow the implementation of predictive manufacturing in the industrial processes. It involves machine learning algorithms which can process large amount of data, discover possible anomalies or system failures, and ensure the product quality. In this scenario, the paper explores data analytics models, particularly for sand casting foundry industries. This work represents an important contribution to the sector at stake and to the impact of the SIM 4.0 demonstrations.

The proposed method encompasses: (*i*) prediction of sand parameters using a MLR; and (*ii*) prediction of casting rejections using M5-rules. In terms of performance, we compare the error between a constant value (baseline approach) and the real values (proposed approach). The results show that the proposed approaches improve the predictive accuracy in both models. By applying the models reported here, it is possible to insert low cost humidity sensors along the circuit line and, in conjunction with the process data already available, enable real-time quality control of the green sand anywhere in the circuit, predict defects attributed to sand preparation and, ultimately, reject sand with parameters out of the optimal range.

As future work, we intend to increase the dataset in order to refine the models. Concretely, the rule-based model requires further exploitation once the error for extreme values tends to be higher.

ACKNOWLEDGENTS

This work is financed by FEDER funds through COMPETE 2020 in the context of the SIAC system (Collective Actions Support System): 03/SIAC/2016; investment project nº 026752.

REFERENCES

Aramide, F., Aribo, S., & Folorunso, D. (2009). Optimizing the Moulding Properties of Recycled Ilaro Silica Sand. *Leonardo Journal of Sciences*, 19: 93–102.

Chandran R.M., U. (2013). Optimization of Process Parameters to Minimize the Casting Defects. *International Journal of Advances in Engineering Science and Technology*, 3(2): 105–111.

Fletcher, J. (2009). Multiple linear regression. *BMJ*, 308: b167.

Guharaja, S., Noorul Haq, A., & Karuppannan, K. M. (2006). Optimization of green sand casting process parameters by using Taguchi's method. *International Journal of Advanced Manufacturing Technology*, 30: 1040–1048.

Holmes, G., Hall, M., & Prank, E. (1999). Generating rule sets from model trees. In *Lecture Notes in Computer Science (including subseries Lecture Notes in Artificial Intelligence and Lecture Notes in Bioinformatics)*.

Ihom, A. P., Ogbodo, J. N., Allen, A. M., Nwonye, E. I., & Ilochionwu, C. (2014). Analysis and prediction of green permeability values in sand moulds using multiple linear regression model, *2*(February): 8–13.

J G Professor, Rajkolhe, R., & Khan. (2014). Defects, Causes and Their Remedies in Casting Process: A Review. *International Journal of Research in Advent Technology*, 2(3): 375–379.

Karunakar, D. B., & Datta, G. L. (2007). Controlling green sand mould properties using artificial neural networks and genetic algorithms - A comparison. *Applied Clay Science*, 2-3: 58–66.

Kiers, H. A. L., & Smilde, A. K. (2007). A comparison of various methods for multivariate regression with highly collinear variables. *Statistical Methods and Applications*, 16: 193–228.

Kozłowski, J., Sika, R., Górski, F., & Ciszak, O. (2019). Modeling of foundry processes in the era of industry 4.0. In *Lecture Notes in Mechanical Engineering*.

Kumar, S., Satsangi, P. S., & Prajapati, D. R. (2011). Optimization of green sand casting process parameters of a foundry by using Taguchi's method. *International Journal of Advanced Manufacturing Technology*, 55: 23–34.

Lewis, M. (2016). Industry 4.0 and what it means to the foundry industry. In *72nd World Foundry Congress, WFC 2016*.

Mittal, S., Khan, M. A., Romero, D., & Wuest, T. (2018). A critical review of smart manufacturing & Industry 4.0 maturity models: Implications for small and medium-sized enterprises (SMEs). *Journal of Manufacturing Systems*, 49: 194–214.

Ober, P. B. (2013). Introduction to linear regression analysis. *Journal of Applied Statistics*, 40(12): 2775–2776.

PATIL, G. G., & INAMDAR, D. (2014). Prediction of Casting Defects Through Artificial Neural Network, 02(05): 298–314.

Sai, T. V., Vinod, T., & Sowmya, G. (2017). A Critical Review on Casting Types and Defects. *IJSRSET* 3(29): 463–468.

Saikaew, C., & Wiengwiset, S. (2012). Optimization of molding sand composition for quality improvement of iron castings. *Applied Clay Science*, 67-68: 26–31.

Singha, S. K., & Singh, S. J. (2015). Analysis and Optimization of Sand Casting Defects With the Help of Artificial Neural Network. *International Journal of Research in Engineering and Technology*, 4: 24–29.

Upadhye, R., & Keswani, I. (2012). Optimization of Sand Casting Process Parameter Using Taguchi Method in Foundry. *International Journal of Engineering*, 1(7): 1–9.

Willmott, C. J., Ackleson, S. G., Davis, R. E., Feddema, J. J., Klink, K. M., Legates, D. R., O'Donnell, J., & Rowe, C. M. (1985). Statistics for the evaluation and comparison of models. *Journal of Geophysical Research*, 90(C5): 8995–9005.

Zhong, R. Y., Xu, X., Klotz, E., & Newman, S. T. (2017). Intelligent Manufacturing in the Context of Industry 4.0: A Review. *Engineering*, 3(5): 616–630.

Industry 4.0 – Shaping The Future of The Digital World – da Silva Bartolo et al. (eds)
© 2021 Taylor & Francis Group, London, ISBN 978-0-367-42272-1

The role of "JavaMach Cluster" to training for Industry 4.0

S. Dobrotvorskiy, L. Dobrovolska, Y. Basova & E. Sokol
National Technical University «Kharkiv Polytechnic Institute», Kharkiv, Ukraine

M. Edl
University of West Bohemia, Pilsen, Czech Republic

ABSTRACT: Currently, there is an intensive expansion of the ideas and solutions of Industry 4.0 for many industries. At the same time, the complexity of the tasks being solved is growing and the need for training specialists capable of synthesizing various parts of Industry 4.0 into a coherent whole becomes obvious. On the other hand, we need tools and platforms to connect the various branches of Industry 4.0 into one whole. This article discusses the concept of using a "JavaMach cluster" to solve these problems based on the use of Java technologies. At the core of the JavaMach Cluster, ideology is the creation of a single educational, scientific, practical platform core industry 4.0 based on the use of common standards and open source technologies. A new approach to the training of specialists both within individual universities and on the basis of international university relations has been considered (Java Europe Education).

1 INTRODUCTION

It is well known (Alcácer & Cruz-Machado 2019; Lu et al. 2016; Mavrikios et al. 2017; Hamrol et al. 2018; Kozłowski et al. 2018; Edl 2013; Peraković et al. 2019; Prinz et al. 2017) that the main directions of development of Industry 4.0 are: Horizontal and Vertical System Integration, Big Data, Simulation, System integration. IoT, cloud computing, Additive manufacturing, Augmented Reality, Autonomous Robots. The development of all these areas is based on the development of information and software. However, as the analysis of the current state of the issue in this area shows, there is no single view on the creation and use of the software. In each case, specific tasks are solved using a convenient for their solution of various software platforms. As a result of this approach, good local solutions are obtained, but as a rule, they conflict with each other when trying to unite them, and most often are generally incompatible. Modern approaches to the training of Specialists mainly focused on specific subject areas. This is an important task. But narrow specialists will not be able to solve the problems of industry 4.0. Therefore, on the one hand, we need specialists as system integrators of Industry 4.0 tasks, and on the other, a unified software platform for solving these problems.

2 RESEARCH METHODOLOGY

2.1 *"JavaMach Cluster" within the framework of Industry 4.0*

One approach to solving these problems may be the ideology underlying the international "JavaMach Cluster". JavaMach Cluster is a cluster that unites on avoluntary basis universities from different countries, industrial organizations and just individual enthusiasts to one principle: using Java and Java technologies for programming open-source software to solve industry problems 4.0 (Sokol et al. 2017; Dobrotvorskiy et al. 2018a; Kagermann et al. 2013). Therefore, the main goal of JavaMach Cluster is to create a unified educational, scientific, practical platform and core engineering on java and java open compatible technologies. In other words, the specialists among themselves and the machines must speak the same "language".

Currently, in the world, the share of software created exclusively in Java can be estimated at 60-70%. The main part of it has closed codes. However, there is some part of the software with an open code that needs to be systematized to solve the specified tasks. But it is obvious that we need to create a lot of new software.

2.2 *The main common cluster tasks in training*

Therefore, in the field of training, we pursue the following main objectives.

The end-to-end cycle of training students from 1 to 6 the course on a single Java platform.

The teamwork of students, both within universities and at the inter-university level, on various academic and real projects.

Attractiveness for Enrollees. Qualitative Preparation of Students.

Acceleration of Adaptation to European standards and practical learning of English.

Free Exchange of educational materials and programs between departments.

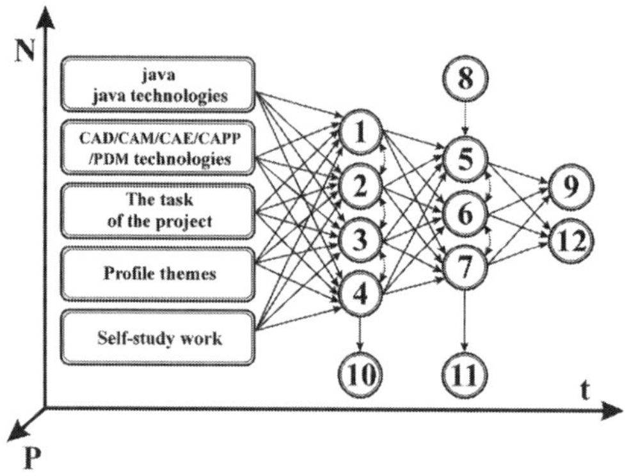

Figure 1. N - quantity of students; P - space; t - time; 1, 2, 3, 4 - coursework project; 5, 6, 7, 8 - bachelor's work; 9 - master's work; 10 - software coursework; 11 - bachelor's software coursework; 12 - master's software coursework.

Creating a single open technology core to integrate JavaMach-core enterprises with CAD/CAM/CAE/CAPP/PDM../ERP systems.

The system of training Bachelors and Masters for Industry 4.0 in the field of mechanical engineering can be simplified as a modernized artificial neural network with the architecture of 5- (4-3) -2 placed in the system of space-time coordinates (Figure 1).

Consider his features on the example of the technological preparation of machine-building production (CAPP).

The ultimate goals of this network are the preparation of bachelors 8, masters 9 and the development by students of ready-made programs (microservices) 10,11, 12, which solve the problems of CAPP. This is reflected in the graph on the tasks for the project.

3 RESULTS

Students learn in-depth the basics of programming on an object-oriented Java language Learn programming environments NetBeans, IntelliJ idea. When choosing any programming environment, special attention is paid to its availability, openness and free of charge and, above all, compatibility with the Java platform. At the next stage, students study methods of storing and processing information from a database point of view.

Big Data, that is a set of technologies designed to perform the storage and processing of large amounts of data, having the ability to work with fast data in very large volumes, to be able to work with structured and poorly structured data.

Here, depending on the complexity of the tasks, the databases of MySql, Oracle, Cloud computing — the provision of computing services (servers, storage, databases, network equipment, software, analytics, etc.) are studied through the Internet.

Cloud storage, convenient for use with Java platform. In this area, students learn cloud storage technology and information sharing on GitHub, where they store the source code of the programs they create. The distinctive features of the GitHub are free of charge, open access to the codes and their adjustment to all members of the project without restrictions. This is the main teamwork tool and is most convenient for the project to be carried out by an international team. The presence of the Internet, a single programming language and GitHub solve a spatial problem in solving the problems of interaction between different universities or developers. This is reflected in the structure of the neural network with additional connections (1.2.3.4) and (5.6.7), in the form of a dash of dotted vectors

4 DISCUSSION

Using Java programming also solves many of the problems of the IoT industry. IoT is a concept of space in which all analog and digital worlds can be combined. This is not just a collection of various devices and sensors combined with wired and wireless communication channels connected to the Internet. This is a closer integration of real and virtual worlds, in which there is a connection between people and devices.

Examples of the successful use of this language are found in the household, industrial appliances, and management of the largest companies in Google, Amazon.

One of the most tangible aspects of the fourth industrial revolution is the idea of "service-oriented design" (Ganesh et al. 2014). It can vary from users who use factory settings to manufacture their products and for companies that supply individual products to individual consumers.

Having a full range of essential knowledge in the field of cybernetics, design, programming, etc.this will require a completely new engineering approach - a digital model, as well as many engineers and programmers who will work with it.

Smart devices that can communicate with each other and with a digital model are required. In this case, control system simulation becomes the most important part of this process. In this plan 3D modeling is a unique method of creating a large array of physical objects based on a single platform. This makes it possible to abandon heterogeneous approaches to solving various problems in favor of a unified approach implemented in digital form. therefore

In the CAD/CAM/CAE cycle, traditional 3D-modeling software such as Solid works, Nx, CITIA with CAM and CAE applications are naturally studied (Dobrotvorskiy et al. 2018b; Dobrotvorskiy et al. 2019). However, for the implementation of projects for microservices for CAPP, students make the most of open desktop or cloud applications (for example,

Onshapé with CAM and CAE applications). This stimulates the ability of students to solve complex problems with minimal time and money.

Practical experience in training specialists in accordance with this algorithm shows that it is necessary to begin training at junior courses. It is desirable from the first course. So, we have transferred to the curriculum of the first course the subject "Computer Science", which studies the basics of programming in the Java language, and "Computer-integrated technologies", where CAD/CAM/CAE systems and their connection with Java technologies are studied. Modular applications in the form of complete programs that have real practical value are able to create bachelors and masters, since they manage, in addition to programming, to study the subject area and understand the problems in their subject area.

Starting the last stage of training (last year) was ineffective. Students had time to study the theoretical part, but did not have time to develop good programs. Complex projects are effective when the group is working on the same task, and it is preferable if students belong to different years of study, which ensures continuity in the further development of the programs after the departure of the masters.

For example, theses of bachelors and masters carried out such developments of the element CAPP.

"Optimization of the selection of blanks from rolled products for machining centers".

"Information and software for the preparation of the production of shafts for a mobile application using Java technology".

"Information and software for production of mechanical engineering products from rolled products using Java technology".

5 CONCLUSIONS

It should be noted that to train specialists of this level for industry 4.0 in universities, new high-level teachers with an unconventional systems approach and enormous knowledge in programming and various subject areas of industry 4.0 are needed. This is an important part of the general problem of training specialists and its solution is no less relevant.

The initiative to create JavaMach Cluster belongs to the Department of Engineering Technologies and Machine Tools NTU "KhPI". There are several reasons for the urgency of a cluster. The first is related to the need for an interactive presentation of the departments of various Ukrainian and European universities.

There are several reasons for the urgency of creating a cluster. The first is due to the need for interactive representation of the Department in the Internet space, i.e. the creation of a virtual Internet Department. The second is an attempt to unite the efforts of machine-building departments of universities in Ukraine and Europe to improve training at the interface of information and engineering technologies on a single Java platform. In addition, the third is important to move from the path of consumer programs to the way they are created. The development of this cluster not only solves the problem of training qualified scientific and industrial personnel.

Maximizing the benefits of the fourth industrial revolution requires significant collaboration that does not limit corporate boundaries, especially when it comes to all machines speaking the same language. Otherwise, the production process will turn into chaos. Thus, the definition of common platforms and languages in which the machines of different corporations freely interact remains one of the main tasks of the spread of cyber-physical systems. JavaMach Cluster can also allow us to solve this problem. Because all software products will be developed on one platform. JavaMach Cluster proposes the development of a single standard for a set of solutions (Dobrotvorsky et al. 2017).

On the other hand, with this approach, excessive homogeneity is not possible when several powerful companies can take on an unnatural advantage in Industry 4.0.

Also partially solved another serious security problem: the openness of the system makes it unattractive for cyber-attacks, since the creation of secure networks is a difficult task, and the integration of physical systems with the Internet makes them more vulnerable to cyber-attacks (Turban et al. 2009.).

Industry 4.0 requires the integration of various technologies and fundamentally new system specialists. Therefore, we propose the concept of using JavaMach Cluster as the center of a single educational, scientific, practical platform of the main industry 4.0 based on the use of a single standard and open source technologies. Creating and operating such a cluster to solve industry problems 4.0 requires the combined efforts of many European countries.

REFERENCES

Alcácer, V. & Cruz-Machado, V. 2019. Scanning the Industry 4.0: A Literature Review on Technologies for Manufacturing Systems. *Engineering Science and Technology, an International Journal* 22: 899–919.

Dobrotvorsky, S.S. et al. 2017. JavaMach Cluster - virtual engineering. In Gennadiy Kostyuk (ed.), *27th international conference "New Technologies in Mechanical Engineering"; Intern. Confer, Koblevo, 3-8 September 2017*. Kharkiv: National Aerospace University H.E. Zhukovsky "Kharkiv Aviation Institute". [Ukrainian].

Dobrotvorskiy, S. et al. 2018a. The role of "JavaMach Cluster" in Industry 4.0. *MMS 2018–3rd EAI International Conference on Management of Manufacturing Systems; Intern. Confer Dubrovnik, 6–8 November 2018*. Croatia: EAI. https://doi.org/10.4108/eai.6-11-2018.2279655.

Dobrotvorskiy, S. et al. 2018b. Effect of the application of microwave energy on the regeneration of the adsorbent, *Acta Polytechnica* 58(4): 217–225. https://doi.org/10.14311/AP.2018.58.0217.

Dobrotvorskiy, S. et al. 2019. The Use of Waveguides with Internal Dissectors in the Process of Regeneration of Industrial Adsorbents by Means of the Energy of Ultra-high-Frequency Radiation. In Ivanov V. et al. (eds), *Advances in Design, Simulation and Manufacturing. DSMIE 2018. Lecture Notes in Mechanical Engineering*: 433–442. Cham: Springer. https://doi.org/10.1007/978-3-319-93587-4_45.

Edl, M. 2013. Educational Framework of Product Lifecycle Management Issues for Master and PhD Study Programmes. In Emmanouilidis C. et al.(eds), *Advances in Production Management Systems. Competitive Manufacturing for Innovative Products and Services. APMS 2012. IFIP Advances in Information and Communication Technology*: 614–621. Berlin, Heidelberg: Springer. https://doi.org/10.1007/978-3-642-40352-1_77.

Ganesh, K. et al. 2014. *Enterprise Resource Planning: Fundamentals of Design and Implementation*. Switxerland: Springer.

Hamrol, A. et al. 2018. Weber M. Analysis of the Conditions for Effective Use of Numerically Controlled Machine Tools. In Hamrol A. et al. (eds), *Advances in Manufacturing. Lecture Notes in Mechanical Engineering*: 3–12. Cham: Springer.

Kagermann, H. et al. 2013. *Recommendations for implementing the strategic initiative INDUSTRIE 4.0. Final report of the Industrie 4.0*. Germany: National academy of science and engineering.

Kozłowski, J. et al. 2018. Modeling of Foundry Processes in the Era of Industry 4.0. In Ivanov V. et al. (eds), *Advances in Design, Simulation and Manufacturing. DSMIE 2018. Lecture Notes in Mechanical Engineering*: 62–71. Cham: Springer. https://doi.org/10.1007/978-3-319-93587-4_7.

Lu, Y. et al. 2016. *Current Standards Landscape for Smart Manufacturing Systems*. Gaithersburg: National Institute of Standards and Technology.

Mavrikios, D. et al 2017. A Web-based Application for Classifying Teaching and Learning Factories. *Procedia Manufacturing* 9: 222–228.

Peraković, D. et al. 2019. Information and Communication Technologies Within Industry 4.0 Concept. In Ivanov V. et al. (eds), *Advances in Design, Simulation and Manufacturing. DSMIE 2018. Lecture Notes in Mechanical Engineering*: 127–134. Cham: Springer. https://doi.org/10.1007/978-3-319-93587-4_14.

Prinz, C. et al. 2017. Implementation of a learning environment for an Industrie 4.0 assistance system to improve the overall equipment effectiveness. *Procedia Manufacturing* 9: 159–166.

Sokol, E.I. et al. 2017. JavaMach Cluster – edinaja platforma obrazovanija, nauki i proizvodstva. Suchasni tehnologiï v mashinobuduvanni [Modern technologies in mechanical engineering], *Bulletin of NTU "KhPI", Series: Techniques in a machine industry* 12: 3–4. [Ukrainian].

Turban, E. et al. 2009. *Information Technology for Management, Transforming Organizations in the Digital Economy*, 4th ed. Massachusetts: John Wiley & Sons.

Industry 4.0 – Shaping The Future of The Digital World – da Silva Bartolo et al. (eds)
© 2021 Taylor & Francis Group, London, ISBN 978-0-367-42272-1

Machine learning approach for biological pattern based shell structures

E. Giannopoulou, P. Baquero, A. Warang & A.T. Estévez
iBAG-UIC Barcelona, Universitat Internacional de Catalunya, Barcelona, Spain

ABSTRACT: Following previous research towards the subject of digital fabrication of thin shell structures, architectural generative design processes sharing similar physical and geometrical characteristics with biological processes were translated to fabrication processes, blurring the lines between physical, digital and biological, and allowed to examine the structural efficiency of segmented stripes arrangements of complex surfaces with less material usage. The goal of this paper is to examine the efficiency of implementing a machine learning approach into an already established design workflow and to develop a creative methodology for decision making. In order to specify the appropriate features, we look at related work that integrates machine learning inside the design and fabrication process.

1 THE CHALLENGE

Integrating intelligent design systems capable to analyze, process and predict results is challenging how we perceive our practice. To Tamke & Thomsen (2018), "the model becomes a creative-analytical engine into which external data can be imported or internally generated to create the basis of an intelligent design practice". These methods aim to expand our ability to work across knowledge domains and to explore potential for innovating existing practice instead of continuing with the early performative architecture practices of interfacing external data and employing generative logics for design exploration and evaluation.

Automated processes are already integral to design; we've just labelled them differently. According to Stoddart, "the idea of automation taking that human agency in design out of the problem is something that I have no interest in exploring because I think you lose the value of design at that point," he says. But we have to address our hubris in understanding our ability to predict solutions to increasingly complex problems (Muklashy 2018).

Recent advances in contemporary biology shows that it has largely become computational biology. The same arise for architectural trans-disciplinary practice which merge computational and biological and fabrication processes inside the design to construction workflow. Menges (2015), argues that the introduction of cyber-physical production systems in the manufacturing industry will also have a major impact on architecture and will not only challenge our understanding of how buildings are made, but more importantly how we think about the genesis of form, tectonics and space.

One of the first objectives of this article is to analyze how recent architectural design projects have creatively implemented machine learning strategies into the design process and understand how those strategies assisted in the design to fabrication process. Finally, a case study demonstrates a creative approach of implementing machine learning and Artificial Neural Networks in an architectural design to fabrication workflow.

2 MULTI OBJECTIVITY & DECISION

Intersections between machine learning and simulation can enable a practice of structural intuition giving rise to a performance based design methodology. Using parametric as well as generative design tools with structural, energetic or other simulation tools is today state-of-the-art practice for experienced practitioners which rely in these situations on intuition. But machine learning can act similarly and predict simulation results out of precedent, how new systems would behave (Tamke et al. 2018). "Moreover, solution spaces are always multi-objective bringing together divergent criteria that don't map to a single optima. As a result, solutions are assessed not absolutely as true or false, but rather qualitatively as better or worse." Therefore, to apply machine learning strategies in architecture "necessitate methods by which results can be evaluated holistically" (Tamke & Thomsen, 2018). To Faircloth et al. (2018) multi-scale architectural models inherently depend on the knowledge of multiple disciplines, and hence multiple methods in order to attempt to describe and predict multi-scale modelling behaviors. These complexities require the profession to develop its own methods, combining models borrowed from other disciplines and validating the handshakes across their respective system boundaries to discover potential for trans-disciplinary, trans-scalar, and trans-temporal practices.

As Tamke et al. says (2017) "the linkage of machine learning with a database enables the memorization of solutions, which build up a kind of experience over time", as in case of Lace Wall (see Figure 1). The case demonstrates how artificial neural networks can categorize the shape of complex geometries based on discretization algorithms, instead of other classification methods. The machine learning-based approach is offering flexibility and precision of previously unseen data and reusing the optimized solutions database and the trained network in multiple iterations of the design. Finally, the kind of intuition that a designer build upon to make design decisions for both complex structural performance choices and behavior can be effectively supported by machine learning providing the means to select alongside a linked database of previously evaluated solutions, which provides the experience on which the selection is based (Tamke et al. 2017).

Fabrication-aware models to Tamke et al (2018) "typically incorporate fabrication limits and material behavior, informed by descriptions of process constraints and predictions of expected behavior. Sourcing information from material data sheets, machine limitations or directly from empirical testing, they seek to incorporate material and fabrication limits inside the design process" to avoid mistakes during fabrication.

However, such models do not include any actual behavior while occurs during the fabrication process (Tamke et al. 2018). As an example, the application of a machine learning approach in the case of Bridge Too Far (see Figure 2), indicates a wider feasibility in architecture, and especially to predict the metal behavior and the forming process results in advance (Zwierzycki et al 2018, Nicolas et al. 2017).

Some cases prohibit traditional computational design optimization approaches. Combining machine learning with the inputs and outputs of a generative design process "one possible strategy for advancing the flexibility of parametric models as performative instruments is tied to rethinking how parameters (or features) are treated in the modelling space" (Stasiuk & Thomsen 2014). To Tamke et al (2018) "this offers the designer an alternative means to more effectively understand, search and discover the complex and varied design outputs that high-dimensional multi-objective optimization algorithms produce".

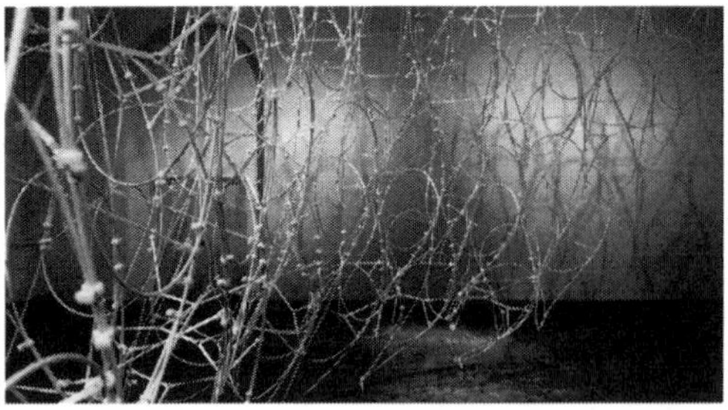

Figure 1. Lace wall detail during the complex modelling exhibition at the Royal Danish academy of fine arts, school of architecture Copenhagen, September–December 2016. (Tamke et al. 2017).

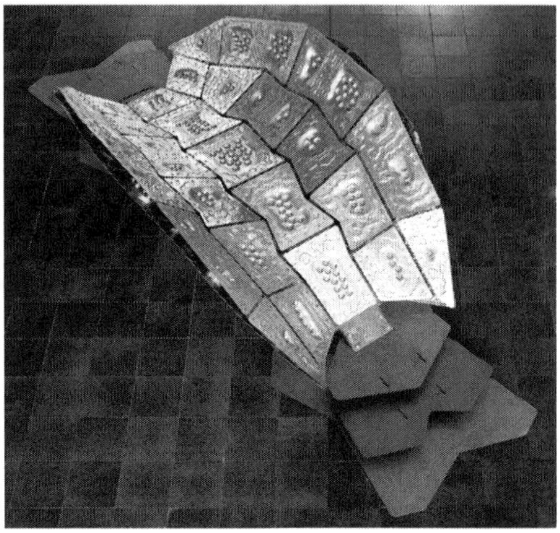

Figure 2. A bridge too far exhibited at the Royal Danish academy of fine arts, school of architecture Copenhagen (Image: © Anders Ingvartsen retrieved from Nicolas et al. 2017).

Although it is still in the experimental stage, there have been multiple attempts by researchers to apply machine learning approaches into building performance prediction and building optimization process. Geyer & Singaravel (2018) were able to develop an artificial neural network model for thermal performance prediction for a building as an example of performance-based design. They have used training data from a physical simulation software for the energy performance for the buildings, and were able to get satisfactory results.

3 CASE STUDY

This research is the evolution towards the subject of digital fabrication of thin shell structures (Bechthold 2008), focusing on the search of a machine learning algorithmic methodology and subsequent computational design techniques which will allow to produce data sets of segmented pattern arrangements in order to help decision making, based on intuition, structural performance and less material usage in one single design workflow. Following the line of previous research (Giannopoulou et al. 2019a,b), also conducted during the Biodigital Architecture Master courses and fabrication workshops, it has been examined how the evolution of architectural generative design processes aimed to apply similar physical and geometrical principles of biological processes, in analogy to biological morphogenesis (Beloussov 2012), translating them to fabrication processes and to test the structurality of branching shell topologies of bifurcating or multi-furcating trees which would not require extra support and withstand an additional weight apart from the material itself. As Stach says (2010), instead of post-rationalizing complex geometrical structures the goal is to "pre-rationalize" the design method.

The logic of stripes has been used as a pre-rationalization construction system (see Figure 3). Some of the most recent examples in practice related to this research is the work of Mark Fornes, who invented a unique approach to describing and building a form: "Structural Stripes" (Schumacher & Fornes 2016).

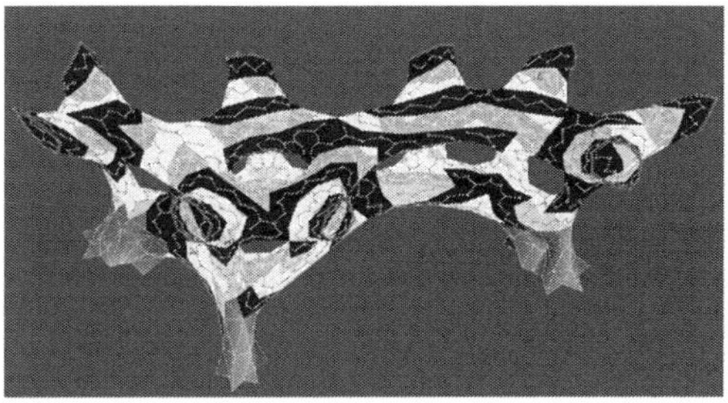

Figure 3. Project analysis of stripe distribution, in order to unroll for fabrication (Biodigital fabrication studio 2018, University master in biodigital architecture, ESARQ-UIC Barcelona).

Patterns were examined for structural behavior in several physical prototypes. Mesh relaxation processes (Piker 2013) and weighted mesh graphs representations (Nejur & Steinfeld 2016) were employed parallel as design tools for the development of minimal structural skins integrating the whole process into the form. The relaxation process, which allowed to arrive close to minimal surfaces, was linked with the segmentation process, which divided the mesh into stripes which in turn was linked with the fabrication process, that integrated material properties, tolerances, constraints, capacities, machine limitations and interactivity, to give eventually the desired structural stripe effect as one unified system in equilibrium as shown in the physical model (see Figure 4). The construction logic, of stripes is conceptualized as a ruled surface (developable, with zero Gaussian curvature).

However, a discussion has been raised upon the stripes topology, if they are closed, or open, and their direction, in relation with the branching topology of the shell structure and its performance. Also, a relation has been observed between stress lines (curves that at each point are tangent to one of the principal stress directions), and the deflected areas, although this could not guarantee that leads to usable structural patterns (Tam & Mueller 2015). Besides, the dual graph concept implemented as a data object, called MeshGraph (Nejur & Steinfeld 2016), is capable of generating a vast amount of patterns with stripe connectivity, to choose from.

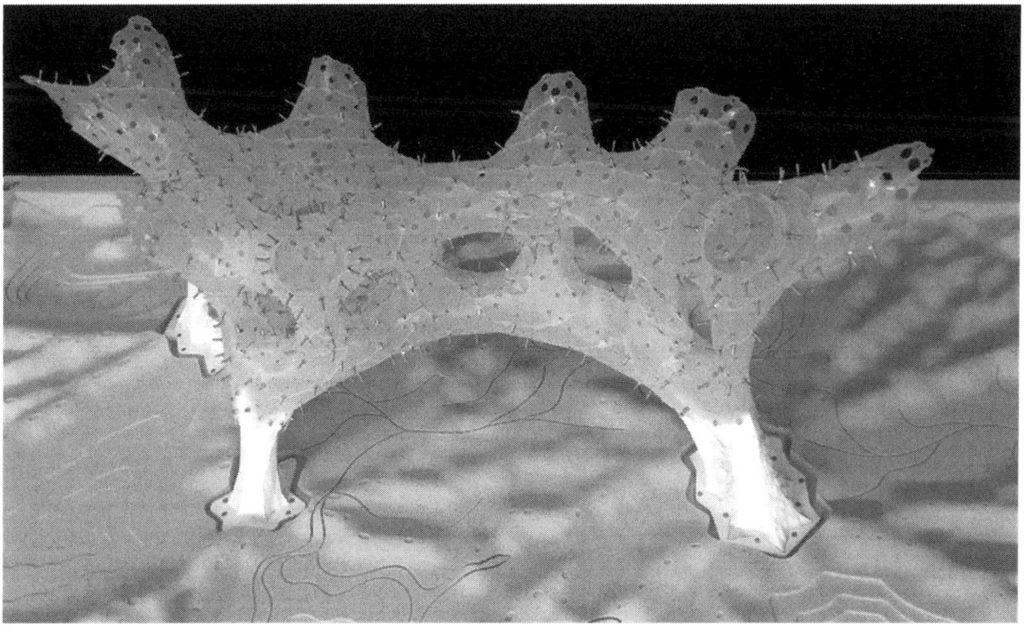

Figure 4. Prototype (Biodigital Fabrication Studio 2018, University Master in Biodigital Architecture, ESARQ–UIC, Barcelona).

Standard topological, shape, size, structural optimization methods nowadays, give standard results but without allowing creativity or the designers interference. It is here that the idea of using a machine learning and artificial neural network approach was introduced.

3.1 *Objectives*

This paper serves as a first attempt and linkage to study intelligent design processes based on ad-hoc machine learning approaches for trans-disciplinary architectural paradigms, which will provide alternative or additional tools to help decision making based on multi-task criteria. Oriented to fabrication, ultimately could allow the designer to interfere/select manually or by intuition that which fits most to his/her desires. The objective is to develop a creative methodology for decision making based partly on the intuition, designer skills and experience, and partly on the machine intelligence and to explore the potentials of state-of-the-art machine learning approaches for the selection process between many geometrical configurations of stripe patterns, for the one with the best structural performance, and which consequently makes the fabrication process more effective.

3.2 *Methodology*

To bridge this gap the idea is to re-examine and modify/extend an already established methodological design workflow of the parametric model. The model should be capable to iterate and to generate datasets in comma-separated values (CSV) file format with the corresponding 3d models which will ultimately serve to predict unexpected outcomes and get insights in order to make better decisions. As a matter of fact, the most interesting part of the method is to determine those sets of attributes/features/behaviors inside the design workflow with which we train the artificial neural network to predict.

The general points of reference/criteria are stated: structurality, minimizing waste of material, less configurations of stripes, less connectivity which means less assembly time and less weight of connection elements (such as screws). Based on the above criteria and looking at the parametric model and geometry, the attributes/inputs for the neural network may defined as:

1. Bottom And Top Number Of Points
2. Bottom And Top X Location Of Points
3. Bottom And Top Y Location Of Points
4. Structures' Height
5. Mesh Subdivision Value
6. Springs Strength Value
7. Segmentation Algorithm
8. Material Thickness

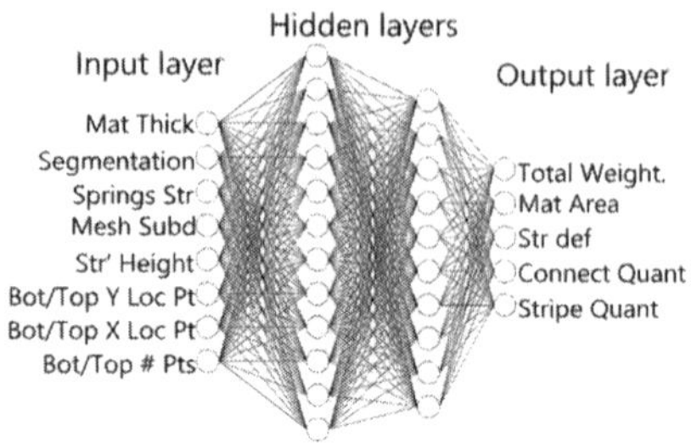

Figure 5. The artificial neural network structure, showing the 7 nodes on the inputs layer, the 2 hidden layers and the 5 nodes on output layer.

One of the attributes introduced in an early design stage is the initial geometry/topology which consists of a bottom and a top system of points joined with a network of columns/lines with boundary conditions which later is converted into quad meshes, like a network of tubules, or hollow vessels referring to a cytoskeleton. In contrary to a static input geometry, the approach allows the generation of a dynamic input geometry which permits the selection between various options that fit best the above criteria. A series of datasets consisting of geometrical outputs are generated, based on the above criteria, to help the designer to have a visual judgement of the numerical values. Those outputs are defined as:

1. Stripes Quantity
2. Connections Quantity
3. Structures' Deflection
4. Material Area Needed
5. Total Weight Of The Structure

The distribution of deflection is generated across the structure using the finite element analysis. The database can give the branching connectivity which defines the initial input geometry, the quantity of stripes and material, the deflection and weight values. The proposed structure of the network is shown in Figure 5.

4 CONCLUSIONS & DISCUSSION

The research re-examines and upgrades an already established generative design workflow to develop a methodology for decision making which allows to produce concatenated taxonomies in the form of datasets. These datasets, which are geometrically segmented patterns and deflected outputs help in the assessment, manipulation and control of the design workflow. The generative method gives the designer the possibility to choose between alternatives which respond to either discrete or associated preset requirements. The vital benefit of creating databases is to be utilized specifically in machine learning to train an artificial neural network and to be able to predict a new building information based on combination of desired parameters. As a result, the algorithm generates an extended amount of possible outputs based on the geometrical, machine and material criteria.

The design strategy extracts processes from fundamental principles that govern both the biological and the fabrication machine. By doing such a geometrical abstraction, one can capture the essence and reveal the rules underlying the apparent complexity. Although biological skin patterns (Kondo 2002) and segmentation in fabrication open a new field of interdisciplinary investigation and architectural applications, there are still many key issues to be further developed, such as the structurality, different material usage and apparently the fabrication technique used, especially for real scale projects.

It is an advantage to use artificial neural network for the prediction of new input values for the thin shell structure and its fabrication from the early stage of design. These methods are very useful especially when the designer wants to have an approximation of the connections and pieces during the design process. Machine learning drastically affects the field of design and architecture through its direct link to computational design and industry, however its applications are still in an experimental stage.

Apart from the many applications and future directions that this generative approach can lead into, the fact that it preserves the intuitiveness of a generic design process is the most important landmark to note. This can be further expanded by saying that the implementation of machine learning and artificial neural network in a design workflow can enhance a designer's inventory by providing control over the scalability of a design process or a design outcome.

Experimentations could also be done in assessing if a machine learning algorithm can predict, assess and analyze input/output parameters for structurally different elements. For example, consider an algorithm trained for analysis of a column structure, and predicting similar attributes for a beam, slab, pedestal or bridge. This research and its explorations could then advance in generating machine learning libraries for different components in architecture with their individual families interacting in a giant, symbiotic, synchronous system, while managing and monitoring a humongous database.

REFERENCES

Bechthold M. 2008. *Innovative surface structures: Technologies and applications*. 1st Edition, Taylor & Francis.

Faircloth, B., Welch, R., Tamke, M., Nicholas, P., Ayres, P., Sinke, Y. & Thomsen, M. R. 2018. Multiscale modeling frameworks for architecture: Designing the unseen and invisible with phase change materials. *International Journal of Architectural Computing* 16(2): 104–122.

Geyer, P. & Singaravel, S. 2018. Component-based machine learning for performance prediction in building design. *Applied Energy*. Elsevier, 228(C): 1439–1453.

Giannopoulou, E., Baquero, P., Warang, A., Orciuoli, A., Estévez, A.T. & Brun-Usan, M.A. 2019a. Biological pat- tern based on reaction-diffusion mechanism employed as fabrication strategy for a shell structure. *IOP Conference Series: Materials Science and Engineering*, 471(10).

Giannopoulou, E., Baquero, P., Warang, A., Orciuoli, A. & Estévez, A.T. 2019b. Employing mesh segmentation algorithms as fabrication strategies: pattern generation based on reaction-diffusion mechanism, *FME Transactions*, 47(2): 379–386.

Kondo, S. 2002. The reaction-diffusion system: a mechanism for autonomous pattern formation in the animal skin. *Genes to Cells* 7(6): 535–541.

Menges, A. 2015. The new cyber-physical making in architecture: computational construction. *Architectural Design* 85 (5): 28–33.

Muklashy W. 2018. How machine learning in architecture is liberating the role of the designer. Redshift by Autodesk. Viewed at 30 February 2019 <https://www.autodesk.com/redshift/machine-learning-in-architecture/>

Nejur, A. & Steinfeld, K. 2016. Ivy: Bringing a weighted-mesh representation to bear on generative architectural design applications. *Proc. 36th ACADIA*, Michigan, US: University of Michigan.

Piker, D. 2013. Kangaroo: Form finding with computational physics. *Architectural Design* 83(1),136–137.

Schumacher, P. & Fornes, M. 2016. The Art of the Prototypical. *Architectural Design* 86 (2): 60–67.

Stach, E. 2010. Structural morphology and self-organization. *Journal of the International Association for Shell and Spatial Structures* 51(165): 217–231.

Stasiuk, D. & Thomsen M.R. 2014. Learning to be a vault. Implementing learning strategies for design exploration in inter-scalar systems. In Emine Mine Thompson (ed.), *Fusion; Proc. 32nd eCAADe Conference*. Newcastle upon Tyne, England, UK, 10-12 September 2014.

Tam, K.M. & Mueller, C.T. 2015. Stress line generation for structurally performative architectural design. In *Computational Ecologies; Proc. 35th ACADIA*, University of Cincinnati, School of Architecture and Interior Design, Cincinnati, Ohio, US.

Tamke, M., Zwierzycki, M., Deleuran, A.H., Baranovskaya, Y. S., Tinning, I. F. & Thomsen, M.R. 2017. Lace wall: Extending design intuition through machine learning. In A. Menges, B. Sheil, R. Glynn, & M. Skavara, *Fabricate 2017*: 98–105. London: UCL Press.

Tamke, M., Paul N. & Zwierzycki M. 2018. Machine learning for architectural design: practices and infrastructure. *International Journal of Architectural Computing* 16 (2): 123–43.

Tamke, M. & Thomsen M.R. 2018. Complex modelling. *International Journal of Architectural Computing* 16 (2): 87–90.

Zwierzycki, M., Nicholas, P. & Thomsen, M.R. 2018. Localized and learnt applications of machine learning for robotic incremental sheet forming. In *Humanizing Digital Reality; Proc. Design Modelling Symposium*, Paris 2017. Springer.

Nicholas, P. et al. 2017 Adaptive robotic fabrication for conditions of material inconsistency: Increasing the geometric accuracy of incrementally formed metal panels. In A. Menges, B. Sheil, R. Glynn, & M. Skavara, *Fabricate 2017*:114–121. London: UCL Press.

Industry 4.0 – Shaping The Future of The Digital World – da Silva Bartolo et al. (eds)
© 2021 Taylor & Francis Group, London, ISBN 978-0-367-42272-1

Machine learning, the Internet of Things, and the seeds of un-behavioral operations: Conceptual bases for the industry of the future

M. Fracarolli Nunes & C. Lee Park
NEOMA Business School, Mont-Saint-Aignan, France

ABSTRACT: With the consolidation of the fourth industrial revolution, machines assumed an even more prominent role in operations, answering for a considerable and increasing portion of companies' output. More than a passing trend, automatization seems to be a solid business tendency, encompassing nearly all sorts of processes. At the same time more complex and efficient learning algorithms are developed, new internet of things technologies significantly diminish machines' dependence on human action. Along with empirical research, the transition towards this 'machine-driven world' demands the construction of robust conceptual bases, particularly around its emerging benefits and challenges. In that direction, the present article addresses some of the potential advantages of technology over human-based activity. Building on the perspectives of two distinguished Brazilian figures, the absence of ego-defense strategies is then argued to be one of the key factors supporting the development of machine learning, stretching the performance gap between creators and creation.

1 INTRODUCTION

In a recent debate with Brazilian philosopher Luiz Felipe Pondé, the journalist, communicator and innovation professor Marcelo Tristão Athayde de Souza (as described in his personal website: http://www.marcelotas.com.br) – also Brazilian and better known as Marcelo TAS – exposed some intriguing views over the nature and functioning of learning processes (the debate is Available at: https://www.youtube.com/watch?v=oYI2oU3tjhM). Accordingly, the main advantage of machines over humans would relate to the capacity of the former to (1) recognize their own mistakes, and (2) share information about these mistakes with peers (i.e. other machines connected to their networks). In reaction, Pondé suggests that the analysis of such characteristics would pertain to a moral sphere, as, in opposition to humans, machines would be deprived from vanity (condition allowing them to proceed in such manner). This *modus operandi* would then dramatically increase the dissemination of empirical data across connected players.

Considered the capacity of machines and data networks to store information indefinitely, these conditions would allow for either the permanent avoidance or the fast treatment of any previously documented failure, even of those initially experienced by a geographically or temporally distant machine/occasion. Boosted by the exponential growth of internet of things technologies, and with numerous machines feeding these databases, information quality would be continuously improved.

In fact, machines' patterns would be contrasting to those of humans in a number of aspects, ranging from the most elementary relationships we keep with errors, until the complex social interactions that may compromise our capacity or willingness to share them. From this angle, it may be argued that humans tend to offer forms of psychological and sociological resistance to the development of individual and collective learning. In some cases, psychology literature classifies individuals' refusal to acknowledge their personal failures as ego or self-protection strategies (e.g. Sommer, 2001; Hepper et al. 2010; Petersen, 2014). Along with the use of denial (Cramer, 1999), other practices such as excuses (Mehlman & Snyder, 1985), justification (Jost & Banaji, 1994), and scapegoating (Allport et al., 1954) would be commonly employed by those seeking to preserve the perception they have about themselves, as well as the image they intend to have in the eyes of observers. In that way, in the absence of typical human mechanisms, the acquisition of abilities would be considerably simpler for machines, with these characteristics standing as the main pillars of what came to be known as machine learning.

In face of the distinct approaches of machines and humans to errors, a formal comparison between them shall be useful. More specifically, the joint analysis of the reactions of each group to unconformities and failures must be revealing of the potentially positive impacts that the elimination, or, at least, the minimization of peoples' psychological limitations may cause. Aiming to contribute to the construction of the conceptual and theoretical bases of machine learning processes, the present article concentrates then on the following research questions: (1) Does the absence of psychological constraints maximize the potential of machine learning processes?; and (2) What would be the main factors involved?

Building on the intersections of the literatures on machine learning, on the internet of things, and on ego-protection strategies, the study presents five distinct but interrelated propositions. Beyond formalizing different perspectives on the issue, these propositions are intended to contribute to the development of a conceptually robust literature of unbehavioral operations, here defined as the operations in which humans are absent, or have their presence minimized. Section 2 ahead contemplates then the main aspects of the consulted literature, followed by the development of our propositions in section 3, discussion and conclusion on section 4, and limitations and suggestions for future research on section 5.

2 LITERATURE REVIEW

2.1 *Machine learning*

As discussed by Jordan & Mitchel (2015:255) in addressing the question of "how to build computers that improve automatically through experience", machine learning would lie "at the core of artificial intelligence and data science". Still after the authors, it would represent a method of choice for a series of developments, including robot control, computer vision, language processing, and speech recognition, in a way that many developers of artificial intelligence recognize as considerably more practical to train systems by the presentation of examples than by their manual programming. On that matter, LeCun et al. (2015) adds that the most common form of machine learning would be what they classify as supervised learning, or the processes through which learning algorithms minimize deviations from desired and actual outputs when provoked by human stimuli. These characteristics would contribute to the perception that machine learning is currently one of the fastest growing technical fields (Jordan & Mitchel, 2015).

From authors' reasoning, it must be argued that the construction of machines' capacity to acquire forms of knowledge is inserted in a functionalist perspective (Campos et al, 1994), representing means for the achievement of major goals. Differently put, the development of machine learning applications – as well as all other form of artificial intelligence – would not represent an end on itself, but rather a manner or a way to meet broader objectives. Such characteristics would be aligned with most of Operations Management concerns, which, broadly, would be intended to minimize the cost and the time of processes (e.g. Hum & Sim, 1996; Guide Jr. & Van Wassenhove, 2001). Indeed, as a Science, Management *per se* seems to be structured around the building of knowledge capable to serve specific interests, either relating to individual classes of stakeholders – e.g. employees (Guest, 2002); shareholders (Friedman, 1970); customers (Fornell, 1992) or society in general (Freeman, 1994), in the potentially non-realistic assumption that trade-offs shall be ignored in the last case (Jensen, 2010). In this direction, along with the consideration of technological challenges, literature around machine learning must be supported by a robust moral and ethical debate, particularly due to its capacity to lead to a complete substitution of human workforce in a not so distant future. Beyond the possibly contrasting economic consequences of such phenomenon – expressive gains of operational efficiency on the one hand, and potential mass unemployment on the other – psychological, social, and political outcomes must be closely followed, under penalty of a major disorganization of social tissue. As previously discussed, the ability to recognize and own mistakes is apparently one of the main advantages of machines over humans. Equally important seems to be their capacity to share related information within network schemes. The following subsection approaches thus the main technologies that make the fast and precise dissemination of data possible.

2.2 *Internet of things*

The now widely employed term "internet of things" would have been coined by Kevin Ashton in a presentation for Procter & Gamble back in 1999 (Ashton, 2009:97). Originally employed to link applications of RFID (radio frequency identification) technologies to the 'then-hot-topic of the internet', the concept would be often misunderstood. As stressed by the author, it would be mainly anchored on the idea that computers – and therefore the internet – are nearly totally dependent on people for the acquisition of information. That would include diverse human actions such as typing, pressing buttons, taking digital pictures, and scanning bar codes, among others. Still, despite the importance of ideas and information, relevant aspects of material life (e.g. economy, society, survival needs) would not be structurally based on them, but, rather, on physical things. Given the dependence of computers on data issued from people, they would be more adequate to deal with abstract information and ideas than with the real world, what would ultimately stand as a limitation to their usefulness.

In that way, through the reduction of computers' demands on humans for informational input, applications using internet of things technologies would maximize machines' potential, as they would have direct access to a range of data issued directly from things. As pointed by the author, offering computers such possibility "so that they can see, hear, and smell the world for themselves, in all its random glory" would represent a sort of empowerment, as, in the limit, these machines would be nearly independent.

On what relates to the present study, the technologies supporting the exchange of information between inanimate objects represent a key point allowing machines to rapidly and precisely share error information among them, enhancing the overall potential of machine learning. In adding the respective dimensions of human nature to this discussion, the next subsection introduces the main aspects of ego and self-protection strategies which may prevent optimal learning by humans, contributing to an eventual advantage of machines in terms of learning capacity, as follows.

2.3 *Ego-protection strategies*

In investigating the behavior of low performing students in environments of high social comparison, Skaalvick (1993) claims that, upon the disclosure of their underachievement, individuals' identities (i.e. self-perception or evaluation) would be somehow threatened, characterizing a situation of high ego-involvement. In response to such psychological perils, self-protective responses, such as non-involvement, disruptive behavior, and withdraw would be expected. As highlighted by the author, these attitudes would be triggered by students' necessity to avoid the exposition of deviances from normal performance, with self-protective behaviors being more accentuated to disguise students' outcomes in those subjects for which they had greater difficulties. In that line, Horton & Sedikides (2009) discuss ego-protection responses of both narcissists and non-narcissists in face of the different social status of ego aggressors. While the first group would employ defensive behaviors regardless the characteristics of the interlocutor, non-narcissists would show some level of mercy to individuals counting on low social status in the application of self-defense actions. Likewise, Sommer (2001) point to two distinct categories of behavioral defenses employed by individuals, either in anticipation, or in response to interpersonal rejection: self-enhancement and self-protection, both characterized as defenses to chronic self-esteem.

From the examination of these positions, it follows that humans apparently have urgency in protecting the perception they have around themselves, refuting perils to such conscience. Among the different types of potentially damaging circumstances to the ego or self-esteem of individuals would be those situations in which they are confronted with failures, errors or deviations from expected behavior and/or performance. In those cases, it seems that actions destined to attenuate, deform, deny, or hide such realities would be close to survival strategies, as ultimately ego, conscience, and mental health would be closely related and mutually dependent. When it comes to the acquisition and dissemination of learning, these facets of human behavior must not be neglected, as they shall cause negative impacts in the conduction of basically any human activity. Within the limits of this investigation, that would be present in the loss of operational performance due to intrinsic difficulties for the acquisition and sharing of knowledge by humans. To some degree, the study of human aspects, either individually or in the interaction with other humans, may be seen as the bases of what is understood as behavioral operations, with factors related to the influence of cultural dimensions, for example, being particularly relevant in supply chain management relationships (Lee Park et al. 2018), as well as in the formulation and implementation of operations strategy (Lee Park & Paiva, 2018).

Following the presentation of the mains ideas over which our arguments are built, the next section presents five propositions, which, when simultaneously considered, form the initial conceptual bases of what is here denominated un-behavioral operations.

3 PROPOSITIONS

As discussed, the incapacity or the unwillingness of humans to recognize and assume mistakes may prevent optimal learning. By attenuating their responsibility over errors and failures, people would preserve the perception they have around themselves, once the acceptance of imperfections must seriously compromise the self-esteem of an individual. If the mental sanity of those applying ego-defense strategies may be reasonably preserved in such occasions, the fact that a failure does not come to be properly recognized and accepted prevents the development of an ideal solution to be implemented in the future. From this apparent trade-off between the preservation of the self and the capacity to learn comes our first proposition:

Proposition 1 – Ego-protection strategies compromise human capacity to learn. As also discussed, ego-protection strategies are not exclusively applied in the preservation of the perception individuals have around themselves. Instead, it seems that they are also employed in the administration of the perspectives believed to be held by third parties. In reality, it must be argued that the perception of observers is also a factor contributing to the esteem people develop around themselves. In this mutually impacting cycle, ego-protection strategies would be employed in the building and maintenance of external facades, which, in the case of errors, would be driven to cultivate images of competence, accuracy, and trustworthiness, among others. In the context of the present study, this need would be reflected in human incapacity or unwillingness to share information around the mistakes they have eventually made, with such constraint negatively impact the creation of collective knowledge. In this sense emerges our second proposition:

Proposition 2 – Ego-protection strategies compromise human capacity to share error information, preventing collective learning. As also pointed, in opposition to humans, machines are deprived from the need to preserve their identities or perceptions they hold about themselves. That would be due to a lack of conscience, and, incidentally, to the unnecessity to protect it. While the presence and the comparatively high level of conscience may be one of the most distinctive characteristics of humans in relation to all other creatures, both animate and inanimate, these factors must also work as powerful sources of weaknesses, among which, the necessity to protect its arguably fragile constitution.

Within this set, the absence of a proper conscience would liberate machines to protect it. Ego-protection strategies would be then unnecessary, leading them to offer absolutely no obstacles to the recognition of mistakes. Being this the initial steps in the construction of solutions, learning process would be more favorable among machines than among humans, as translated in the third proposition of the study:

Proposition 3 – Incidental to the absence of conscience and of the respective need to employ ego-protection strategies, machine learning is more favorable than human learning. Similarly, the absence of conscience would allow machines to share information to other machines, with absolutely no concerns in terms of the impacts this may have in their images among peers. This would represent a possible dimension of machines' lack of vanity, as discussed by the philosopher Luiz Felipe Pondé. Just like ego-protection strategies may prevent the building of collective human knowledge, the fact that they are not necessary among machines may be seen as an additional facilitator of learning, as framed in our fourth proposition:

Proposition 4 – Incidental to the absence of conscience and of the respective need to employ ego-protection strategies, the sharing of knowledge among machines is more favorable than the sharing of knowledge among humans. In parallel with the capacity to recognize data emerging from mistakes, the ability to properly store them (along with solutions) in a way they can be accessed to avoid or treat similar situations in the future stands as an additionally favorable dimension of machines' functioning. In analogy to the memory of humans, the potential of machines on that matter would be superior both in terms of the quantity of data passive to be stored by them, as well as by the theoretically infinite period through which such data must be stored. While human life is inevitably destined to cease, machines can be eternally preserved, or have their "memories" transferred to other machines if necessary. That would guarantee the perennialism of anything learned to be used for an indefinite period of time. These perspectives are identified in out fifth proposition.

Proposition 5 – Machines' capacity to store higher quantities of data indefinitely makes machine learning more favorable than human learning.

4 DISCUSSION AND CONCLUSION

Based on the debate between Brazilian philosopher Luiz Felipe Pondé and Marcelo Tas, the present study conceptually formalizes the conditions and contexts allowing machines to learn in a potentially more efficient way than humans. Departing from initial philosophical considerations, we concentrate on aspects of the human psyche that may work as obstacles to ideal learning processes. Once they would be absent in machines – just like vanity – the consideration of the impacts of such difference may add to the understanding of machine learning as a whole.

We seek to offer then conceptual bases for the comprehension of what seems to be a major tendency in industrial operations. Even if the substitution of human workforce by machines is not new, it seems to have considerably accelerated in the last decades. Following important disruptive technological advancements in the fields of telecommunications and data processing, structural modifications in production processes took place, pushing companies towards significant adaptations in the use of their resources and capabilities. Among the main developments would be the creation and consolidation of the internet, as well as the numerous technologies allowing the exchange of data (e.g. RFID). From a more comprehensive viewpoint, all the applications related to the so-called internet of things would be equally included on that classification.

In this way, the gap between the operational efficiency of humans and that of machines seems to have stretched, meaning that the choice for the use of the second in detriment of the first would be increasingly more likely. In face of this still unstable environment, a solid body of empirical, conceptual, and theoretical knowledge must be built or adapted to treat the specificities of the operations conducted in the total absence or in the minimized presence of humans. In contrast to the already developed field of behavioral operations – which investigates how human condition may impact and be managed so that performance is preserved or improved –, we refer to the administration of machine-driven schemes as un-behavioral operations, as analyses of the absence of human factors must be one or the most relevant elements in the comprehension of this evolving situation.

Our considerations represent then the adjustment of some of the most relevant psychological issues to hypothetical contexts in which machines are expected to hold most of production and service processes, being, if not capable to fully govern themselves, considerably more active in the planning, execution, and control of their activities. Key in that direction is the capacity of machines to develop, stock, access, and share knowledge, justifying our interest in approaching machine learning specifically. As discussed throughout the text, we concentrate in analyzing how the absence of conscience and the consequent unnecessity to employ ego-protection strategies must favor learning process in the building of artificial intelligence. In doing that, we consider important developments of the literatures on machine learning, internet of things, and psychology (ego-protection strategies). The intersection of these often-distant debates allowed for the construction of five distinct propositions which are expected to offer initial bases for future developments, as discussed next.

5 LIMITATIONS AND SUGGESTIONS FOR FUTURE RESEARCH

As it is proper of scientific literature, our study contains limitations that must be identified, so that the validity of our contributions shall be more adequately accessed. Initially, it must be considered that our discussion is solely based on the joint perspectives offered by the literature, and not on original empirical data. While the consideration of evidence is expected in studies positioned within more objectivist and positivist traditions (ontologically and epistemologically, respectively), it must be noted that we do not intend to propose confirmations of cause and effect relations. Given the still incipient status of the literature supporting the developments of the fourth industrial revolution – also called industry 4.0 –, we believe that the building of an initial conceptual debate is particularly relevant. In this vein, even if not based on empirical evidence, our propositions must be useful in the building of theoretical reasonings capable to frame the evolution of technological applications in industrial contexts as a whole. Thus, future research dedicated to test our propositions would be particularly welcome.

Comparative studies investigating the actual distinction between human and machine learning outcomes could be notably useful in the consolidation of the ideas proposed here. Also, qualitative investigation on the functioning of ego-protection strategies adopted by employees would add to the understanding of their inexistence when it comes to processes of machine learning. Along with that, observations on the nature and the speed of information sharing by machines inserted in networks shall allow for the contrast with similar attributes of human exchange. For such task, the analysis of human behavior in social networks may offer insights.

Also, the impact of the nearly immediate communication among machines in the dissemination of negative events in supply chains must be addressed. As shown by Fracarolli Nunes (2018, 2019), negative corporate events of different natures may contaminate both upstream and downstream partners. The comparison between human and machine-driven operations in the speed and impact of this process must add new perspectives on the collateral effects of negative corporate events, particularly in supply chain contexts. Finally, additional debate on the validity of adapting human conditions to machines are necessary.

REFERENCES

Allport, G.W., Clark, K. & Pettigrew, T. 1954. *The nature of prejudice*.

Ashton, K. 2009. That 'internet of things' thing. *RFID Journal*, 22(7): 97–114.

Campos, J.J., Mumme, D., Kermoian, R. & Campos, R.G. 1994. A functionalist perspective on the nature of emotion. *Japanese Journal of Research on Emotions*, 2(1): 1–20.

Cramer, P. 1999. Ego functions and ego development: Defense mechanisms and intelligence as predictors of ego level. *Journal of Personality*, 67(5): 735–760.

Fornell, C. 1992. A national customer satisfaction barometer: the Swedish experience. *Journal of Marketing*, 56(1): 6–21.

Fracarolli Nunes, M. 2018. Supply chain contamination: An exploratory approach on the collateral effects of negative corporate events. *European Management Journal*, 36(4): 573–587.

Fracarolli Nunes, M. 2019. The impact of negative social/environmental events on the market value of supply chain partners. In L. de Boer & P.H. Andersen (eds), *Operations Management and Sustainability*: 151–178. Cham: Palgrave Macmillan.

Freeman, R.E. 1994. The politics of stakeholder theory: Some future directions. *Business Ethics Quarterly*, 4(4): 409–421.

Friedman, M. 1970. The social responsibility of business is to increase its profits. *The New York Times Magazine*, 13: 32–33.

Guest, D. 2002. Human resource management, corporate performance and employee wellbeing: Building the worker into HRM. *The Journal of Industrial Relations*, 44(3): 335–358.

Guide Jr., V.D.R. & Van Wassenhove, L.N. 2001. Managing product returns for remanufacturing. *Production and Operations Management*, 10(2): 142–155.

Horton, R.S., & Sedikides, C. 2009. Narcissistic responding to ego threat: When the status of the evaluator matters. *Journal of Personality*, 77(5): 1493–1526.

Hepper, E.G., Gramzow, R.H. & Sedikides, C. 2010. Individual differences in self-enhancement and self-protection strategies: An integrative analysis. *Journal of Personality*, 78(2): 781–814.

Hum, S.H. & Sim, H.H. 1996. Time-based competition: literature review and implications for modelling. *International Journal of Operations & Production Management*, 16(1): 75–90.

Jensen, M.C., 2010. Value maximization, stakeholder theory, and the corporate objective function. *Journal of Applied Corporate Finance*, 22(1): 32–42.

Jordan, M.I., & Mitchell, T.M. 2015. Machine learning: Trends, perspectives and prospects. *Science*, 349(6245): 255–260.

Jost, J.T., & Banaji, M.R. 1994. The role of stereotyping in system-justification and the production of false consciousness. *British Journal of Social Psychology*, 33(1): 1–27.

LeCun, Y., Bengio, Y., & Hinton, G. 2015. Deep learning. *Nature*, 521(7553): 436–444.

Lee Park, C., Fracarolli Nunes, M., Muratbekova-Touron, M. & Moatti, V. 2018. The duality of the Brazilian

jeitinho: An empirical investigation and conceptual framework. *Critical Perspectives on International Business*, 14(4): 404–425.
Lee Park, C. & Paiva, E.L. 2018. How do national cultures impact the operations strategy process? *International Journal of Operations & Production Management*, 38 (10): 1937–1963.
Mehlman, R.C. and Snyder, C.R. 1985. Excuse theory: A test of the self-protective role of attributions. *Journal of Personality and Social Psychology*, 49(4): 994–1001.
Petersen, L.E. 2014. Self-comparison and self-protection strategies: The impact of self-comparison on the use of self-handicapping and sandbagging. *Personality and Individual Differences*, 56: 133–138.
Skaalvick, S. 1993. Ego-involvement and self-protection among slow learners: Four case studies. *Scandinavian Journal of Educational Research*, 37(4): 305–315.
Sommer, K. 2001. Coping with rejection: Ego defensive strategies, self-esteem and interpersonal relationships. In M. R. Leary (ed), *Interpersonal Rejection*: 167–188. New York: Oxford University Press.

Industry 4.0 – Shaping The Future of The Digital World – da Silva Bartolo et al. (eds)
© 2021 Taylor & Francis Group, London, ISBN 978-0-367-42272-1

Two-phase methodology for smart reconfigurable assembly systems

F. Sunmola, H. Alattar, K. Trueman & L. Mitchell
School of Engineering and Computer Science, University of Hertfordshire, Hatfield, Hertfordshire, UK

ABSTRACT: Increasing need for responsive manufacturing in competitive global markets presents a challenge for assembly systems. Smart reconfigurable assembly systems offer a response to this challenge, addressing uncertainties in manufacturing when subjected to dynamic product mix and volumes amongst others. Reconfigurability is a purposeful change to assembly systems through the addition, removal, or rearrangement of assembly operations, processes, functions and system components. Following an empirical study of assembly reconfigurations in a real use case, a two-phase methodology for engineering smart reconfigurable assembly systems is developed and presented in this paper. The two-phase approach consists of primary reconfiguration and secondary reconfiguration. The primary reconfiguration allows a strategic approach to reconfigurable manufacturing in which strategic changes are made to an assembly system. Operational changes are accommodated in the secondary reconfiguration. The proposed methodology is piloted in a contract electronics manufacturing environment and results demonstrate its potential benefits for the industry.

1 INTRODUCTION

Increasing need for responsive manufacturing in competitive global markets presents a challenge for assembly systems as they need to better manage changes in products and volume whilst meeting manufacturing strategy expectations. An assembly system should not only produce high-quality products at the lowest possible price, it should also be able to react quickly to market changes and consumers preferences. Current assembly systems are known to be limited under uncertainty, particularly when additional processes, changes in process sequence or processing time are required at short notice.

Smart reconfigurable assembly systems offer a response to this challenge. Uncertainties in manufacturing subjected to dynamic product mix and volumes can be addressed through the concept of smart reconfigurable assembly systems. In smart reconfigurable assembly systems intelligence is embedded in the products, workstations and advances in technology is creating more opportunities for increased autonomy in communication between entities in the system with more adaptable control of assembly flow and system performance (ElMaraghy & ElMaraghy, 2016).

Reconfigurability in this context is a purposeful change to assembly systems through the addition, removal, or rearrangement of assembly operations, processes, functions and system components. The benefits of re-configurability include economies of scale, opportunities to optimally accommodate product / component change, product variety handling capability, and lead time reduction. Reconfigurable manufacturing systems have key characteristics that are suited for addressing the identified challenges. The characteristics are (Bortolini et al. 2018): a) modularity, the compartmentalisation of operational functions into units, b) integrability, the ability to connect cells/modules rapidly and precisely, c) diagnosability, the ability to provide a correct diagnosis so as to rapidly attend to the diagnosis, d) convertibility, the ability to easily transform functionalities towards meeting new requirements, e) customisation, the ability to modify system to suit a particular individual or task leading to customised flexibility, and f) scalability, the capacity of the manufacturing system to be changed in size or scale.

Design of the reconfigurable manufacturing system (RMS) represents a significant challenge compared to the design of traditional manufacturing systems, as it should be designed for efficient production of multiple variants, maintaining cost-efficiency and environmental sustainability, as well as multiple product generations over its lifetime. Thus, critical decisions regarding the degree of scalability and convertibility of the system must be considered in the design phase, which affects the abilities to reconfigure the system in accordance with changes during its operating lifetime (Andersen et al. 2017). Computational Design Synthesis tools to support decision-making in the design of reconfigurable manufacturing systems and to show potential benefits are emerging (e.g. Gronau et al. 2016, Colledani & Tolio 2005).

In this paper, a two-phase methodology for engineering reconfigurable assembly systems is developed and presented with a case study. The two-phase approach consists of primary reconfiguration that focuses on strategic aspects and secondary reconfiguration that accounts for operational change requirements. The remainder of the paper is structured into four sections. A literature review of related work is presented in Section 2 and this is followed in Section 3 by a description of the proposed methodology. A case study insight into the proposed methodology is contained in Section 4. The paper ends in Section 5 with conclusions and areas of future work.

2 OVERVIEW OF RELATED WORK

Reconfiguration of manufacturing systems is recognised in the literature and such systems are often required in meeting demands for a diverse set of individualized products to be manufactured in small quantities and with very short delivery lead time. Reconfigurable manufacturing system (RMS) is designed for rapid adjustment of production capacity and functionality, in response to new circumstances, by rearrangement or change of its components. In these responsive systems, reconfigurations are done to provide appropriate capacity for processing part families with ability to accommodate new and unanticipated changes in product design, processing needs and constraints.

Extensive reviews of reconfigurable manufacturing systems exist e.g. Bortolini et al. (2018), Huettemann et al. (2016) and Bi et al (2008). Key findings indicate that conventional manufacturing system design methods do not support the design of an RMS and that a systematic RMS design method is lacking (Andersen et al. 2017) hence the need for newer methodologies. Also highlighted (Bortolini et al. 2018) is the increasing strong links between reconfigurable manufacturing systems and industry 4.0, particularly smart technologies. Huettemann et al. (2016) concluded that reconfigurable manufacturing systems are well suited for small-lot and small to medium series assembly systems.

The utility of RMS is greatly increased if it is designed for multiple configurations in which the configurations form the product. Configurations specify common units for creating product variants. The number of different parts to be manufactured for a product family may be significantly reduced through configurations and combination of different configuration allows manufacturing of sufficient product variety.

The major strategy of modelling the RMS is to handle the varying functionality and capacity demands over the planning cycle which can consist of multiple time horizons. RMS modelling objective varies but mostly centred on the reduction in cost and reconfiguration effort. Design for reconfiguration is an essential part of new manufacturing system design particularly in global competitive market environments. For existing manufacturing systems, there is often the need for improvements requiring adaptations to handle variations. In general, implementation of reconfigurable manufacturing systems often calls for an analysis of the current system to understand the extent to which they can fulfil re-configurability characteristics.

Design processes for RMS are usually conceptualized either as cyclic problem solving methods or phased methods. A comparison of the methods is contained in Andersen et al. (2017). In phased methods for RMS design, focus is primarily on the structural aspects related to the design process as a sequence of decisions progressing through different sequential phases that represent groups of related activities e.g. phases of design specification, concept development, preliminary design, and detailed design.

An initial step in RMS design is typically requirements analysis, sometimes expressed as identification of need for re-configurability, identification of change drivers, definition of requirements for re-configurability and the system, and defining need for scalability and convertibility (Bi et al 2008). In addition, there is often the need to assess the extent to which the requirements can be satisfied (Andersen et al. 2017, Rösiö 2019, Spena et al 2016). For existing systems this will also include an assessment of the current ability to reconfigure the existing manufacturing system.

A variety of measures of manufacturing system re-configurability exists. For example, Rösiö et al, (2019) proposed an assessment criterion for measuring re-configurability of existing manufacturing systems and the readiness of such system to accommodate change with respect to products and volume variations. The criterion they proposed can also be used for establishing the set of parameters to focus attention on in order to achieve a higher degree of re-configurability. Beauville dit Eynaud et al. (2019) presented an Analytic Hierarchy Process to identify weights of characteristics of configurability based on experience of decision-makers.

Another common phase in the design of reconfigurable manufacturing systems is the configuration phase. This phase is characterized by the need for evaluating a large number of system alternatives often in a large solution space. This often calls for computational and performance evaluation tools that are able to quickly analyse and explain the behavior and performance of the system. Such tools are presented in, for example, Colledani & Tolio (2005) and Gyulai et al. (2014).

Gyulai et al. (2014) studied a long-term decision problem relating to relocation of assembly system for a product with decreasing demand, changing product manufacture from a dedicated to a reconfigurable line based on investment and operational costs. They used an integrated computational methodology that combined discrete-event simulation and machine learning techniques. A key point of interest in their work is the distinction between dedicated and reconfigurable assembly lines for product manufacture and the complications inherent in identifying the set of products worth assembling in a reconfigurable line as opposed to dedicated ones.

3 THE TWO-PHASE METHODOLOGY

The two-phase methodology presented in this paper consists of primary reconfiguration and secondary reconfiguration. The primary reconfiguration allows for a more strategic approach to reconfigurable manufacturing in which strategic changes are made to an assembly system. Operational changes are accommodated in the secondary reconfiguration.

A key concept underlying the methodology is manufacturing cell reconfiguration stability. The notion of reconfiguration stability is not new in the engineering and computer science literature. For example, in the context of autonomous and adaptive systems in distributed parallel applications, reconfiguration stability is expressed in terms of the average time between consecutive reconfigurations of the same component and optimizing the reconfiguration amplitude (number of allocated/deallocated resources) (Mencagli et al. 2014).

Mencagli et al. (2014) used the Model-based Predictive Control (MPC) technique to create a trade-off between reconfiguration stability and amplitude. It is important to also respect the stability implications of reconfigurations. In this paper, it is asserted that the more the reconfigurations a manufacturing cell requires the lower is its cell reconfiguration stability. A reconfiguration stability threshold is used for discriminating between primary and secondary reconfigurations. The threshold may be linked to strategic and operational change requirements. The threshold may vary depending on the manufacturing system's environment and requirements.

The steps in the two-phase methodology are broadly described as follows (see Figure 1 above). First, reconfiguration requirements are established, and

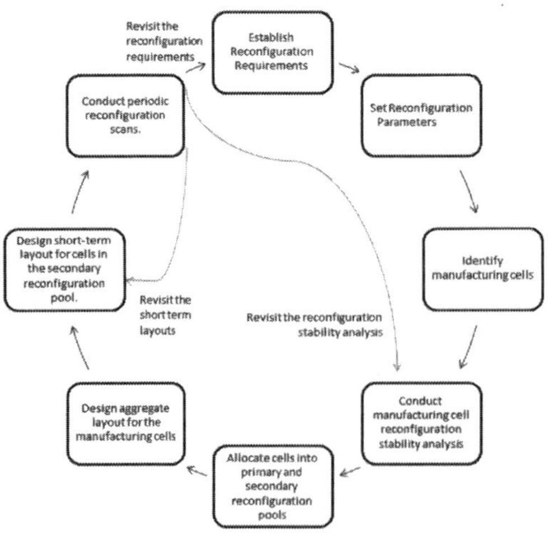

Figure 1. Outline of the two-phase methodology.

parameter values set. Product demand requirements are established and situated within a pre-specified reconfiguration design horizon. Second, manufacturing cell reconfiguration stability parameter is set. Third, manufacturing cell configurations are formed based on product characteristics and demand requirements.

Fourth, manufacturing cell stability for each cell is analysed and compared to the established threshold. Fifth, cells above the threshold are assigned to a primary reconfiguration pool and the others to secondary reconfiguration pool. Sixth, an aggregate layout of the process is designed. The aggregate layout defines the layout organisation of the manufacturing cells for the plant. It aims to improve efficiency by optimally arranging cells on the plant in order to eliminate waste in material flows, inventory handling and management. Seventh is the reconfiguration design of manufacturing cells in the secondary reconfiguration pool. Finally, periodic re-configuration scans are carried out to evaluate the performance of the cells in the secondary reconfiguration pool which may trigger follow-up actions.

4 CASE STUDY

Company X is one of the UK leading contractor manufacturers. The company is specialized in the contract electronics manufacturing (CEM) which focuses on manufacturing of PCBs, cables and Box build i.e. General Assembly. In the arena of CEM, company X focuses on manufacturing low to medium volumes, which could vary in physical sizes as well as batch sizes.

As competition is rising in the CEM market, it was essential for company X to explore new methods of manufacturing to be able to compete in the market and produce products more efficiently. First, the company is facing a lack of adequate footprint on shop floor as cell allocations are customer associated which can become redundant at times. This issue then leads to the difficulty of flexibility on shop floor for accommodating large size products that requires rearranging some areas to accommodate the current needs. Furthermore, batch sizes could also create a challenge, in which large batch sizes could consume a high portion of shop floor footprint which could then lead to flexibility issues.

Second, the visibility on shop floor is significantly limited as jobs are allocated on shop floor to be assembled based on customers, demand and size. High quantity products are built in the high-volume section and, on the other hand, other products are cell based on products type as well as customers. Operators are then located to jobs and, in some cases, operators are responsible for the whole operation or job i.e. a one-person operation. At some other times, jobs are allocated on shop floor regarding the nearest test point or any vacant place available.

Company X is facing limitations with the approach used in allocating jobs. Company X is also affected by fluctuating customer demands and short of product life cycles. Due to the dynamic of the nature of contract manufacturing regular adjustments of the assembly area setup are necessary and hence the reason behind Company X looking into smart reconfigurable manufacturing.

The production line in Company X consists of four main sections PCBs, Cables, GA and Dispatch (See Figure 2 below). Our focus in this case study is the General Assembly (GA) area of the manufacturing production line General Assembly (GA) is combined of both electrical and mechanical assembly processes in which products are manufactured, tested, configured and packaged.

The first step in the analysis process was establishing reconfiguration requirements as well as setting reconfiguration parameters for GA. The analysis of the GA line indicates the two-phased methodology for assembly will be beneficial for Company X as it will allow flexibility in producing a variety of products including the variances of size and batches. This will allow the benefit of mass customization and thus improving productivity of the assembly line. This is achieved by reducing setup time as some identified areas/cells will require little to no configuration. This will also increase the efficiency of production as required tools and machines are located at the right place which will reduce waste time reflecting a lean philosophy.

The second step is identifying manufacturing cells; the cells are shown in Figure 3 below.

A reconfiguration stability analysis of the cells with a threshold of stability was used to discriminate between primary reconfiguration and secondary reconfiguration of cells in the GA. The threshold was established based on experience and brain storming. Two specific areas were identified to require short term reconfigurations and the other areas are found to require little or no configuration in the short term. These areas identified for short term reconfigurations are general fitting area and fitting assembly. These two areas are significantly affected by changing operator's skill levels. On the other hand, housing assembly and cables assembly require little or no configuration in the short term based on the frequent use of similar machines and processes.

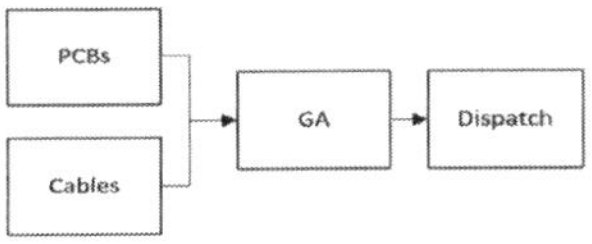

Figure 2. An overview of the production line.

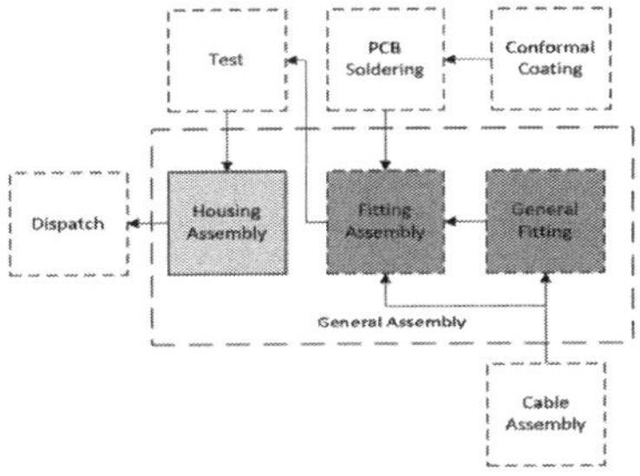

Figure 3. An overview of the proposed layout.

The operations covered in the general fitting is explained from the term, in which, general activities and operations are carried out and affected by the size of products as well as the batch size. These operations are mainly carried out on chassis of the product and require short time to fit, thus the need of frequent change and frequent semi configurations. Fitting assembly also benefits from the two-phased configuration, since fitting assembly operations includes complex operations of fitting PCBs and cables and thus variant operations are required. Fitting assembly is also affected by size of products and volume, and therefore, secondary configuration is done through operational re-configuration such as skill level changes. Designing the General Assembly this way derives benefits from the two-phase reconfiguration methodology thus presenting opportunities for assembling products more efficiently particularly through reduction in preparation and setup time.

Several other sections in the assembly only require strategic changes, as it is set with the required tools. An example is Conformal Coating area. Conformal coating area aims to finish the build of PCB by applying a coat on the PCB, this can either done by hand or a smart machine. However, the setup of the area is fixed, and no configuration is required. Similar sections that could also require little to no configuration is PCB soldering as well as test sections. PCB soldering requires the same tools and a possibly little configuration will be required dependent on the size of product. Furthermore, test section will only require configuration with new products which requires new test machines and jigs, otherwise the configuration of the workplace is standardised.

Further analysis showed that housing assembly is requires somewhere between strategic and operational changes. As a result, it is classified differently. Aggregate layout for the manufacturing cells to establish the overall layout of the areas/cells was carried out. This is followed by short term layout for cells requiring operational changes. This proposed layout also aims to streamline the manufacturing of General Assembly products and decrease the cost of build by reducing waste as well as using the appropriate skill for the operations required.

An initial pilot project for a product, Product A, was carried out. The aim of the pilot was to investigate the indicative performance of the new methodology for General Assembly, concentrating on both the process and time of the build to create a deeper understanding of the configurations required. After analysing the results from the pilot project, Product A was found to be efficiently built using the new methodology and resulted in some noticeable improvements in build time. This reinforced the desirability of the two-phase methodology as well as an indication of the required configurations for GA. As a result, two-phase methodology is found to beneficial to company X and further investigation and pilot project is planned.

5 CONCLUSION AND FUTURE WORK

Manufacturing systems are increasingly required to be responsive to increasing variety of products, volume changes and shortening lead times. To address these challenges, manufacturers are turning to reconfigurable assembly systems and seeking to exploit emerging opportunities associated with smart technologies.

A two-phase methodology for engineering reconfigurable assembly system is presented in this paper to contain a mix of steps and cycles. The initial step establishes reconfiguration requirements including information regarding product demand requirements. Central to the approach is the manufacturing cell reconfiguration stability whose value is used to discriminate between cells requiring reconfiguration at strategic levels and those that require operational changes. A case study implementation of the methodology in a contract electronic manufacturing company indicates that the methodology is promising. Future work is required to develop computational algorithms and tools for establishing the methodology. This should include analytical tools for computing manufacturing cell reconfiguration stability in the context of the methodology. The phases of the methodology can extend beyond two, to incorporate for example a medium-term phase. Future work could explore other phases that may be of interest.

REFERENCES

Andersen A., Brunoe T. D., Nielsen K., Rösiö C., 2017. Towards a generic design method for reconfigurable manufacturing systems: Analysis and synthesis of current design methods and evaluation of supportive tools, Journal of Manufacturing Systems, 42: 179–195.

Beauville dit Eynaud A., Klement K., Gibaru O., Roucoules L., & Durville L. 2019. Identification of reconfigurability enablers and weighting of reconfigurability characteristics based on a case study, Procedia Manufacturing, 28: 96–101.

Bi Z, Lang S, Shen W, Wang L. 2008. Reconfigurable manufacturing systems: the state of the art. International Journal of Production Research 46:967–92.

Bortolini M., Galizia F. G., & Mora C. 2018. Reconfigurable manufacturing systems: Literature review and research trend. Journal of Manufacturing Systems, 49:93–106.

Carin Rösiö C, Aslam T., Sri-kanth K. B, & Shetty S. 2019. Towards an assessment criterion of reconfigurable manufacturing systems within the automotive industry. Procedia Manufacturing, 28: 76–82.

Colledani M. & Tolio T. 2005 A Decomposition Method to Support the Configuration/Reconfiguration of Production Systems, CIRP Annals, 54,1: 441–444.

ElMaraghy H. & ElMaraghy W. 2016. Smart Adaptable Assembly Systems, Procedia CIRP, 44: 4–13.

Gronau N., Vogel-Heuser B., Weber E, Ullrich A., & Schütz D. 2016 Modellability of System Characteristics - Using Formal Mark-up Languages for Change Capa-bility by Design, Procedia CIRP, 52: 118–123.

Gyulai D., Kádár B., Monostori L. 2014 Capacity Planning and Resource Allocation in Assembly Systems Consisting of Dedicated and Reconfigurable Lines, Procedia CIRP, 25: 185–191.

Huettemann G., Gaffry C., Schmitt R. H. 2016. Adaptation of Reconfigurable Manufacturing Systems for Industrial Assembly – Review of Flexibility Paradigms, Concepts, and Outlook. rocedia CIRP, 52: 112–117.

Mencagli G., Vanneschi M., & Vespa E. 2014. A Cooperative Predictive Control Approach to Improve the Reconfiguration Stability of Adaptive Distributed Parallel Applications. ACM Transactions on Autonomous and Adaptive Systems 9, 1: 27 pages.

Spena P. R., Holzner P., Rauch E., Vidoni R., & Matt D. T. 2016. Requirements for the Design of Flexible and Changeable Manufactur-ing and Assembly Systems: A SME-survey, Procedia CIRP, 41: 207–212.

Industry 4.0 – Shaping The Future of The Digital World – da Silva Bartolo et al. (eds)

A screw unfastening method for robotic disassembly

J. Huang, D.T. Pham, R. Li, K. Jiang, M. Qu, Y. Wang, S. Su & C. Ji
Department of Mechanical Engineering, School of Engineering, University of Birmingham, Birmingham, UK

ABSTRACT: Disassembly automation is critical to enabling autonomous remanufacturing of the returned end-of-life products in a circular economy. Screw unfastening is a disassembly method that still poses difficulty to robotic disassembly systems. This paper presents a new method of automated screw unfastening for robotic disassembly by combining the strategies of torque control, position control and active compliance. The paper outlines the setup, process and control procedure of the proposed robotic screw unfastening technique. It then presents an experiment to show a practical implementation of the proposed method. The experiment involved the dismantling of screws from an automotive turbocharger by an industrial collaborative robot. The test results demonstrate the effectiveness of the proposed method.

1 INTRODUCTION

Remanufacturing, as an efficient way of adding value to end-of-life (EoL) products, is increasingly important to delivering economic, environmental and social benefits, enabling a circular economy (Kalmykova et al., 2018). Only in Europe, the remanufacturing industry is estimated to generate billions of euros yearly (Kurilova-Palisaitiene et al., 2018). Disassembly is a step critical to handling the returned EoL products for remanufacturing. Disassembly involves high workloads and low efficiency with high labour costs (Wolff et al., 2017). However, currently, most disassembly operations are carried out manually as they are more difficult to be fully automated than assembly operations due to the uncertain shapes, sizes and physical conditions of used products (Vongbunyong & Chen, 2015). Robots have unique advantages in handling repetitive disassembly tasks such as screw unfastening.

Screws are one of the most common mechanical fasteners, as they could be dismantled non-destructively. This makes products suitable for future maintenance and rework (Jia et al., 2019). According to a survey of more than 400 products, unscrewing operations account for around 40% of all mechanical disassembly operations (Pham, 2018). However, screw unfastening is one of the most difficult disassembly methods fully to automate, especially for small screws in the consumer electronics industry (Jia et al., 2019). An understanding of screw unfastening, possible failure modes and control strategies is important to enhance the robustness and efficiency of automated screw unfastening systems.

This paper presents a novel method suitable for screw unfastening using industrial robots. The proposed method employs the strategies of torque control, position control and active compliance to achieve automated screw unfastening. The paper is organised as follows. Section 2 gives a short literature review covering robotic disassembly and automated screw unfastening. Section 3 introduces the proposed methodology including system setup, unfastening process and control procedure. Section 4 describes an experiment with implementing the proposed method to unfasten screws from an automotive turbocharger. Section 5 concludes the paper.

2 RELATED WORK

2.1 *Robotic disassembly*

A Cartesian robot was designed for automated assembly and disassembly (Danišová et al., 2012). A platform demonstrated how human knowledge at planning and operational levels was transferred to cognitive robots for automated disassembly (Vongbunyong et al., 2017). A robotic disassembly system with a cooperative task planner was presented for PC disassembly (ElSayed et al., 2012). A systematic framework was developed for disassembling and recycling EV comments using robots (Li et al., 2014). A prototype system with a translational motion robot was reported for the automated disassembly of batteries (Schumacher & Jouaneh, 2013). A vision-based intelligent disassembly cell was outlined for dealing with E-waste (Weyrich & Wang, 2013). A robot manipulator on a mobile platform was presented to perform disassembly tasks in a production line (Filipescu et al., 2012).

A disassembly sequence planning method using a Genetic Algorithm was investigated for robotic disassembly of EoL products (ElSayed et al., 2010). Human-robot collaboration was implemented for disassembling EoL EV batteries to overcome uncertainty and unpredictability with their conditions (Gerbers et al., 2018). A systematic framework of human-robot collaborative disassembly was presented for sustainable manufacturing (Liu et al., 2019). A flexible gripper was designed for automated disassembly of lithium-ion battery systems (Schmitt et al., 2011).

2.2 *Automated screw unfastening*

The state of the art in threaded fastening automation was surveyed and open questions for further research were discussed (Jia et al., 2019). Screw unfastening cannot be merely considered as the reverse of screw fastening. The procedure and control strategies of screw unfastening are different from screw fastening. It is critical to detect when the screw has been completely removed from a screw hole and torque control of the unscrewing tool cannot work.

An initial investigation was carried out for unscrewing tasks using a compliant robot in a hybrid disassembly work station (Chen et al., 2014). A robotic system combining force and vision sensing was reported for automated removal of screws from laptops (DiFilippo & Jouaneh, 2018). A computer vision module was employed to locate screw holes automatically and an accelerometer was used to determine when the screw has been completely loosened. However, the performance of the proposed system was affected by camera brightness levels. Screwing and unscrewing robots and tools were developed for automated disassembly of electronic devices (Mironov et al., 2018). An algorithm for diagnosing the condition of the unscrewing process was presented and demonstrated (Apley et al., 1998). An image processing algorithm was investigated to detect screws automatically, enabling robotic disassembly of electric vehicle motors (Bdiwi et al., 2016). A robot equipped with an electric screwdriver was implemented for unscrewing tasks in a battery disassembly workstation (Wegener et al., 2015). The method of detecting screws on a metal-ceiling structure was based on a multi-frame matching process, which was investigated for screw removal tasks using a robot (Cruz-Ramirez et al., 2008).

The current research mainly focuses on automated screw fastening and there has not been much reported research into robotic screw unfastening for autonomous disassembly. It is still a challenge for automated screw unfastening systems to be efficient and robust.

3 METHODOLOGY

3.1 *System setup*

As shown in Figure 1, the system to demonstrate the proposed method consists of an industrial collaborative robot (KUKA LBR iiwa 14 R800) with its controller, an electrical nutrunner (torque spindle, Georges Renault SAS MC51-10) and its controller, and a geared offset attachment (LUBBERING basic line 4-22-15). The collaborative robot has active compliance control with configurable stiffness and damping using a Cartesian impedance controller. Sensor information including the torque and position of the robot is employed to control the screw unfastening process.

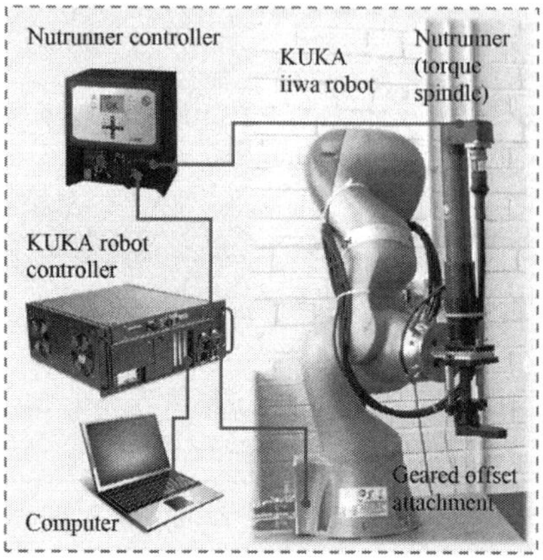

Figure 1. System setup of automated screw unfastening.

The nutrunner can be controlled by the robot which can start and stop its operation. The nutrunner equipped with a geared offset attachment is mounted on the robot's flange using an adapter plate.

3.2 *Automated screw unfastening process*

Screw unfastening involves multiple operation stages. Figure 2 illustrates the proposed process of automated screw unfastening, which includes four stages: approach, search and engage, unfasten, and assess and leave.

1) Approach. The robot equipped with a nutrunner approaches a set position below screw head (Figure 3(a)). Then, the robot triggers the nutrunner to rotate in the fastening direction of the targeted screw. The nutrunner moves to the screw head until it touches its surface under active compliance control (Figure 3(b)).
2) Search and engage. The robot searches for the screw head along a spiral path to align and engage the screw head (Figure 3(c)). The torque information measured by the nutrunner and the robot is used to detect whether the nutrunner engages with the screw head (Figure 3(d)).
3) Unfasten. Once the nutrunner engages with the screw head, it rotates in the unfastening direction to remove the screw from the screw hole. At the same time, the robot follows the movement of the screw, which is pushed by the screw flange.
4) Assess and leave. The robot employs its position information to assess the end of the unfastening stage. Due to the geometry of the screw, when a screw thread ends and continues to rotate, it will fall downward every complete rotation (DiFilippo & Jouaneh, 2018). Finally, the robot leaves its current position with the unfastened screw.

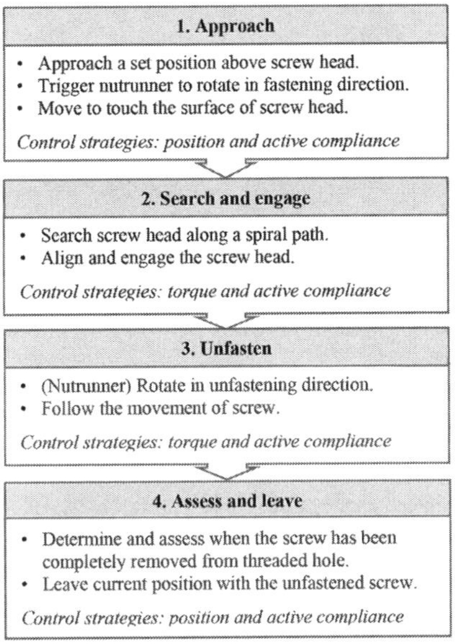
1. Approach
Approach a set position above screw head.
Trigger nutrunner to rotate in fastening direction.
Move to touch the surface of screw head.
Control strategies: position and active compliance
2. Search and engage
Search screw head along a spiral path.
Align and engage the screw head.
Control strategies: torque and active compliance
3. Unfasten
(Nutrunner) Rotate in unfastening direction.
Follow the movement of screw.
Control strategies: torque and active compliance
4. Assess and leave
Determine and assess when the screw has been completely removed from threaded hole.
Leave current position with the unfastened screw.
Control strategies: position and active compliance

Figure 2. Automated screw unfastening process.

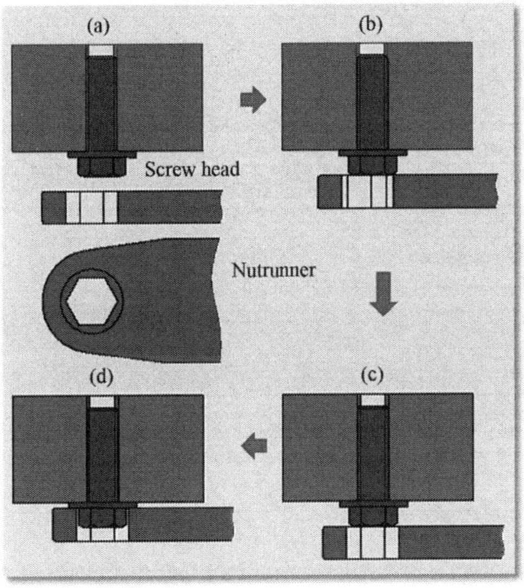
(a)
(b)
Screw head
Nutrunner
(d)
(c)

Figure 3. Screw unfastening process (a) approach, (b) search, (c) align, and (d) engage.

3.3 *Control procedure*

Figure 4 illustrates the control procedure of the robot and nutrunner in the proposed screw unfastening process. First, the robot equipped with a nutrunner moves to a set position below the screw head and sends a signal to the nutrunner to trigger it to rotate at a low speed in the fastening direction. After the robot turns on its compliant mode, it will move to touch the surface of the screw head. Then, the robot begins to search for the screw head along a spiral path until the socket in the nutrunner aligns and engages with the screw head. As the nutrunner rotates in the fastening direction, the torque applied on the nutrunner will increase once it engages with the screw head. When the nutrunner's torque reaches 10.0 Nm (a set value), the torque on the robot arm in the z-direction will exceed 5.0 Nm, simultaneously. At that point, the nutrunner begins to rotate in the unfastening direction with a high speed and the robot stops its searching operation. After that, the robot will be pushed by the screw flange to follow the movement of the screw until the screw is completely removed from the threaded hole. When the position value of robot in the z-direction stops decreasing and oscillates periodically, this means that the screw is free from the threaded hole. The robot will switch off its compliant mode and move away with the unfastened screw. Finally, the robot sends a stop signal to the nutrunner.

4 EXPERIMENT AND RESULTS

To validate the proposed method for automated screw unfastening, an experiment was conducted in the authors' Autonomous Remanufacturing Laboratory. The test involved the disassembly of screws from an automotive turbocharger using an industrial collaborative robot with active compliance control.

4.1 *Experiment setup*

Figure 5 shows the layout of the robotic disassembly cell used. A pneumatic clamping vice (Schunk, TANDEM KSP-LH PLUS 250) was employed to secure an EoL turbocharger (BorgWarner 54359710029) for disassembly on a workbench. There are four flanged hexagon head screws (M6) on the compressor housing of the turbocharger. A robot equipped with an electrical nutrunner was implemented to disassemble the screws automatically using the proposed method. The robot program was written in Java on the KUKA Sunrise Workbench.

4.2 *Testing results*

The control procedure described in Section 3.3 was tested in the robotic disassembly cell. The

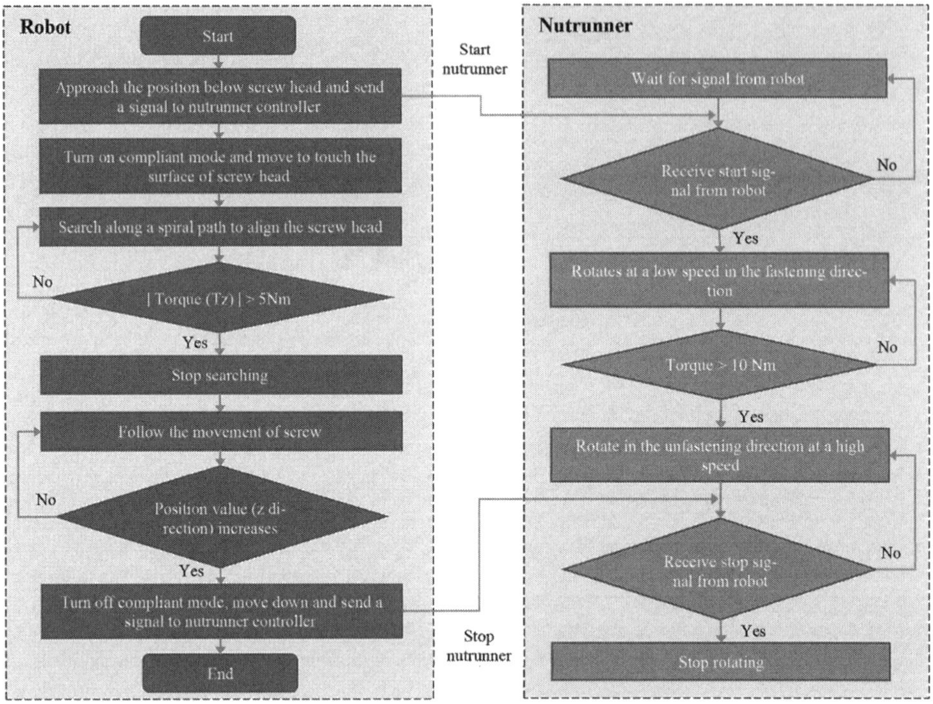

Figure 4. Flow chart of control procedure.

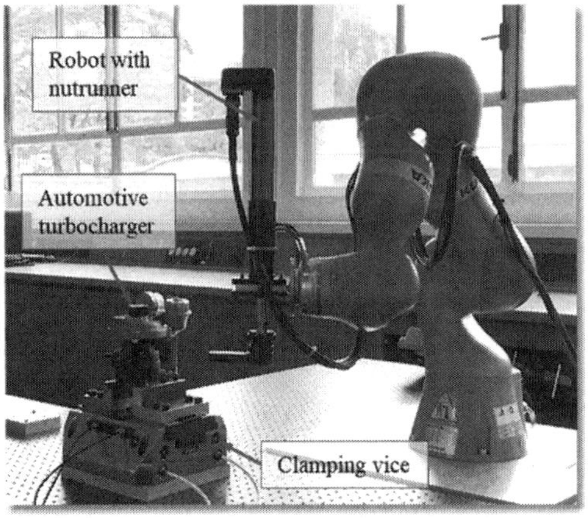

Figure 5. Experiment setup.

screws were unfastened successfully with a total time of approximately 17 seconds, including 7 seconds for searching and engaging, and 6 seconds for unfastening. A video, which could be found by visiting the link in the Appendix, recorded the whole screw unfastening process.

Figure 6 shows selected image frames captured from the process of automated screw unfastening. The nutrunner approached a set position below the screw head (Figure 6(b)) from its initial position (Figure 6(a)). The nutrunner rotated in the fastening direction and the robot turned on its compliant mode. Then, the nutrunner moved to touch the surface of the screw head (Figure 6(c)). After that, the robot searched for the screw head (Figure 6(d)) along a spiral path to align the non-runner socket with the screw head (Figure 6(e)). Once it had engaged with the screw head, the nutrunner rotated in the unfastening direction to remove the screw from the threaded hole. Pushed by the flange of the screw, the robot followed the movement of screw (Figure 6(f)). When the screw had been completely unfastened from the threaded hole (Figure 6(g)), the robot left its current position with the unfastened screw (Figure 6 (h)). Finally, the robot finished the screw unfastening process by sending a stop signal to the nutrunner and carrying the unfastened screw away (Figure 6(i)).

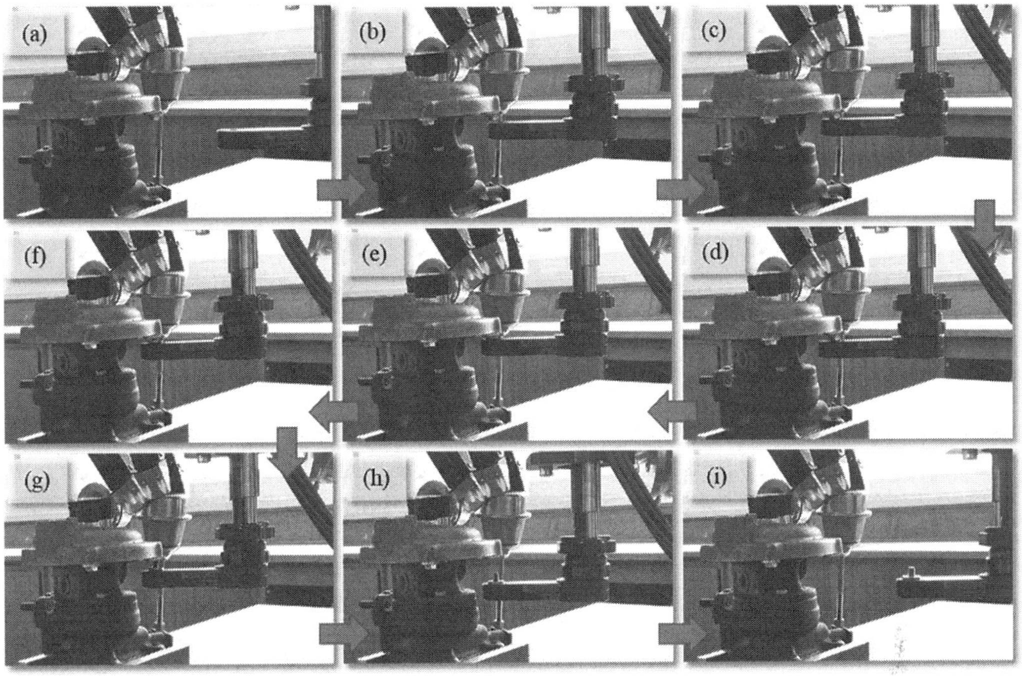

Figure 6. Automated screw unfastening of a screw from a turbocharger: (a) initial position, (b) approaching the screw head, (c) touching the surface of screw head, (d) searching, (e) engaging, (f) unfastening and following the movement of screw, (g) judging the end of unfastening process, (h) leaving current position, and (i) carrying the unfastened screw away.

5 CONCLUSION

Screw unfastening is difficult to be fully automated in robotic disassembly. This paper has presented a new method for automated screw unfastening using an industrial collaborative robot equipped with an electrical nutrunner. The method employs strategies combining torque control, position control and active compliance. The paper has provided details of the setup, process and control procedure. An experiment was carried out to validate and demonstrate the application of the proposed method and system. The experiment involved the disassembly of hexagon head screws from an automotive turbocharger. The results obtained show the proposed method and system have potential application in robotic disassembly.

Future work will focus on possible failure modes of the proposed screw unfastening method and strategies to mitigate them. More tests will be carried out to investigate the proposed system to enhance its robustness and increase its efficiency.

ACKNOWLEDGMENTS

This research was supported by Innovate UK (Contract Ref. 103667) and the EPSRC (Grant No. EP/N018524/1). We appreciate the provision of the automotive turbochargers by Reco Turbo Ltd. for the experimental work.

APPENDIX

The case study described in Section 4 can be viewed at:

https://drive.google.com/open?id=1PXovqCGZXDWFCYfyCjy5bjMAva9ev6Cd.

REFERENCES

Apley, D. W., Seliger, G., Voit, L. & Shi, J. 1998. Diagnostics in disassembly unscrewing operations. *International journal of flexible manufacturing systems*, 10, 111–128.

Bdiwi, M., Rashid, A. & Putz, M. 2016. Autonomous disassembly of electric vehicle motors based on robot cognition. IEEE International Conference on Robotics and Automation (ICRA), 2016. IEEE, 2500–2505.

Chen, W. H., Wegener, K. & Dietrich, F. 2014. A robot assistant for unscrewing in hybrid human-robot disassembly. *IEEE International Conference on Robotics and Biomimetics (ROBIO 2014)*, 2014. IEEE, 536–541.

Cruz-Ramirez, S. R., Mae, Y., Takubo, T. & Arai, T. 2008. Detection of Screws on Metal-Ceiling Structures for Dismantling Tasks in Buildings. *IEEE/RSJ International*

Conference on Intelligent Robots and Systems, 2008. 4123–4129.

Danišová, N., Ružarovský, R. & Velíšek, K. 2012. Robotics System Design for Assembly and Disassembly Process. *International Scholarly and Scientific Research & Innovation*, 6, 1211–1217.

DiFilippo, N. M. & Jouaneh, M. K. 2018. A System Combining Force and Vision Sensing for Automated Screw Removal on Laptops. *IEEE Transactions on Automation Science and Engineering*, 15, 887–895.

ElSayed, A., Kongar, E. & Gupta, S. M. 2010. A genetic algorithm approach to end-of-life disassembly sequencing for robotic disassembly. *Northeast Decision Sciences Institute Conference*, 2010. 402–408.

ElSayed, A., Kongar, E., Gupta, S. M. & Sobh, T. 2012. A robotic-driven disassembly sequence generator for end-of-life electronic products. *Journal of Intelligent & Robotic Systems*, 68, 43–52.

Filipescu, A., Filipescu, S. & Minca, E. 2012. Hybrid system control of an assembly/disassembly mechatronic line using robotic manipulator mounted on mobile platform. *7th IEEE Conference on Industrial Electronics and Applications (ICIEA)*, 2012. IEEE, 447–452.

Gerbers, R., Wegener, K., Dietrich, F. & Dröder, K. 2018. Safe, Flexible and Productive Human-Robot-Collaboration for Disassembly of Lithium-Ion Batteries. *Recycling of Lithium-Ion Batteries*. Springer.

Jia, Z., Bhatia, A., Aronson, R. M., Bourne, D. & Mason, M. T. 2019. A Survey of Automated Threaded Fastening. *IEEE Transactions on Automation Science and Engineering*, 16, 298–310.

Kalmykova, Y., Sadagopan, M. & Rosado, L. 2018. Circular economy–From review of theories and practices to development of implementation tools. *Resources, Conservation and Recycling*, 135, 190–201.

Kurilova-Palisaitiene, J., Sundin, E. & Poksinska, B. 2018. Remanufacturing challenges and possible lean improvements. *Journal of Cleaner Production*, 172, 3225–3236.

Li, J., Barwood, M. & Rahimifard, S. 2014. An automated approach for disassembly and recycling of Electric Vehicle components. *IEEE International Electric Vehicle Conference (IEVC)*, 2014. IEEE, 1–6.

Liu, Q., Liu, Z., Xu, W., Tang, Q., Zhou, Z. & Pham, D. T. 2019. Human-robot collaboration in disassembly for sustainable manufacturing. *International Journal of Production Research*, 1–18.

Mironov, D., Altamirano, M., Zabihifar, H., Liviniuk, A., Liviniuk, V. & Tsetserukou, D. 2018. Haptics of Screwing and Unscrewing for Its Application in Smart Factories for Disassembly. *Haptics: Science, Technology, and Applications*, 10894, 428–439.

Pham, D. T. 2018. Is Smart Remanufacturing a Tautology? *2nd International Workshop on Autonomous Remanufacturing*, Wuhan University of Technology, Wuhan, China, 2018.

Schmitt, J., Haupt, H., Kurrat, M. & Raatz, A. 2011. Disassembly automation for lithium-ion battery systems using a flexible gripper. *15th International Conference on Advanced Robotics (ICAR)*, 2011. IEEE, 291–297.

Schumacher, P. & Jouaneh, M. 2013. A system for automated disassembly of snap-fit covers. *The International Journal of Advanced Manufacturing Technology*, 69, 2055–2069.

Vongbunyong, S. & Chen, W. H. 2015. Disassembly automation. *Disassembly Automation*. Springer.

Vongbunyong, S., Vongseela, P. & Sreerattana-aporn, J. 2017. A process demonstration platform for product disassembly skills transfer. *Procedia CIRP*, 61, 281–286.

Wegener, K., Chen, W. H., Dietrich, F., Dröder, K. & Kara, S. 2015. Robot assisted disassembly for the recycling of electric vehicle batteries. *Procedia CIRP*, 29, 716–721.

Weyrich, M. & Wang, Y. 2013. Architecture design of a vision-based intelligent system for automated disassembly of E-waste with a case study of traction batteries. *IEEE 18th Conference on Emerging Technologies & Factory Automation (ETFA)*, 2013. IEEE, 1–8.

Wolff, J., Kolditz, T., Günther, L. & Raatz, A. 2017. Development of a Methodology for the Determination of Conceptual Automated Disassembly Systems. *Tagungsband des 2. Kongresses Montage Handhabung Industrieroboter*. Springer.

Industry 4.0 – Shaping The Future of The Digital World – da Silva Bartolo et al. (eds)
© 2021 Taylor & Francis Group, London, ISBN 978-0-367-42272-1

Using active adjustment and compliance in robotic disassembly

R. Herold, Y.J. Wang, D.T. Pham, J. Huang, C. Ji & S. Su
University of Birmingham, Birmingham, UK

ABSTRACT: With a greater demand for waste reduction, the need to remanufacture products has increased and the industry is beginning to move to disassembly automation in the same way that assembly had previously. This research investigates the use of active adjustments and compliance to improve the efficiency of the disassembly of products where one component is fixed into another component through a slot or channel. The aim is to cater for uncertainties in the relative positioning of the components, thus minimising damage and enabling fast separation. Several strategies were employed to identify the most effective method of separation. Experiments reveal that adjusting the position proportionally to the forces measured could provide good results. It was found that using an oscillating motion path rather than a linear motion as a part of an active adjustment and compliance strategy can greatly reduce resistance forces.

1 INTRODUCTION

Remanufacturing brings used products to a 'like-new' state. It can be a useful endeavour as the sale of the lower-priced remanufactured goods can increase a product's market share. Adding incentive for a customer to return faulty goods allows for greater understanding of failures in the product which can be channelled into the development of new products. Re-use of the products gives savings in both energy and raw materials while reducing waste to landfill.

To remanufacture a product, it must first be disassembled so that parts can be inspected, cleaned and where necessary replaced. Disassembly, the step that determines the product recovery and therefore the potential return (D. H. Lee, Rang, & Xirouchakis, 2001), is replete with uncertainties. There could be uncertainties about the level of disassembly for maximum profit (H. B. Lee, Cho, & Hong, 2010), the best disassembly sequence (Smith, Smith, & Chen, 2012) or even the cost-effectiveness of disassembling a product for remanufacture (Vadde, Kamarthi, & Gupta, 2006). Other uncertainties include the unknown condition, quality, materials and structure of the parts, along with variance of the same product between manufacturing years and manufacturers (Zussman, 1995). Not only is there a lack of knowledge, but the knowledge itself is acquired only by remanufacturers (Zhu, Gu, Wen, & Yu, 2008) who often are not inclined to share that which might give them a competitive edge. Due to variability in the condition of the returned products, disassembly tends to be manually carried out. It is labour intensive, given the complexity of the operations involved (Liu et al., 2017).

Developments in automated disassembly systems started in the mid-1990s with the robotic disassembly of a PC (Kopacek & Kronreif, 1996), followed by several successful attempts at dismantling electrical devices and automotive components (Barwood, Li, Pringle, & Rahimifard, 2015; Gil et al., 2007; Vongbunyong & Chen, 2015). Those reported studies were mostly product-orientated and have not offered any fundamental insight into the process of disassembly.

This research focuses on a specific operation in disassembly, the separation of a component slotted into a channel in another component. Slotted connections are common in mechanical products (Figure 1). Retaining parts using channels is a simple method of locating two parts relative to each other. Interlocking parts in this manner prevents all motion except in the direction of assembly or disassembly. It is often used for locating parts for further fixation via screws or other methods.

The main challenge in the automated separation of a slotted product is collision caused by uncertain start and end points, as shown in the example in Figure 2. As the degree of freedom is limited, uncertain movement along an incorrect axis can cause two components to collide. Either deformation in long-term usage or inaccurate point teaching during programming can cause uncertainties.

This paper explores methods to optimise the motion such that the forces, time and collisions are reduced thus increasing the efficiency of disassembly and reducing the chance of damage. Four strategies based on active adjustment and compliance are proposed and experimentally investigated.

2 METHODOLOGY

This research experimentally investigates the use of active adjustment and compliance to improve the efficiency of disassembly. The slotted product

Figure 1. Examples of slotted products.

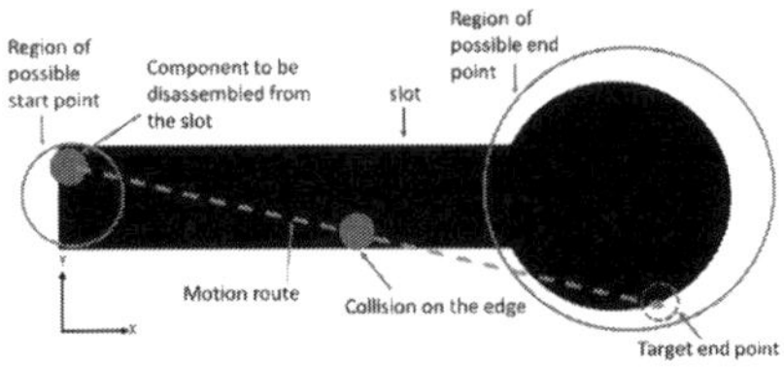

Figure 2. An example of a collision in the disassembly of a slotted product.

example in Figure 3 was adopted. A KUKA LBR robot (Figure 4) was used to move the pin along the slot to reach the detachable point.

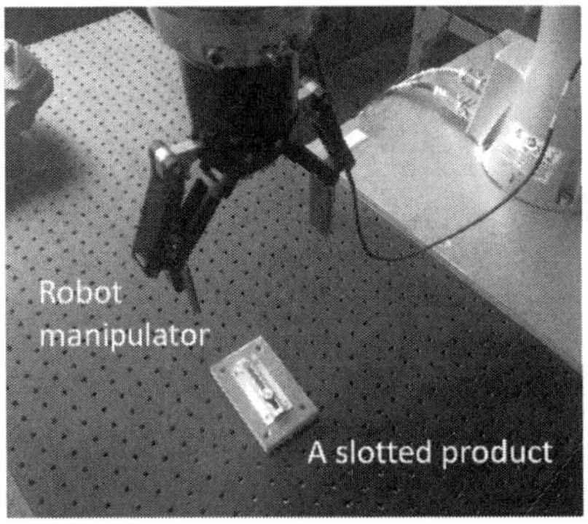

Figure 3. Experimental set up.

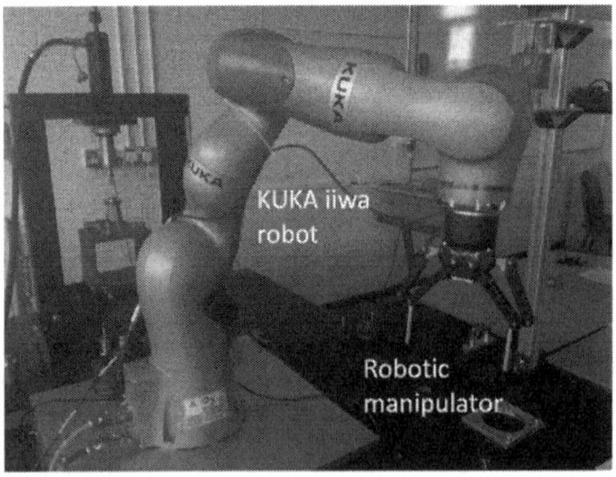

Figure 4. KUKA LBR iiwa robot.

The internal sensors in the robot were employed to detect forces and use the signal to guide adjustment and compliance. In total, four strategies were investigated as explained in Section 2.1. The experimental procedure and setup are described in Section 2.2.

2.1 *Strategies for active adjustment and compliance*

Four strategies were proposed to reduce the forces and time taken in the disassembly of a slotted product: 3axis, ROT, propF and Osc.

- 3axis. The compliant robot detects resistance forces along the X, Y, and Z axes. Reaching a force limit on an axis would cause a fixed small adjustment of the position along the axis accordingly.
- ROT. The compliant robot detects not only the forces along the three axes but also moments at the tip of the gripper. The fractional adjustment involves position shift along or rotation around X, Y and Z.
- propF. This uses both the motions in X, Y and Z as well as the rotational elements as in ROT. However, instead of adjusting each parameter by a fixed amount should forces and moments be above the threshold, this strategy adjusts all the elements proportionally to the fraction of the sum of the forces or torques for which it accounted.
- Osc. Instead of moving linearly, the compliant robot is programmed to adopt an oscillating movement. This would involve position adjustment in the same way as in ROT. The oscillating direction is perpendicular to the direction of the slot.

2.2 *Experimental set up and procedure*

The general procedure is given in Figure 5. The robot would first grip the pin and data recording begins. Then the robot would start moving the pin along the channel, in either linear (3axis, ROT and propF) or oscillating motion (OSC), during which the robot make adjustments as necessary. Once the pin reaches the end point, the test would be completed. Each strategy would be tested using 6 sets of points with uncertainties. Each set of points would be tested 3 times.

For every test, the time taken for the movement, forces along X, Y, Z and number of 'collisions' (adjustments caused by force thresholds being reached) were recorded. X was the direction of the slot. Y was perpendicular to X with the XY plane being parallel to the test table. Z was vertical. A test was considered a failure when it showed no signs of the components having been separated after 100 collisions. The parameters adopted in the experiment are given in Table 1.

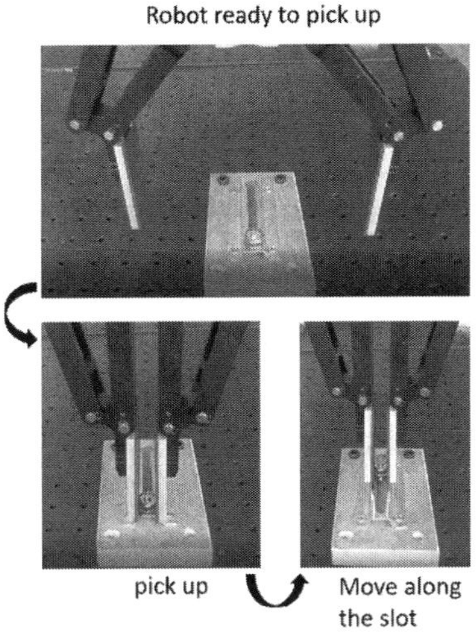

Figure 5. Test procedure.

Table 1. Parameters set up for active adjustment and compliance.

Variables	Values
Cartesian stiffness (XYZ)	300 N/m
(3axis and ROT) adjustment values X&Y	0.75mm
(3axis and ROT) adjustment values Z	1mm
(ROT) adjustment values ABC	0.5°
Adjustment value for propF = N*((X or Y or Z)/(total force))	N= 1
Force condition Z	25N/m
Force condition X &Y	15N/m
Recording sample rate	25ms
Adjustment Cartesian velocity	1mm/s
(Osc) Oscillation frequency	3Hz
(Osc) Oscillation amplitude	2 mm

3 RESULTS AND DISCUSSION

3.1 *Successful rate*

Using all strategies can achieve reliable separation of the two components, as summarised in Table 2. Although using propF failed to complete one test, the successful rate is high. Also, the propF can be more efficient in terms of time and force, to be discussed in Section 3.2.

3.2 *Successful rate*

As shown in Figure 6, the maximum and average forces of the first three strategies are very similar. Interestingly, based upon only the collected force data, it would seem that including rotations into the adjustment in ROT did not provide an improvement in reducing the forces when compared to the normal 3 axis adjustment. As the data shows, based solely upon the forces the rotational program actually performed slightly worse. Although propF introduced minor improvements, they are statistically insignificant. It is apparent that using oscillating motion resulted in greater force reductions. The average force was around half of that of that obtained without oscillations and the maximum force measured was 15% lower.

Figure 7 gives an example of how forces change during the operation. At the beginning, the force was high because of initial misalignments. The robot adjusts the pin position to reduce forces to enable the movement of the pin in the slot. The first three strategies demonstrated similar performances and propF showed a slight improvement in time consumption. The use of oscillating motion paths allowed rapid reduction of resistance forces and thus easier disassembly.

Table 2. Successful rate summary.

Strategies	Success rate	Description
3axis	100%	-
ROT	100%	-
propF	94.4%	Failed to complete one among the 18 tests (6 sets of points with uncertainties, and each set repeated 3 times)
Osc	100%	-

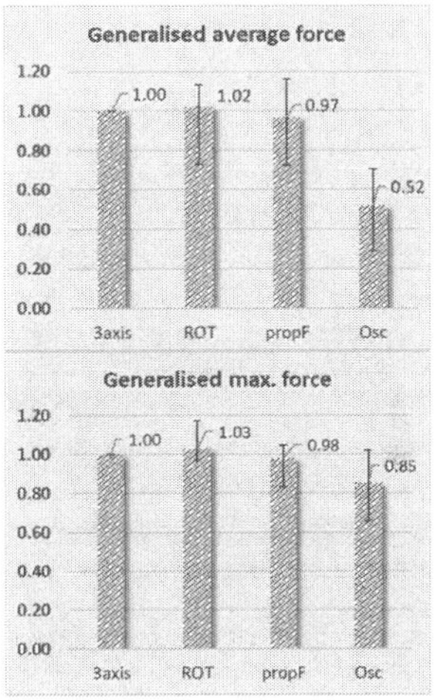

Figure 6. Effects of the four strategies on disassembly forces.

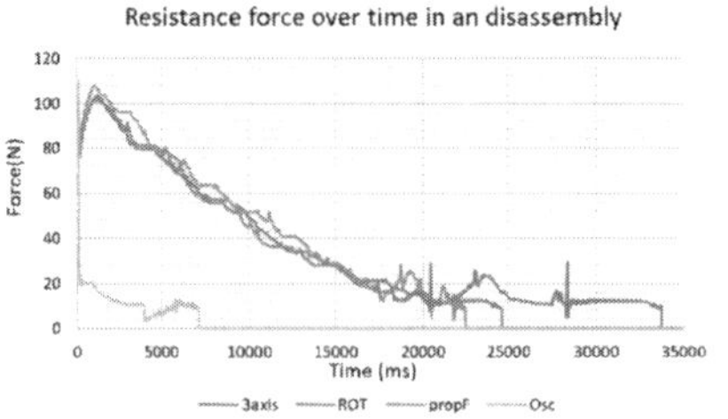

Figure 7. An example of a force-time plot in a disassembly.

Figure 8 shows the disassembly times required using the four strategies. ROT and propF proved to be more helpful than 3axis. Although propF appeared to perform well, the result may be skewed by the fact that it failed to complete one test, as in Table 2. What is clear is that the addition of a rotational component in the adjustment method did greatly reduce the time consumption. Comparing Osc to the other three strategies, the oscillating motion was far quicker than its linear counterparts. This was mostly due to the relatively low number of collisions that the robot experienced in this motion path and thus the reduced need to stop and adjust. The other three strategies, however, were often so "stop-start" that the robot never accelerated to full speed before stopping again. On average, both 3axis and ROT incurred almost the same average number of collisions (8.10 and 8.17 respectively). The propF program was marginally better with an average of 7.53 collisions. Osc showed significant improvement with 5.41 collisions.

3.3 *Evaluation & discussion*

The data presented in this paper has shown that, in every respect, using an oscillating motion path is superior to adopting linear movements. Using an oscillating motion path when separating two slotted components has many benefits, the key reason being the reduction in the forces experienced by the parts and the robot. The strategies mimic the "wriggling" movements humans often adopt when trying to separate parts. This is very useful information for the subject of disassembly for remanufacture, as one of the key problems with disassembly is the unknown condition of parts. Knowing how to reduce the forces experienced by the parts will help to reduce damage thus allowing more parts to be salvaged.

It is expected that similar benefits could be derived for other disassembly operations such as removing a peg from a hole. In the next stage of the research, such operations would be investigated. Also, a scientific explanation for the experimental observation reported here would be explored. Another focus for research could be the investigation of the effects of amplitude and frequency on the force measured by the robot. During experimentation, an amplitude that matched the clearance between the pin and the slot seemed to work the best. However, it is unknown if there is a more optimum setting.

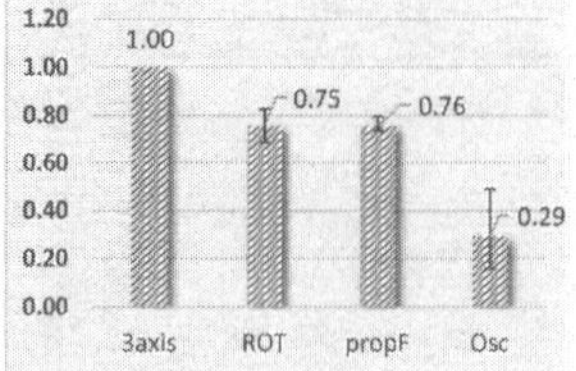

Figure 8. Effects of the four strategies on disassembly time.

4 CONCLUSION

This research experimentally investigated the effects of four strategies using active adjustment and compliance on the efficiency of slotted components disassembly. From the data collated in the experiments, it can be concluded that there is a definite advantage to using an oscillating motion in the disassembly of slotted components, both in separation time and force. Applying oscillating motion to other similar scenarios in disassembly to improve efficiency can be very promising. With regard to the position adjustment of the robot, using a proportional method seems to yield the best results compared to a fixed adjustment.

ACKNOWLEDGEMENT

This research was supported by the EPSRC (Grant No. EP/N018524/1).

REFERENCES

Barwood, M., Li, J., Pringle, T., & Rahimifard, S. (2015). Utilisation of reconfigurable recycling systems for improved material recovery from e-waste. In Procedia CIRP (Vol. 29, pp. 746–751).

Gil, P., Pomares, J., Puente, S. V. T., Diaz, C., Candelas, F., & Torres, F. (2007). Flexible multi-sensorial system for automatic disassembly using cooperative robots. International Journal of Computer Integrated Manufacturing, 20(8), 757–772.

Kopacek, P., & Kronreif, G. (1996). Semi-automated robotic disassembling of personal computers. In EFTA '96 - IEEE Conference on Emerging Technologies and Factory Automation (Vol. 2, pp. 567–572). IEEE.

Lee, D. H., Rang, J. G., & Xirouchakis, P. (2001). Disassembly planning and scheduling: Review and further research. Proceedings of the Institution of Mechanical Engineers, Part B: Journal of Engineering Manufacture, 215(5), 695–709.

Lee, H. B., Cho, N. W., & Hong, Y. S. (2010). A hierarchical end-of-life decision model for determining the economic levels of remanufacturing and disassembly under environmental regulations. Journal of Cleaner Production, 18(13), 1276–1283.

Liu, J., Zhou, Z., Pham, D. T., Xu, W., Ji, C., & Liu, Q. (2017). Robotic disassembly sequence planning using enhanced discrete bees algorithm in remanufacturing. International Journal of Production Research, 1–18.

Smith, S., Smith, G., & Chen, W. H. (2012). Disassembly sequence structure graphs: An optimal approach for multiple-target selective disassembly sequence planning. Advanced Engineering Informatics, 26(2), 306–316.

Vadde, S., Kamarthi, S. V., & Gupta, S. M. (2006). Pricing decisions for product recovery facilities in a multi-criteria setting using genetic algorithms. In Proceedings of SPIE - The International Society for Optical Engineering (Vol. 6385).

Vongbunyong, S., & Chen, W. H. (2015). Disassembly Automation. Springer, Cham.

Zhu, Z., Gu, J., Wen, L., & Yu, S. (2008). The challenges and importance of knowledge management in remanufacturing. In 4th World Congress on Maintenance (pp. 1–4).

Zussman, E. (1995). Planning of disassembly systems. Assembly Automation, 15(4), 20–23.

Industry 4.0 – Shaping The Future of The Digital World – da Silva Bartolo et al. (eds)
© 2021 Taylor & Francis Group, London, ISBN 978-0-367-42272-1

An ontology for sensors knowledge management in intelligent manufacturing systems

M. Mabkhot, M. Krid, A. M. Al-Samhan & B. Salah
Industrial Engineering Department, King Saud University, KSU, Riyadh, Saudi Arabia

ABSTRACT: Industry 4.0 (I4.0) promises new type of intelligent factories; it enables connection between the physical and virtual elements via cyber physical systems. Sensors are important enablers of a cyber-physical system (CPS) as they facilitate the observation and measurement of physical elements. Intelligent factories rely on numerous sensors to perform different functions in different system resources. Sensors also require substantial amounts of information and expert knowledge to manage their activities during their life-cycle. System users need frequent access to sensor information and knowledge to take better decisions. Existing software related to sensors do not provide such knowledge for different levels of user's experience. However, to succeed in such a system, sensors knowledge should be captured in understandable form for machines and humans, and accessible for different levels of user's experience. This study addresses this gap by suggesting an ontology for capturing knowledge about sensors in intelligent manufacturing system.

1 INTRODUCTION

Several technologies such as CPSs, Internet of Things (IoT), big data, cloud manufacturing and virtualization evolved rapidly (Adamson et al. 2017). These technologies formalize the vision of future manufacturing systems (MSs) under the umbrella of I4.0. The emphasis for this vision is laid on intelligent factories, which characterized by autonomous behavior and dynamic and cooperative interactions (Lee 2015). In these systems, sensors are embedded in physical components including machines, robots, AGV, heavy equipment (e.g., cranes and hoists), or building components that have the ability to connect to the internet (Oesterreich and Teuteberg 2016). Systems contains very large number of sensors to perform different functions. Deloitte (2017) estimates billions of sensors will exist to enable the CPS.

To be successful in intelligent system, data and information generated in the system and the knowledge required to manage the system resources should be available and accessible to different autonomous devices and humans (Vogel-Heuser and Hess 2016). Sensors are heterogeneous and serve different resources in different shop floor locations, which make it difficult for users to manage such diverse and huge knowledge (Legat *et al.* 2014). Legacy software in MSs related to sensors do not offer the required knowledge about sensors and its related activities. In addition, users have different levels of experience and expertise in using such software.

Storing sensor knowledge in a meaningful form is mandatory (Compton *et al.* 2012). Semantic Web is benefitted to make the knowledge understandable to machines and humans (Mabkhot *et al.* 2018). Arbitrary components in the system can understand the knowledge from heterogeneous resources. The powerful reasoning capabilities offered by Semantic Web, and more particularly ontologies, make modeling using such technologies very advantageous over other concept modeling technologies in classification, formalization, and sharing.

Since its emergence, ontology has been adopted in MSs in order to speed up manufacturing processes by facilitating access to knowledge encoded in a way that makes it reusable (Ramos 2015). Dasgupta & Dey (2013) suggested a semantic representation of sensor data. Eid, Liscano & El Saddik (2007) suggested an ontology to describe heterogeneous sensor knowledge. Corcho & García-Castro (2010) reviewed such works and discussed the need to interpret, manage, and integrate data derived from heterogeneous sensor networks in a meaningful way. W3C Semantic Sensor Network Incubator group (the SSN-XG) produced a Semantic Sensor Network (SSN) ontology that describes sensors in terms of capabilities, measurement processes, observations, and deployments. This ontology was later deployed in Dey, Jaiswal, Dasgupta, & Mukherjee (2016) to semantically present the energy meter as a sensor. The aim of the deployment was to use semantic technology to manage and improve the energy efficiency in a building.

However, existing works on sensor knowledge are general and not suitable for MSs environments that have different characteristics and interact with different components such as machines and robots. Silva *et al.* (2012) suggested a knowledge-based system to detect, identify, and disambiguate various sensor and system faults in an electro-mechanical actuator system. This ontology was dedicated to actuator faults monitoring. To the best of our knowledge there is no previous works considered developing ontologies to capture, define, and represent sensor knowledge in a MS. This study addresses this gap.

2 ONTOLOGY-BASED CONCEPT MODEL FOR SENSORS KNOWLEDGE

This section presents the concept of the knowledge base. Sensor concepts in MSs can be classified into two types. The first type relates to the details of sensors' capabilities and features as provided from the manufacturers. For example, information related to function, characteristics, input and output signals, type of energy required, response time, expected lifetime, recommended calibration schedule, etc. The second type relates to life sensors' cycle, starting from installing the sensor until its end of life. Such information is related to the behavior of sensors, their relationship with another system elements (e.g., actuators, modules), and the activities performed on the sensors (i.e., installation, calibration, and formatting). Figure 1 shows the main elements for capturing this information. In the following subsections, we will elaborate on how the concept model represents sensors knowledge in MSs.

2.1 *Sensors class*

MSs consist of many physical elements, and sensors are one of these constituents. A sensor is a device that detects and responds to some form of input from the physical environment. To build complete knowledge about sensors in MSs, this study focuses on device sensors and some MSs constituents that have direct relations to sensors knowledge. Figure 1 shows these constituents, which include sensor, actuator, module, and worker. Knowledge about manufacturer and energy types will also be captured. The main class of this ontology is the *Sensor* class, which is defined to represent sensors concepts. *Sensor* class is a subclass of *MfgSysElement* and defined in a hierarchy of subclasses with a set of properties as follows:

2.1.1 *Hierarchy of sensors class*

Sensor class captures all the information about the sensors. To facilitate the capturing of this information, we classified this class into four subclasses, namely, *SensCharacteristic*, *SensFunction*, *SensFaultFailures&Malfunction*, *SensActivity*, and *SysSensor*, as shown in Figure 2.a. *SensCharacteristic* describes sensor features such as accuracy, precision, and repeatability. *SensFunction* class captures knowledge about the types of function the sensor can perform in the system. *SensFaultFailures&Malfunction* records deviation in sensor performance events during the sensor lifecycle. *SensActivity* captures knowledge about tasks that are performed on the sensor during its lifecycle. *SysSensor* describes all existing system sensors. The following subsections illustrate these subclasses.

2.1.2 *Sensors characteristic class*

SensCharacteristic class is designed to capture knowledge about the features of the sensor. Sensor features describe the specification and capability of the sensors as provided by the manufacturers and the performance of a sensor in the system. Based on this, we classified *SensCharacteristic* class into two subclasses:

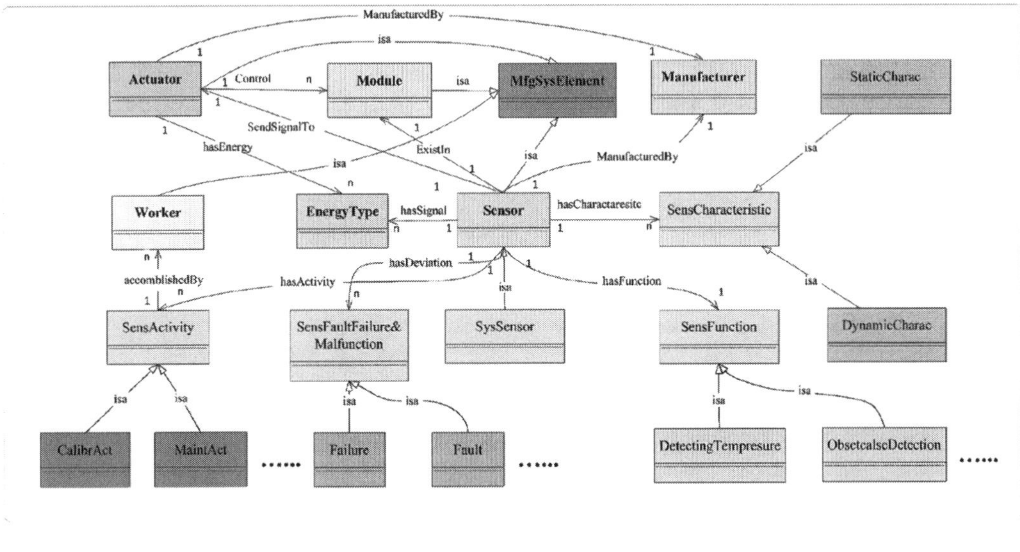

Figure 1. An ontology-based concept for manufacturing sensors knowledge.

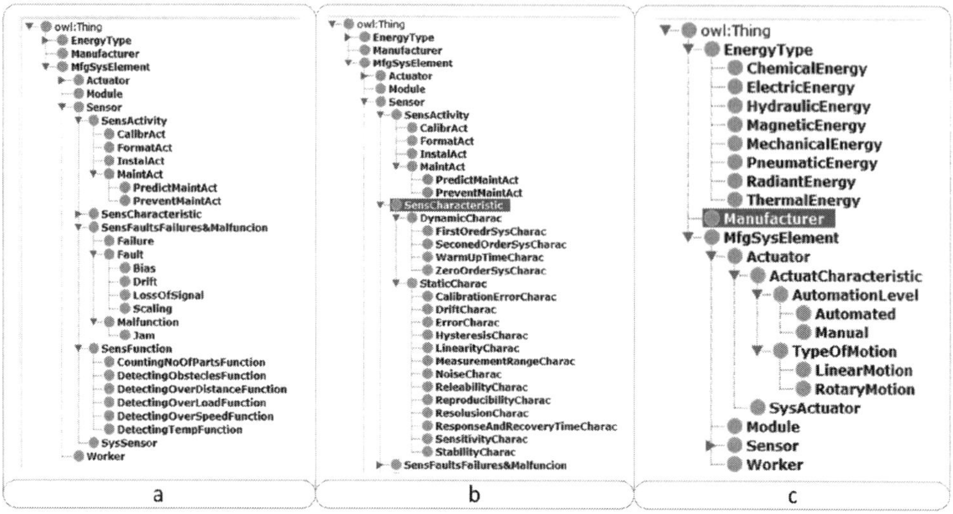

Figure 2. Hierarchy of ontology classes.

StaticCharac and *DynamicCharac*. *StaticCharac* subclass includes knowledge about accuracy, precision, repeatability, reproducibility, stability, error, noise, drift, resolution, calibration error, sensitivity, linearity, hysteresis measurement range, response and recovery time. During its lifecycle, sensor may not respond with perfect accuracy and varies over the time. It occurs because both sensor and its coupling with the stimulus source cannot always respond instantly.

This behavior is time-dependent and can be described as dynamic characteristics, which includes warm up time, zero order, first order, and second order. These characteristics knowledge is described in *DynamicCharac* subclass.

2.1.3 *Sensors function class*

SensFunction class is designed to capture knowledge about the role of the sensor device in the system. Sensors are used in MSs to perform different functions (cf. Figure 3). Mostly, it is used for detection purposes, that is, for discovering and identifying the presence of deviations in working conditions, deviation in system constituents (e.g., machine overload), product specifications (e.g., product weight), or in system conditions (e.g., temperature and humidity). In addition, sensors are used for other functions such as counting numbers of products.

2.1.3.1 SENSOR FAULT, FAILURE, & MALFUNCTION CLASS

A history of sensor faults, failures, and malfunction events is captured during its life cycle in the *SensFaultFailure&Malfunction* class (cf. Figure 4). Sensor fault is an unpermitted deviation of at least one characteristic property (feature) of the sensor from the acceptable, usual, standard operating condition.

Sensor faults are classified into four types: *Bias, Drift, Scaling*, and *LossOfSignal*. Sensor failure is a permanent interruption of the specified operating conditions.

Sensor's ability to perform a required function. Sensor malfunction is an intermittent irregularity in the fulfilment of the sensor's desired function such as Jam. An instance (fault, failure, or malfunction individual) will be created to capture all relevant knowledge. The name of the instance will be coded

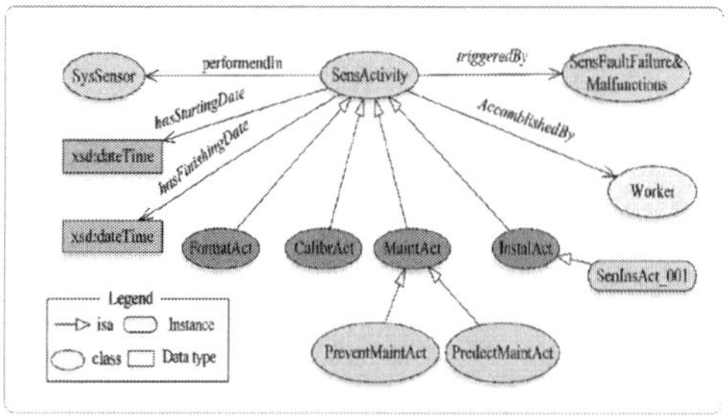

Figure 3. Sensor Activity class properties.

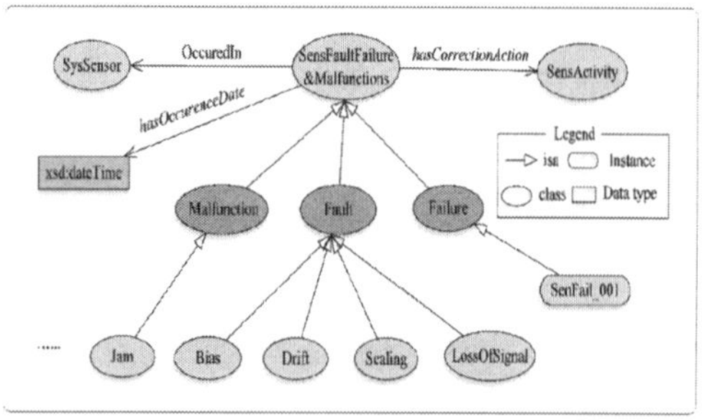

Figure 4. Sensor fault, failure, and malfunction events class properties.

as a unique identifier such as *SenFail_001* instance, which means sensor failure event number 001. This knowledge is described using a set of objects and data properties as shown in Figure 2.

2.1.4 *Sensors activity class*

SensActivity class captures knowledge about system activities performed on a sensor (cf. Figure 4). It captures the history of activities on the sensor, starting with the installation activity up to the uninstallation. It includes calibration, format, as well as predictive and preventive maintenance activities. For each activity, an instance will be created to capture all related knowledge. Name of instance will be coded as a unique identifier such as *SenInsAct_001*. This knowledge is described using a set of objects and data properties as shown in Figure 3.

2.1.5 *System sensors class*

SysSensor class captures knowledge related to existing sensors in the system. An instance will be created in this class to represent each physical sensor.

The knowledge captured in all ontology classes will be used to provide a complete knowledge of sensor instances. *SysSensor* class is described using a set of data and object properties as shown in Figure 3. Sensor instance name will be coded as a unique identifier in the system such as *Sensor_001*. More details on how to capture sensor instance knowledge will be explained in section 3.1.

2.2 *Actuator class*

Actuator class is designed to capture knowledge about actuators in the system, which is required to capture complete sensors knowledge. An actuator is a device that reacts to information provided from sensors. Reaction can be to move or control system or components such as to open or close a valve. Actuator class is classified into two main subclasses: *ActuatorCharacteristic* and *SysActuator* (cf. Figure 4.b).

ActuatorCharacteristic captures knowledge about the characteristics of the actuator relating to automation level (automated or manual) and motion type (linear or rotary). *SysActuator* class captures knowledge about existing actuators in the system. Each system actuator has a unique identifier such as Actut_001. *SysActuators* are described using a set of object and data properties. *hasEnergy* object property describes energy type (i.e., pneumatic or electric). *hasLevelOfAtumation* object property describes the level of automation (manual or automated). *hasMotionType* object property describes motion type (linear, rotary). *hasPhysicalLocation* object property describes the Module the actuator is located in. *hasMaxExtension* data property describes the maximum distance of the actuator extension. *hasMaxSpeed* data property describes the maximum actuator speed and *hasOutputTorque* data property describes the output torque of the actuator.

2.3 *Other ontology classes*

In addition to capturing a complete picture of sensors knowledge in the system, we created classes for each system constituent that has a relation with the sensor, including energy type, module, worker and manufacturer (cf. Figure 2c). *EnergyType* class describes the energy input and output needed for the sensor and actuator. Module class captures all system workstations, machine tools or material handling. Sensor is located in shop floor on a Module, and sensor function is needed to accomplish the overall Module function. For example, a sensor that measures the distance a machine stroke traveled is necessary to inform the machine controller the distance moved. Worker class captures knowledge of the work force who accomplished the activities related to the sensors described in section 2.1.5. Manufacturer class captures knowledge about system sensor suppliers such as *DeltaOH*, *Phillips*, *Fastron* and *Kamtrup*. Knowledge about all these classes is necessary to enable different types of queries about sensors.

3 KNOWLEDGE INSTANCES AND RULES

The above ontology provides a semantic representation and hierarchal classification of sensors knowledge. Web Ontology Knowledge (OWL) is an effective tool to manage such a large body of knowledge. We built the suggested ontology in Protégé 5.2 editor (Stanford Center for Biomedical Informatics Research 2015). In MSs, many sensors exist so that thousands of classes are required to define such knowledge. However, not all user levels know how to retrieve the required information from the ontology. To enable use of different interfaces for different levels of MS users, we use Semantic Query-enhanced Web Rule Language (SQWRL) along with the ontology. SWRL-based query language has been widely used in the past (Zhang *et al.* 2015). SQWRL is similar to the SWRL antecedent and effectively employed it as a query specification. SQWRL is developed in the SWRLTab plug-in in Protégé-OWL. Hence, this query is run in the SQWRLTab plug-in.

3.1 *Sensors knowledge instance*

In ontology-based semantic representations, an instance is a specific realization of an ontology class, and a class is a pool of instances. For example, *SysSensor* class is a pool of sensors such as *Sensor_001*, *Sensor_014*, and *Sensor_127* (cf. Figure 5). Instances capture knowledge of each individual in the system (i.e., physical sensor, actuator, worker, module, and manufacturers). For clarification purpose, we will explain how to capture knowledge of a sensor instance. As we described earlier, each specific sensor has specific function, characteristics, physical location, activities history, etc. Figure 6 shows the definition of *Sensor_001* instance.

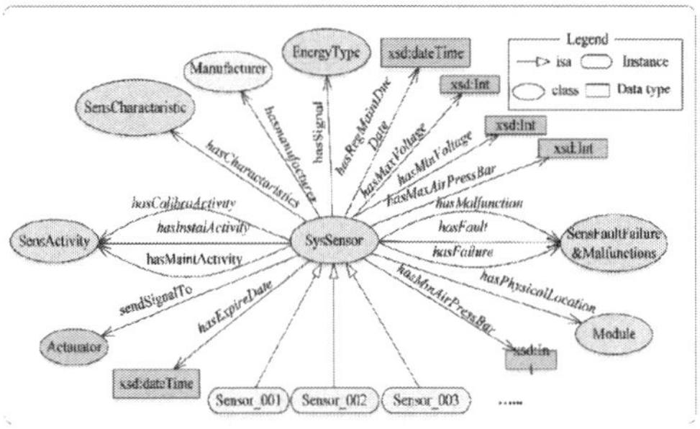

Figure 5. System sensor class properties.

3.2 *Knowledge rules for sensors*

To enable retrieving the required information for different levels of MS users, the ontology is combined with SQWRL. SQWRL is a concise, readable, and semantically robust query language for OWL. SQWRL is expressed such as SWRL rules in the form: Antecedent → Consequent. Antecedent part refers to the body, and the consequent part refers to the head. The antecedent part is expressed as the conjunctive formula of atoms a_1^a_2 ^, …,^a_n. Atom a_i ($1 \leq i \leq n$)) can be in either of the forms C(?x) or P (?x, ?y), in which if x is an instance of class C, then C (?x) holds; if x is related to y by property P, then P(?x, ?y) holds. The consequent part is different from SWRL and consist of the primary operator sqwrl: select. The consequent part can include many secondary operators to organize the enquired knowledge such as tabulating enquired knowledge or ordering the information. Table 1 shows two examples of SQWRL queries explained in the following:

- *Due calibration sensors query*: retrieves sensors that have scheduled maintenance on a specified date.
- *Expire sensors query:* each sensor has an expected life. Running sensors more than the expected life increases the chances of failures and malfunctions. The second SQWRL query in Table 1 retrieves all sensors that will expire in a specific date.

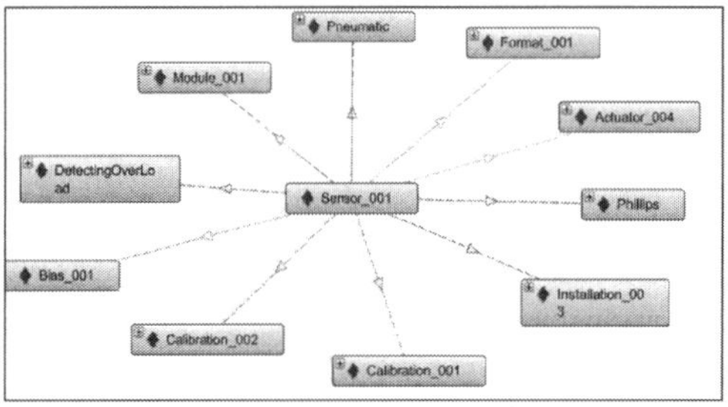

Figure 6. Sensor_001 definition.

Table 1. Example of SQWRL queries.

Antecedent	Consequent
SysSensors(?x) ^hasCalibrationDDate (?x,?value)^swrlb:equal(?value, 20171226)^hasPhysical-Location(?x, ?y)	sqwrl:select(?x, ?value, ?y)^ sqwrl:columnNames ("Sensor name", "Calibration date", Physical location")
SysSensors(?x)^hasExpireDate (?x, ?value)^swrlb:lessThan(?value, 20171216)^hasPhysical-Location(?x, ?y)	sqwrl:select(?x, ?value, ?y)^ sqwrl:columnNames ("Sensor name", "Expire date", "Physical location")

3.3 *DL query*

DL query tab in Protégé-OWL is an interface for querying OWL ontologies. Similar to OWL, DL query is based on Description Logic (DL), which is a family of knowledge representations (Lehmann 2016). The

DL query interface is powerful and easy to use for searching for a classified ontology. Firstly, we classify the ontology using a reasoners (classifier), which is an artificial intelligence piece of software used in ontology to infer logical consequences from a set of asserted facts or axioms. For reasoning in this paper, we use Pellet 3.0 because it has some required capabilities (such as complex data-type reasoning and support for SWRL rules) not present in other reasoners. Table 2 gives a set of DL queries, explained as follows:

- *Energy type of sensor query*: Sensors need a form of energy to transform input signal to output. Sometimes, the source of energy has a shortage (e.g., pneumatic and electrical) in the system and the Decision maker (DM) has to know what sensors work on a specific energy type. DL query number one in Table 2 lists all pneumatic sensors that exist in the system.
- Deviation type of sensor query: To improve predictive maintenance schedule and reduce faults. Failure or malfunction occurrences, it is necessary to study the history of the sensors' deviation types that occurred in the system in the past. DL query number two in Table 2 lists all bias events that occurred for sensors that exist in the system.

Table 2. Example of DL queries.

#	DL query
1	SysSensor and hasEnergy some PneumaticEnergy
2	SysSensor and hasDeviation some Bias
3	SysSensor and hasDeviation some LossOfSignal and SendSignalTo only {Actuator_001}

Moreover, DL query can retrieve more specific information. For example, if we want to study the loss of signal deviation that occurred to sensors connected with an actuator named *Actuator_001*, DL query number three in Table 2 lists these sensors.

4 CONCLUSION

Today's MSs are going to be intelligent and will rely on a vast number of sensors. This article suggested a knowledge base to support the management of sensors in an intelligent MS. An ontology was built to structure to the knowledge and cover a wide range of sensor knowledge. The structure of the suggested knowledge base is modular and can be adapted to build similar knowledge bases for other MS components. Moreover, the structure of the ontology is generic enough and can be merged with upper MS ontologies. The SQWRL query is employed to retrieve organized information from the ontology while three types of query interfaces are explained to serve different user levels in the MS. Finally, the potential of the suggested approach motivates to build another knowledge base for MS components such as machines and robots. Such work initiates a platform structure for CPS and this research direction represents the focus for our future work.

REFERENCES

Adamson, Göran, Lihui Wang, and Philip Moore. 2017. "Feature-Based Control and Information Framework for Adaptive and Distributed Manufacturing in Cyber Physical Systems." Journal of Manufacturing Systems 43:305–15.

Compton, et al. 2012. "The SSN Ontology of the W3C Semantic Sensor Network Incubator Group." Journal of Web Semantics 17:25–32.

Corcho, Oscar and Raúl García-Castro. 2010. "Five Challenges for the Semantic Sensor Web." Semantic Web 1 (1–2):121–25.

Dasgupta, Ranjan and Sounak Dey. 2013. "A Comprehensive Sensor Taxonomy and Semantic Knowledge Representation." Pp. 791–99 in Seventh International Conference on Sensing Technology.

Deloitte. 2017. Sensor Technology.

Dey, Sounak, Dibyanshu Jaiswal, Ranjan Dasgupta, and Arijit Mukherjee. 2016. "Organization and Management of Semantic Sensor Information Using SSN Ontology: An Energy Meter Use Case." Pp. 468–73 in Proceedings of the International Conference on Sensing Technology, ICST.

Eid, Mohamad, Ramiro Liscano, and Abdulmotaleb El Saddik. 2007. "A Universal Ontology for Sensor Networks Data." Pp. 59–62 in Proceedings of the 2007 IEEE International Conference on Computational Intelligence for Measurement Systems and Applications, CIMSA.

Lee, Jay. 2015. Smart Factory Systems.

Lehmann, Jens. 2016. DL-Learner Manual.

Mabkhot, Mohammed M., Abdulrahman M. Al-Ahmari, Bashir Salah, and Hisham Alkhalefah. 2018. "Requirements of the Smart Factory System: A Survey and Perspective." Machines 6(2):23.

Oesterreich, Thuy Duong and Frank Teuteberg. 2016. "Understanding the Implications of Digitisation and Automation in the Context of Industry 4 . 0: A Triangulation Approach and Elements of a Research Agenda for the Construction Industry." Computers in Industry 83:121–39.

Ramos, Luis. 2015. "Semantic Web for Manufacturing, Trends and Open Issues. Toward a State of the Art." Computers & Industrial Engineering 90:444–60.

Stanford Center for Biomedical Informatics Research. 2015. "Protégé."

Vogel-Heuser, B. and D. Hess. 2016. "Guest Editorial Industry 4.0 Prerequisites and Visions." IEEE Transactions on Automation Science and Engineering 13(2):411–13.

Industry 4.0 – Shaping The Future of The Digital World – da Silva Bartolo et al. (eds)

Long range RFID indoor positioning system with passive tags

J.S. Pereira
Polytechnic Institute of Leiria, School of Technology and Management, Leiria, Portugal
Instituto de Telecomunicações, Portugal
Centre for research in Informatics and Communications - CIIC, Portugal

H.M.C. Gomes
Polytechnic Institute of Leiria, School of Technology and Management, Leiria, Portugal
Instituto de Telecomunicações, Portugal

S.P. Mendes
Polytechnic Institute of Leiria, School of Technology and Management, Leiria, Portugal
Centre for research in Informatics and Communications - CIIC, Portugal

R.B. Santos & S.D.C. Faria
Polytechnic Institute of Leiria, School of Technology and Management, Leiria, Portugal

C.F.C.S. Neves
Polytechnic Institute of Leiria, School of Technology and Management, Leiria, Portugal
INESC Coimbra, Portugal

ABSTRACT: The Industrial Internet of Things (IIoT) brings together machines, advanced analytics and people. It is a network of monitoring devices and sensors that can supervise processes; collect, exchange and analyse data and use that information to continually adjust the manufacturing process. Passive radio frequency identification (RFID) tags are well-known electronic devices that could contribute to help the IIoT purpose, by providing an easy wireless identification way, normally used in short range applications. In this work, the use of these devices for longer range positioning is investigated. More specifically, one of the goals of the project is building an RFID indoor positioning system (IPS) prototype to be used in the context of the mould industry that allows the location of passive RFID tags in a range of 15 metres from the reader. One UHF RFID reader was mounted and tested in an industrial crane that can cover a large industrial zone.

1 INTRODUCTION

The problem of indoor location of parts, products and materials is being studied in several research centres around the world (Xiaolin 2012), (Accenture 2015). For example, the Institute of Telecommunications and the Polytechnic Institute of Leiria have been developing some research in the area of indoor positioning, by radio frequency, through Portuguese projects like LINPOSYS (2015), TOOLING4G (2018-2020) and RINPOSYS (2019), (Bagarić *et al.* 2015), (Bagarić 2016), (Pereira *et al.* 2015), (Ferreira *et al.* 2015), (Lanza-Gutierrez *et al.* 2015), and (Flores *et al.* 2017).

Nowadays, some companies are producing commercial indoor positioning system (IPS) solutions: (Wifarer 2019) Canada; (IndoorAtlas USA Inc. 2019) Finland, United States, China, and Japan; (Pozyx Labs 2019) British Indian Ocean; Phillips' LED-based IPS (Netherlands); and (Nextome SRL 2019) Italy. Currently, the best solution seems to be the Pozyx solution with an estimation error of around 10 centimetres.

In addition, there are also some commercial solutions of national and international companies on the market. In the Portuguese reality, we highlight the company RCSoft (Rcsoft 2019) that implements solutions based on radio frequency identification. Currently, RCSoft is present in the industry, with solutions to control and trace certain types of objects in the specific context of each client's activities through RFID technology. RCSoft has recently implemented its RFID tag location solution in Portuguese companies such as Revigrés (ceramic tiles industry) and Raporal S.A. (meat industry). In the international scene, the View Technologies consortium (a joint venture of Stanley Black & Decker, Inc. and RF Controls, formed in 2014) holds a vast portfolio of location patents (View 2019). Their products cover various solutions through RFID control. Briefly, such systems use a set of transmitting and receiving antennas with transmission direction control by phase variation of the radio frequency signals.

Nevertheless, the various national and international commercial solutions for locating existing RFID tags are not adapted to the real problem of location in industrial environments where electromagnetic interferences generated by electric machines predominate. It is well known (Pereira 2015) that the electromagnetic multiple-path interference is one of the main responsible for the inhibition of long range readings of RFID tags. In industrial mould environments, where electromagnetic wave-reflecting metallic materials prevails, the multiple-path interference is the main barrier to the implementation of a reliable long-range RFID locating solution with passive tags (Wu 2012) and (Lazaro *et al.* 2009).

Several companies such as RCSoft, Link Technologies Inc, Jiangmen Shenghua Science And Technology Co have started to market passive RFID tags (ISO 18000 6C, EPC CLASS1 GEN2) with metal presence protection, high temperature resistance and immunity to moisture. These anti-metal RFID tags will be used in our mould industry real-world scenario.

We propose a radio frequency location solution that mitigates the unreliability of current RFID locating systems in industrial environments. In our proposal, we have implemented passive anti-metal RFID tags (ISO 18000 6C) paired with an innovative long range IPS built with a set of directional antennas and UHF RFID readers. The proposed novel solution will not infringe the patents of View Technologies and will attempt to outperform it, using patents pending from IPLeiria and the Telecommunications Institute (Pereira *et al.* 2015), (Pereira *et al.* 2016) and (Pereira *et al.* 2017)

The structure of this paper is as follows: Section 2 describes the hardware and software used on our new long-range RFID indoor positioning system. Section 3 presents a new mathematical model of an IPS using short and long-range antennas. The experimental results are shown in section 4 and some conclusions are provided in section 5.

2 RFID IPS HARDWARE AND SOFTWARE

Our long range IPS solution with passive tags has been tested with an UHF RFID reader CH-MU904 and two different antennas. An antenna type 1 (Figure 1) is a 5 dBi short range directional antenna and antenna type 2 (Figure 2) is a 12 dBi long range directional antenna. Major electrical and mechanical specifications are as follows:

The passive tags used are anti-metal tags for ISO 18000-6C and Alien Passive Tags 860 MHz. The RFID reader was tested using the Chafon SDK (Software Development Kit) for Windows operating system. The RFID software has a parameter that defines the reading quality (received power) of the

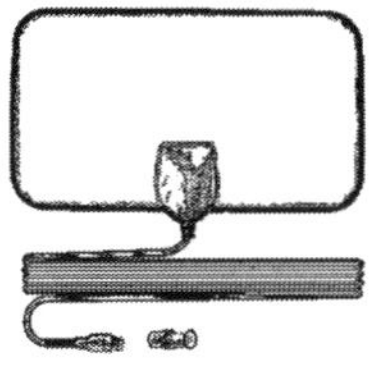

Figure 1. Short-range antenna type 1 (Ultra-thin Digital 2019); Digital Indoor Antenna TV HDTV model; 5 dBi Gain; 50 Ω Impedance; Linear polarization; 120 mm x 210 mm x 0.6 mm; 433 MHz - 918 MHz.

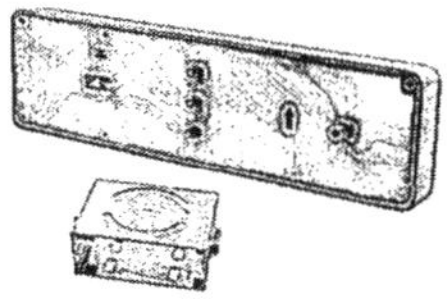

Figure 2. Long-range antenna type 2 (Chafon 2019a); CF-RA1202 model; 860 MHz - 960 MHz; 100 MHz bandwidth; 12 dBi Gain; Horizontal and vertical beam width 90° and 32°; F/B Ratio ≧ 23; VSWR ≦ 1.3; 50 Ω Impedance; Linear polarization; 580 mm x 160 mm x 45 mm.

Figure 3. UHF RFID Reader Module (Chafon 2019b); CF-MU904 model; 865 MHz - 868 MHz; ISO 18000-6C (EPC GEN2) protocol; 26 dBm (adjustable) RF power; 50 tags/s readings; ~15 m reading range (depending on antenna/tag pair).

passive tags defined as rssi. When the passive tag is at the maximum range of the antenna, the rssi value drops to 185. This value will be used in the next section to find a mathematical model to estimate the tag location on a bi-dimensional surface.

3 MATHEMATICAL MODEL

This section proposes a mathematical model that estimates the rssi values by an equation for both types of antennas (short and long-range). Equation 1 is used in (2) to define the rssi surface.

3.1 *Rssi surface*

Equation 1 has been estimated based on a set of rssi measurements of different RFID antennas. The equation computes the rssi value in a bi-dimensional space (x, y, z), and is given by:

$$y = \frac{L}{\alpha - \beta}\sqrt{(\alpha - z)^2 - \gamma^2(x - x_0)^2} \quad (1)$$

where x and y represent the Cartesian coordinates from the tag to the reader referential (x_0, 0) in metres; z represents the rssi value obtained by the RFID reader; α is the maximum rssi value at the coordinate (x_0, 0); β is the rssi value at the coordinate (x_0, L), where L is the maximum range of the RFID antenna; γ is an integer value depending on the type of the antenna used (short and long range); x_0 is a shift value of the antenna distance over the x axis.

After solving (1) it is possible to find the RFID antenna rssi sensitivity z. This rssi surface is given by:

$$z = \alpha - \sqrt{\left(\frac{\alpha - \beta}{L}\right)^2 y^2 - \gamma^2(x - x_0)^2} \quad (2)$$

For the two antenna types (short and long range) the rssi surface plots are presented in the next section.

3.2 *Rssi surface of two different RFID antennas*

For the short-range antenna, Figure 4 represents the real measurements taken with an 0.08 m step until the maximum of 0.62 m has been reached. Figure 5 represents the values obtained with (2).

For the long range antenna the measurements were taken with an 0.5 m step until the maximum of 15 m. Figure 6 represents the measurements taken up to 5 m. Figure 7 represents the values obtained with (2) within the long range scenario.

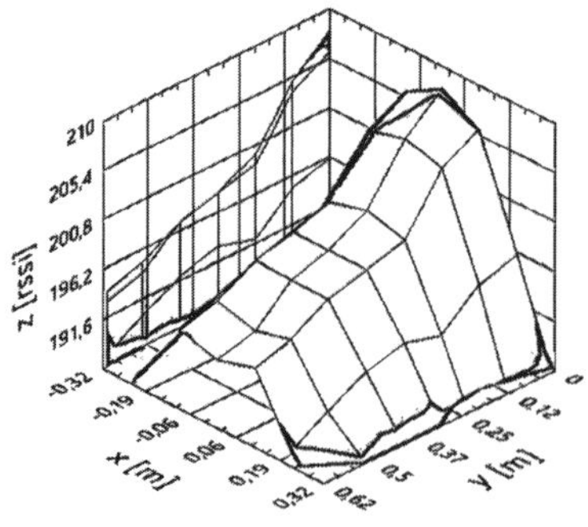

Figure 4. The real rssi surface of the short RFID antenna type 1.

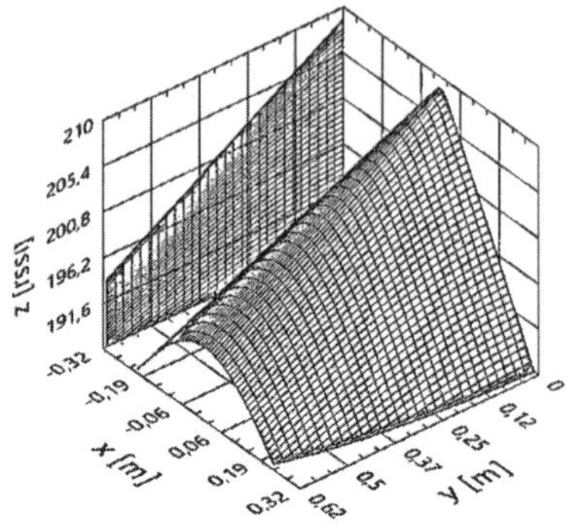

Figure 5. The theoretical rssi surface of the short RFID antenna type 1.

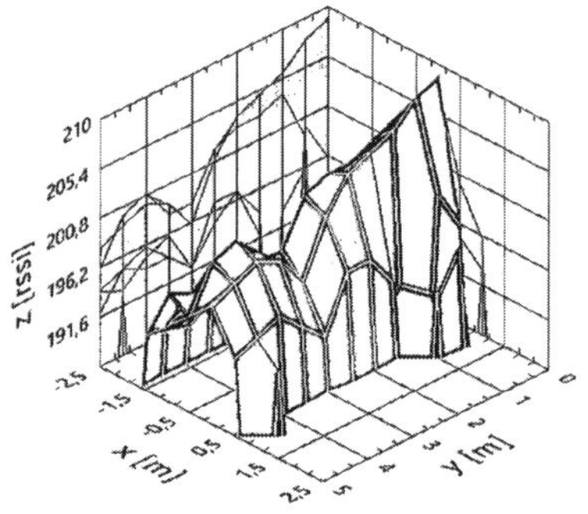

Figure 6. The real rssi surface of the long-range RFID antenna type 2.

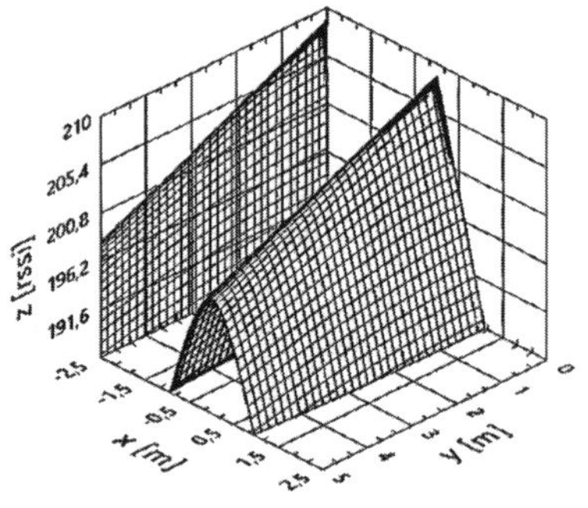

Figure 7. The theoretical rssi surface of the long-range RFID antenna type 2.

Figures 5 and 7 were obtained using the antenna parameters presented in Table 1.

The mathematical rssi surface of the two antennas is used to estimate the location of passive tags, when two identical antennas are placed side by side, in front of the passive RFID tags (as presented in Figure 8).

Table 1. Parameters of the two different RFID antennas.

z parameter	Antenna Type 1 (Short range)	Antenna Type 2 (Long range)
α	210 rssi	rssi
β	195 rssi	198 rssi
L	0.56 m	5 m
γ	75	20
x_0	0 m and 0.08 m	0 m and 0.5 m

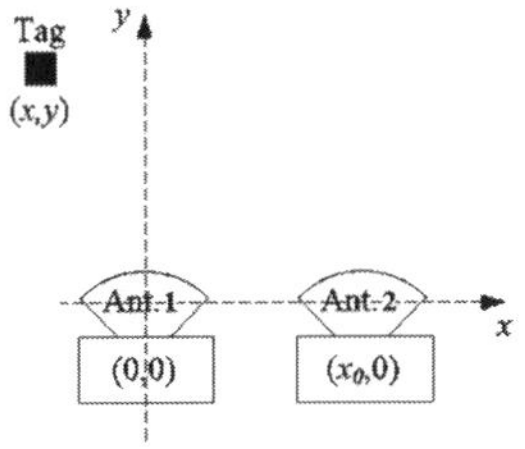

Figure 8. Scenario of two antennas side by side.

The next section shows an intersection method to compute the tag location.

3.3 *Rssi intersection curves for tag location*

By using (2) it is possible to discover the intersection of both curves y (of two antennas placed side by side) at a distance of x_0. The first antenna, located at coordinate (0, 0) with an rssi equal to z_1, will be equal to:

$$y = \frac{L}{\alpha - \beta}\sqrt{(\alpha - z_1)^2 - {\gamma_1}^2 x^2} \quad (3)$$

The second antenna, located at coordinate (x_0, 0) with an rssi equal to z_2, will be as follows:

$$y = \frac{L}{\alpha - \beta}\sqrt{(\alpha - z_2)^2 - \gamma^2 (x - x_0)^2} \quad (4)$$

The intersection of the two previous curves gives us:

$$x = \frac{z_1(z_1 - 2\alpha) + z_2(2\alpha - z_2) + \gamma^2 x^2}{2\gamma^2 x_0} \quad (5)$$

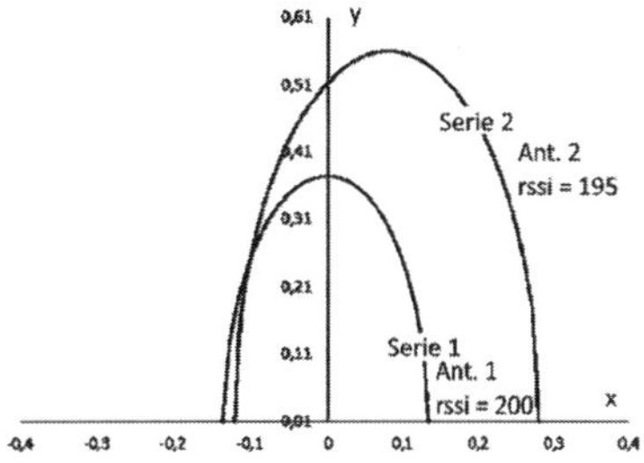

Figure 9. The intersection method of two rssi curves (for same antenna type).

To obtain the y coordinate of the previous intersection we replace (5) in (3) or (4).

Figure 9, "Serie 1" is the rssi curve of Antenna 1 (any type) located at (0, 0), when rssi z_1= 200. "Serie 2" is the rssi curve of Antenna 1 located at (0, 0.08), when rssi z_2= 195. The tag location is (-0.10, 0.25) in metres, within this specific scenario, as shown by the intersection of both curves in Figure 9.

After defining the mathematical model for the RFID IPS location, we tested and validated it, using two antenna types presented in Table 1. Results are discussed in the following section.

4 EXPERIMENTAL IPS RESULTS

Table 2 presents the RFID IPS results when two Antennas of Type 1 are placed side by side at a distance of 0.08 m. The coordinates (x', y') are the estimated values of (5) and (3). The real tag location is (x, y). The distance between the estimated point and the real tag location is the error value. The RFID IPS error average, obtained with 5 readings, is 0.2 metres when two Antennas type 1 are used.

Table 3 presents the RFID IPS results when two Antennas type 2 are placed side by side at a distance of half a metre. Here, the second RFID IPS error average, obtained with 5 readings, is 0.5 metres when two Antennas type 2 are used.

Table 2. RFID IPS results with antenna type 1.

Tag reading	Ant. 1 at (0, 0)	Ant. 1 at (0.08, 0)	x'	y'	X real	Y real	Error
1	198	193	-0.1	0.3	0.1	0.3	0.0
2	197	192	-0.1	0.3	0.8	0.4	0.7
3	196	200	0.1	0.4	0.1	0.2	0.1
4	196	204	0.1	0.2	0.1	0.2	0.0
5	194	195	0.1	0.6	0.0	0.5	0.1
RFID IPS error average [m] = 0.2							

Table 3. RFID IPS results with antenna type 2.

Tag reading	Ant. 2 at (0, 0)	Ant. 2 at (0.5, 0)	x'	y'	X real	Y real	Error
1	201	202	0.3	0.3	0.1	2.5	0.3
2	198	196	0.1	0.3	0.8	5.0	0.2
3	194	199	0.6	0.4	0.1	3.5	1.0
4	197	204	0.5	0.2	0.1	2.0	0.2
5	201	195	-0.1	0.6	0.0	4.0	0.6
RFID IPS error average [m] = 0.5							

Each antenna type 2 can read RFID tags up to 15 metres. However, we performed our experimental tests up to a maximum of 5 metres.

5 CONCLUSIONS

A long-range RFID indoor positioning system with passive tags has been designed, implemented and experimented with promising results. The solution will be further tested in a harsher industrial environment to enable the location of any object in the context of the Industrial Internet of Things. The detection of passive RFID tags is currently performed up to a range of 15 metres with a small error precision that can turn the IIoT a full reality.

In the future we will experiment the long-range antenna between 5 and 15 metres and we will place the IPS system prototype on an industrial crane that can cover a large industrial zone.

ACKNOWLEDGEMENT

The authors would like to thank the Polytechnic Institute of Leiria, School of Technology and Management of Leiria, the Centre for research in Informatics and Communications, and the "Instituto de Telecomunicações" of Portugal, and INESC-Coimbra.

This work is partially financed by Portuguese funds through FCT – "Fundação para a Ciência e a Tecnologia", I.P., under the project grants UID/CEC/04524/2019 and UID/EEA/50008/2019.

Additionally, this works is partially supported by the Tooling4G project (24516), funded by the Compete 2020 program, within the context of the PORTUGAL 2020 program through FEDER.

REFERENCES

Accenture (2015) *Winning with the Industrial Internet of Things*. Available at: https://www.accenture.com/t00010101t000000z__w__/it-it/_acnmedia/pdf-5/accenture-industrial-internet-of-things-positioning-paper-report-2015.pdf.

Bagarić, J. (2016) *Indoor Positioning System for Mobile Devices using Radio Frequency and Perfect Sequences*. Polytechnic Institute of Leiria.

Bagarić, J., Ferreira, M., Pereira, J. S. & Mendes, S. P. (2015) 'On estimating indoor location using wireless communication between sensors', in *Conf. on Telecommunications - ConfTele 2015*.

Chafon (2019a) *UHF 12 dBi linear antenna for highway road*. Available at: http://www.chafon.com/productdetails.aspx?pid=590.

Chafon (2019b) *UHF RFID Reader Module*: Available at: http://www.chafon.com/productdetails.aspx?pid=565.

Ferreira, M., Bagaric, J., Lanza-Gutierrez, J., Mendes, S. P., Pereira, J. & Gomez-Pulid, J. (2015) 'On the Use of Perfect Sequences and Genetic Algorithms for Estimating the Indoor Location of Wireless Sensors', *International Journal of Distributed Sensor Networks*, 2015(Article ID 720574), 1–12.

Flores, D., Pereira, J. S. & Marcillo, D. (2017) '3D Localization System for an Unmanned Mini Quadcopter based on Smart Indoor Wi-Fi Antennas', in *WorldCIST 2017*.

IndoorAtlas USA Inc. (2019) *IndoorAtlas*. Available at: https://www.indooratlas.com/(Accessed: 19 March 2019).

Lanza-Gutierrez, J., Gomez-Pulido, J., Mendes, S. P., Ferreira, M. & Pereira, J. (2015) 'Planning the Deployment of Indoor Wireless Sensor Networks Through Multiobjective Evolutionary Techniques', in *Proc EvoStar European Conf. on the Applications of Evolutionary Computation - EvoApplications*. Copenhagen, Denmark, 128–139.

Lazaro, A., Girbau, D. & Villarino, R. (2009) 'Effects of Interferences in Uhf Rfid Systems', in *Prog. Electromagn. Res.*, 425–443.

Nextome SRL (2019) *Nextome*. Available at: https://www.nextome.net/en/indoor-positioning-technology.php (Accessed: 19 March 2019).

Pereira, J. (2015) *Sequências perfeitas para sistemas de comunicação*. Edited by Novas Edições Acadêmicas. Saarbrücken.

Pereira, J., Pujari, P., Manjunath, G. & Gasparovic, M (2016) 'Standing Wave Cancellation and Shadow Zone Reducing Wireless Transmitter, System and Respective Method and Uses'. Pending patent, Portugal.

Pereira, J., Ferreira, M. P. M. & Gasparovic, M. (2017) 'Indoor Positioning System and Method'. Pending patent, Portugal.

Pereira, J., Mendes, S. & Bagaric, J. (2015) 'Standing Wave Cancellation – Wireless Transmitter, Receiver, System and Respective Method'. Pending patent, Portugal.

Pozyx Labs (2019) *Pozyx*. Available at: http://pozyx.io/ (Accessed: 19 March 2019).

Rcsoft (2019) *Rcsoft*. Available at: https://www.rcsoft.pt/ (Accessed: 19 March 2019).

Ultra-thin Digital (2019) *Indoor Antenna TV HDTV Antenna High Signal Capture Cable Signal Amplifier Antenna*. Available at: https://www.aliexpress.com.

View (2019) *View*. Available at: https://www.innovationleader.com/why-tools-giant-stanley-formed-joint-venture-big-data-iot/(Accessed: 19 March 2019).

Wifarer (2019) *Wifarer*. Available at: http://www.wifarer.com/(Accessed: 19 March 2019).

Wu, J. (2012) 'Three-Dimensional Indoor RFID Localization System', 199.

Xiaolin, J. (2012) 'RFID technology and its applications in Internet of Things (IoT)', in *2012 2nd International Conference on Consumer Electronics, Communications and Networks (CECNet)*. China, 1282–1286.

Additive manufacturing and virtual prototyping

Industry 4.0 – Shaping The Future of The Digital World – da Silva Bartolo et al. (eds)
© 2021 Taylor & Francis Group, London, ISBN 978-0-367-42272-1

3D printing for sustainable construction

Y.W.D. Tay, B.N. Panda, G.H.A. Ting, N.M.N. Ahamed, M.J. Tan & C.K. Chua
Singapore Centre for 3D Printing (SC3DP), School of Mechanical and Aerospace Engineering, Nanyang Technological University (NTU), Singapore

ABSTRACT: Sustainability is interpreted as the effective use of resources as well as the preservation of the environment. A sustainable building is giving back to the environment more than it takes and ensuring that the resources is being used in an effective way that would benefit the community. A combination of smart design, efficient technology and designing buildings with sustainability in mind from the start of the designing phase is therefore necessary. 3D concrete printing can be a sustainable solution because of its ability to manufacture complex shapes to enable passive design thus reducing energy consumption. This article presents an overview on the sustainability of 3D concrete printing in terms of printable green materials as well as sustainable architectural design to achieve passive system.

1 INTRODUCTION

3D concrete printing, also known as additive manufacturing in the building industry, is expected to be a true game-changer in Industry 4.0. The potential of this technology when it reaches maturity can revolutionize the construction market and make major changes such as shorter building time, cheaper construction, freedom of shape and integration of functionality (Kothman & Faber 2016). In addition, the correct use of green materials and selection of sustainable complex architectural design can amplify the sustainability of this technology. s

1.1 *3D concrete printing and printable material*

3D concrete printing has two main different techniques to create complex structure: Binder jetting and material deposition method (Tay et al. 2019a). Both of these methods create a complex structure by adding small layers of material over the previous layer. However, the latter technique seems to be more favourable in the field of research in terms of publication (Tay et al. 2019a). Its paradoxical rheological property is one of the challenges to be addressed before successful printing can be possible.

The rheological performance of a printable concrete material is different from the conventional casted material. For material to be printable, it has to be flowable so that it can be delivered to the nozzle by a pump (Roussel 2018). Furthermore, after extrusion, it has to be stiff enough to hold its shape and the weight of the subsequent layers (Tay et al. 2019a). Whereas, in the conventional casting technique, the formwork will hold the fresh concrete in place. Lastly, the interlayer bonding between the layers, which is a distinct feature compared with conventional casted concrete, should be strong enough to sustain the structures. These interfaces between layers tend to be the weakness of the whole structure (Tay et al. 2019b).

1.2 *Sustainable construction*

Sustainable construction is the aim to meet present-day needs for infrastructure, housing and working environments without compromising the ability of future generations to meet their own needs in times to come. This means ensuring that resources are being used in an efficient way that would benefit the community and the world.

Several sustainable materials that have a low carbon footprint can be used for 3D printing. Fly ash, geopolymer and recycled glass are some of the green material that has been used by the industry (Panda et al. 2018). Although these sustainable materials have been used in conventional casting methods, the rheological behaviour for printing is different. The mixtures have to be tailored to this new manufacturing process for printing to be successful. Apart from printing sustainable materials, printing passive design maximizes the potential of 3D printing to create comfortable space for the users. A passive system is a combination of energy-efficient design to take advantage of the climate to maintain the comfort level in an infrastructure. Such an approach reduces energy consumption during operation. However, the passive design has to be implemented during the design phase.

Sustainable construction is a broad term and can involve different types of issues (Tan et al. 2011). The focus of this paper is the sustainability of 3D concrete printing in terms of construction material and architectural design to improve the way people build and live in the building sector.

2 SUSTAINABLE PRINTABLE MATERIALS

The materials presented below are the green materials currently under research at SC3DP, NTU, Singapore. The usage of these materials is considered sustainable since dumping to a landfill will cause a negative environmental impact. These researches revolve around investigating a suitable mixture ratio to fulfil the required behaviour for printing.

2.1 *High volume fly ash concrete*

Fly ash-based materials are one of the possible alternatives for printing sustainable concrete structures. Fly ash is a by-product from the coal industry and is considered as a waste product. It contains some toxic metals that will degrade the soil and will cause air pollution. As such, a research carried out SC3DP, NTU offers suitable high-volume fly ash based formulation for 3D printing application which can reduce the environmental impact instead of disposing them to an open environment (Panda et al. 2018). Hence, a high volume of fly ash was incorporated in the formulation and it was found to improve long term strength performance of the building materials.

The rheological properties of the printable material in its fresh state is crucial. High thixotropy behaviour allows the material to become less viscous when stress is applied and allow it to return to its more viscous state when at rest. It can be measured in terms of structural breakdown and recovery. A shear-thinning protocol was used to calculate a structural index, λ, by shearing the mortar at 300 s^{-1} for 300 seconds. In general, the higher the λ, the higher the thixotropy. The result is shown in Figure 1a. It can be seen that the structural index of the material increased with resting time. This phenomenon was linked to both the physical interaction of the particles and the hardening process (Panda et al. 2019).

Additionally, the viscosity recovery was measured to investigate the fresh property of the mortar after the extrusion process. If the original viscosity of the initially deposited layer has not recovered before the deposition of the second layer, it may cause deformation in the printed structure. the protocol for this second test is as follows: (i) $0.01 s^{-1}$ for 60 seconds; (ii) $300 s^{-1}$ for 30 seconds and; (iii) $0.01 s^{-1}$ for 60 seconds. These three different intervals correspond to the material state (i) initially at rest; (ii) extrusion and (iii) at rest after extrusion. The result is shown in Figure 1b. Almost 80% of fly ash was utilized to formulate the mix design that exhibits excellent thixotropic behaviour and achieving 35Mpa mechanical compressive strength, which is suitable for non-structural application (Biranchi Panda et al. 2019).

Furthermore, this mixture is used to print a modular toilet shown in Figure 2a. The toilet was printed in three parts and later assembled on site. Comparing to the conventional method of creating concrete structures, 3D concrete printing can save the material wastage, production time, cost and ultimately fetch sustainability in our built environment.

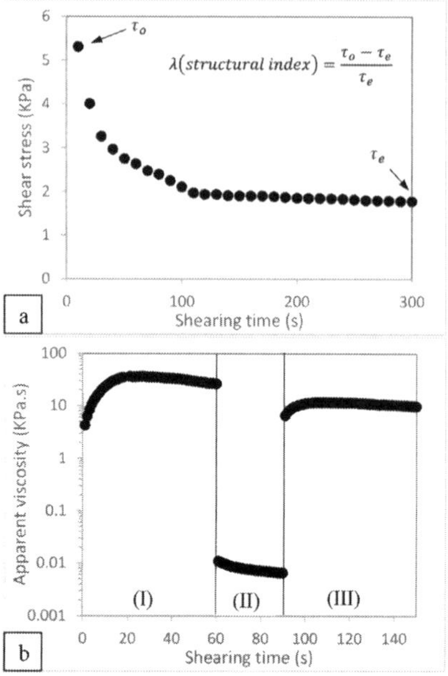

Figure 1. Thixotropy behaviour of 3D printed concrete (a) structural breakdown (b) structural recovery.

2.2 *Geopolymers*

3D printing with high volume fly ash is a challenge as the strength development in an early age is not quick enough to support subsequent layers. To avoid this problem, fly ash was activated with an alkali solution according to geopolymerization mechanism. Figure 2b shows an example 3D printing of geopolymer mortar extruded through a rectangular orifice of a 4-axis gantry printer. To enhance the reaction process, 5-10% slag, which is also regarded as one of the by-products of steel power plant industries, was used.

2.3 *Recycled glass aggregates*

Despite the abundant supply of sand from the desert or the seabed, the world is facing a shortage of construction sand. This is mainly due to the nature of the desert sand and sea sand are not suitable for applications in construction, thus, only river sands are used in the construction industry. Hence, with the limited resource for construction sand, an alternative solution is needed to meet the increasing demand for construction materials. Singapore generates over 70,000 tons of glass waste annually, in which only less than 20% is being recycled (National Environment Agency 2018). The remaining waste glass is usually disposed of in landfill where it is

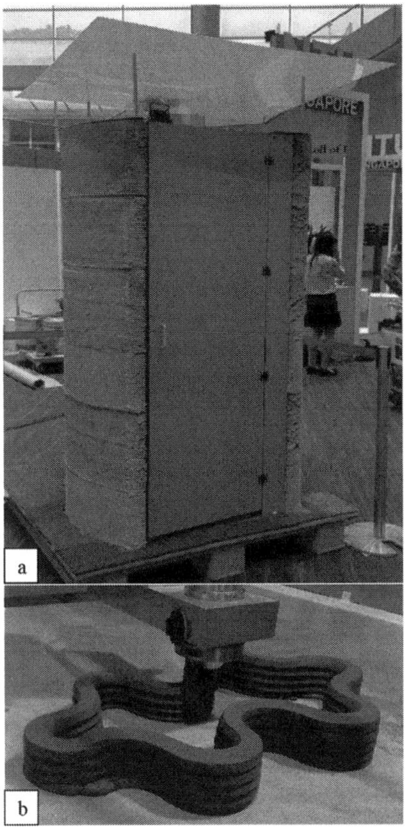

Figure 2. (a) 3D printing of modular toilet with high volume fly ash at SC3DP, NTU. (b) 3D printing of geopolymer.

not suitable due to its non-biodegradable nature. The study of replacing river sand by the recycled glass in cementitious materials has been established for decades.

At SC3DP, NTU the research on using recycled glass aggregates for 3D concrete printing focuses on the formulation of mix design and the effect of recycled glass aggregate gradation on printability (Ting et al. 2018). The comparison of the river sand and recycled glass aggregates was studied to evaluate the printability performance of the materials. The gradation study of the recycled glass particles was also conducted to optimize the material performance for 3D concrete printing application. Furthermore, the alkali-silica reaction which is a commonly known issue in the recycled glass aggregates is currently being investigated.

3 SUSTAINABLE ARCHITECTURAL DESIGN

Construction sustainability does not necessarily relate only to the selection of green materials. The energy used in the fabrication process, and the energy used in its operation after construction is equally important. The implementation of passive designs can dramatically reduce energy consumption and is an area where 3D concrete printing can potentially make a significant contribution because of the ease of creating a complex structure.

3.1 *Management of passive cooling or heating effect*

In warm climates, maximizing the use of natural airflow can contribute significantly to the energy load of the building. Excessive usage of air conditioning due to poor cooling efficiency design is not desirable. Clever designs such as “Cool Brick” can help in passive cooling (Fratello & Rael 2015) as shown in Figure 3a. This use of natural cooling effect must be incorporated in the design stage. In cold climates, the same principle should be applied in designing the insulation that could retain and recirculate the heat with minimum energy used.

The complexity of the shapes that can be printed by 3D printing is endless. The modular wall shown in Figure 3b is made up of modular blocks. These modular blocks can allow passive cooling and maximize the use of natural airflow while providing shade from the sunlight. Furthermore, such a design improves the appearance of the wall.

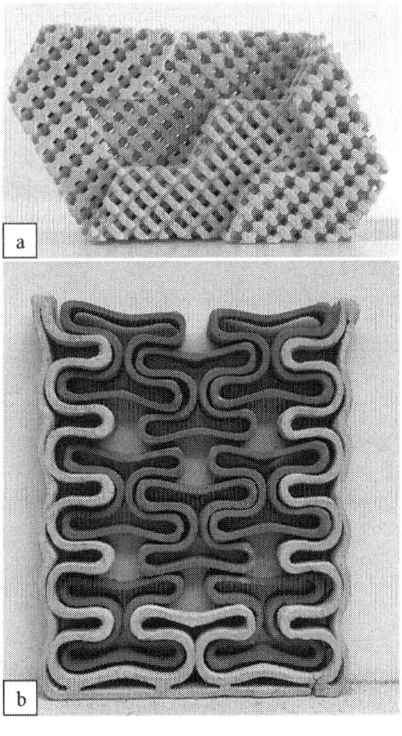

Figure 3. (a) Brick that provides passive cooling made by additive manufacturing (Fratello et al., 2015) (b) 3D printed modular wall printed in SC3DP, NTU.

3.2 *Management of natural lighting effect*

Management of natural lighting is another important aspect. While the type of material and building orientation can achieve natural lighting, using additive manufacturing can control the lighting permeability while improving the appearance of the wall. Small modular blocks can be printed to control the amount of light allowed to be pass through as shown in Figure 4 (Stott 2015). Indoor spaces that are naturally brightened can reduce the need for artificial lighting at a later stage. Such designs of using natural lighting need to be considered in the early stages of the designing process.

Extrusion over a doubly curved surface can be achieved with 3D printing without the need for formworks. Research work carried out in SC3DP, NTU printed the framework and was later assembled to form the curve bed for printing. The wireframe is then covered with a flexible textile to provide a surface for printing as shown in Figure 5. This printing serves as a proof of concept to build larger modular curved facades and complex architectural shape which can then be used for controlling the natural light.

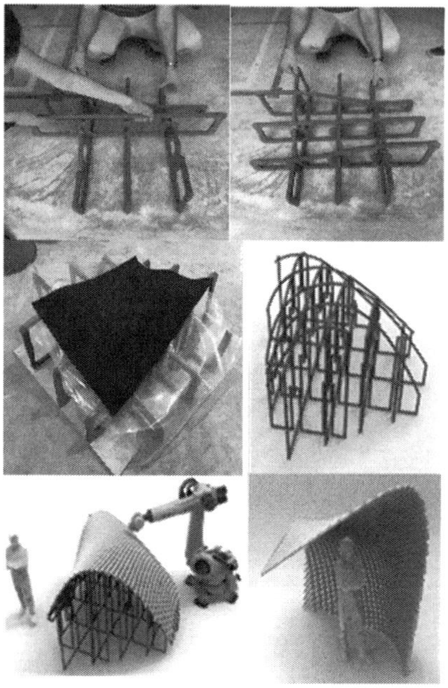

Figure 5. Curved bed printing with printed frames printed in SC3DP, NTU.

3.3 *Management of acoustic effect*

Dampening the sound surrounding the building or within the premises can improve the comfort of the user. Using passive designs can reduce the need for additional material for soundproofing. Figure 6 shows an element that was designed to enhance the soundproofing capability of a wall (Gosselin et al. 2016). The generic elements were stacked together to form a complete wall. The geometries of the holes dampen the acoustic waves passing through.

Figure 6. Concrete printed acoustic wall element (Valente et al. 2019).

Figure 4. Structure printed in modular that have different light permeability (Stott 2015).

3.4 *3D printing optimized concrete structure*

Beyond economical and architectural benefits, 3D concrete printing can be used to reduce the environmental footprint of the industry. The increased level of control offered by 3DCP enables the use of the advanced computational algorithm to reduce the density of the structure by creating a lattice structure. These optimized structures (See Figure 7) not only reduce the overall weight but also improve the use of resources effectively. In addition, it can incorporate structural members

Figure 7. 3D printed lattice shape wall element printed in SC3DP, NTU.

such as rebar, pre-stressing cable to produce structural concrete (Lim et al. 2018). 3D printed pedestrian bridge (Salet et al. 2018) is one of the recent examples, where the printed structure was post-tensioned after assembling the individual 3D printed sections.

It is worthy to note that, the traditional optimization software only considers isotropic material properties during the analysis, however for 3D printing projects the part design needs to be finalized whilst taking the properties and limitations of a 3D concrete printer and the material properties into account.

4 CONCLUSIONS

3D concrete printing is a promising technology to address the sustainability challenges in today's construction industry and open up new opportunities for design possibilities. The future of construction is most likely to be an integrated process that allows the organization to take advantage of both conventional and additive manufacturing technologies.

With rapid urbanization in many developing countries, there is an urgent need to come up with clever ideas that optimize the sustainable performance of the buildings that we live and work in.

Construction firms today need to recognize that sustainable construction is becoming a greater concern. Builders who invest in the latest sustainable technologies in the construction process can recoup those costs over time in the form of decreased building operation costs as a result of greater energy efficiency. Regulators also play a significant role in sustainable construction by creating the right incentives for companies that choose to build sustainably. Finally, the government can legislate and create mandates that require firms to build in a sustainable way.

ACKNOWLEDGEMENT

This research is supported by the National Research Foundation, Prime Minister's Office, Singapore under its Medium-Sized Centre funding scheme, Singapore Centre for 3D Printing, and Sembcorp Design & Construction Pte Ltd.

REFERENCES

Fratello, V. S., & Rael, R. (2015). *Cool brick* [Online]. Emerging Objects. Available at: http://www.emergingobjects.com/2015/03/07/cool-brick/(Accessed: 31 January 2019).

Valente, M., Sibai, A., & Sambucci, M. (2019). Extrusion-Based Additive Manufacturing of Concrete Products: Revolutionizing and Remodeling the Construction Industry. *J. Compos. Sci. 3*, 88.

Kothman, I., & Faber, N. (2016). How 3D printing technology changes the rules of the game: Insights from the construction sector. *J. Manuf. Technol. Manag. 27*, 932–943.

Lim, J. H., Panda, B., & Pham, Q.-C. (2018). Improving flexural characteristics of 3D printed geopolymer composites with in-process steel cable reinforcement. *Constr. Build. Mater. 178*, 32–41.

National Environment Agency. (2018). *Waste management statistics and overall recycling measures* [Online]. Available at: https://www.nea.gov.sg/our-services/waste-management/waste-statistics-and-overall-recycling (Accessed: 31 January 2019)

Panda, B., Tay, Y. W. D., Paul, S. C., & Tan, M. J. (2018). Current challenges and future potential of 3D concrete printing. *Mater. Sci. Eng. Technol. 49*, 666–673.

Panda, Biranchi, Ruan, S., Unluer, C., & Tan, M. J. (2019). Improving the 3D printability of high volume fly ash mixtures via the use of nano attapulgite clay. *Compos. Part B Eng. 165*, 75–83.

Roussel, N. (2018). Rheological requirements for printable concretes. *Cem. Concr. Res. 112*, 76–85.

Salet, T. A. M., Ahmed, Z. Y., Bos, F. P., & Laagland, H. L. M. (2018). Design of a 3D printed concrete bridge by testing. *Virtual Phys. Prototyp. 13*, 222–236.

Stott, R. (2015). *Emerging objects creates "Bloom" pavilion from 3D printed cement* [Online]. Arch Daily. Available at: https://www.archdaily.com/613171/emerging-objects-creates-bloom-pavilion-from-3-d-printed-cement (Accessed: 31 January 2019).

Tan, Y., Shen, L., & Yao, H. (2011). Sustainable construction practice and contractors' competitiveness: A preliminary study. *Habitat Int. 35*, 225–230.

Tay, Y. W. D., Li, M. Y., & Tan, M. J. (2019a). Effect of printing parameters in 3D concrete printing: Printing region and support structures. *J. Mater. Process. Technol. 271*, 261–270.

Tay, Y. W. D., Qian, Y., & Tan, M. J. (2019b). Printability region for 3D concrete printing using slump and slump flow test. *Compos. Part B Eng. 174*, 106968.

Ting, G. H. A., Tay, Y. W. D., Annapareddy, A., Li, M., & Tan, M. J. (2018). Effect of recycled glass gradation in 3D cementitious material printing. *Proc. 3rd Int. Conf. Prog. Addit. Manuf. (Pro-AM 2018)* 50–55.

Industry 4.0 – Shaping The Future of The Digital World – da Silva Bartolo et al. (eds)
© 2021 Taylor & Francis Group, London, ISBN 978-0-367-42272-1

Galvanometer calibration using image processing for additive manufacturing

B. King, A. Rennie, J. Taylor & S. Walsh
Lancaster University, Lancaster, UK

G. Bennett
Euriscus Ltd., London, UK

ABSTRACT: The process of calibrating a laser scanner system in additive manufacturing applications can be expensive and lengthy if the standard build materials and powders are used. A method is presented to calibrate the process parameters for a laser scanner. Patterns scanned on thermally reactive paper are analysed using image processing to ascertain the optimal parameter value. Samples were tested at varying mark speed and analysed using an image-processing algorithm. The ideal process values were obtained and an approximation curve was created to predict the delay parameters any mark speed. The data contained in the approximation curve was verified by selecting an alternative mark speed and then retesting using the image processing technique found that delay parameters can accurately be predicted within 20μs.

1 INTRODUCTION

Laser-based Additive Manufacturing (AM) technologies, including: photopolymerisation processes, where resins are laser cured to produce solid objects, and some Powder Bed Fusion (PBF) methods, such as, Selective Laser Sintering (SLS) and Selective Laser Melting (SLM) (Brandt, 2017) where polymer or metal powders are sintered or melted using the application of a laser (Gibson et al. 2010), all utilise galvanometer scanners to drive the toolpath. The laser is mounted above the build bed is directed onto a position on the build surface using a galvanometer scanner which reflects the laser beam downwards in addition to providing movement in the X and Y directions, Figure 1.

Galvanometer scanners are opto-mechatronic systems that offer a high accuracy solution for laser beam positioning in a wide range of applications (Yoo et al, 2016). In cases where actuation is required in Two Dimensions (2D), two mirrors are employed, the rotation of each mirror referring to a movement in the X or Y direction. The rotational movement of the mirrors is actuated by servo motors, allowing for an extremely quick response to a step input, and very accurate positioning (Mnerie et al. 2013).

Galvanometer scanners are not perfect systems, due to the processing time required between issuing and execution of that command. There also exists inherent mechanical faults that come from using servomotors to operate within a 2D space; there can exist a degree of "bow-tie" distortion caused by the limited 180° range of motion of using a combination of reflective mirrors. Secondly, due to the requirement for rapid movement of the mirrors, there is a period of oscillation of the servomotor as it comes to rest at its set point, which can be seen on the resultant marked vector.

The Laser Scanner process variables (Matsuka et al, 2015) include:

- Mark speed (m/s), the speed at which the laser beam travels while switched on, determined by the energy density required to alter the state of the material, laser beam diameter and laser power;
- Jump speed (m/s), the speed at which the predicted position of the laser beam travels while switched off between marking vectors;
- Tracking error (μs), the delay between the scanner being issued with a command and it's execution;
- Mark delay (μs), a delay between finishing a jump command and executing a mark command;
- Jump delay (μs), The delay between finishing the jump command and executing a sintering vector;
- Polygon delay (μs), a delay between two marking vectors.

In laser-based AM processes, the standard calibration of the laser system consists of producing and analysing parts for characteristics of a suboptimal parameter (Ha et al. 2015; Wang 1999; Senthilkumaran et al, 2009). This process is subject to inaccuracies as there is an inability to maintain consistency in the thermal environment necessary for laser-based AM. Further to this, producing parts

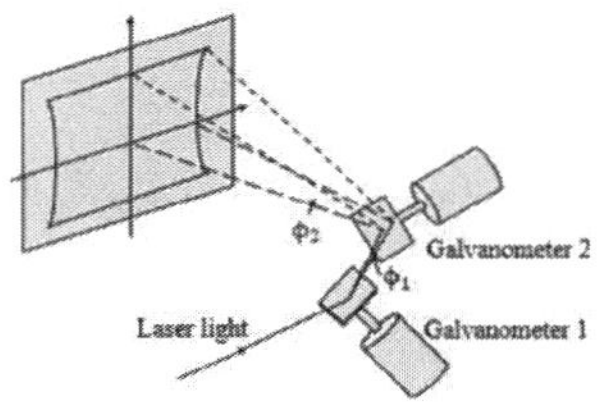

Figure 1. Galvanometer Scanner (Nguyen, 2018).

is lengthy, expensive and wasteful, with respect to the unnecessary use of materials and the time taken to run a build. The need for a calibration method that will eliminate the need to produce parts was developed accordingly.

The objectives of this paper was, firstly, to introduce and verify a method of calibrating process parameters for galvanometer scanners in AM, without utilising any build materials. Secondly the development of a data model capable of predicting process parameters, that can correctly predict process parameters at untested values.

2 METHODOLOGY

To calibrate the laser scanner, thermally reactive paper was identified as a method to record the movement of the laser beam on the build bed. When Infrared radiation from the laser is directed onto the sheet, the paper colours, due to the reaction of Bisphenol-A in the coating of the paper (Geens et al. 2012). The laser scanner is programmed to run test patterns directed onto the thermal paper and is subsequently analysed using image processing techniques.

2.1 *Experimental set up*

A Raylase AxialScan 30 galvanometer scanner controlled by and SP-ICE3 controller card (Raylase, 2019) was used in conjunction with a Coherent Diamond C70 CO_2 laser (Coherent, 2009), operating at 25 kHz with a maximum power output of 70W, to mark the test patterns on the thermal paper. The laser scanner unit was mounted in a test rig with an adjustable platform, at a Z height of between 250-750mm above the base of the rig where the thermal paper was secured for scanning, to simulate the setup of a range of AM machines, Figure 2.

The control data for the laser scanner was sent through an Ethernet connection to the SP-ICE3 controller card, which controlled the position of the laser beam through the scanner and the PWM output signal to the scanner. The remaining features including power supplies, laser chiller and safety features were managed by an Allan Bradley PLC (Rockwell Automation, 2013).

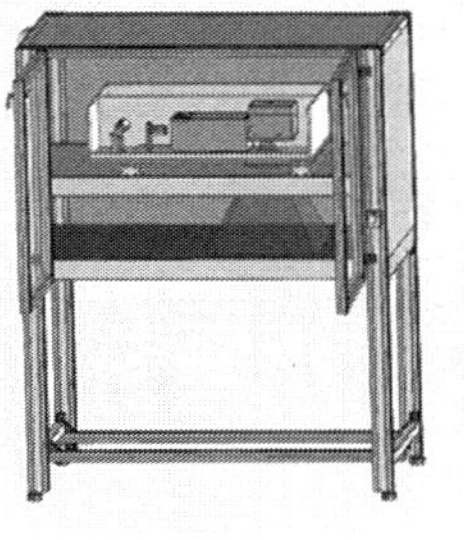

Figure 2. CAD rendering of laser scanner test rig.

2.2 *Test patterns*

The effect of incorrect delay values must first be defined before a calibration method can be ascertained. When mark delay is badly adjusted, there is a reduced length and an oscillatory quality at the start of a marking vector due to the settling period of the servomotors that drive the galvanometer mirrors, Figure 3a. When jump delay is too short, the resultant vectors overextend beyond their intended position when transitioning to a jump vector, Figure 3b. Lastly, Figure 3c shows the curved quality of adjoining vectors when poly delay is too short, when any delay parameter is too long a burned mark is present at the area of effect.

Three test patterns were designed to assess the values of jump, polygon and mark delay. Figure 4 shows the test patterns selected, with solid lines, representing marking vectors and dashed lines indicate a jump vectors. Each test pattern was selected to test for each of the parameters to be calibrated: pattern a) consists of two vectors at 90° which tests for jump delay; test pattern b) involves two connected vectors at a 45°, which can show the pronounced effect of a misaligned mark delay; and test pattern c) a series of long and shorter vectors marking an S pattern, which can show the curved effect to too short poly delay or vector burnout in the corners when poly delay is set to be too long.

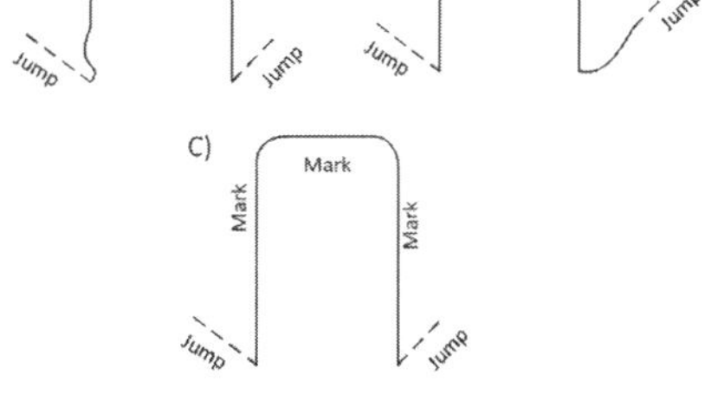

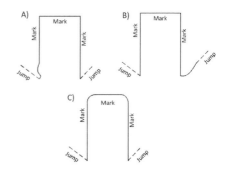

Figure 3. Effect of too short delay parameters.

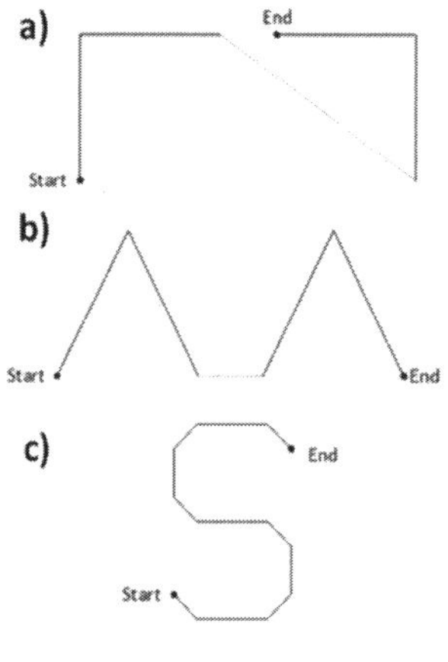

Figure 4. Parameter calibration test patterns.

2.3 *Scanning parameters*

The ideal delay parameters will be different depending on the mark and jump speeds, as there may be a call to vary the mark speed of any given AM system, due to energy density requirements of materials, it is suggested that calibration for a galvanometer scanner is run for a variety of mark speeds. The fastest scanning speeds recorded in SLS are 5m/s (Beard et al, 2010) and the slowest are recorded in SLM, as low as 0.005m/s (Yap & Sing, 2015). These were selected due to the range of experiment values; the aim was generation of an extrapolation curve that could then be used to determine any delay values between these mark speeds. Jump speed can remain consistent at any selected value, in this case a jump speed of 10m/s was chosen.

Prior to scanning the test patterns, the correct laser power had to be determined to mark the paper without burning it. A series of lines were scanned at each speed, increasing the pulse width slightly on each line. The scanned paper was then photo-scanned and analysed using MATLAB testing for an average pixel saturation value of the line of 95% black. The appropriate pulse width to mark the paper for each mark speed is shown in Table 1. Once the correct marking speed had been determined, all test patterns were marked using the combinations of speed and pulse width in Table 1.

Table 1. The pulse width to mark the thermal paper.

Mark speed (m/s)	Pulse Width (μs)	Laser Power (W)
0.005	0.02	0.035
0.1	0.9	0.158
0.5	1.7	2.975
1	2.3	4.025
1.5	3	5.250
2	3.5	6.125
3	5	8.75
4	6.7	11.725
5	8.4	14.700

To ascertain the correct delay parameter values using image processing, the test patterns in Figure 4 were scanned varying either jump, mark or polygon delay from 0 to 1000μs, then analysed for the optimum value using the technique described in Section 3, this process was repeated three times on separate sheets and an average value taken. A region of 200μs was then identified, ±100 of the previously identified value, for more precise testing; in this case, each of the test patterns was rescanned a further three times, varying the delay parameter by 20 μs. This scheme was repeated for all of the mark speeds, and the thermal paper was photo-scanned to a 600 x 600 dpi JPEG file.

3 IMAGE PROCESSING

On visual inspection of the scanned sheets, at lower speeds the implementation of any delay parameter has a negative impact, resulting in burnout at the start, end or join of vectors, as suggested by Raylase (2018). For this reason, it was decided that patterns scanned at a speed of 0.1m/s would be used as control patterns that are used to compare against scan patterns with uncelebrated delays. Figure 5 shows the test patterns scanned at 0.1m/s. The flowchart in Figure 6 shows the general procedure of the image-processing algorithm used to calibrate each parameter.

All photoscans were processed using the MATLAB image processing toolbox. The first stage is to load the image, and then convert it into a binary edge matrix, using the "*edge*" function, which returns a matrix of 1's and 0's, where a 1 represents the boundary of a region of black laser markings compared to the white background of the thermal paper. The data is stored in a series of 2D points, where each point is the location of a pixel on the edge boundary. Following this, the area in pixels of each identified region is calculated and the region rejected if it is under a minimum threshold value of 50 pixels, indicative of an outline that is too small to be a feature of a continuous test pattern. The plotted accepted region for test pattern B is shown in Figure 7.

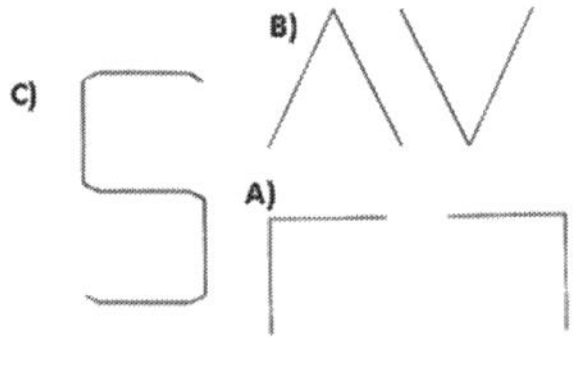

Figure 5. Test patterns A, B and C from Figure 3 scanned at 0.1m/s.

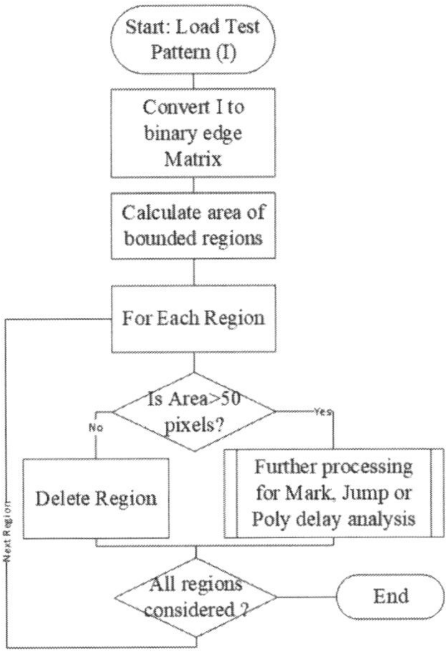

Figure 6. Image processing flowchart.

Figure 7. Highlighted accepted region for test pattern A.

3.1 *Jump delay*

To obtain the correct jump delay parameter, test pattern A in Figure 4 is considered. It is necessary to compute the number of vectors present in the scanned sample in comparison to the control sheet scanned at 0.1 m/s. The change in gradient of the perimeter of the scan path is calculated, and if the gradient is consistent over a certain threshold derived from analysis of the control sample, in this case 20 pixels, a vector is identified. The effect of jump delay on the inverted matrix is shown in Figure 8, where the typical tick

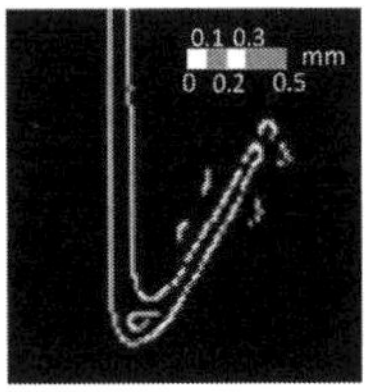

Figure 8. Inverted binary pixel matrix of too short mark delay, Test pattern A.

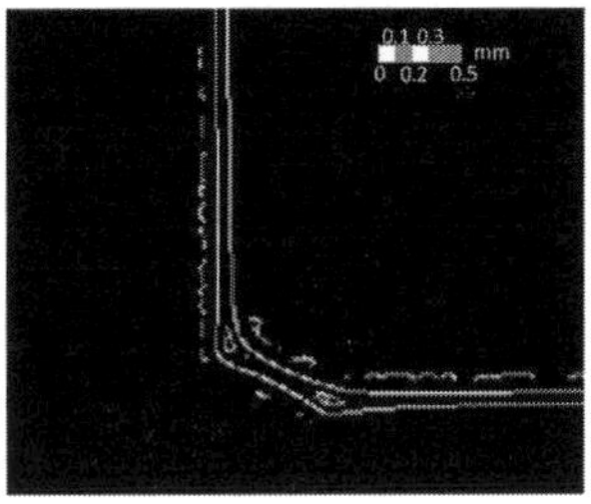

Figure 9. Inverted binary matrix of Test pattern C when Polygon delay parameter too long.

pattern at the end of a vector where marking delay is too short, would be recognised as a third vector in one right angled sample.

When any delay value is too long, the effect as shown in Figure 9 is witnessed where the scan path balloons at the start, end or join of vectors depending on which delay is being calibrated. To identify this, the transition between vectors is considered at the join between the two right-angled vectors should take 15 ±2 pixels. For this reason, if the scan path gradient has not transitioned from 1 to 0 within 17 pixels, then the burnout has occurred due an excessive delay parameter value.

3.2 *Mark delay*

To obtain the correct mark delay value, the length of the vectors has to be considered. When mark delay is too short, in addition to the oscillatory quality at the start of a marking vector after a jump command, the vector is also shorter than it needs to be. The location of the ends of the vectors in test pattern B can be compared to see if they are parallel. If they are aligned within 2 pixels, mark delay can be deemed to be perfectly calibrated. If mark delay is too short, the first vector after the jump command will be too short and if mark delay is too long, then the vector will be longer due to the balloon effect that is seen on the paper when the laser.

3.3 *Polygon delay*

To test for too long of a polygon delay, two methods can be employed: firstly, the nested region inside the test pattern can be identified, as corner burnout from the laser lingering too long on joins between vectors; secondly, the region boundary expands as thermal radiation spreads through the paper as the laser dwells, Figure 9. As the nested region inside the scan pattern is so small, it is rejected by the filtering portion of the algorithm. The increase in diameter of the perimeter boundary where vectors meet is the focus of the algorithm.

The inverted binary image in Figure 10 shows the effect of too short a Poly delay and when compared with the pattern in Figure 5, it can be seen that the

Figure 10. Inverted binary matrix of too short polygon delay, test pattern C.

shorter vectors have become elliptical. In this instance, the same vector identification method is used as described in Section 3.1. The number of vectors recorded in the scanned image will be much lower than the number present in the control sample.

4 RESULTS & ANALYSIS

Table 2 shows the ideal delay values derived from the image processing for each mark speed; the results are plotted in an approximation curve, using a standard regression model in Figure 11. In the case of this galvanometer scanner, mark delay is not necessary until mark speeds of almost 2m/s, but polygon and mark delays are necessary even at the lowest speeds. Jump delay requires the largest value due to the compounded effect of the mark speed and the jump speed together, however, if the jump speed were to decrease so would the delay value.

Table 2. Delay values at varying mark speeds.

Mark Speed (m/s)	Jump Delay (μs)	Mark Delay (μs)	Poly Delay (μs)
1	80	0	40
1.5	120	0	60
2	160	20	120
3	260	80	180
4	320	120	240
5	480	180	320

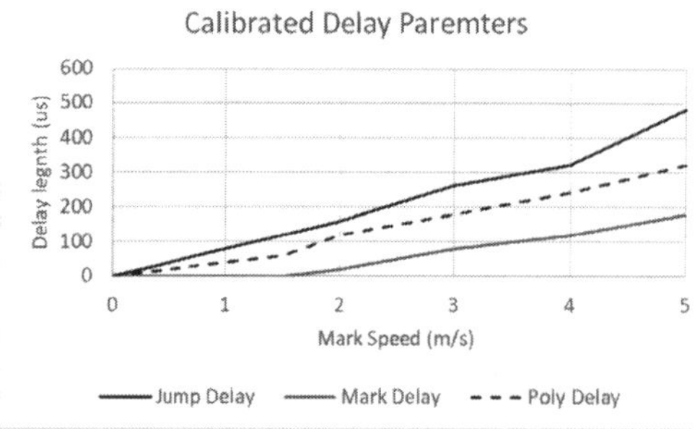

Figure 11. Approximation curve for delay values at varying mark speeds.

Figure 12. Correctly calibrated scan patterns at 3.5m/s.

To validate the calibration method, a further test was undertaken at a mark speed of 3.5m/s, with data read from the graph in Figure 12 to the nearest 20μs, with a mark delay of 100μs a jump delay of 300μs and a poly delay of 200μs. The result on the thermal paper was photo scanned and analysed. The scanned sheet can be seen in Figure 11 and met all conditions for accurately calibrated delay parameters, when reanalysed by the image processing software.

5 CONCLUSION

A galvanometer laser scanner delay parameter calibration method using image processing has been designed and tested using photoscans of laser marked thermal paper at different mark speeds with varied parameter values. Image processing techniques are used to find the 2D coordinates of the perimeter outline of the scan path. Using standard vector analysis, the value of the parameter can be deemed too short or too long. By repeating the experiments within a 100μs window, the delay parameters can be accurately calibrated within 20μs.

REFERENCES

Beard. M, Ghita. O, Evans. K. 2010. Evaluation of laser energy density effects on mechanical properties of selective laser sintering components using Raman spectroscopy, 21st Danube Adria Association For Automation & Manufacturing (DAAM) Conference, 10-23rd October 2010, Zadar: Croatia

Bradt. M. 2017. Laser additive manufacturing: materials, design, technologies and applications, Elsevier, 21–42,

Geens. T, Goeyens. L, Kannan. K, Neels. H & Covaci. A. 2012. Levels of bisphenol-A in thermal paper receipts from Belgium and estimation of human exposure, Science of the Total Environment, 435: 30–33

Gibson. I, Rosen. D.W & Stucker. B 2009, Additive Manufacturing Technologies: Rapid Prototyping to Direct Digital Manufacturing, Springer, 1–35

Ha. S, Han. H, Kwon. D, Kim. N, Hwang. C, Shin. H & Park. K. 2015. Systematic Dimensional Calibration Process for 3D Printed Parts in Selective Laser Sintering (SLS), ASME 2015 International Design Engineering Technical Conferences and Computers and Information in Engineering Conference, August 2–5 2015 Boston: USA

Matsuka. D, Fukushima. S & Iwaski. M. 2015. Compensation for torque fluctuation caused by temperature change

in, fast and precise positioning of galvanometer scanners, 2015 IEEE International Conference on Mechatronics (ICM), 642–647,
Mnerie.C, Preitl. S & Duma.V. 2013. Performance Enhancement of Galvanometer Scanners Using Extended Control Structures, 8th IEEE International Symposium on Applied Computational Intelligence and Informatics, May 23–25, 2013, Timisora: Romania
Ngyen., Dinh. V. L & Kim. H 2018. A High-Definition LIDAR System Based on Two-Mirror Deflection Scanners, IEEE Sensors Journal, 19: 559–568,
Raylase, AxialScan 20/30 datasheet, Raylase. 2018. Rockwell Automation: SLC 500 systems selection guide, Rockwell Automation.
Wang. X 1999, Calibration of shrinkage and beam offset in SLS process, Rapid Prototyping Journal, 5(3): 129–133,
Senthilkurmaran. K, Pandey. P.M & Rao. P.V.M 2009. Influence of building strategies on the accuracy of parts in selective laser sintering, Materials and Desingn, 30(8): 2946–2954,
Yap. C & Sing. S 2015, Review of selective laser melting: Materials and applications, Applied Physics Reviews, 2: 41101
Yoo. H.W, Ito. S, Schitter. G. 2016: High speed laser scanning microscopy by iterative learning control of a galvanometer scanner, Control Engineering Practice, 5: 12–21.

Industry 4.0 – Shaping The Future of The Digital World – da Silva Bartolo et al. (eds)
 ISBN 978-0-367-42272-1

A multi-material extrusion nozzle for functionally graded concrete printing

F. Craveiro & H. Bártolo
CIAUD, Lisbon School of Architecture, Universidade de Lisboa, Portugal
School of Technology and Management, Polytechnic of Leiriat, Portugal

J.P. Duarte
Stuckeman Center for Design Computing, Stuckeman School of Architecture and Landscape Architecture, Penn State University, USA

P.J. Bártolo
School of Mechanical, Aerospace and Civil Engineering, The University of Manchester, UK

ABSTRACT: 3D printing (3DP) is one of the breakthrough innovations for architecture, engineering and the construction industry. However, most 3DP technologies are limited to the design and fabrication of physical parts with homogeneous material properties, assuring structural safety but with no efficient use of material resources. A different strategy must be used to produce building components with varying material constituent or distribution. A Computational Fluid Dynamics (CFD) analysis was carried out to simulate the behavior of two 3D printing print-heads designed to combine two different materials, one composed by a simple "Y" shape, and the other integrating a dynamic mixing rod. The dynamic one presents a good mixing behavior, allowing to produce building components with locally optimized performance. Future work will include experimental tests and the analysis of new materials to compare simulated and experimental material behaviors.

1 INTRODUCTION

According to an independent expert report from the European Commission (European Commission 2019), 3D printing (or additive manufacturing, AM) of large objects is one of the major innovations for the future of construction. This technology will transform the construction sector, enabling producing complex geometrical shapes without formwork (Craveiro et al 2019b, Roussel 2018), providing a unique advantage over conventional construction, allowing architects and engineers to convert a 3D digital model into a physical part, by laying down successive layers of materials. In recent years, several types of printing systems (i.e. robots and cranes), as well printable materials, such as steel, clay, glass, polymers, geopolymers and concrete, were proposed (Tay et al. 2017). Different AM technologies (fusion, jetting or material extrusion) can be used to build each layer. Among these techniques, the extrusion of concrete through print-heads is the most widely used (Craveiro et al. 2019b).

On the other hand, most layered concrete elements, as well precast ones, are entirely produced with the same material, assuring structural safety, but not allowing an efficient use of resources and not tackling increasing materials costs (Río et al. 2015). Several strategies, such as topology optimization (Dillenburger et al. 2017) and functionally graded materials (FGM) (Craveiro et al 2018, Herrmann and Sobek 2017, Pietras and Sadowski 2019) can be used to enhance the material distribution, reducing the total mass of manufactured elements, this way minimizing concrete usage.

FGM allows changing properties over the volume, thereby enabling the production of more efficient composite materials with multifunctional properties. Functionally graded components can be obtained by either varying porosity or changing the volume fraction of two or more materials. They can be manufactured with stepwise changes of material properties (i.e. a different volume fraction for each new layer) or by continuously change the properties of the material.

Thus, the production of concrete elements with a continuous gradient using extrusion technologies, requires the development of print-heads capable of constantly varying the mixing ratio of different materials.

2 PRINT-HEADS

A print-head or extrusion head is the element of a 3D printing machine, which delivers the material on a desired location, previously pressurized by the pumps. The end part of the print-head is the nozzle, which profiles the material in a continuous filament.

Nozzle shape for concrete-like materials is commonly round or rectangular/square (Paul et al. 2018b), though elliptical (Mingyang Li et al. 2018) and trapezoid (Lao et al. 2018) nozzles have also been explored. Non-circular shapes, typically used to create contours, involve an additional degree of freedom (DOF) to keep the orientation of the nozzle tangent to the printing path, this way preserving the filament shape (Bos et al. 2016). The nozzle can also be equipped with side trowel guides (Khoshnevis et al. 2006) designed to improve the surface finishing (Craveiro et al. 2011).

In addition, to produce continuous graded elements, the print-heads may have two or more inlets to collect homogeneous materials, previously mixed and pressurized by individual concrete pumps (Craveiro et al. 2018), as well one outlet to deliver the FGM. Through the printing process, the pumping ratio varies continuously, though the total output volume is constant.

In this work, two different extrusion heads are proposed. The first one is a simple "Y" shape print-head with 15 mm of internal diameter (Figure 1a), while the second one includes a dynamic mixer (mixing rod) ensuring a good mixture. The second print-head is composed of a metal body with two inlets and one outlet (nozzle) with an internal diameter of 15 mm (Figure 1b). An internal mixing rod is composed of 8 mixing blades disposed with an angle of 15° relative to the flow, assisting in material flowing. The spin is assured by a 12V DC high torque motor, which varies the speed by tuning the voltage from 0V to 12V.

In both cases, circular nozzle shapes were considered, which gives more freedom in printing internal filaments, simplifying the movements' printing system (requiring no extra DOF).

This study aims to simulate the behavior of fine-aggregate concretes flowing through both print-heads, as well assessing the quality of the extruded material, which is part of an ongoing work comprising the development of cementitious printing materials together with the development of a novel printing system to design and produce functionally graded building components (Craveiro et al 2018, 2017).

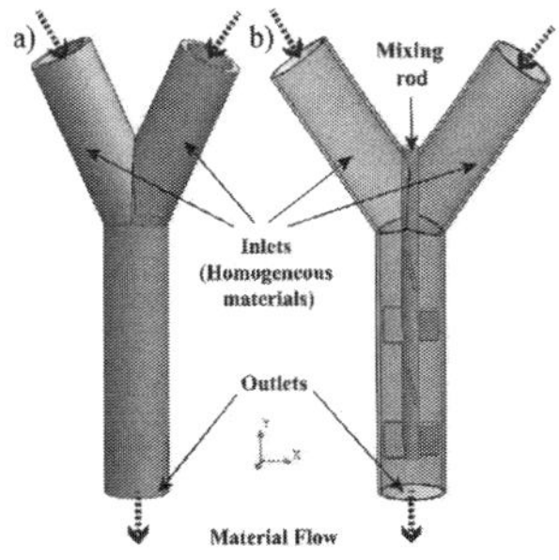

Figure 1. CAD models of the print-heads: a) "Y" shape and b) dynamic.

3 MATERIALS AND FLOW SIMULATION

A Computational Fluid Dynamics (CFD) analysis was performed to predict material flow fields and the mixing process for each extrusion head. Firstly, the 3D CAD program SolidWorks was used to create the geometries, while the numerical analysis was performed by Finite Volume Method (FVM), based on Navier-Stokes equations to simulate the interaction of fluids with surfaces, using SolidWorks Flow Simulation tool (Sobachkin and Dumnov 2013). CFD analysis can be used for cement pastes containing aggregate particles with a maximum size of 2 mm (Roussel and Gram 2014), so the fine-aggregate concretes used in this work can be considered fluids.

The properties of the cementitious materials are important to obtain a consistent numerical result. Printable materials must be able to support its own weight and subsequent layers, presenting a rigid body behavior at low stress and, when submitted to stresses higher than a critical value (yield stress), flowing as a viscous fluid (Roussel 2018). The Bingham model behavior, commonly used to describe the rheological behavior of cementitious materials (Ferraris 1998). It presents a linear relationship between shear stress (τ) and shear rate (γ), characterized by a plastic viscosity (μ), which applies to a material that flows only when submitted to stresses higher than the yield stress (τ_0), according to the following equation:

$$\tau = \tau_0 + \mu\gamma \tag{1}$$

where τ is the shear stress (*Pa*), τ_0 is the yield stress (*Pa*), μ is the plastic viscosity (*Pa×s*) and γ is the shear rate (*1/s*). The materials' properties used for simulation are presented in Table 1.

In this work, the rheological properties were considered similar for two concrete mixtures containing different contents of cork. The thermal conductivity and specific heat were obtained by transient plane source through an Hotdisk TPS2500S equipment, and the density was calculated by measuring the

Table 1. Materials used for the simulation (Craveiro et al. 2019a; Paul et al. 2018a).

	Material	
Specifications	0 vol% cork	10 vol% cork
Density (fresh) – *kg/m³*	1920	1770
Consistency coefficient – *Pa×s*	144*	144*
Yield stress – *Pa*	3622*	3622*
Specific heat – *J/(kg×K)*	888.889	865.591
Thermal conductivity – *W/(m×K)*	0.82	0.73

* adapted from (Paul et al. 2018a).

weight of a constant volume for each composition (Craveiro et al. 2019a). The compressibility of the materials was neglected, no roughness and adiabatic wall were considered.

The Reynolds number (*Re*) was used to assess if the flow is laminar or turbulent. For a yield stress fluid, such as concrete, *Re* is calculated according to the following equation (Roussel and Gram 2014).

$$R_e = \frac{\rho V^2 D}{D\tau_0 + \mu V} \tag{2}$$

where ρ is the density of the material, D is the typical dimension of the flow, V is the flow velocity, τ_0 is the yield stress of the concrete and μ is the plastic viscosity.

Whenever a fluid flows in parallel layers with no disruption between layers (no mixing), it can be named as a laminar flow. For cylindrical pipes, the flow is considered laminar when the Reynolds number is less than 2300 (Çengel and Cimbala 2014). In this study the flow is laminar.

The properties of both materials and the boundary conditions are key parameters. Due to hardware limitations of the printing system, the extrusion velocity was fixed to 25 *mm/s*. This study considered 50% of the extrusion velocity in each inlet (12.50 *mm/s*) and the following boundary conditions:

- The temperature in the internal face of the inlet lids (20 °C);
- The material concentrations were set as a volume fraction;
- The pressure of the outlet was set to be equal to the environment pressure (internal face of the outlet lid);
- The inlet velocity in the internal face of both inlet lids is 12.5 mm/s.

The specific convergence goal "volume flow rate" was selected and determined on the outlet lid surface, in order to direct the solver to the desirable results, minimizing errors (Matsson 2017). Time-dependent analysis with 20 seconds was selected to simulate the complete movement of the material flowing from the inlets to the outlet.

The mesh, in the computational domain, was created by Solidworks and consists of elements in the form of rectangular parallelepiped. Additional refinement levels were added in order to improve the mesh inside the tube, resulting in a total mesh of about 100000 elements. All simulations were performed with uniform processing conditions. The flow velocity and the rotational speed of the mixer were the key parameters investigated.

4 RESULTS AND DISCUSSION

The computed results of both print-heads were analyzed and discussed in detail. For the simple

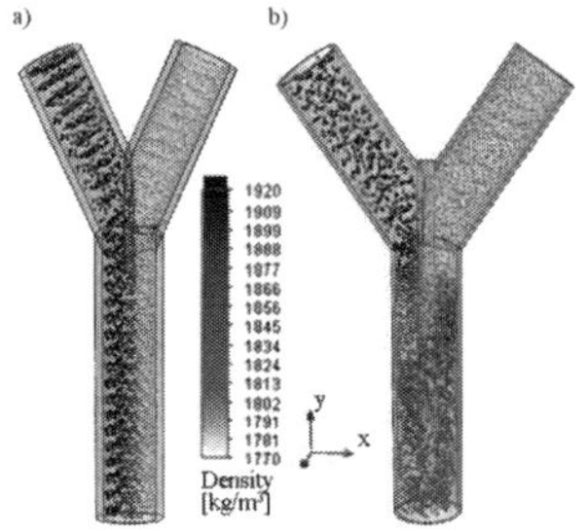

Figure 2. Particles showing flow distribution and material density: a) Simple "Y" print-head, b) dynamic print-head at 40 RPM.

print-head, the target was to identify the distribution of the inflowing materials both along the tube and on the outlet, while for the dynamic one, the main goal was to find the optimum number of the mixing rod revolutions, allowing good homogeneity of the material. In Figure 2, particles represent flow distribution and material density (color gradation) inside each print-head.

4.1 *Simple "Y" print-head*

Figure 3 shows the material density distribution along the extrusion head and on the outlet. As predicted, the materials flowing from the inlets to the outlet will not mix properly, each one filling half filament. The homogeneous materials can be easily identified by the maximum and minimum density patterns along the cross-section XY, where only a small area in the interface of both materials assumes an average density (cross-section XZ).

The velocity distribution, observed in Figure 4, indicates that materials will flow out of the nozzle with different velocity along the cross section of the filament. In this case, the velocity is higher in the interior (~43 mm/s) and slows down to 7 mm/s in the filament walls.

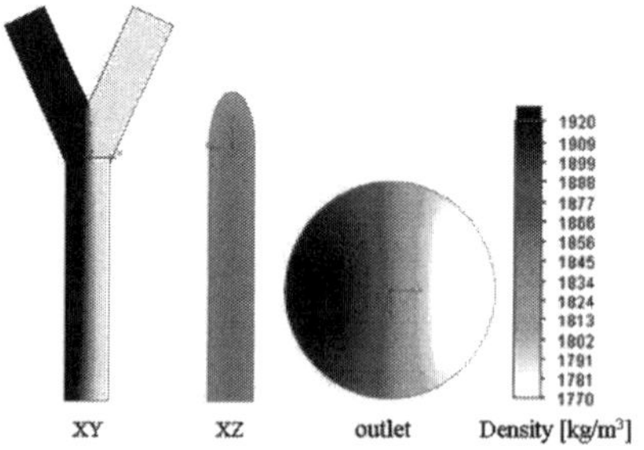

Figure 3. Simple "Y" print-head calculation results: density.

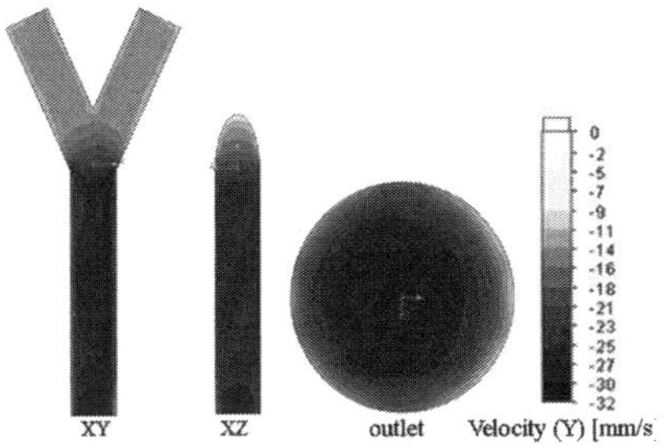

Figure 4. Simple "Y" print-head calculation results: velocity.

4.2 *Dynamic print-head*

The main objective of the dynamic print head simulation was to find the minimum optimal rotating speed to obtain a good mixing, since high speeds can segregate the particles. Perpendicular cross-sections of the nozzle (XY and XZ) and outlet density maps were obtained at different speeds (from 0 to 100 RPM) and are illustrated in Figure 5. At 0 RPM, an inefficient mixture is obtained, comparable to the outcome of the simple nozzle. At each increase of 10 RPM, a better mix can be observed.

Figure 6 Quantifies the density distribution on the outlet of the dynamic extrusion head. The minimum and the maximum density values occurring on the outlet were plotted, and the variation percentage was then obtained for each rotational speed. At 60 RPM, a difference of 8.8% can be observed, which seems to be a good mixed material, almost homogeneous.

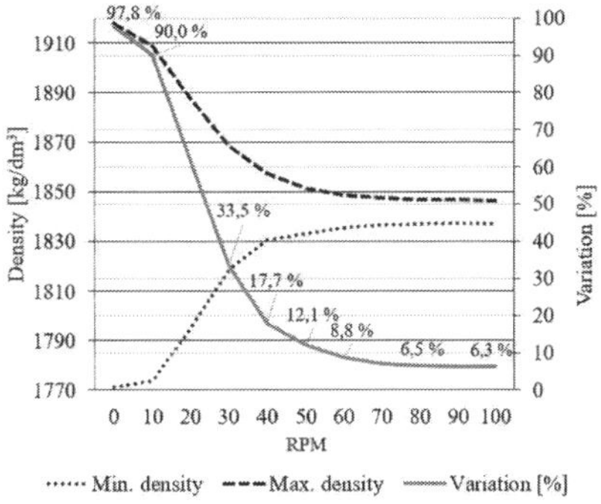

Figure 6. Maximum vs minimum outlet density, according RPM.

5 CONCLUSIONS

The integration of digital technologies into building processes allows to produce complex shapes composed of new materials with multi-performance capabilities. In this study, a flow simulation of two 3D printing extrusion heads to produce continuous graded materials was presented, using Solidworks flow simulation tool.

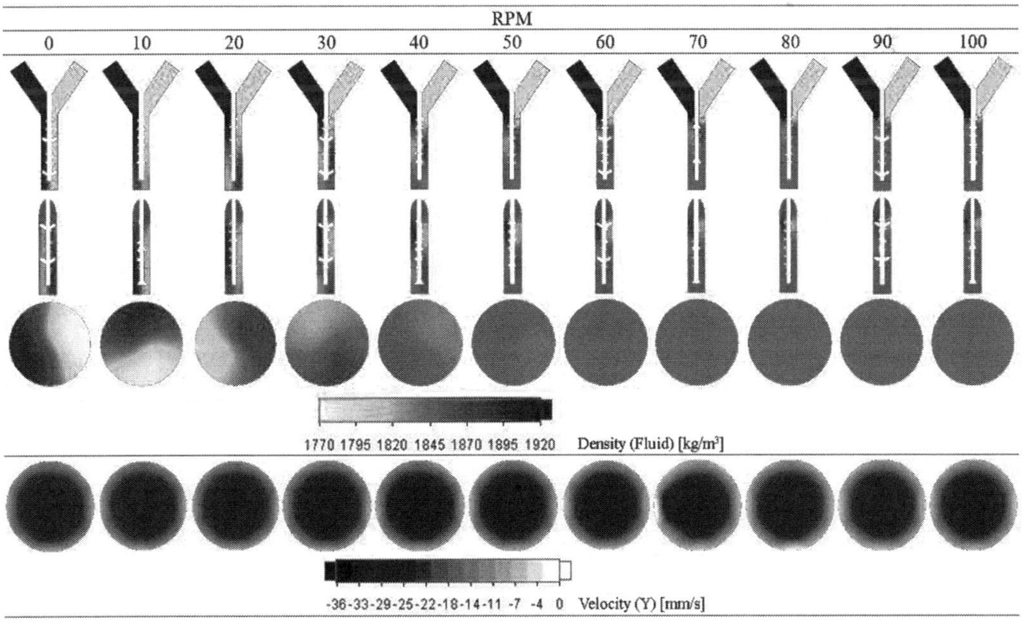

Figure 5. RPM vs density distribution (XY & XZ dynamic print-head cross-sections and outlet) and velocity (outlet).

Some conclusions can be drawn from the numerical analysis:

- Simple "Y" nozzle allows no good mixing;
- The dynamic nozzle presents a good density distribution at 60 RPM (mixing rod), enough to get less than 10% of density difference and obtain an almost homogeneous material;
- It was observed that, for both print-heads, the velocity of the materials is not constant along outlet planes (high velocity in the interior of the filament and low velocity near the exterior walls), which needs further investigation to get a better understanding of its potential impact on the printing process.

Future work will include experimental tests and the analysis of new materials to compare simulated and experimental material behaviors.

ACKNOWLEDGMENTS

This research work was supported by grant SFRH/BD/105404/2014 from the Portuguese Foundation for Science and Technology (FCT).

REFERENCES

Bos, F., Wolfs, R., Ahmed, Z., Salet, T., 2016. Additive manufacturing of concrete in construction: potentials and challenges of 3D concrete printing. Virtual and Physical Prototyping 11, 209–225. https://doi.org/10.1080/17452759.2016.1209867

Çengel, Y.A., Cimbala, J.M., 2014. Fluid mechanics: fundamentals and applications, 3rd edition. ed. McGraw-Hill, New York, NY.

Craveiro, F., Bártolo, H., Bártolo, P.J., Duarte, J.P., 2019a. Characterization of concrete/cork materials for functionally graded additive manufacturing. [Manuscript submitted for publication].

Craveiro, F., Bártolo, H., Bártolo, P.J., Nazarian, S., Duarte, J.P., 2018. Additive Manufacturing of Functionally Graded Building Parts: Towards Seamless Architecture, in: 4th Biennial Residential Building Design & Construction Conference Proceedings. Pennsylvania Housing Research Center, State College, pp. 529–540.

Craveiro, F., Bártolo, H., Gale, A., Duarte, J.P., Bártolo, P.J., 2017. A design tool for resource-efficient fabrication of 3d-graded structural building components using additive manufacturing. Automation in Construction 82, 75–83. https://doi.org/10.1016/j.autcon.2017.05.006

Craveiro, F., Duarte, J.P., Bártolo, H., Bártolo, P.J., 2019b. Additive Manufacturing as an enabling technology for digital construction: a perspective on Construction 4.0. Automation in Construction 103, 251–267. https://doi.org/10.1016/j.autcon.2019.03.011

Craveiro, F., Matos, J.P., Bártolo, H., Bártolo, P.J., 2011. Automation for building manufacturing, in: P. J. Bártolo, et al. (Ed.), Innovative Developments in Virtual and Physical Prototyping. CRC Press, London.

Dillenburger, B., Jipa, A., Aghaei-Meibodi, M., Bernhard, M., 2017. The Smart Takes from the Strong, in: Sheil, B., Menges, A., Glynn, R., Skavara, M., Lee, E. (Eds.), Fabricate Rethinking Design and Construction. UCL Press, pp. 210–217.

European Commission, 2019. 100 Radical Innovation Breakthroughs for the future (No. KI-04-19-053-EN-N), Research and Innovation. Publications Office of the European Union, Luxembourg. https://doi.org/10.2777/24537

Ferraris, C.F., 1998. Testing and modelling of fresh concrete rheology (No. NIST IR 6094). National Institute of Standards and Technology, Gaithersburg, MD. https://doi.org/10.6028/NIST.IR.6094

Herrmann, M., Sobek, W., 2017. Functionally graded concrete: Numerical design methods and experimental tests of mass-optimized structural components. Structural Concrete 18, 54–66. https://doi.org/10.1002/suco.201600011

Khoshnevis, B., Hwang, D., Yao, K.T., Yeh, Z., 2006. Mega-scale fabrication by Contour Crafting. IJISE International Journal of Industrial and Systems Engineering 1, 301–320. https://doi.org/10.1504/IJISE.2006.009791

Lao, W., Tay, D., Quirin, D., Tan, M.J., 2018. The effect of nozzle shapes on the compactness and sternght of structures printed by additive manufacturing of concrete, in: Proceedings of the 3rd International Conference on Progress in Additive Manufacturing. Nanyang Technological University, Singapore. https://doi.org/10.25341/D4V01X

Matsson, J.E., 2017. An Introduction to SOLIDWORKS Flow Simulation 2017. SDC Publications, Mission, KS.

Mingyang Li, Wenxin Lao, Lewei He, Ming Jen Tan, 2018. Effect of rotational trapezoid shaped nozzle on mateiral distribution in 3D cementitious material printing process, in: Proceedings of the 3rd International Conference on Progress in Additive Manufacturing. Nanyang Technological University, Singapore, pp. 56–61. https://doi.org/10.25341/D4ZP4W

Paul, S.C., Tay, Y.W.D., Panda, B., Tan, M.J., 2018a. Fresh and hardened properties of 3D printable cementitious materials for building and construction. Archives of Civil and Mechanical Engineering 18, 311–319. https://doi.org/10.1016/j.acme.2017.02.008

Paul, S.C., van Zijl, G.P.A.G., Tan, M.J., Gibson, I., 2018b. A review of 3D concrete printing systems and materials properties: current status and future research prospects. Rapid Prototyping Journal 784–798. https://doi.org/10.1108/RPJ-09-2016-0154

Pietras, D., Sadowski, T., 2019. A numerical model for description of mechanical behaviour of a Functionally Graded Autoclaved Aerated Concrete created on the basis of experimental results for homogenous Autoclaved Aerated Concretes with different porosities. Construction and Building Materials 204, 839–848. https://doi.org/10.1016/j.conbuildmat.2019.01.189

Río, O., Nguyen, V.D., Nguyen, K., 2015. Exploring the Potential of the Functionally Graded SCCC for Developing Sustainable Concrete Solutions. Journal of Advanced Concrete Technology 13, 193–204. https://doi.org/10.3151/jact.13.193

Roussel, N., 2018. Rheological requirements for printable concretes. Cement and Concrete Research 112, 76–85. https://doi.org/10.1016/j.cemconres.2018.04.005

Roussel, N., Gram, A., 2014. Physical Phenomena Involved in Flows of Fresh Cementitious Materials, in: Roussel, N. (Ed.), Simulation of Fresh Concrete Flow, State-of-the-Art Report of the RILEM - Technical Committee 222-SCF. Springer, pp. 1–24.

Sobachkin, A., Dumnov, G., 2013. Numerical Basis of CAD-Embedded CFD, in: Proceedings of NAFEMS World Congress 2013. Presented at the International Conference on Simulation Process and Data Management, Salzburg, Austria.

Tay, Y.W.D., Panda, B., Paul, S.C., Noor Mohamed, N.A., Tan, M.J., Leong, K.F., 2017. 3D printing trends in building and construction industry: a review. Virtual and Physical Prototyping 12, 261–276. https://doi.org/10.1080/17452759.2017.1326724

Industry 4.0 – Shaping The Future of The Digital World – da Silva Bartolo et al. (eds)
© 2021 Taylor & Francis Group, London, ISBN 978-0-367-42272-1

Analysis of 3D printed 17-4 PH stainless steel lattice structures with radially oriented cells

P.S. Ginestra, L. Riva, G. Allegri, L. Giorleo, A. Attanasio & E. Ceretti
Department of Mechanical and Industrial Engineering, University of Brescia, Brescia, Italy

ABSTRACT: Additive Manufacturing was initially born as a way to rapid prototyping but the potential of this process made it spread really fast as a production technique. Among all the materials, polymers are the most used for 3D printing but metals are expected to have the biggest growth rate: 38% by 2020. The interest on lattice geometries is widely in-creasing mainly due to the possibility of obtaining light-weight, well performing and multifunctional products for different fields of application like aerospace, automotive and biomedical. In this work, 17-4PH stainless steel lattice samples with different geometry of cells were manufactured by Selective Laser Melting and tested under uniaxial compression tests. In particular, both the orientation of the cells in relation to the sample and the building orientation of the structures with respect to the building plate were analyzed.

1 INTRODUCTION

Additive Manufacturing (AM) of pure metals can be di-vided in different sub-categories according with the material charging method: powder bed fusion systems, where materials are added layer-by-layer and direct energy deposition systems, where materials are added through a nozzle or a wire fed system (Lowther et al. 2019). Cellular structures are usually divided in two major groups: stochastic porous structures and cellular lattice structures (Yan et al. 2014). Stochastic porous structures are characterized by pores located randomly throughout the entire volume of the object, thus, the mechanical properties result not uniform and difficult to control. Cellular lattice structures consist in a unit cell repeated in all directions, making the mechanical properties controllable and repeatable. Consequently, lattice structures with a certain volume fraction present better mechanical properties than stochastic porous structures. As an example, in the aerospace field, the production of a component with the same level of performances but a reduced weight consists in a huge economical advantage. On the biomedical side, lattice structures can be advantageous enhancing the proper-ties of implanted prostheses and functional orthoses (Ginestra et al. 2016). Moreover, the surface properties of a metal lattice structure can improve the interaction with the physiological environment by stimulating cells through their morphology which was found to be a key factor for the biointegration of a scaffold (Gastaldi et al. 2015, Ginestra et al. 2017a,b). Furthermore, an open cell structure makes a sample suitable even for non-structural application like acoustic insulation, energy ab-sorption and filtration.

Selective Laser Melting (SLM) of lattice structures is a relatively new process so there is no complete characterization of their performances in the literature yet. The properties of the lattice structures depend on several factors: cell geometry, material, struts dimension, loading directions and boundary conditions (Leary et al. 2018). Although the lattice structures have a lot of advantageous properties there are some issues that have to be taken into account. The majority of the possible geometries is not isotropic, consequently the mechanical properties of an object show an anisotropic trend. Moreover, only few materials and geometries have been studied and the comparison between different results is often difficult due to the differences in tests procedures, process parameters and printing quality (Marbury 2017).

For example, the orientation of the cells within the sample changes the number of struts in the direction of compression causing a change in the strength of the sample. It also changes the thermal gradient that the part is subjected to, and consequently also the residual stresses within the structure. The study of different orientations is then fundamental for a complete analysis of the potential of lattice geometries. Moreover, considering the widespread use of lattice structures for load-bearing prostheses design in the biomedical field, a radial orientation of the cells within a lattice sample would replicate the structure of the Harvesian lamellae of the cortical bone (Rho et al. 1998). In this work, 17-4 PH Stainless Steel was used to manufacture cylindrical lattice samples via SLM. The high tensile stress and hardness combined with a good corrosion resistance below 315°C make this material appealing for many ap-plications (Mahmoudi et al. 2017). The geometry of three different cells was chosen considering the most representative unit cells for cellular lattices available in the majority of the printers' design platforms and currently analyzed in literature (Jin et al. 2019, Maskery et al. 2018). The building orientation of the samples was modified to evaluate the effect of these parameters on the mechanical performances of the structures.

2 MATERIALS AND METHODS

17-4PH stainless steel powder was used as printing material. All the powders were selected from the same batch to exclude the influence of the material. The powder exhibits spherical shape particles with a size in the range of 5-35 µm (Zhang et al. 2017). The chemical com-position is reported in Table 1.

The low content of impurities is crucial to avoid side effects of embrittlement. The samples were manufactured using a laser-based powder bed fusion machine (ProX 100, 3D SYSTEMS). The laser melting process occurred in a protective Nitrogen atmosphere with O2 content less than 0.1 vol.% and the processing parameters were set as reported in Table 2. Three replicas of each produced substrate have been analyzed for the compression tests. When the SLM process was finished, the samples were removed from the plate through a band saw.

2.1 *Design and production of the lattice structures*

The lattice structures studied in this paper were gene-rated by means of the integrated software 3DXpert. In particular, three different geometries were selected: diagonal, diamond and face centric cubic cells (FCC) (Figure 1).

The dimensions of the unit cells were kept to 2(X) x 2(Y) x 2(Z) mm3. A diameter of struts (Dc) of 0,5 mm and spherical nodes (Dn) of 1 mm were selected. Following the ISO standard 13314, cylindrical samples with a 24 mm diameter and a 30 mm height were produced. Two building orientations were considered: 0° and 90° in relation to the building plate direction (Figure 2).

Table 1. Chemical composition of 17-4PH stainless steel powder.

17-4 PH	Cr	Ni	Cu	Mn	Mo	Nb	Si
Wt (%)	16.71	4.09	4.18	0.8	0.19	0.23	0.53

Table 2. Process parameters used in the SLM process.

Process Parameter	Value
Laser power (W)	50
Spot diameter (µm)	80
Scan speed (mm/s)	300
Hatch spacing (µm)	50
Layer thickness (µm)	30

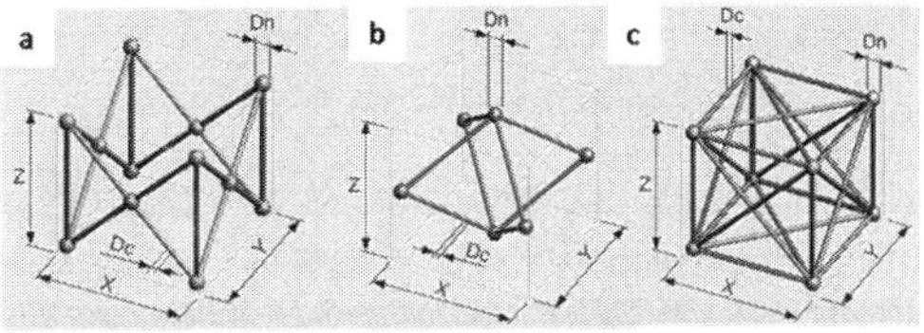

Figure 1. Diagonal (a), Diamond (b) and FCC (c) unit cells.

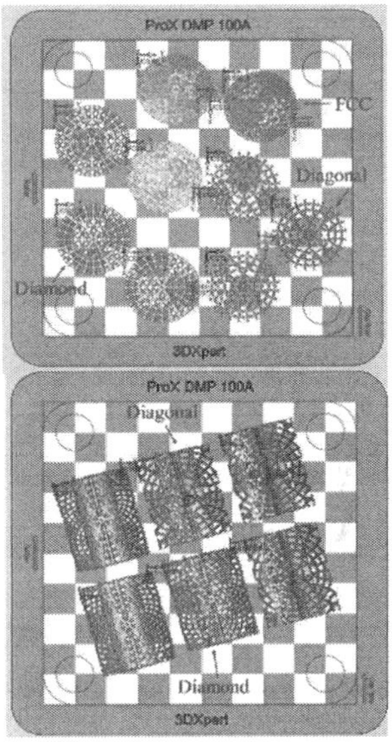

Figure 2. Building plates design in 3DXpert. The direction of the roller is up-down.

In order to analyze the effect of the cells orientation on the mechanical properties of the samples, the cells were arranged with a radial distribution, in which the cells are not lined up one next to the other but form concentric features (Figure 3).

Figure 3. Samples manufactured with 0° (left) and 90° (right) building orientation: Diagonal (a), Diamond (b) and FCC (c).

2.2 *Characterization of the lattice samples*

In order to obtain the relative density of the samples, the geometry and weight of the lattices were analyzed. The height and diameter of the cylinders were measured with a digital caliber while a microbalance was used to obtain the weight and calculate the relative density of the samples. The surface morphology of the samples was observed under an Hirox RH-2000 optical microscope.

The mechanical response of the lattice samples was evaluated by uniaxial compression tests carried out at 23° C by means of a hydraulic press equipped with a 1000kN load cell. The specimens were subjected to a compressive ramp under force control up to a displacement of 15 mm without intermittence. The samples oriented at 0° were subjected to a load parallel to the building direction while the samples built at 90° were subjected to a load perpendicular to the building direction. The results are reported as nominal stress vs. nominal strain curves, where the nominal stress is the force normalized on the overall specimen cross-section (about 452 mm2), while the nominal strain is the crosshead displacement normalized on the overall specimen thickness (Ceretti et al. 2017). This representation allowed an easier comparison of the various structures, avoiding effects ascribed to the differences in the cross-section of the samples. The elastic modulus (E) was considered as the slope of the linear fit of the stress-strain curve.

3 RESULTS

3.1 *Morphology of the samples*

The geometry of the samples was analyzed to evaluate the comparison between the designed models and the as-built samples and revealed a good consistency of the production process (Table 3).

The higher dispersion of data in relation to the diagonal cells is due to the geometry of the samples that may influence the right positioning of the caliber. As for the other cells geometries, the slightly higher standard deviation on the diameter of the samples built at 0° is probably due to the cutting edges. The relative density was analyzed considering that the volume fraction is one of the key parameters controlling the mechanical properties of porous parts (Table 4).

Table 3. Geometry data of the as built samples.

Sample	Building angle (°)	Diameter (mm)	Height (mm)
Diagonal	0	23.8±0.06	29.4±0.07
	90	23.4±0.01	30.0±0.05
Diamond	0	24.1±0.04	29.4±0.06
	90	23.4±0.02	30.0±0.01
FCC	0	24.1±0.06	29.5±0.02
	90	23.8±0.01	29.9±0.01

Table 4. Weight and density values calculated on the as built samples.

Sample	Building angle (°)	Weight (g)	Relative density (g/mm^3)
Diagonal	0	27.3±0.37	2.1E-03
	90	29.6±0.04	2.3E-03
Diamond	0	30.0±0.07	2.2E-03
	90	29.5±0.16	2.3E-03
FCC	0	39.5±1.97	3.0E-03
	90	44.0±1.19	3.3E-03

The building orientation of the samples has a reduced impact on the relative density of the parts while the geometry of the cells, especially for the FCC configuration, may cause the presence of unmelted powder trapped inside the structure causing an increase of the relative density.

The surface morphology of the produced samples was observed to identify the presence of defects and collapsed struts (Figure 4).

The struts and the nodes show the typical cylindrical and spherical shapes but their surfaces are covered with partially melted powder on the top layer. At higher magnifications (Figure 4), it is possible to see the path followed by the laser while scanning and melting the powder.

The results of the compression tests are reported in terms of the nominal stress (i.e. stress over the initial section) vs. the engineering strain (i.e. the displacement over the overall specimen thickness) and are compared in Figures 5 and 6, as the most representative nominal stress vs. nominal strain curves for each material group. In this representation, the use of normalized force versus displacement allows the comparison between specimens with different cross sections and thickness.

All the curves show the occurrence of a multi stage compression history, as suggested by the three

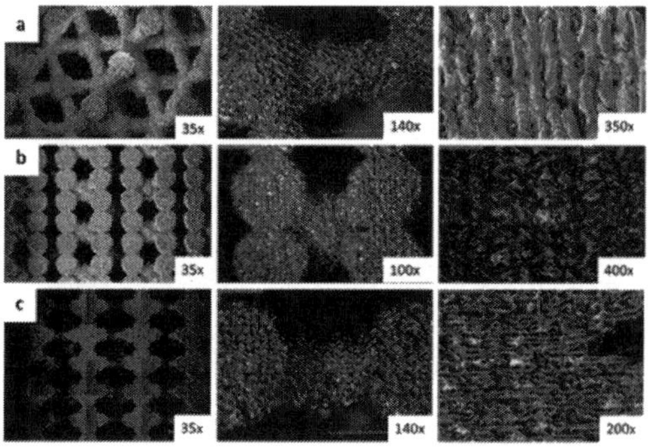

Figure 4. Optical microscope images at different magnifications of the samples manufactured with 0° building orientation: Diagonal (a), Diamond (b) and FCC (c).

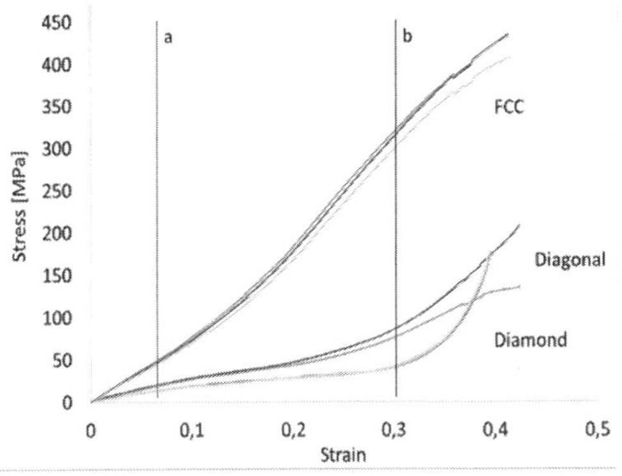

Figure 5. Nominal stress vs. nominal strain curves obtained from the compression tests performed on the samples built at 0° of orientation.

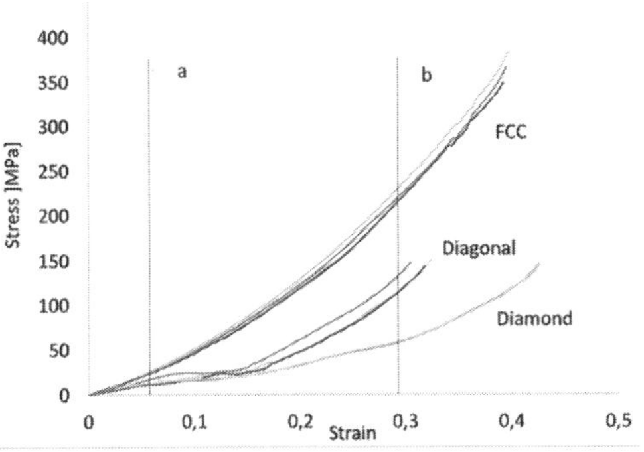

Figure 6. Nominal stress vs. nominal strain curves obtained from the compression tests performed on the samples built at 90° of orientation.

different slopes of the curves: the early deformation of the structure allows to evaluate the material stiffness at small strains (Figure 7). This trend is followed by a reduction of the slope as a possible consequence of the col-lapse of the structure trough local collapses of the lattice layers. Finally, a subsequent increase of the slope is found after the collapsed structure has approached a more compact state. In particular, the early elastic de-formation stage is followed by a plastic deformation region at higher values of strain (Figure 5 and 6 line a), a subsequent collapse stage and a final hardening stage (Figure 5 and 6 line b).

The difference in the performances between the two orientations for the diagonal and FCC samples are probably caused by an anisotropy due to different number of struts in the direction of the compression, especially for the FCC configuration. In particular, the obtained curves for the samples oriented at 90° are influenced by an effect of densification of the structure that requires higher levels of stress compared to the other cells geometries. As for the

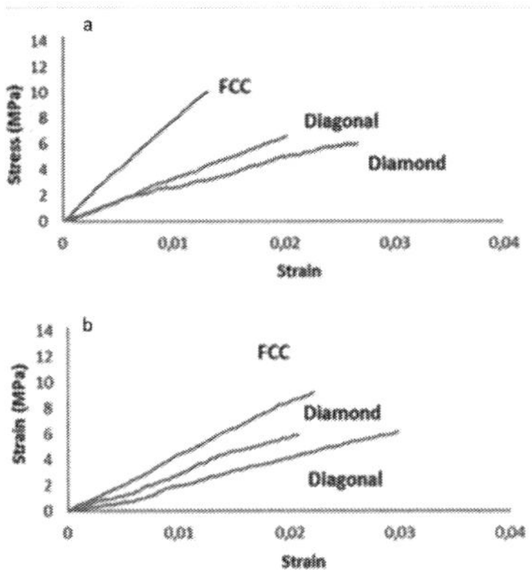

Figure 7. Detail of the early elastic deformation stage on the samples built at (a) 0° and (b) 90° of orientation.

diamond geometry, it can be observed that the building direction is not affecting the mechanical response in the same way, leading to a more isotropic behavior of these cells highlighted by the evident replicability of the compression trend. With regard to the diagonal geometry, the difference found between the trends observed at different building orientations is probably due to the less presence of struts parallel to the compression direction. The nominal stress vs. nominal strain curve stiffness (i.e. the slope of the curve in the linear initial trend) is reported in Table 5.

The reason of the observed difference between the stiffness calculated for the FCC configuration at different building orientations is probably the radial distribution of the cells that caused the two orientations to have very distinct internal structures and therefore different mechanical properties compared to the typical iso-tropic behavior of these cells.

The compressed samples are shown in Figure 8.

The images of the compressed samples show the layer-by-layer buckling deformation, typical of the lattice structures, that was accompanied by a shear band (45°) clearly visible for the diagonal configuration as already observed for this type of geometry (Liu et al. 2018).

Table 5. Stiffness (E) calculated on the samples tested under compression.

Sample	Building angle (°)	E (MPa)
Diagonal	0	391.1±51.9
	90	261.9±36.1
Diamond	0	225.8±33.3
	90	286.0±29.3
FCC	0	782.5±35.6
	90	416.0±23.3

Figure 8. Compressed samples: Diagonal (a), Diamond (b) and FCC (c).

4 CONCLUSIONS

In this work, 17-4PH stainless steel lattice cylinders with a radial distribution of cells were built using a 3D printer. Three cell geometries, diagonal, diamond and FCC, and two different building orientations, 0° and 90°, were chosen. The 3DXpert software was used to design the samples, produced using the ProX100 printer (3D SYS-TEMS). The structures were characterized and subject-ed to compression tests by means of an hydraulic press. The geometry of the designed samples was reproduced by the process with an error of less than 1% and the geometries of the struts and nodes have been produced without defects. Regarding the mechanical performances of the samples under compression tests, the results suggested that when a diagonal geometry of the cells is used, the building orientation can strongly in-fluence the mechanical properties. In particular, the stiffness of the samples is higher in relation to a building orientation of 0° probably due to the higher number of struts along the compression direction. The diamond geometry showed isotropic mechanical properties al-most independent from the building orientation of the structures. As for the FCC cells, the stiffness of the samples calculated in correspondence to a building orientation of 0° was significantly higher than the value calculated for the samples built at 90° with respect to the building plate. Although the geometry of the cells is isotropic, the performances resulted very different. Considering the relationship between the volume fraction and the mechanical properties of these samples, the results suggest that the radial orientation of the cells can influence the mechanical response of the lattice structures when the geometry of the cells can be varied by giving a radial distribution in relation to the building orientation.

The anisotropy of the mechanical properties found in SLM parts can be attributed to the oriented layer-by-layer growth process along the direction of the substrate. As expected, an increase of the overall system stiff-ness is found as the density increases. This suggests the possibility to easily tune the stiffness of the lattices, and to vary the object structure without significant changes in the overall stiffness. In particular, the optimization of the orientation of the cells can improve the mechanical properties of the lattice structures and enhance the performance efficiency of the produced cellular parts and, as a consequence, enhance the performances of the final product.

REFERENCES

Ceretti, E., Ginestra, P., Neto, P.I., Fiorentino, A., Da Silva, J.V.L. 2017. Multi-layered scaffolds production via Fused Deposition Modeling (FDM) using an open source 3D printer: process parameters optimization for dimensional accuracy and design reproducibility. *Procedia. CIRP* 65: 13–18.

Gastaldi, D., Parisi, G., Lucchini, R. et al. 2015. A Predictive Model for the Elastic Properties of a Collagen-Hydroxyapatite Porous Scaffold for Multi-Layer Osteochondral Substitutes, *International Journal of Applied Mechanics* 7(4): 1550063.

Ginestra, P.S., Ceretti, E., Fiorentino, A. 2016. Potential of modeling and simulations of bioengineered devices: Endoprostheses, prostheses and orthoses. *Proceedings of the Institution of Mechanical Engineers, Part H: Journal of Engineering in Medicine* 230(7): 607–638.

Ginestra, P., Fiorentino, A., Ceretti, E. 2017. Microstructuring of titanium collectors by laser ablation technique: a novel approach to produce micro-patterned scaffolds for tissue engineering applications. *Procedia CIRP* 65:19–24.

Ginestra, P.S., Pandini, S., Fiorentino, A. et al. 2017. Microstructured scaffold for guided cellular orientation: Poly(ε-caprolactone) electrospinning on laser ablated titanium collector. *CIRP Journal of Manufacturing Science and Technology* 19:147–157.

Jin, N., Wang, F., Wang, Y., et al. 2019. Failure and energy absorption characteristics of four lattice structures under dynamic loading, *Materials and design* 169:107655.

Leary, M., Mazur, M., Williams, H., et al. 2018. Inconel 625 lattice structures manufactured by selective laser melting (SLM): Mechanical properties, deformation and failure modes. *Materials and design* 157:179–199.

Liu, F., Zhang, D.Z., Zhang, P., Zhao, M., Salman J. 2018. Mechanical Properties of Optimized Diamond Lattice Structure for Bone Scaffolds Fabricated via Selective Laser Melting *Materials* 11: 374. doi:10.3390/ma11030374.

Lowther, M., Louth, S., Davey, A. et al., 2019. Clinical, industrial, and research perspectives on powder bed fusion additively manufactured metal implants, *Additive Manufacturing* 28:565–584.

Mahmoudi, M., Elwany, A., Yadollahi, A. et al., 2017. Mechanical properties and microstructural characterization of selective laser melted 17-4 PH stainless steel, *Rapid Prototyping Journal* 23:280–294.

Maskery, I., Sturm, L., Aremu, A.O., et al. 2018. Insights into the mechanical properties of several triply periodic minimal surface lattice structures made by polymer additive manufacturing, *Polymer* 152: 62–71.

Marbury, F., 2017. Characterization of SLM Printed 316L Stainless Steel and Investigation of Micro Lattice Geometry. PhD dissertation.

Rho, J.Y., Kuhn-Spearing, L., Zioupos, P., 1998. Mechanical properties and the hierarchical structure of bone, *Medical Engineering and Physics* 20: 92–102.

Yan, C., Hao, L., Hussein, A., et al. 2014. Advanced lightweight 316L stainless steel cellular lattice structures fabricated via selective laser melting, *Materials and Design*, 2014 55: 533–541.

Zhang, Z., Peng, T., Xu, S. 2017. The influence of scanning pattern on the part properties in powder bed fusion processes: an experimental study. *Procedia CIRP* 61: 606–611.

Industry 4.0 – Shaping The Future of The Digital World – da Silva Bartolo et al. (eds)
© 2021 Taylor & Francis Group, London, ISBN 978-0-367-42272-1

Control of process parameters for directed energy deposition of PH15-5 stainless steel parts

Y.Y.C. Choong
HP–NTU Digital Manufacturing Corporate Lab, School of Mechanical & Aerospace Engineering, Nanyang Technological University, Singapore

K.H.G. Chua & C.H. Wong
Singapore Centre for 3D Printing, School of Mechanical & Aerospace Engineering, Nanyang Technological University, Singapore

ABSTRACT: Directed energy deposition (DED) is an additive manufacturing (AM) process that utilizes a concentrated heat source, which may be a laser or electron beam, with in situ delivery of powder material for subsequent melting to accomplish layer-by-layer fabrication. It has received high attention in recent years because of its potential to provide complex prototyping and strong metallic coating for high-value component repairing. However, there is a lack of published studies with the necessary parameters to draw conclusions on precipitation hardening stainless steel such as PH15-5. This work focused on the effects of the laser parameters such as laser power, powder feed rate and scanning strategies on the mechanical properties. A relationship between the internal features of the printed parts and their mechanical properties is established where X-ray computed tomography (CT) analysis was performed to visualize and quantify the volume fraction of pores and inclusions in selected samples prior to testing.

1 INTRODUCTION

Additive manufacturing (AM) has developed rapidly in modern manufacturing for a range of applications in the manufacturing, medical and industrial sectors (Choong et al. 2017; Choong et al. 2016; Eng et al. 2017; Tan et al. 2019). AM manufactures the part by depositing material layer-by-layer. It differs from the conventional manufacturing such as subtractive processes, formative processes and joining processes (Choong et al. 2016; Choong et al. 2020; Eng et al. 2017). It is well-known for its short lead time and ability to create complex parts which can be challenging via conventional manufacturing methods. Directed energy deposition (DED) has received high attention in recent years because of its potential to provide complex prototyping and strong metallic coating for high-value component repairing. DED technology can provide higher deposition rates and wider fabrication space to manufacture larger workpieces as compared to the other metal-based AM methods (Selcuk, 2011). It utilizes a concentrated heat source, which may be a laser or electron beam, with in situ delivery of powder material for subsequent melting to accomplish layer-by-layer part fabrication. A small heat affected zone (HAZ) is created to enable metallurgical bonding, hence producing high density part and is suitable for single-to-multi layer cladding/repair (Costa & Vilar, 2009).

Similar to powder bed fusion (PBF) processes, DED is a means to build metallic parts. However, instead of a separate material and selective energy delivery process, DED combines the material and energy delivery for simultaneous deposition and part forming. The powder preform is typically blown through nozzles and can result in non-used powder accumulation which leads to partial melting. Moreover, due to the layer-by-layer method, lack-of-fusion defects are often introduced at the interfaces between each laser pass and layers. As a result, it becomes difficult to fabricate fully dense or defect-free components that consistently produce components with repeatable mechanical properties (DebRoy et al., 2018). To accomplish effective material joining, the successful combination of material, or feedstock, and energy delivery is required. The process of determining the printable parameters for materials using DED has been challenging due to high occurrence of over-melting or under-deposition that affect the mechanical properties of printed parts. Some of the critical process parameters are laser power, feed rate, powder feed rate, layer thickness and stepover distance, which imparted significant variations on the properties of the printed parts.

On the other hand, numerous studies have shown that the quality of a printed part by DED process is dependent on the different deposition patterns. These patterns are controlled by the scanning strategies that dictates the building direction of the nozzle on the workpiece. There are various common scanning strategies used for DED which include raster, zigzag and spiral. A variation in these patterns and strategies will affect the mechanical and geometrical quality of the final printed parts. Dai *et al.* discovered that using spiral pattern would be possible to reduce the residual stress to one third of the one printed with the zigzag pattern (Dai & Shaw, 2002). Kristler *et* al. also reported that using the zigzag pattern with 90° rotation in between layers would result in a workpiece with higher porosity level as compared to one printed without rotation in between layers (Kistler et al. 2019). These studies reveal that the choice of scanning strategy plays an important role in the quality of the final printed product.

Many researchers have performed characterizations on the mechanical properties of metallic materials such as 316L stainless steel, Inconel, titanium alloy and etc (Choong et al. 2018; Chua et al. 2018; Tan et al. 2019), however fabrication of PH15-5 using DED is infrequently reported in literature. PH15-5 is a precipitation hardening stainless steel that is known for its ability to gain high strength through heat treatment and exhibits the corrosion resistance of austenitic stainless steel. Hence, it is widely used in aerospace and other high technology applications such as high strength shafts, turbine blades and nuclear waste casks (Hongmin, 2003). Researchers are shifting their attention to using PH15-5 in DED to repair high-value components.

Therefore, this work investigates on the control of process parameters such as laser power, powder feed rate following a spiral and zigzag tool path to enable the fabrication of PH15-5 parts using DED. The aim of this study is not AM process optimization, but rather characterization of the internal structure and mechanical properties of the DED printed PH15-5 components. As such, the components were examined for porosity and internal defects using X-ray computed tomography (CT). Based on the results of the mechanical and microstructural characterization of samples, a control of the process parameters on the quality of the printed parts is established.

2 EXPERIMENTAL SECTION

2.1 *Material and printing process*

The metallic powder material used in this work is PH15-5 precipitation hardening stainless steel, which was purchased from Dura-Metal Pte Ltd. The particle size distribution in DED ranges from 45 to 105 μm, which is significantly larger than particles used in SLM processes so as to prevent powder cloud formation when the powder exits the nozzle. If the particle size is too small, problems with powder delivery can occur causing a reduction in deposition height and efficiency (Kong et al. 2007). Table 1 shows the chemical composition of PH15-5 used in this experiment.

LASERTEC 65 3D hybrid from DMG MORI (Germany) was used to carry out the DED process of PH15-5 samples. The machine's chamber was backfilled with argon for deposition with oxygen content maintained below 20 ppm. During deposition, PH15-5 powder was injected into the melt pool through four radially symmetrical nozzles with orifice diameters of 3 mm at a powder feed rate of 10 to 14 g/min and power ranging from 1000 to 1400 W. A stainless-steel substrate was positioned at a working distance of 11 mm such that the beam diameter is kept at a processing distance (i.e. its layer thickness) of 1 mm.

Three processing parameters were varied to assess their effects on microstructural and mechanical properties: laser power (*P*), powder feed rate (*PFR*) and scanning strategies (spiral and zigzag patterns as shown in Figure 1). During each run, the processing parameters were varied while the other parameters were held at nominal values as illustrated in Table 2. The layer deposition height and stepover distance are fixed at 1 mm and 50% of the nozzle size respectively.

2.2 *Density measurement*

The density of the samples was measured by a densitometer (Mettler Toledo; XS204) using Archimedes' method. Three repetitive measurements were taken for each sample to increase confidence of the data. In practice, the Archimedes' method usually gives lower density values. Therefore, computed tomography (CT) scans

Figure 1. Spiral and zigzag scanning strategy.

Table 1. Chemical composition, % by wt, of PH15-5.

Cr	Ni	Cu	Mn	Si	C	Mo	Nb	Fe
14.0–15.5	3.5–5.5	2.5–4.5	1.0	1.0	0.07	0.5	0.15–0.45	Balance

Table 2. Process parameter combinations used in experiment.

Process parameters			
	Input variables		
Scanning strategy	Power, *P*	Feed rate, *FR*	Powder feed rate, *PFR*
	W	mm/min	g/min
Spiral	1400	1000	14.3
Spiral	1300	1000	14.3
Spiral	1200	1000	14.3
Spiral	1100	1000	14.3
Spiral	1000	1000	14.3
Zigzag	1400	1000	14.3
Zigzag	1300	1000	14.3
Zigzag	1200	1000	14.3
Zigzag	1100	1000	14.3
Zigzag	1000	1000	14.3

were carried out on the samples as a complementary method to quantify the porosity and identify internal defects in the samples.

2.3 *X-ray Computed Tomography (CT)*

X-ray computed tomography is a radiographic based non-destructive evaluation technique capable of 10μm spatial resolution and interrogation of internal features in bulk produced samples. CT scans were performed on the samples using a Bruker micro-CT Skyscan 1173 machine to examine internal pores or defects within the build parts, and to quantify the pore volume fraction. The specimen is incrementally rotated to render a three-dimensional representation of its entire volume for analysis.

2.4 *Vickers hardness measurement*

The microhardness test was carried out using FM-300e Vickers microhardness tester on the y-z plane where a load of 1 kg force is applied onto the surface of the polished sample. A loading time of 15 seconds was used in accordance to ASTM E18.

3 RESULTS AND DISCUSSION

3.1 *Effects of laser power on mechanical properties*

The influence of laser power on mechanical properties in terms of ultimate tensile strength, density and hardness were plotted to establish a relationship between the process parameter and part property. We can

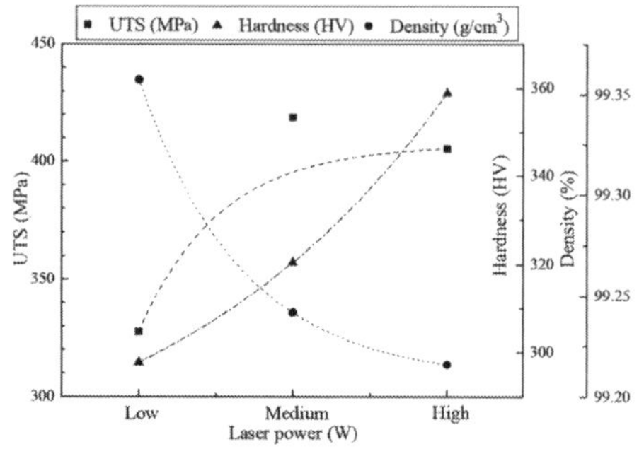

Figure 2. Effects of laser power on mechanical properties.

analyze the relationship based on the profile as appeared in Figure 2.

The results obtained show that with the increasing laser power, the hardness improves greatly from 297.73 HV to 358.87 HV. Microhardness measurements were also made across the layers from the outer edge to the core of the sample. The microhardness analysis showed that there is reduction in the hardness from the outer edge to the core. At the beginning of material deposition, the heat quickly dissipated via heat conduction of the workpiece. The initial thermal transience produced a rapid quenching rate effect at the beginning state of the laser deposition process, which resulted in increased hardness (Amine et al. 2014).

Based on the printed samples' density measurements, printing PH15-5 with laser power ranging from 1000-1400 W yields high density parts. In comparison to the typical density of an annealed PH15-5 bar of 7.78 g/cm^3, the samples obtained high relative density of > 98% which is exceeding the benchmark (95%) set for determining high density printed parts (Sufiiarov et al. 2017). There is also a direct relationship between density and hardness such that higher density is a result of closer packing of atoms, leading to higher hardness.

Figure 2 also demonstrates a trend where the printed parts become denser as the laser power increases. This is attributed to the powder being fully melted, enabling effective layer-wise bonding of material. The mechanical properties of the DED printed parts were analyzed to be associated with the internal features. CT images of the cross-sections printed with low power and high power revealed the internal features of the parts as shown in Figure 3.

It is observed in Figure 3(a) that there were multiple microcracks present in parts that were printed using lower laser power. During the DED build process, residual stresses were developed within the material (Wu et al. 2014). The rapid heating and cooling cycles associated with the laser deposition process increase the thermal stress and as a result, the dislocation density. Moreover, lower laser power results in insufficient energy to fully melt and

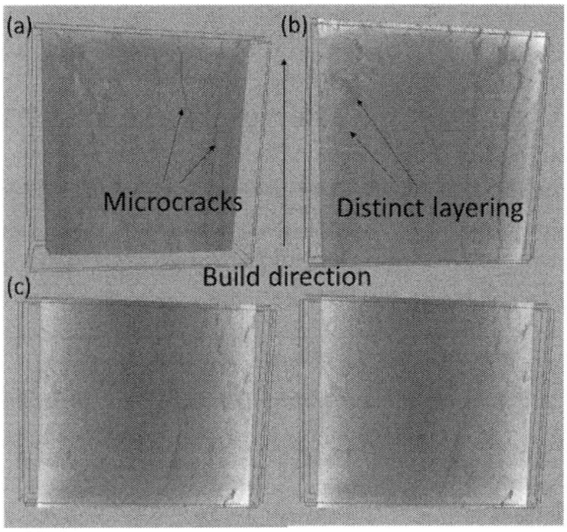

Figure 3. CT images of cross-section showing (a) microcracks present in parts printed with low laser power; (b) distinct layering was observed due to partial melting; (c) significantly fewer microcracks in parts printed with higher laser power.

Table 3. Density and hardness results due to variation in PFR.

Process parameters			Part properties	
	Input variables		Non-destructive test	Destructive test
Scanning strategy	Power, *P*(W)	Powder feed rate, *PFR* (g/min)	Density (%)	Hardness (HV)
Spiral	1400	14.3	99.22	354.69
	1400	13	99.49	300.00
	1400	12.1	99.23	329.98
	1400	11	99.00	285.70
	1400	10	99.20	358.87
Zigzag	1400	14.3	99.61	342.19
	1400	13	99.50	338.65
	1400	12.1	99.65	332.88
	1400	11	99.40	297.23
	1400	10	99.64	325.81

inadequate penetration of the molten pool of the upper layer into the previously deposited layer. Consequently, this generates lack of fusion voids which causes distinct layering to be seen from the CT images in Figure 3(b). Microcracks, porosity and lack of fusion defects are the most common internal defects seen in parts built by DED. The morphology, location, size and volume fraction of internal defects are the primary variables that have significant impact on the mechanical properties.

On the other hand, increasing the laser power reduces the formation of microcracks and lack of fusion voids. Higher heat input effectively minimizes the lack of fusion voids by increasing the pool depth with sufficient penetration into the previously deposited layer, thus ensuring proper bonding as seen in Figure 3(c). As a result, it was reiterated in Figure 2 that the ultimate tensile strength (UTS) increases with increasing laser power due to minimization of internal defects that bring adverse effects on mechanical properties.

3.2 *Effects of powder feed rate on mechanical properties*

Powder feed rate (*PFR*) is also considered as one of the critical process parameters in DED process that determines whether there is effective material joining. Table 3 shows the density and hardness results when PFR is varied.

According to the results in Table 3, the variations in powder feed rate affect the mechanical properties in terms of density and hardness. This is attributed to the amount of powder deposited that determines the deposition efficiency of the printing process. As the amount of powder deposited increases, there is insufficient energy input to melt the powder fully given the same amount of laser power. Laser attenuation occurs as a result of increasing powder feed rate that causes a slight decrease in the mean temperature of the powder stream. This is substantiated by the CT images of parts printed with higher powder feed rate as shown in Figure 4(a). There is a significant presence of lack of fusion defects and pores which can be detrimental to the mechanical properties.

Conversely, no lack of fusion defects was found in samples that were printed with lower powder feed rate (as seen from Figure 4(b), but at the resolution of the X-ray CT scanning parameters used, a distribution of small spherical pores and inclusions with high contrast were detected. The porosity level was found to be in the range of 0.09 to 0.11%, which is considerably low. Hence, parts printed with lower powder feed rate exhibit denser property and higher hardness as compared to parts printed with higher powder feed rate. In particular, as the need arises for large structures to be

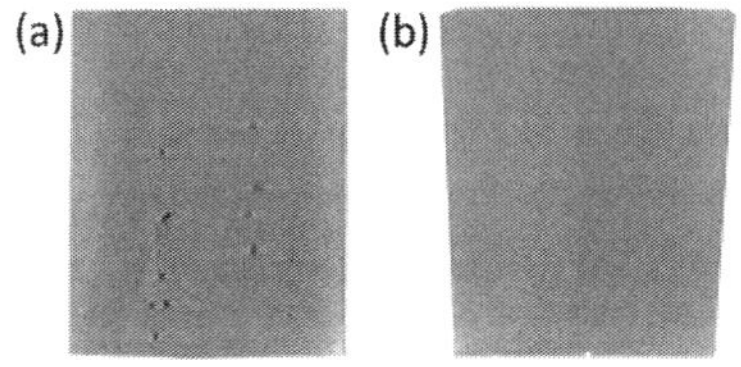

Figure 4. CT images of cross-sections for samples built with (a) high powder feed rate; (b) low powder feed rate.

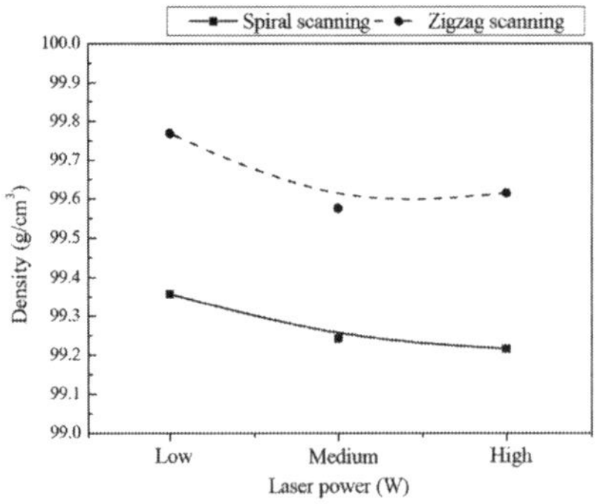

Figure 5. Density comparison between parts printed with spiral and zigzag scanning strategies.

fabricated by DED, higher heat inputs and laser powers will be required corresponding to higher powder feed rate/deposition rates.

3.3 *Effects of scanning strategies on mechanical properties*

There are several scanning strategies that are available for DED process, out of which only the spiral and zigzag patterns were investigated. They are the common scanning strategies that operators usually employ during their printing (Yu et al., 2011). A comparison in terms of density was plotted between parts printed with spiral and zigzag scanning strategies as shown in Figure 5.

Figure 5 shows that the zigzag scanning strategy produces parts of higher density as compared to parts that were fabricated using the spiral scanning strategy. The values were discovered to be relatively high (> 99% density values) with negligible changes when laser power increases. The slightly lower values of parts printed with spiral scanning may be a result of slightly higher porosity levels. Scanning strategies did not have any considerable effect on the hardness of the samples as well as they held similar martensitic microstructure due to similar laser power and powder feed rate processing parameters. Hence, changing the deposition path does not have a significant influence in the melting or deposition efficiency unlike the effects of laser power and powder feed rate.

4 CONCLUSIONS

PH15-5 stainless steel parts were fabricated using laser-based directed energy deposition additive manufacturing in order to elucidate the effects of processing parameters on the internal features and mechanical properties of the printed parts. Based on the effects of the three influencing factors, the laser power has the largest impact, followed by powder feed rate and scanning strategies. Mechanical properties in terms of density and hardness are highly dependent on whether the powder has been being fully melted, enabling effective layer-wise bonding of material. High laser power is necessary to minimize the presence of internal defects such as porosity, microcracks and lack of fusion voids in the printed parts. The CT images revealed that low laser power causes inadequate penetration of the molten pool of the upper layer into the previously deposited layer, hence generating distinct layering that adversely affect the mechanical properties. Moreover, high powder feed rate also contributes to laser attenuation that reduces the effective bonding between layers. Conversely, scanning strategies showed minimal effects on the mechanical properties, but using zigzag scanning produces slightly denser parts than spiral scanning.

ACKNOWLEDGEMENTS

This research was supported by Nanyang Technological University Singapore and Y.Y.C Choong also wishes to acknowledge HP-NTU Digital Manufacturing Corporate Lab.

REFERENCES

Amine, T., Newkirk, J. W., & Liou, F. (2014). An investigation of the effect of direct metal deposition parameters on the characteristics of the deposited layers. *Case Studies in Thermal Engineering*, *3*, 21–34.

Choong, Y. Y. C., Chua, G. K. H., & Wong, C. H. (2018). Investigation on the integral effects of process parameters on properties of selective laser melted stainless steel parts. *Proceedings of the 3rd International Conference on Progress in Additive Manufacturing (Pro-AM 2018)*, 274–279.

Choong, Y. Y. C., Maleksaeedi, S., Eng, H., Su, P.-C., & Wei, J. (2016). Curing characteristics of shape memory polymers in 3D projection and laser stereolithography. *Virtual and Physical Prototyping*, 1–8.

Choong, Y. Y. C., Maleksaeedi, S., Eng, H., Wei, J., & Su, P.-C. (2017). 4D printing of high performance shape memory polymer using stereolithography. *Materials & Design*, *126*, 219–225.

Choong, Y. Y. C., Maleksaeedi, S., Eng, H., Yu, S., Wei, J., & Su, P.-C. (2020). High speed 4D printing of shape memory polymers with nanosilica. *Applied Materials Today*, *18*, 100515.

Choong, Y. Y. C., Saeed, M., Eng, H., & Su, P.-C. (2016). Curing behaviour and characteristics of shape memory polymers by UV based 3D printing. *Proceedings of the 2nd International Conference on Progress in Additive Manufacturing (Pro-AM 2016)*, 349–354.

Chua, G. K. H., Choong, Y. Y. C., & Wong, C. H. (2018). Investigation of the effects on the print location during selective laser melting process. *Proceedings of the 3rd International Conference on Progress in Additive Manufacturing (Pro-AM 2018)*, 613–618.

Costa, L., & Vilar, R. (2009). Laser powder deposition. *Rapid Prototyping Journal*, *15*(4), 264–279.

Dai, K., & Shaw, L. (2002). Distortion minimization of laser-processed components through control of laser scanning patterns. *Rapid Prototyping Journal*, *8*(5), 270–276.

DebRoy, T., Wei, H., Zuback, J., Mukherjee, T., Elmer, J., Milewski, J., … Zhang, W. (2018). Additive manufacturing of metallic components–process, structure and properties. *Progress in Materials Science*, *92*, 112–224.

Eng, H., Maleksaeedi, S., Yu, S., Choong, Y., Wiria, F., Tan, C., … Wei, J. (2017). 3D Stereolithography of Polymer Composites Reinforced with Orientated Nanoclay. *Procedia engineering*, *216*, 1–7.

Eng, H., Maleksaeedi, S., Yu, S., Choong, Y. Y. C., & Wiria, F. E. (2017). Development of CNTs-filled photopolymer for projection stereolithography. *Rapid Prototyping Journal*, *23*(1), 129–136.

Hongmin, L. (2003). Technology Research on the Vacuum Ageing Treatment of PH15-5 Precipitated Hardening Stainless Steel [J]. *Development and Application of Materials*, *2*.

Kistler, N. A., Corbin, D. J., Nassar, A. R., Reutzel, E. W., & Beese, A. M. (2019). Effect of processing conditions on the microstructure, porosity, and mechanical properties of Ti-6Al-4V repair fabricated by directed energy deposition. *Journal of Materials Processing Technology*, *264*, 172–181.

Kong, C., Carroll, P., Brown, P., & Scudamore, R. (2007). *The effect of average powder particle size on deposition efficiency, deposit height and surface roughness in the direct metal laser deposition process*. Paper presented at the 14th International Conference on Joining of Materials.

Selcuk, C. (2011). Laser metal deposition for powder metallurgy parts. *Powder Metallurgy*, *54*(2), 94–99.

Sufiiarov, V. S., Popovich, A., Borisov, E., Polozov, I., Masaylo, D., & Orlov, A. (2017). The effect of layer thickness at selective laser melting. *Procedia engineering*, *174*, 126–134.

Tan, H. W., An, J., Chua, C. K., & Tran, T. (2019). Metallic nanoparticle inks for 3D printing of electronics. *Advanced Electronic Materials*, *5*(5), 1800831.

Tan, H. W., Saengchairat, N., Goh, G. L., An, J., Chua, C. K., & Tran, T. (2019). Induction Sintering of Silver Nanoparticle Inks on Polyimide Substrates. *Advanced Materials Technologies*, 1900897.

Wu, A. S., Brown, D. W., Kumar, M., Gallegos, G. F., & King, W. E. (2014). An experimental investigation into additive manufacturing-induced residual stresses in 316L stainless steel. *Metallurgical and Materials Transactions A*, *45*(13), 6260–6270.

Yu, J., Lin, X., Ma, L., Wang, J., Fu, X., Chen, J., & Huang, W. (2011). Influence of laser deposition patterns on part distortion, interior quality and mechanical properties by laser solid forming (LSF). *Materials Science and Engineering: A*, *528*(3), 1094–1104.

Industry 4.0 – Shaping The Future of The Digital World – da Silva Bartolo et al. (eds)
 ISBN 978-0-367-42272-1

Investigation into post-processing electron beam melting parts using jet electrochemical machining

T. Kendall & P. Bartolo
University of Manchester, Manchester, UK

C. Diver
Manchester Metropolitan University, Manchester, UK

D. Gillen
University of Manchester, Manchester, UK & Blueacre Technology, Dundalk, Ireland

ABSTRACT: This paper presents the findings of using jet electrochemical machining (Jet-ECM) as a post-processing technique on electron beam melting (EBM) parts. Poor surface finish is a negative characteristic, typically found in parts produced by additive manufacturing, and is prominent in EBM. Jet-ECM is a non-contact machining technique that uses anodic dissolution to locally remove material from electrically conductive parts. Jet-ECM has been proven to improve the surface finish on metallic parts. This study aims to further investigate the post-processing capability of Jet-ECM on a Ti6Al4V part produced by EBM. The influence of supply current, machining time and feed rate were assessed using a Jet-ECM system under development by Blueacre Technology and the University of Manchester. The study shows surface finish is improved from a starting value of 25 μm Ra and that Jet-ECM is a viable technique for the post processing of metallic printed parts.

1 INTRODUCTION

Additive manufacturing (AM) is a technology that is developing at a rapid pace, especially in the manufacture of metal parts. Unlike traditional manufacturing, AM provides a freedom of design, producing single parts that were unfeasible with traditional techniques, or required the assembly of several sub-parts. Due to the additive nature, AM provides an economical manufacturing option with very little or no material wastage when compared to traditional subtractive manufacturing processes.

While AM can deliver parts of very intricate and complex geometries with a minimum need for post-processing, it is still hampered by low productivity, poor quality and uncertainty of final mechanical properties (Rombouts et al. 2013). Due to this uncertainty in surface quality and final quality, in some cases AM parts require a form of post-processing.

One form of AM is Electron Beam Melting (EBM) or electron-beam additive manufacturing, a powder bed process in which powder material is fused together by an electron beam in a vacuum environment (Murr et al. 2012). EBM has the ability to process a wide variety of metallic alloys, including Ti6Al4V as used in this study. Ti6Al4V is widely used in aerospace and automotive industries as it offers excellent mechanical properties from low density material with strong corrosion resistance caused by the oxide layer. The alloy is also used in biomedical application due to its biocompatibility. Like other powder bed additive manufacturing processes, EBM can produce complex parts with minimal waste which are not feasible with conventional manufacturing techniques.

Parts produced by EBM are associated with a coarser resolution and higher surface roughness when compared to similar laser powder bed techniques (Algardh et al. 2016) and are orders of magnitude higher than conventionally machined surfaces (Nicoletto et al. 2018). For example, a study on the roughness values of Ti-6Al-4V produced by EBM reported Ra values in the order of 20 to 30 μm (Safdar et al. 2012). Surface roughness on EBM parts is dependent on a range of parameters including the raw material size, the system parameters and build orientation as well as other mechanisms such as balling and stepping (Gong et al. 2013). The surface roughness values, as well as the mechanical properties of the produced components are of the upmost importance to the wear characteristics and the usability of parts for certain applications and industries, with properties of EBM parts being widely investigated (Nicoletto et al. 2018, Wang et al. 2017, Cunningham et al. 2017, Klingvall Ek et al. 2016, Algardh et al. 2016, Gong et al. 2013, Safdar et al. 2012, Murr et al. 2012).

For surface modification of AM parts, laser remelting has been widely investigated, including the roughness values, hardness and wear behavior (Xu et al. 2006, Pinto et al. 2003, Lamikiz et al. 2007). Lamikiz et al. found an 80.1% reduction in mean roughness in selective laser sintered (SLS) parts, an alternative AM technology to LMD, with a final surface roughness of 1.49μm Ra (Lamikiz et al. 2007). Yasa & Kruth (2011) performed a study in to laser remelting as a method to enhance surface roughness on parts produced using selective laser melting (SLM), finding the average roughness, Ra, of AISI 316L stainless steel samples decreases from 12 μm down to 1.5 μm, a surface roughness enhancement of about 90%, but at a cost of longer production times (Yasa & Kruth 2011).

Jet-ECM is a non-contact machining technique in which a free jet of electrolyte locally removes material from an electrically conductive workpiece via anodic dissolution. Although the process shares several similarities with conventional electrochemical machining (ECM), unlike conventional ECM, Jet-ECM does not use a tool to dissolve the desired shape on the workpiece, but instead uses a pressurised jet of electrolyte expelled through a nozzle to locally dissolve material. In Jet-ECM, the electrically conductive workpiece is the positively charged electrode (anode), the metallic nozzle is the negative electrode (cathode) and the electrolyte allows the transmission of electrons from the workpiece. Only material exposed to the jet is removed (Natsu, Ikeda & Kunieda 2007), therefore highly specific material removal can be achieved through varying the feedrate, current, working gap, electrolyte concentration, temperature and flow rate. Figure 1 provides an illustration of the process.

Several aspects of Jet-ECM have been investigated over the past two decades, including studies on the generation of 3D surfaces (Natsu, Ikeda & Kunieda 2007, Kawanaka et al. 2014, Schubert et al. 2016), nozzle geometries (Kunieda et al. 2011), surface finishes (Kawanaka & Kunieda 2015, Kawanaka et al. 2014, Hackert-Oschätzchen et al. 2012, Natsu, Ikeda, & Kunieda 2007) and various materials (Liu et al. 2017, Jain et al. 2016, Hackert-Oschätzchen et al. 2016, Mizugai, Shibuya, & Kunieda 2013, Hackert-Oschätzchen et al. 2013). However, no full investigation into using Jet-ECM as a post processing technique on AM parts has been carried out.

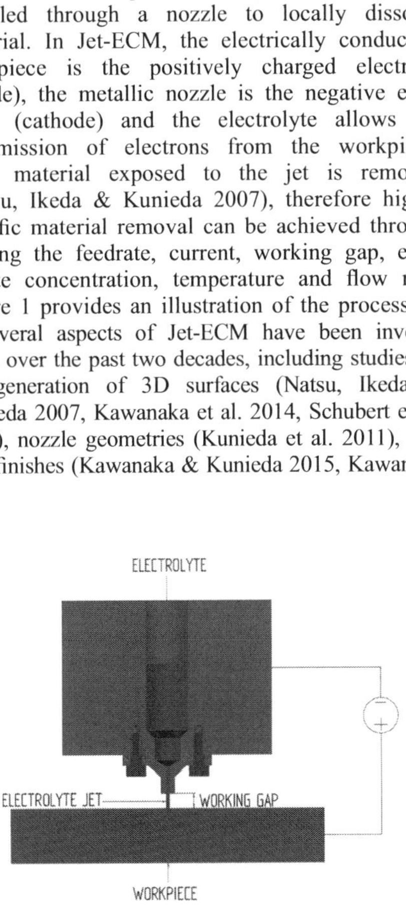

Figure 1. Illustration of jet-electrochemical machining.

Studies have shown that the Jet-ECM process can be used to machine titanium alloys (Speidel et al. 2016, Liu et al. 2017, Mitchell-Smith & Clare 2016), with sodium chloride electrolyte solutions being more effective than sodium nitrate electrolytes in the machining process on TB6 alloy (Liu et al. 2017) and Ti-6Al-4V alloy (Speidel et al. 2016) due to lower passivation caused by sodium chloride electrolytes and higher machining efficiencies achieved (Speidel et al. 2016).

Jet-ECM has been proven as a technique to provide high surface finishes from machining applications, with studies from various research groups producing surface finishes on stainless steel up to 0.1 μm Rz (Natsu, Ikeda, & Kunieda 2007), 1.1 μm Rz and 0.1 μm Ra (Hackert-Oschätzchen et al. 2012), 14.1 nm Ra (Kawanaka et al. 2014) and 0.2 μm Rz (Kawanaka & Kunieda 2015), with mirror finishes being reached in most surface roughness studies. A benefit of JET-ECM over physical ablation and laser processes is that no heat-affected layers, cracks, nor burrs generated as a result of the process (Kai et al. 2012). Jet-ECM also has the ability to alter and produce complex surfaces (Schubert et al. 2016, Kawanaka et al. 2014, Natsu, Ooshiro, & Kunieda 2008, Hackert, Meichsner, & Schubert 2008) which lends the process to post processing EBM parts.

In further comparison to other techniques, Jet-ECM has the ability to machine weak, thin and brittle parts where traditional techniques would struggle, has the ability to remove surface undulations unlike electro polishing, and has the ability to produce a better surface finish than electro-discharge machining (EDM). A combination of all these factors leads to a belief that Jet-ECM is worth further investigation as a post-processing technique.

This investigation will use a Jet-ECM system to post-process Ti6Al4V parts produced by EBM. The applied current, feed rate and machining time will be varied, and the resulting machined profiles and surface qualities compared. The results will be used to indicate the potential suitability of Jet-ECM as a post-processing technique on titanium alloy parts.

2 METHODOLOGY

2.1 *Experimental equipment*

This investigation was performed on the Jet-ECM R500 system currently under development by Blueacre Technology in collaboration with the University of Manchester School of Mechanical, Aerospace and Civil engineering.

The system is comprised of a three axis gantry system providing a working envelope of 400 x 400 x 150 mm with a positional accuracy of between 20 and 40 microns dependent on axis, and maximum repeatability of 12 microns. The z-axis houses the small diameter nozzle (0.3-1 mm diameter) and is electrically isolated from the axis via an insulating block. The electrolyte is circulated from the reservoir to the nozzle via a near pulseless, variable flow micropump and filter system. A flow controller circuit varies the flow provided to the nozzle.

The machining current is provided to the nozzle and workpiece by a 1.5 kW programmable power supply from B and K precision. A short circuit detection system is installed to provide an accurate working gap between the nozzle and workpiece. A sensor array monitors pH, temperature, flow rate and pressure. This data is collected, visualized and transferred via National Instruments hardware and LabView software. The path and velocity of the nozzle movement and the electrical output parameters are programmed and controlled via a PC.

2.2 *The workpieces*

The first workpiece used for this study is a Ti6Al4V bone plate manufactured using EBM at The University of Manchester. The EBM Arcam A2 model (Arcam, Sweden) was used for printing the workpieces. The samples were produced at 60 kV under a vacuum pressure of 2.0×10^{-3} mBa, a scanning speed of 4530 mm/s, beam focus offset of 3 mA, line offset of 0.1 mm, substrate plate temperature of 600 °C and build temperature of 750 °C.

The powder used was Ti6Al4V with powder diameter of 45 to 100 μm, composed of 6.04% of aluminum, 4.05% of vanadium, 0.013% of carbon, 0.0107% of iron, 0.13% of oxygen with the balance titanium.

The material layer thickness was 50 μm, and the process used a low beam current and high scanning speed to reduce residual stresses in the component.

The workpieces contain complex internal features and several thin walled sections. For this study, the workpiece will be referred to as the AM part.

Figure 2. Material surface on AM part, x100 magnification.

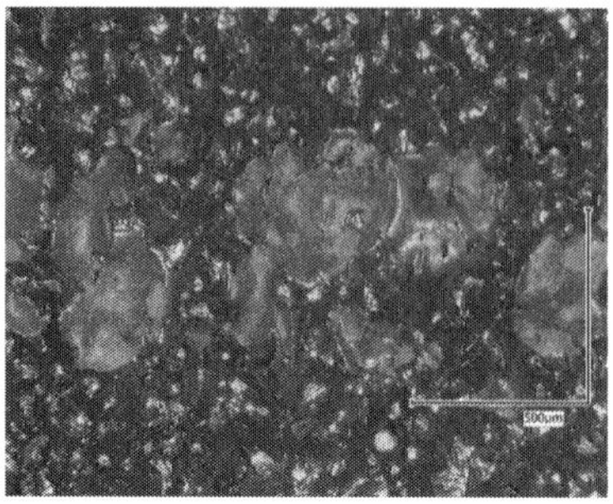

Figure 3. Material surface on AM part from trial 6, x200 magnification.

Using a NanoFocus μscan non-contact optical profilometer the surface roughness on the top surface was found to be approximately 25 μm Ra when averaged from three areas on the surface. The cutoff length used was 0.8mm, measured over a length of 4.5mm. Although the top surface was flat with good dimensional accuracy, the surface, even under no magnification, was visibly rough, with residual powder particles recognisable. Figure 3 shows the top workpiece surface at × 100 magnification, in which the surface irregularities and undulations can be seen.

To compare the part produced by EBM to one produced by traditional methods, a second workpiece of Ti6Al4V plate stock was used of length, width and height of 100 mm, 60 mm and 5 mm respectively. Using the same measurement method as previously outlined, the surface roughness of the plate stock was found to be 1.24 50 μm Ra. Whereas the AM part has a rough, irregular surface as well as a porous structure, the plate stock has a smooth, regular surface with no porosity in the part. For this study, this workpiece will be referred to as the plate stock.

2.3 *Experimental method*

The same sets of experimental parameters were used on both workpieces (Table 1). For the experiments, the same experimental parameters were used to machine channels on the top surface of each workpiece over a length of 7mm, perpendicular to the longest axis of the part. For the first five trials on each workpiece, the feedrate was kept constant at 0.05mm/s, while the current was increased from 0.05 A to 0.25 A in 0.05A intervals so that both the effect of machining current and federate can be analysed. The same five current values were then used at the other feedrates of 0.1, 0.2 and 0.3 mm/s. All other experimental conditions were set to be constant, however fluctuations in pressure and flow from the pump cause a slight variation in the conditions, these experimental conditions can be seen in Table 2.

Table 1. Experimental parameters.

	Current	Feedrate
Trial	Amps	Mm/s
1	0.05	0.05
2	0.1	0.05
3	0.15	0.05
4	0.2	0.05
5	0.25	0.05
6	0.05	0.1
7	0.1	0.1
8	0.15	0.1
9	0.2	0.1
10	0.25	0.1
11	0.05	0.2
12	0.1	0.2
13	0.15	0.2
14	0.2	0.2
15	0.25	0.2
16	0.05	0.3
17	0.1	0.3
18	0.15	0.3
19	0.2	0.3
20	0.25	0.3

Table 2. Experimental conditions.

Electrolyte	Sodium chloride aqueous solution
Electrolyte electrical conductivity	170 mS/cm
Flow rate	1.7-1.8 l/min
Working gap	0.5 mm
Nozzle diameter	0.5 mm
Electrolyte pressure	1.9-2.0 bar

Following the machining trials, the sample was cleaned in an ultrasonic bath of water and soap to remove any machining debris and residual sodium chloride on the surface. Each machined sample was analysed under a Keyence VHX-500 digital microscope to calculate the machined depth and width of the channels, and to provide a visual analysis of the surface On those areas that indicated successful machining, the surface roughness values were measured on a NanoFocus µscan non-contact optical profilometer.

3 RESULTS AND DISCUSSION

The surface profiles, channel widths, depths and the images presented in this section were all taken on a Keyence VHX 500 digital microscope. The values for surface roughness were calculated on a NanoFocus µscan non-contact optical profilometer. A surface roughness reading taken along the centre of the channel, in the direction parallel to the longest edge of the groove, and a reading 0.2 mm either side of the centre were averaged to provide the roughness value for each trial. The cutoff length used was 0.8mm, measured over a length of 4.5mm.

Due to the nature of the surface on the AM part, measuring the depth and width of the machined channel proved difficult if the material removal was very low (< 20 µm depth). At this depth, the powder particles on the surface of the part are larger in height than the depth of machining and therefore an upper surface to measure depth cannot be attempted. In these cases, the machined depth and width are recorded as zero, as seen in trials 6, 11, 16 and 17 in Table 3.

It should be noted in these cases there is still a visible change on the surface of the material, as seen in Figure 3, however the machined areas are not continuous across the length of the channel, with gaps between the machined areas.

The results from trials 6, 11, 16 and 17 indicate either the applied current was too low to produce constant machining, i.e. the current was on the limit of machining, the transitional feedrate was too high to allow prolonged machining, or most likely a combination of both factors.

It can be seen in Figure 5 that for all trials performed on the AM part, the surface roughness was improved from an original value of 25 µm Ra, with the best surface roughness value of 7 µm Ra found at an applied current of 0.15A and 0.05 mm/s feedrate.

The general trends in Figure 5 suggest that the surface roughness improves with increased current, and also with decreased feedrate, with several anomalies in the results. these anomalous results, found in trials 9, 10 and 18 may be due to the measurement technique used, and do not fully represent the entire machined channel in each case. alternatively, the surface roughness post process is also driven by the surface roughness pre-process, where a rougher surface on the pre-machined part leads to a rougher surface after processing. as the surface of the am part was not homogenous, the surface roughness is not a set variable and will differ throughout the surface and impact the roughness value achieved.

in accordance with faraday's law of electrolysis for a constant current, as the applied current is increased, the mass of material removed increases linearly as in equation 1 below:

$$m = ItM/Fz \qquad (1)$$

where m = mass of the substance removed, I = the applied current, M = molar mass of the substance,

Table 3. Experimental results.

	Machined Depth		Machined Width		Surface Roughness	
	μm		μm		μm Ra	
Trial	AM part	Plate Stock	AM part	Plate Stock	AM Part	Plate Stock
1	20	26	690	1072	19.1	0.9
2	102	61	904	1212	12.6	4.3
3	148	103	1008	1230	7.0	2.0
4	176	122	956	1150	7.9	1.0
5	216	181	966	1153	8.2	2.8
6	0	19	0	908	21.5	1.8
7	83	28	740	1111	15.8	3.7
8	96	63	850	1054	12.9	2.8
9	102	90	846	982	13.7	1.7
10	171	119	881	948	12.9	2.1
11	0	12	0	676	23.6	3.1
12	38	23	692	772	20.7	8.4
13	83	31	839	839	18.2	4.6
14	59	58	799	749	13.4	8.7
15	69	63	771	820	8.6	8.1
16	0	0	0	0	22.5	4.2
17	0	13	0	805	19.9	4.8
18	21	30	681	850	21.9	6.7
19	42	50	771	946	13.6	9.4
20	59	54	823	970	13.2	9.8

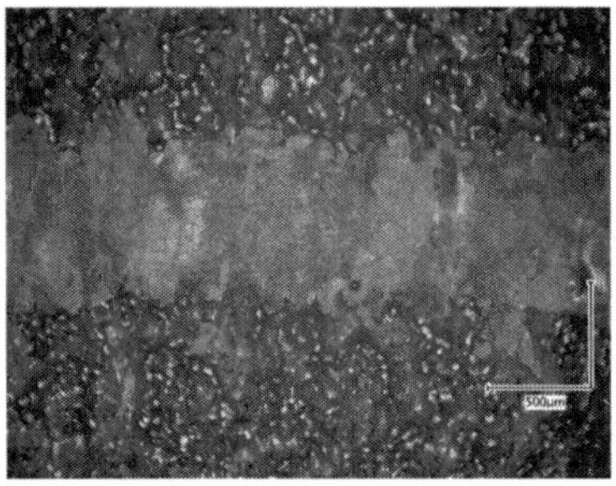

Figure 4. Material surface on AM part from trial 3, x 100 magnification.

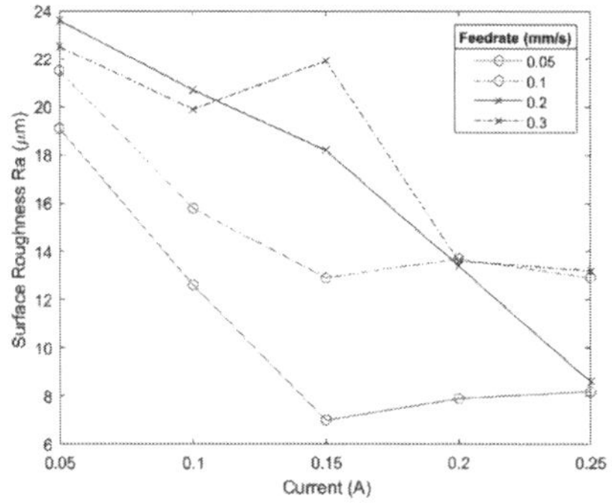

Figure 5. Surface roughness vs current at all feed rates on AM part.

F = the Faraday constant and z = the valency of the substance.

Due to the nature of material removal in Jet-ECM where the current reduces dramatically in the radial direction of the electrolyte jet, an increase in current tends to have a larger effect on the machined depth than the machined width. In turn, the mass of material removed is indicated by the depth of the machined channel.

It can be seen in Figure 6 that the machined depth increases with increasing current for the trials performed at 0.05 mm/s and 0.1 mm/s feedrate on both the AM part and the plate stock. Also in accordance with Faraday's law (Equation 1) the depth of cut decreases with increased feed rate as the machining time is reduced. The largest channel depth achieved using the parameters in this experiment is 216 μm (trial 5), however the results suggest that the surface roughness can be improved with the machined depths of less than 100 μm, which is important for applications where dimensional accuracy is of importance.

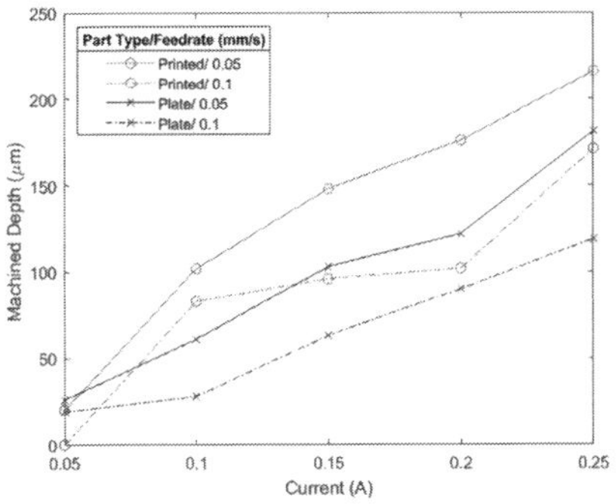

Figure 6. Machined depth vs. current for AM part and plate stock.

The machined depth was consistently higher on the AM part over the plate stock, illustrated in Figure 6. This is due to the density of the machined parts. Although the base material for both parts is of the same density, the parts themselves are not. Whereas the plate stock is homogenous and has no porosity, the AM part is less dense, with porosity throughout the material. As the mass removal is influenced by the current, if the same mass is removed from both parts, the volume of material removed from the AM part will be greater than the plate stock, as is indicated in the found results. This raises an important consideration for the processing of metallic parts produced by additive manufacturing in materials and processes outside of those studied in this paper. If processing is required to a specific tolerance in dimensional accuracy and surface roughness, then the part density and composition must be taken into consideration.

4 CONCLUSION

This study set out to investigate Jet-ECM as a post processing technique for metallic parts produced by EBM. The results indicate that:

- The surface roughness was improved from 25μm Ra on the AM part for all parameters used in this experiment, with the best surface roughness achieved of 7 μm Ra.
- In general, the surface roughness improved with increased current and reduced feedrate, however due to the nature of the part there were some anomalies in the trend.
- The surface roughness was improved on the AM part with low material removal (<100 μm machined depth).
- Machined depth increased with current on both the AM part and the plate stock, however due to the density of the part, for the same machining parameters, the depth was greater on the AM part.

The results from this study indicate that Jet-ECM has a positive effect on the surface roughness of the AM part in this study and should be further investigated as a post-processing technique for metallic parts produced by additive manufacturing. Further work is required to determine the optimum parameters to allow accurate material removal for producing desired profiles and roughness values on parts produced by EBM in a controlled and repeatable manner. Further studies should be performed on AM parts of other materials and produced by other processes such as selective laser sintering and laser metal deposition.

Currently, the process is slow, and is limited to the outer surfaces of components; therefore more development of the process is required for Jet-ECM to be accepted as a viable option for wider use.

ACKNOWLEDGEMENT

The authors would like to acknowledge the support of Blueacre Technology for both their financial assistance and sponsorship in this research, as well as the support of the DTP scholarship through ESPRC.

REFERENCES

Algardh, Joakim Karlsson, Timothy Horn, Harvey West, Ronald Aman, Anders Snis, Håkan Engqvist, Jukka Lausmaa, and Ola Harrysson. 2016. "Thickness dependency of mechanical properties for thin-walled titanium parts manufactured by Electron Beam Melting (EBM)®." *Additive Manufacturing* no.12:45–50. doi: https://doi.org/10.1016/j.addma.2016.06.009.

Cunningham, Ross, Andrea Nicolas, John Madsen, Eric Fodran, Elias Anagnostou, Michael D. Sangid, and Anthony D. Rollett. 2017. "Analyzing the effects of powder and post-processing on porosity and properties of electron beam melted Ti-6Al-4V." *Materials Research Letters* no. 5 (7):516–525. doi: 10.1080/21663831.2017.1340911.

Gong, Haijun, H. Rafi, Thomas Starr, and Brent Stucker. 2013. The Effects of Processing Parameters on Defect Regularity in Ti-6Al-4V Parts Fabricated By Selective Laser Melting and Electron Beam Melting. In *24th Annual International Solid Freeform Fabrication Symposium*. Austin, Tex, USA: University of Texas.

Hackert-Oschätzchen, M., N. Lehnert, A. Martin, and A. Schubert. 2016. "Jet Electrochemical Machining of Particle Reinforced Aluminum Matrix Composites with Different Neutral Electrolytes." *IOP Conference Series: Materials Science and Engineering* no. 118 (1):012036.

Hackert-Oschätzchen, Matthias, André Martin, Gunnar Meichsner, Mike Zinecker, and Andreas Schubert. 2013. "Microstructuring of carbide metals applying Jet Electrochemical Machining." *Precision Engineering* no. 37 (3):621–634. doi: http://dx.doi.org/10.1016/j.precisioneng.2013.01.007.

Hackert-Oschätzchen, Matthias, Gunnar Meichsner, Mike Zinecker, André Martin, and Andreas Schubert. 2012. "Micro machining with continuous electrolytic free jet." *Precision Engineering* no. 36 (4):612–619. doi: http://dx.doi.org/10.1016/j.precisioneng.2012.05.003.

Hackert, M., G. Meichsner, and A. Schubert. 2008. "Generating micro geometries with air assisted jet electrochemical machining." *European Society for Precision Engineering and Nanotechnology, EUSPEN 2008.* no. 2:420–424.

Jain, N. K., A. Potpelwar, Sunil Pathak, and N. K. Mehta. 2016. "Investigations on geometry and productivity of micro-holes in Incoloy 800 by pulsed electrolytic jet drilling." *The International Journal of Advanced Manufacturing Technology* no. 85 (9):2083–2095. doi: 10.1007/s00170-016-8342-9.

Kai, Shoya, Haruo Sai, Masanori Kunieda, and Heikan Izumi. 2012. "Study on Electrolyte Jet Cutting." *Procedia CIRP* no. 1:627–632. doi: http://dx.doi.org/10.1016/j.procir.2012.05.011.

Kawanaka, Takuma, Shigeki Kato, Masanori Kunieda, James W. Murray, and Adam T. Clare. 2014. "Selective Surface Texturing Using Electrolyte Jet Machining." *Procedia CIRP* no. 13:345–349. doi: http://dx.doi.org/10.1016/j.procir.2014.04.058.

Kawanaka, Takuma, and Masanori Kunieda. 2015. "Mirror-like finishing by electrolyte jet machining." *CIRP Annals - Manufacturing Technology* no. 64 (1):237–240. doi: http://dx.doi.org/10.1016/j.cirp.2015.04.029.

Klingvall Ek, Rebecca, Lars-Erik Rännar, Mikael Bäckstöm, and Peter Carlsson. 2016. "The effect of EBM process parameters upon surface roughness." *Rapid Prototyping Journal* no. 22 (3):495–503. doi: 10.1108/RPJ-10-2013-0102.

Kunieda, M., K. Mizugai, S. Watanabe, N. Shibuya, and N. Iwamoto. 2011. "Electrochemical micromachining using flat electrolyte jet." *CIRP Annals - Manufacturing Technology* no. 60 (1):251–254. doi: http://dx.doi.org/10.1016/j.cirp.2011.03.022.

Lamikiz, A., J. A. Sánchez, L. N. López de Lacalle, and J. L. Arana. 2007. "Laser polishing of parts built up by selective laser sintering." *International Journal of Machine Tools and Manufacture* no. 47 (12–13):2040–2050. doi: http://dx.doi.org/10.1016/j.ijmachtools.
2007.01.013.

Liu, Weidong, Sansan Ao, Yang Li, Zuming Liu, Zhengming Wang, Zhen Luo, Zhiping Wang, and Renfeng Song. 2017. "Jet electrochemical machining of TB6 titanium alloy." *The International Journal of Advanced Manufacturing Technology* no. 90 (5):2397–2409. doi: 10.1007/s00170-016-9500-9.

Mitchell-Smith, J., and A. T. Clare. 2016. "ElectroChemical Jet Machining of Titanium: Overcoming Passivation Layers with Ultrasonic Assistance." *Procedia CIRP* no. 42:379–383. doi: http://dx.doi.org/10.1016/j.procir.2016.02.215.

Mizugai, K., N. Shibuya, and M. Kunieda. 2013. "Study on Electrolyte Jet Machining of Cemented Carbide." *International Journal of Electrical Machining* no. 18:23–28. doi: 10.2526/ijem.18.23.

Murr, Lawrence E., Sara M. Gaytan, Diana A. Ramirez, Edwin Martinez, Jennifer Hernandez, Krista N. Amato, Patrick W. Shindo, Francisco R. Medina, and Ryan B. Wicker. 2012. "Metal Fabrication by Additive Manufacturing Using Laser and Electron Beam Melting Technologies." *Journal of Materials Science & Technology* no. 28 (1):1–14. doi: https://doi.org/10.1016/S1005-0302(12)60016-4.

Natsu, W., S. Ooshiro, and M. Kunieda. 2008. "Research on generation of three-dimensional surface with microelectrolyte jet machining." *CIRP Journal of Manufacturing Science and Technology* no. 1 (1):27–34. doi: http://dx.doi.org/10.1016/j.cirpj.2008.06.006.

Natsu, Wataru, Tomone Ikeda, and Masanori Kunieda. 2007. "Generating complicated surface with electrolyte jet machining." *Precision Engineering* no. 31 (1):33–39. doi: http://dx.doi.org/10.1016/j.precisioneng.
2006.02.004.

Nicoletto, G., R. Konečná, M. Frkáň, and E. Riva. 2018. "Surface roughness and directional fatigue behavior of as-built EBM and DMLS Ti6Al4V." *International Journal of Fatigue* no. 116:140–148. doi: https://doi.org/10.1016/j.ijfatigue.2018.06.011.

Pinto, Maria Aparecida, Noé Cheung, Maria Clara Filippini Ierardi, and Amauri Garcia. 2003. "Microstructural and hardness investigation of an aluminum–copper alloy processed by laser surface melting." *Materials Characterization* no. 50 (2–3):249–253. doi: http://dx.doi.org/10.1016/S1044-5803(03)00091-3.

Rombouts, M., G. Maes, W. Hendrix, E. Delarbre, and F. Motmans. 2013. "Surface Finish after Laser Metal Deposition." *Physics Procedia* no. 41:810–814. doi: http://dx.doi.org/10.1016/j.phpro.2013.03.152.

Safdar, Adnan, Wise He, Liu-Ying Wei, Anders Snis, and Luis Chávez de Paz. 2012. *Effect of process parameters settings and thickness on surface roughness of EBM produced Ti-6Al-4V.* Vol. 18.

Schubert, Andreas, Matthias Hackert-Oschätzchen, André Martin, Sebastian Winkler, Danny Kuhn, Gunnar Meichsner, Henning Zeidler, and Jan Edelmann. 2016. "Generation of Complex Surfaces by Superimposed Multi-dimensional Motion in Electrochemical Machining." *Procedia CIRP* no. 42:384–389. doi: http://dx.doi.org/10.1016/j.procir.2016.02.216.

Speidel, Alistair, Jonathon Mitchell-Smith, Darren A. Walsh, Matthias Hirsch, and Adam Clare. 2016. "Electrolyte Jet Machining of Titanium Alloys Using Novel Electrolyte Solutions." *Procedia CIRP* no. 42:367–372. doi: http://dx.doi.org/10.1016/j.procir.2016.02.200.

Wang, Pan, Wai Jack Sin, Mui Ling Sharon Nai, and Jun Wei. 2017. "Effects of Processing Parameters on Surface Roughness of Additive Manufactured Ti-6Al-4V via Electron Beam Melting." *Materials (Basel, Switzerland)* no. 10 (10):1121. doi: 10.3390/ma10101121.

Xu, W. L., T. M. Yue, H. C. Man, and C. P. Chan. 2006. "Laser surface melting of aluminium alloy 6013 for improving pitting corrosion fatigue resistance." *Surface and Coatings Technology* no. 200 (16–17):5077–5086. doi: http://dx.doi.org/10.1016/j.surfcoat.2005.05.034.

Yasa, E., and J. P. Kruth. 2011. "Microstructural investigation of Selective Laser Melting 316L stainless steel parts exposed to laser re-melting." *Procedia Engineering* no. 19:389–395. doi: http://dx.doi.org/10.1016/j.proeng.2011.11.130.

Industry 4.0 – Shaping The Future of The Digital World – da Silva Bartolo et al. (eds)
© 2021 Taylor & Francis Group, London, ISBN 978-0-367-42272-1

Manufacturing of a hollow propeller blade with WAAM process - from the material characterisation to the achievement

G. Pechet, J.-Y. Hascoet & M. Rauch
Centrale Nantes/GeM - UMR CNRS, Nantes, France
Joint Laboratory of Marine Technology (JLMT) Centrale Nantes - Naval Group

G. Ruckert & A.-S. Thorr
Naval Group Research, Technocampus Ocean, Bouguenais, France
Joint Laboratory of Marine Technology (JLMT) Centrale Nantes - Naval Group

ABSTRACT: Currently, blades of maritime propellers are mostly cast and this manufacturing process is well-known. However, large cast parts suffer geometrical constraints, shrinkage cavity, coarse microstructure, important lead times, etc. Additive manufacturing and more specifically the Wire Arc Additive Manufacturing (WAAM) process is a promising alternative to casting. Centrale Nantes and Naval Group have worked together to manufacture additively a hollow blade by taking advantage of the WAAM process capabilities. Hollow blades will make the propeller lighter, overall improving the hydrodynamics capacities and modifying the noise and the vibrations. The final aim of the project is to build a 6 meters propeller. To achieve this objective, a material characterisation on a duplex stainless steel was performed to ensure good mechanical properties. Then, a preliminary step of process engineering was achieved on sensitive areas of a hollow blade to finally successfully manufacture a demonstrative part.

1 INTRODUCTION

Propeller blades traditionally used on ships are cast metallic parts. The usual manufacturing equipment used for the manufacturing of propeller blades (foundry and machining) limit intrinsically the improvement possibilities of propeller performances (such as efficiency and acoustic discretion) because they limit the practicable outer forms, impose the production of massive parts and limit the possibilities of integration of absorbing materials.

Besides, the production of propeller blades by usual techniques can be a long, complex and material consuming because of the important overthicknesses to be eliminated by machining process.

The problem addressed here is the capability of Additive Manufacturing (AM) to be a credible alternative of the conventional production processes and to authorise future improvements of propellers performances. This technology manufactures parts adding layer by layer the materials, allowing a reduction of material wastage and complex manufacturing parts. AM techniques make it also possible to build functionally graded materials (FGM) previously impossible with common manufacturing as forging or casting (Hascoet et al. 2011; Muller al. 2013). Also, hybrid manufacturing (coupling of additive and subtractive method) combines the best features of both of these approaches, thus offering a wide range of feasibility (Laguionie et al. 2011). The technology Wire Arc Additive Manufacturing (WAAM) uses common welding process as a heat source and wire as feedstock. One of the most commonly used is Metal Inert/Activ Gas (MIG/MAG) which suits perfectly to this application because the arc is directly formed with the workpiece and the consumable wire electrode (no need of external feed). This process is an interesting alternative to other applications due to their numerous advantages: high deposition rate, cost competitiveness, unlimited build envelops. WAAM technology already shown its potential to successfully manufacture high-dimensional parts in various materials like steel, aluminium and titanium (Williams et al. 2015; Hascoet et al. 2017).

Additively manufactured blades have already been produced these last few years in WAAM, as some notable projects have demonstrated. For instance, in France, Centrale Nantes and Naval Group printed a full-scale propeller blade (Queguineur et al. 2017) while the company RAMLAB in Netherlands have manufactured a 3D complete marine propeller (Ya & Hamilton, 2018). Both of these demonstrators are in aluminum-copper alloy.

However, none of these demonstrators used the flexibility offered by additive manufacturing in a more efficient way for the design of blades. In contrast, hollow blades could provide many advantages, such as lighten propellers, reduce noise and vibrations of propellers, improve their hydrodynamic efficiency, etc.

Thus, within the European project RAMSSES, Centrale Nantes and Naval Group joined together in order to build the first hollow blade demonstrator. This article relates firstly the material characterisation realised on the chosen material to ensure it follows the requirements. Then, the focus is made on the process engineering for the blade and the manufacturing of the demonstrator is exposed.

2 MATERIAL CHARACTERISATION

The filer material used here is a duplex stainless steel (G 22 9 3 NL). Duplex stainless steels are dual-phase material made up of a ferrite δ matrix in which grows austenite grains. It confers good mechanical properties and corrosion resistance which is suitable for marine applications. Several test blocks have been built to characterise properties of parts made by filler deposition (Figure 1). Preliminary work was made to identify processing parameters, toolpath strategy. The chosen parameters (wire-feed speed, travel speed, layer height, etc.) have been defined in order to ensure layers overlapping, wetting and an important deposition rate (3.5 kg/h), while limiting spatters.

2.1 *Metallographic examinations*

Metallographic samples were taken over the whole thickness of the product. Several cross sections were extracted from the samples with the aim to assess the ferrite number of the duplex block in addition to underline its internal defects and to examine its microstructure via micrographic and macrographic observations.

Ferrite content measurements underline that the measurements are few scattered around the mean value of 33 ± 1 FN and do not significantly fluctuate from an area to another (Figure 2).

Ferrite content is homogeneous over the whole height of the block. Nevertheless, the ferrite content is lower than a duplex stainless steel which usually contains between 40% to 60% of ferrite in the base material and between 25% to 70% in welds and heat affected zones (or in measurable terms between 35 to 100 FN) (World Conference on Duplex Stainless Steels 1997). Besides, following a chemical composition of WAAM samples, WRC diagram predicts a value of around 50 FN in welding conditions.

Figure 1. Test blocks produced for characterisation.

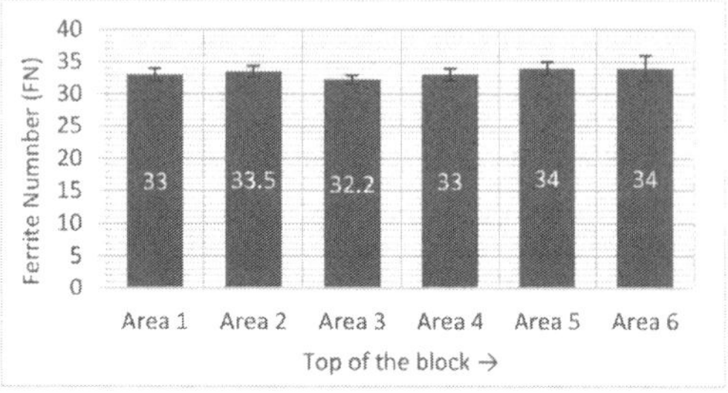

Figure 2. Ferrite number vs the localisation in the block.

Therefore, these unexpected FN values cannot be attributed to the chemical composition. The most plausible hypotheses of this weak ferrite rate are the thermal cycles. Duplex stainless steels solidify in a ferritic field and during cooling, part of the ferrite (δ) transforms (in solid state) into austenite (γ). Additive manufacturing implies the multiplication of layers one above the others. Consequently, heat accumulates layer after layer. Moreover, in welding joints, thermal conduction through the base material is much higher than thermal convection. In WAAM applications it is the opposite, thermal convection prevails. The impact is then a decrease in the cooling rate for WAAM applications compared with welding (Figure 3). Thus, the transformation δ ferrite → austenite (γ) (at the cooling stage) has more time to be achieved.

The macrostructure of the two samples highlights a multilayer structure (Figure 4). Both cross sections do not show any macroscopic volumetric defects. Besides, the layers are well interpenetrated (no lack of fusion or crack is visible). Moreover, no internal discontinuity is visible (inclusions, porosity, etc.).

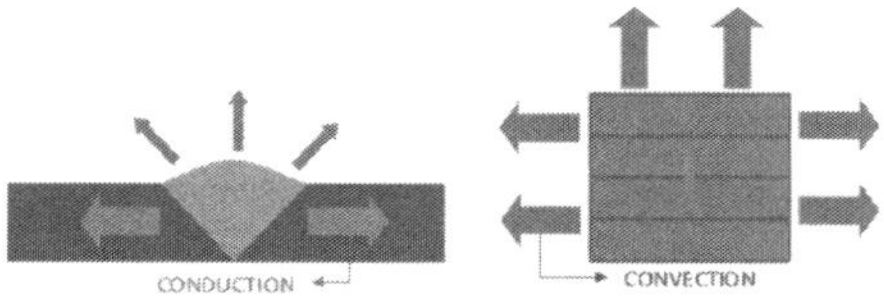

Figure 3. Heat transmission in a) welds, b) WAAM blocks.

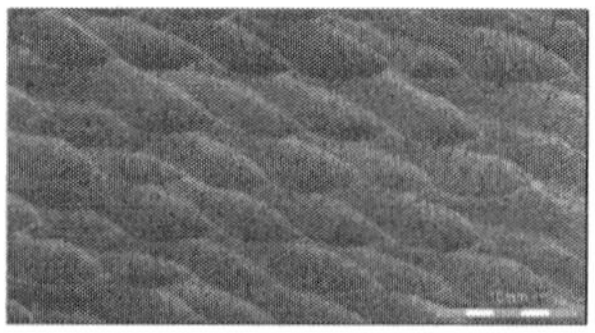

Figure 4. Macrograph of a transverse section – Magnification x4.

Micrographic examinations show classic austeno-ferritic fine microstructure with austenite (γ) in a ferrite δ matrix (Figure 5). Yet, the volume fraction of austenite is predominant and corroborates the ferrite measurements. Figure 6 point out defects such as porosities (less than 60 µm in diameter) and inclusions (few tens of microns) like silicates. If porosities are scarce, microscopic inclusions are recurring. These defects are though microscopic and below quality levels for inspection as defined in the standard (NF EN ISO 5817-1, 2014).

2.2 *Tensile tests*

Tests were realised on 12 samples at room temperature. 6 samples (Longi) were extracted through the longitudinal direction of the block (x-axis) and 6 (Trans) through the long transverse direction (z-axis, see Figure 1). Tensile tests were realised following the standard ISO 6892-1, (2016). As shown in Figure 7, properties measured via tensile tests (yield

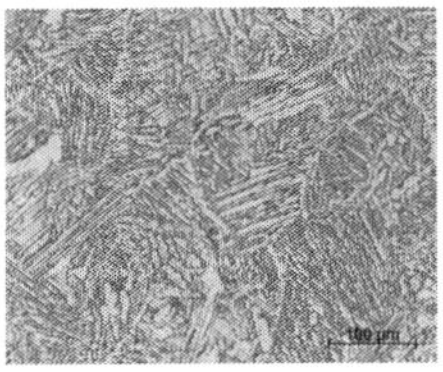

Figure 5. Micrograph of a transverse section – Magnification x200.

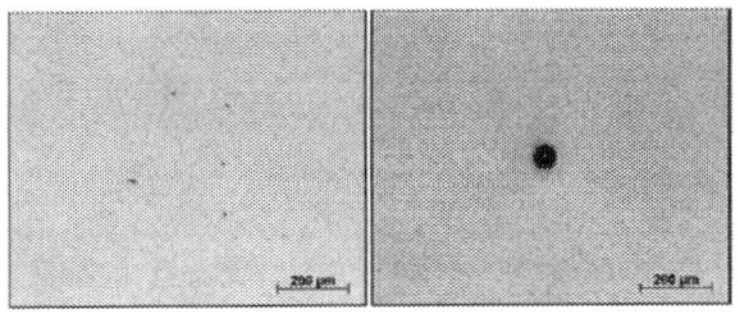

Figure 6. Few defects observed: Inclusions (left) and porosities (right).

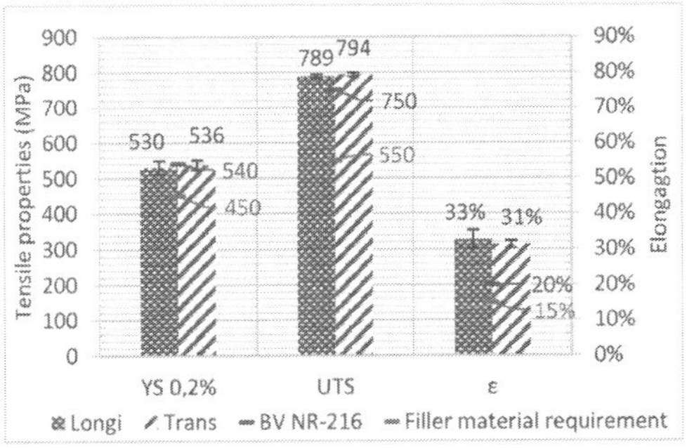

Figure 7. Tensile properties of the WAAM produced tensile samples.

stress, ultimate strength and elongation) are not significantly different whether the sample was extracted through the longitudinal direction of the block or through the long transverse direction of the block especially in considering standard variations.

Moreover, the location of the samples extraction (top, middle or bottom of the block) does not show any particular differences on the mechanical properties.

Tensile tests based on samples extracted from the WAAM produced block indicate that the tensile properties (0.2 offset yield stress, ultimate strength and total elongation) are above minimums required for the filler materials (duplex stainless steel G 22 9 3 NL). Therefore, WAAM process does not deteriorate raw material properties and even override them (e.g. about +40% on the ultimate strength in comparison with requirements, Figure 7). Despite most of the samples show a 0.2% offset yield stress below (but very close) to the requirement of BV rules for a martensitic stainless steel casting (16 Cr 5 Ni), other mechanical characteristics are above BV NR-216 exigencies (Bureau Veritas). Duplex WAAM produced blocks are thereby competitive with martensitic castings even if this material is a priori strengthen.

To conclude, the material characterisation have shown good mechanical properties and validate the use of duplex stainless steel G 22 9 3 NL with the WAAM process for the intended application which is the manufacturing of a hollow propeller blade.

3 MANUFACTURING OF A HOLLOW PROPELLER BLADE

3.1 *WAAM process engineering*

A hollow blade contains specific geometric singularities which lead to key manufacturing difficulties. In order to estimate the feasibility of realisation with the WAAM process, separate test specimens are performed based on each significant area:

1. Cavity opening;
2. Closing cavity;
3. Trailing edge;
4. Leading edge;
5. Monobead edge.

The specimens were all built with a single orientation of the welding torch to rely on the feedback learn with the preliminary experiments and simplify the toolpath generation. Figure 8 shows the identified test parts manufactured. In order to have representative results, the CAD model of the identified singularities are extracted the extent possible from the CAD of the hollow blade. The obtained parts gave excellent results in terms of surface aspects and geometry tolerance compared to the CAD model.

However, some problems were found for the trailing edge. Indeed, the design of the blade has lots of

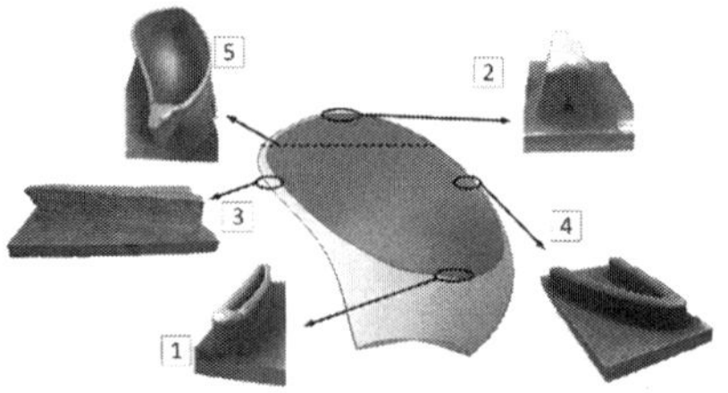

Figure 8. Geometric singularities of a hollow blade. 1: Cavity opening, 2: Closing cavity, 3: Trailing edge, 4: Leading edge, 5: Monobead edge.

areas with slope more or less important. Angles with the WAAM process have consequences on the feasibility of the part. In this document, the terms "positive" and "negative" will refer to the angle of the slope (Figure 9). A part with a negative slope will have a portion of its upper layers without overlapping of previous layers. Depending on the angle of the slope, the manufacturing will not be possible at some point, as the bead will not overlap enough the previous layer. On the other hand, a positive slope will not be a problem to manufacture but it may have an effect on the geometry of the part due to the slicing (staircase effect, common in additive manufacturing).

For the isolated singularity part trailing edge, CAD with the most critical slope was extracted from the blade and reached a 45° negative slope compared to the normal along Z axis. Manufacture of the part showed that the trailing with this too high slope was not possible with a fixed orientation of the torch, see Figure 10. The tip tended to collapse, and the electrical arc was disrupted because the wire did not reach the previous layer, causing wire feeding in the vacuum. Nevertheless, the general aspect of the part is excellent, with good flatness of the layers and correct radius at the extremity of the cavity. Therefore, the cleaning allowance to machine will be limited.

In order to see the maximum permissible angle, two trailing edges with different angles were tested, one with 45° angle and a second with 36° angle. Scans of the two parts showed that the maximum permissible angle for this shape of trailing edge with the chosen parameters with a vertical orientation of the welding torch is around 32° (Figure 11).

To conclude, preliminary experiments showed that the high negative slopes complicate the manufacture of the blade when a single orientation of the torch is used. To ensure the realisation of the part, it will be required to adapt the set-up during manufacturing.

3.2 *Manufacturing of the blade*

First of all, the geometry of the blade comes from a compromise between hydrodynamics properties, strength and mass reduction while being achievable with the WAAM process. The scale of the blade was fixed to 1/3, giving an approximate weight of 320 kg and size of 1 m. The blade was cut with a horizontal plane just above the starting of the hub ensuring to start on a flat baseplate.

Like expose previously, realisation of parts with geometrical singularities and especially the trailing edge showed that a too steep slope leads to lack of matter and non-respect of the geometry model. To overcome these limitations, the built of the blade was divided in two positioning set-ups (Figure 12):

- First set-up starting flat on the baseplate with the CAD model cut with a plane orientated at 30°. The critical point for the change of orientation was determined according to a 30° angle

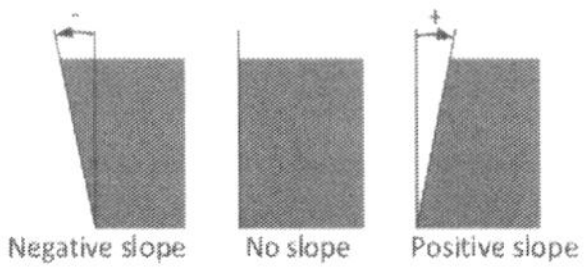

Figure 9. Illustration of the slopes.

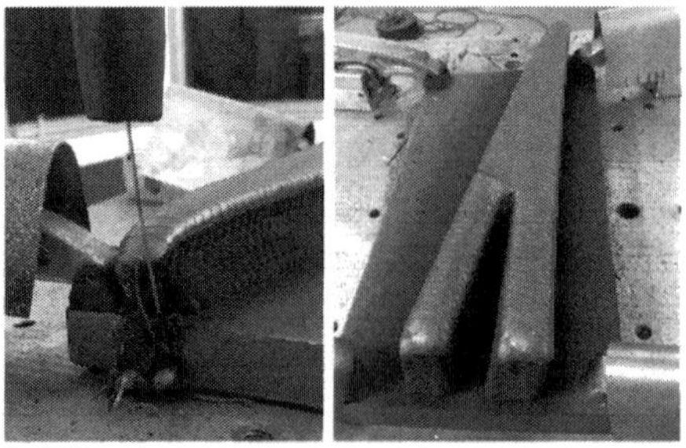

Figure 10. Trailing edge: Arc disruptions (left), General aspect (right).

Figure 11. Scans of trailing edges as-fabricated with 36° (left) and 45° (right) respectively of negative slopes.

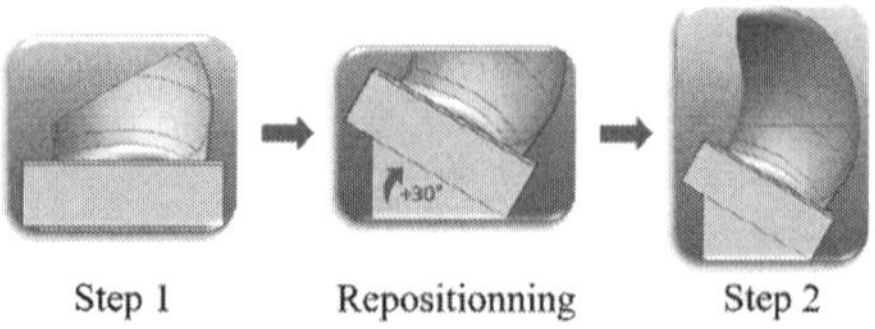

Figure 12. Illustration of the steps of manufacturing.

compared to the Z axis at the leading edge (Figure 12).

- Repositioning with a 30° orientation of the set-up. The previous cut of the CAD model with 30° plane allows to start again on a flat surface. The slopes of the leading and trailing edge are then reduced. Thus, the 45° unfeasible slope (in 3 axis) of the trailing edge tested during the preliminary step become a feasible 15° slope.

The 30° obtained surface was flat and without scale effect. Once the first step was finished, the set-up needed to be orientated to 30°. The new set-up gave excellent result in terms of surface state and flatness (Figure 13).

From a certain height, combination of two phenomenon provoked a change in the state of the melt bead:

- First, slope was increasing near the trailing edge;
- Secondly, thickness of the edges was reducing along the height Z and the toolpath strategy was getting closer to oscillation strategy with one large melt bead.

This large melt bead combined with the increased slope tended to an overflow of the melt on the negative slope and so a smaller layer height in this area of the blade. This caused arc discontinuities as soon the defect increased with the layers (Figure 14).

It was overcome with an adaptation of the parameters in this area. In case of bigger scale of the blade, this phenomenon will probably not happen, as the wall thickness of the blade will be more important.

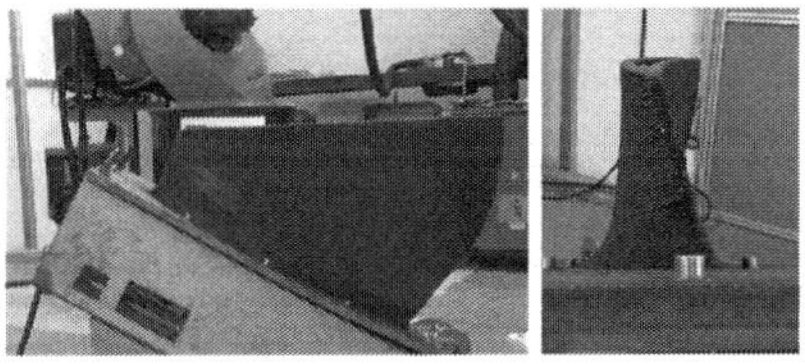

Figure 13. Set-up after the 30° repositioning.

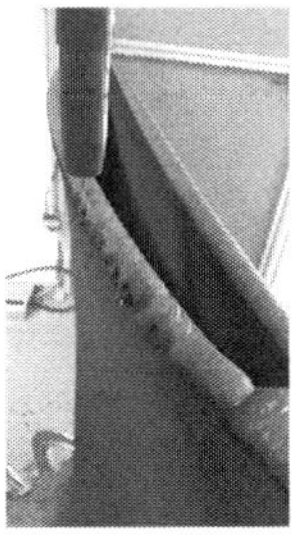

Figure 14. Defect near the trailing edge during manufacturing step 2.

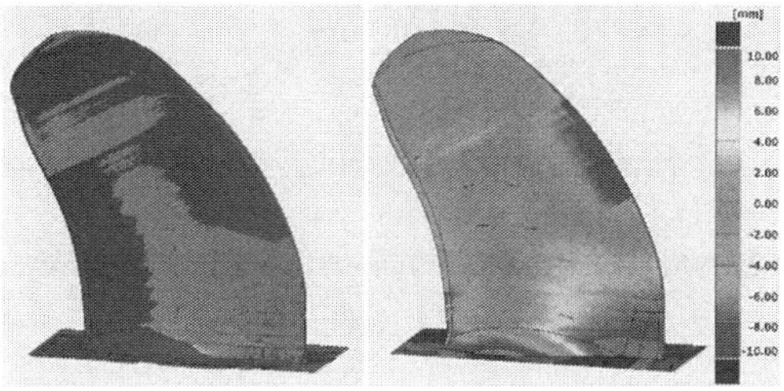

Figure 15. Comparison between CAD model and scan of the blade.

3.3 *Geometrical inspection*

Complete scan of the blade was completed with a handheld laser scanner for 3D inspection. Comparison with CAD model was then realised.

Figure 15 Presents the results obtained: on the right picture, both of CAD model (blue) and scan of the blade (grey) are superposed and the right picture represents the comparison surface on the CAD. There was a deformation of the baseplate. Clamps between the baseplate and the tube were undersized, especially on the leading edge where the plate rose up of several millimetres. This explains the red area with +10 mm of difference with CAD model at the base of the leading edge. These deformations highlight the significant thermic distortions happening during manufacturing of high dimensional parts with WAAM process (baseplate endures high residual stresses which exceed the yield point).

The blue area on the leading edge at the middle height of the blade corresponds to the orientation change of the set-up. This highlight the slight shift which occurred following the change of workplan.

The last third of the blade does not present a big difference with the CAD model. This area matches with the change of strategy and parameters for the closure of the cavity.

4 CONCLUSION

This article shows the potential for added value for industrial applications of the WAAM process, particularly for marine propellers. The realisation of this hollow blade (Figure 16) point out the capacity of additive manufacturing to accomplish new design previously unfeasible and therefore improve hydrodynamics of future propellers. Specific manufacturing methods need to be implemented as well to realise such large components with additive manufacturing.

Mechanical testing demonstrated that the blocks in duplex stainless steel made by filler deposition reach mechanical properties that are above raw materials requirements. No critical defects were

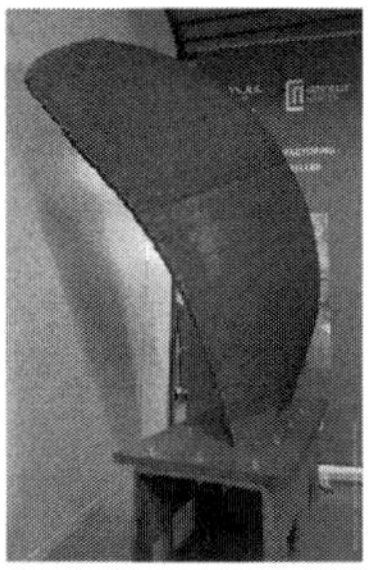

Figure 16. Blade finished.

observed and the implemented WAAM parameters and strategies seem to ensure the integrity of so built metallic parts.

Concerning the realisation of the hollow blade demonstrator, adaptation of the parameters in some areas enabled to manufacture the complete hollow blade. However, multi-axis deposition could facilitate the manufacturing of high offset areas and offer better geometric precision. That is why the next demonstrator (a more than half-scale hollow demo-blade) will be manufacture 5-axis WAAM process and validate the expected benefits.

ACKNOWLEDGEMENTS

The project RAMSSES has received funding under the European Union's Horizon 2020 research and innovation programme under the grant agreement No 723246.

REFERENCES

Bureau Veritas. NR 216.C1 DT R09 E- Bureau Veritas-Rules on Materials and Welding for the Classification of Marine Units. (s.d.).

Hascoet J.-Y., Muller P. & Mognol P. 2011. Manufacturing of Complex Parts with Continuous Functionally Graded Materials (FGM). *Solid Free. Fabr. Symp.*, pp. 557–569, 2011.

Hascoet J.-Y., Querard V. & Rauch M. 2017. Interests of 5 Axis Toolpaths Generation for Wire Arc Additive Manufacturing of Aluminium Alloys. *J. Mach. Eng.*, vol. 13, no. 3, pp. 51–65.

ISO 5817:2014 - Welding — Fusion-welded joints in steel, nickel, titanium and their alloys (beam welding excluded) — Quality levels for imperfections

ISO 6892–1:2016 - Metallic materials - Tensile testing - Part 1: method of test at room temperature

Laguionie R., Rauch M. & Hascoet J.-Y. 2011. A Multi-process Manufacturing Approach Based on STEP-NC Data Model. *Alain Bernard. Global Product Development, Springer, Berlin, Heidelberg*, pp. 253–263.

Muller P., Mognol P. & Hascoet J.-Y. 2013. Modeling and control of a direct laser powder deposition process for Functionally Graded Materials (FGM) parts manufacturing. *J. Mater. Process. Technol.*, vol. 213, no. 5, pp. 685–692.

Queguineur A., Ruckert G., Cortial F. & Hascoet J.-Y. 2017. Evaluation of Wire Arc Additive Manufacturing for large size components in naval applications. *ICWAM Conf.*, pp. 1–6.

Williams S.W., Martina F., Addison A. C., Ding J., Pardal G. & Colegrove P. 2016. Wire + Arc Additive Manufacturing *Mater. Sci. Technol.*, vol. 32, no. 7, pp. 641–647.

World Conference on Duplex Stainless. 1997. Duplex Stainless Steels 5th World Conference, 97th ed., vol. 1. Maastricht, the Netherlands: *KCI Publishing BV.*

Ya W. & Hamilton K. 2018. On-Demand Spare Parts for the Marine Industry with Directed Energy Deposition: Propeller Use Case. *AMPA Conf.*

Industry 4.0 – Shaping The Future of The Digital World – da Silva Bartolo et al. (eds)
© 2021 Taylor & Francis Group, London, ISBN 978-0-367-42272-1

Fully 3D printed horizontally polarised omnidirectional antenna

H.W. Tan & C.K. Chua
Engineering Product Development Pillar, Singapore University of Technology and Design, Singapore

M. Uttamchand & T. Tran
Singapore Centre for 3D Printing, School of Mechanical and Aerospace Engineering, Nanyang Technological University, Singapore

ABSTRACT: The aerospace industry anticipates the use of additive manufacturing (AM) technologies for more significant weight reduction and space utilisation to seek better fuel efficiency and aircraft performances, whereas the electronics industry aims to reduce footprints of electronic devices with increased functionalities. The Computer Simulation Technology (CST) Micro-wave Studio simulation software was used for optimising critical parameters of the printed Alford-loop-structure antenna's designs to achieve an operating frequency in the range 2.4 – 2.5 GHz with good impedance matching. A simulation study was also conducted using frequency domain CST Microwave Solver to find the resonance frequency of Alford-loop-structure antenna and better understand its electrical field distributions. This paper demonstrates a fully 3D printed antenna which is able to achieve a voltage standing wave ratio (VSWR) of approximately 3. The radiation pattern of the antenna in the E-plane is omnidirectional, whereas the radiation pattern in the H-plane is close to omnidirectional.

1 INTRODUCTION

The aerospace and electronics industries express rising interests in direct fabrication of functional antennas and sensors directly onto conformal surfaces through additive manufacturing (AM) technologies (Choong et al. 2017, Choong et al. 2020, Ng et al. 2019, Tan et al. 2016). The aerospace industry anticipates the use of AM technologies for more significant weight reduction and space utilisation (Chua & Leong 2017) to seek better fuel efficiency and aircraft performances, whereas the electronics industry aims to reduce footprints of electronic devices with increased functionalities (Paulsen et al. 2012).

Conventional microwave or radiofrequency (RF) antennas are usually fabricated on 2-dimensional (2D) surfaces using micromachines. These conventional antennas are typically associated with high costs (e.g. capital, materials and labour costs) (Ghazali et al. 2015), and they are also technically challenging to fit into small lattice spaces in electronic devices. In contrast, AM technologies also can allow greater cost-effectiveness, greater materials savings and shorter time bottlenecks for on-demand fabrication of printed antennas on a wide variety of substrates, which potentially can benefit the prototyping process (Tan et al. 2016). In addition, AM technologies allow direct printing of antennas onto conformal surfaces which favour full optimisations of available spaces in electronic devices.

Recent literature has also demonstrated the capabilities of AM technologies for fabricating printed antennas, and they can be categorised in two main approaches: partially additive and fully additive manufacturing. The primary distinction between these two approaches is, as the names suggest, based on whether the antenna in consideration is solely fabricated by material deposition processes. On the one hand, partially additive manufactured antennas generally involve the fabrication of the dielectric layer through AM processes only while the conductive elements were fabricated either by a metallization process (Ghazali et al. 2015, Mäntysalo & Mansikkamäki 2009), or by attaching a conductive layer (Whittow et al. 2014), or by embedding conductive wire on the substrate (Shemelya et al. 2015). The patch antennas or the microstrip antennas are some of the conventional antennas that are fabricated by the partially additive manufacturing approach. A common feature of these types of antennas is that the emitting element is separated from the ground plane by a dielectric layer (Ghazali et al. 2015, Mäntysalo & Mansikkamäki 2009, Pa et al. 2015, Shemelya et al. 2015, Whittow et al. 2014). On the other hand, fully additive manufactured antennas typically involve the deposition of three layers of different materials: a conductive layer as the ground plane, a dielectric layer, and a conductive layer (or pattern) for the emitting element. Patch antennas and inverted-F antenna (IFA) antennas are some of the common antennas design that can be entirely fabricated by the fully additive manufacturing approach (Deffenbaugh et al. 2013, Goh et al. 2016).

Although most wireless systems employ vertically polarised antennas, propagating vertical polarised electromagnetic waves may change their initial polarisation direction after multiple complex scattering and reflections during transmission. There are many research studies showed that by using horizontally polarised antennas, at both receiver and transmitter, can transmit more power than vertically polarised antennas (Lin et al. 2006, Wei et al. 2012). Lin and co-workers (Lin et al. 2006) did a comparison study between a printed antenna with Alford-loop-type structure and a printed IFA. They demonstrated that better H-plane pattern was observed in the former while desiring good horizontally polarised omnidirectional radiation feature. Deffenbaugh and co-workers (Deffenbaugh et al. 2013) proposed a fully 3D printed IFA for Bluetooth and Wi-Fi applications. Their antenna possessed good standing wave ratio (SWR) and had good comparable receiving distance range as compared to a conventional quarter-wave antenna. However, the omnidirectional properties of their antennas were less ideal than the Alford-loop-structure antenna.

In this paper, a fully additive manufactured antenna with Alford-loop-structure is presented in which the dielectric layer (substrate) and the electrically conductive antenna structures are entirely fabricated by AM techniques. This fully additive manufactured antenna aims to achieve horizontal polarised, omnidirectional radiation with low return loss and good impedance matching.

2 ANTENNA DESIGN AND PARAMETRIC STUDY

The Alford-loop-structure antenna design is chosen to be printed on a 3D printed substrate, due to its ability to achieve better omnidirectional radiation with minimum return loss. The Alford-loop-structure design has conductive structures on the top and bottom surfaces of the substrate. The conductive patterns on the top surface and bottom surface resemble a Z-shaped and N-shaped alphabetic letter respectively, with each "arm" at the centre extending out to two "wings" (see Figure 1) (Lin et al. 2006). Both conductive patterns must coincide and aligned perfectly at the "arm" region (top and bottom), with the feed point situated at the centre of the substrate (see Figure 1). A feed (coaxial RF connector) is fitted through the feed point to connect both top and bottom conductive patterns (see Figure 1). The structure symmetry of Alford-loop-structure antenna allows each conductive pattern (top and bottom) to have 180° phase difference from each other with the same magnitude (Lin et al. 2006).

A parametric study was first conducted with the *Computer Simulation Technology (CST) Microwave Studio* simulation software for optimising critical parameters of the printed Alford-loop-structure antenna's designs to achieve an operating frequency in the range 2.4 – 2.5 GHz with good impedance matching. The parameters are v, x and z_1, where v is the width of the "wing", x is the width of the "arm" and z_1 is the length of the "wing" (see Figure 1).

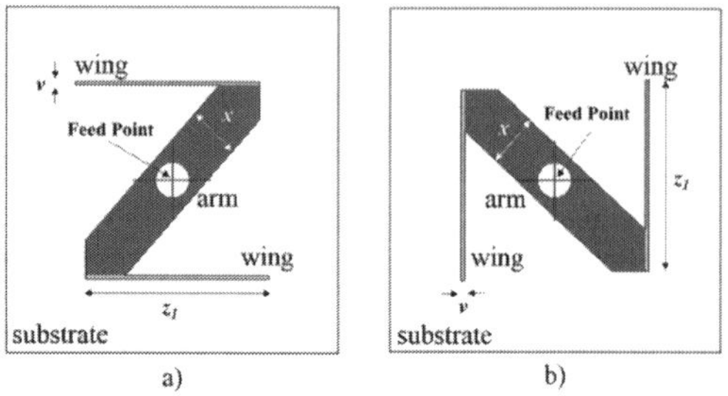

Figure 1. Schematic diagram of the Alford-loop-structure antenna: a) top view, b) bottom view (Lin et al. 2006).

2.1 *Determine the length of the "wing", z_1*

Lin and co-workers (Lin et al. 2006) discussed that the length of the "wing" of the Alford-loop-structure antenna, z_1, is approximately a quarter of the wavelength of the resonance frequency that is effective in the substrate (Sayidmarie & Yahya 2013) due to the absence of ground plate in its design. Hence, the effective dielectric constant can be approximated to be equivalent to the average of the sum of the dielectric constant of the substrate and air (Lin et al. 2006). Thus, the length of the z_1 strip can be expressed as

$$z_1 \approx V 0.25 \times \lambda_e \approx \left[0.25 \times \frac{\lambda_0}{\sqrt{\frac{1+\varepsilon_r}{2}}} \right] \quad (1)$$

where λ_e is the effective wavelength of the resonance frequency in the substrate, λ_0 is the wavelength of the resonance frequency in air, and ε_r is the dielectric constant of the substrate. As the antenna is required to be printed onto the *ULTEM™ 9085* substrate and operated at 2.45 GHz, z_1 is calculated to be approximately 22.2 mm.

2.2 *Determine the width of the "wing", v*

A parametric study was conducted to find the optimal width of the "wing" of the Alford-loop-structure antenna, v, where the frequency (GHz) was plotted against the output reflection coefficient, S_{11}, for four different values of v (i.e. 0.4 mm, 0.6 mm, 0.8 mm, and 1.0 mm). The value of v was varied to match the impedance. From Figure 2, it was observed that the output reflection coefficient decreased as the value of v decreased from 1.0 mm to 0.4 mm. Hence, v = 0.4 mm was selected as the optimal width of the

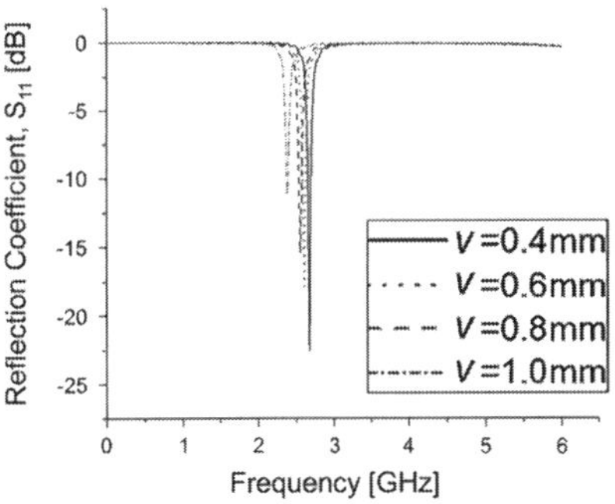

Figure 2. Reflection coefficient, S_{11} of the printed Alford-loop-structure antenna with varying values of the width of the "wing" (v = 0.4 mm, 0.6 mm, 0.8 mm, and 1.0 mm).

"wing" as it gave the lowest reflection coefficient, in which the return loss was minimised.

2.3 *Determine the width of the "arm", x*

A parametric study was conducted to find the optimal width of the "arm" of the Alford-loop-structure antenna, *x*, where the frequency (GHz) was plotted against the output reflection coefficient, S_{11}, for three different values of *x* (i.e. 6.0 mm, 6.22 mm, 6.40 mm). The value of *x* was varied for calculating impedance matching. When *x* was varied between 6.0 mm and 6.20 mm, the operating frequency remained in 2.4 GHz – 2.5 GHz frequency range while the reflection coefficient was between -20 dB and -25 dB (see Figure 3). The reflection coefficient maintained at -25 dB when x was further increased to 6.40 mm, but its operating frequency was shifted to 2.65 GHz, which is out of the desired operating frequency range. However, when the value of *x* was slightly varied from 6.2 mm to 6.22 mm, the impedance was matched at the desired 2.4 GHz frequency with a reflection coefficient of -41 dB. Hence, *x* = 6.22 mm was selected as the optimal width of the "arm" as it gave the lowest reflection coefficient, in which the return loss was minimised.

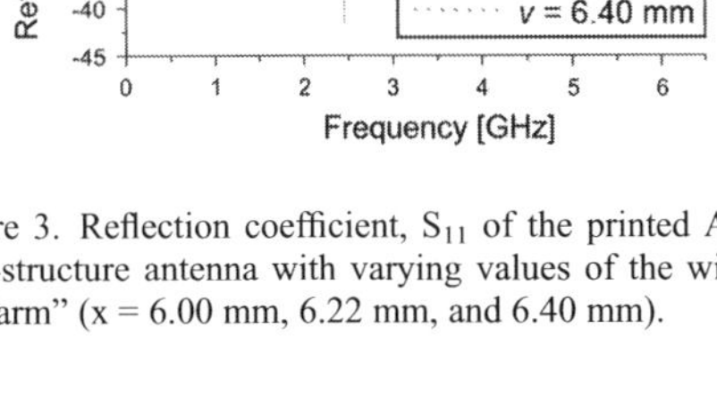

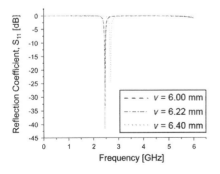

Figure 3. Reflection coefficient, S_{11} of the printed Alford-loop-structure antenna with varying values of the width of the "arm" (x = 6.00 mm, 6.22 mm, and 6.40 mm).

Table 1. Geometric parameters of the printed Alford-loop-structure antenna's conductive patterns.

Geometric parameters of the printed Alford-loop-structure antenna's conductive patterns	Dimensions (mm)
Length of the "wing", z_1	22.2
Width of the "wing", *v*	0.4
Width of the "arm", *x*	6.22

From the parametric study, the optimised parameters of the printed Alford-loop-structure's conductive patterns were obtained and tabulated in Table 1. The parametric and simulation studies of the printed Alford-loop-structure antenna's designs indicate that it operates at 2.45 GHz with desirable performance characteristics. Hence, the designs were then materialised into a fully 3D printed Alford-loop-structure antenna, with both substrates and conductive antenna structures fabricated through additive manufacturing technologies.

3 SIMULATION RESULTS

An Alford-loop-structure antenna was designed with the geometric parameters mentioned in Table 1. A simulation study was conducted using frequency domain *CST Microwave Solver* to find the resonance frequency of Alford-loop-structure antenna and better understand its electrical field distributions.

3.1 *Reflection coefficient*

Reflection Coefficient, S_{11} is a parameter or scattering matrix that provides a complete description of the network (Pozar 2012). It reflects the relationship between incident and reflected voltage and determines the amount of voltage that has been reflected back to the source. Hence, the amount of voltage received by the antenna can be calculated. The reflection coefficient, S_{11} of the Alford-loop-structure antenna was simulated using CST Microwave Solver (see Figure 4). The simulations showed that the resonance peaks at 2.45 GHz, with a reflection coefficient of -41 dB.

3.2 *Electric field distributions*

The simulated electrical field distributions of the Alford-loop-structure antenna after impedance matching are shown in Figure 5, where the strength of the electric field is proportional to the current distribution along the antenna. The radiation pattern of this antenna design is omnidirectional and horizontally polarised, which looks like a "doughnut" shape. The power received by the antenna was dependent on the projected length of the dipole, which was

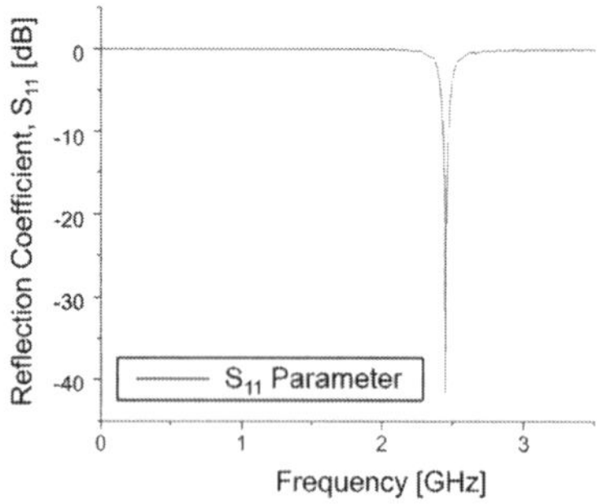

Figure 4. Reflection coefficient, S_{11} of the printed Alford-loop-structure antenna.

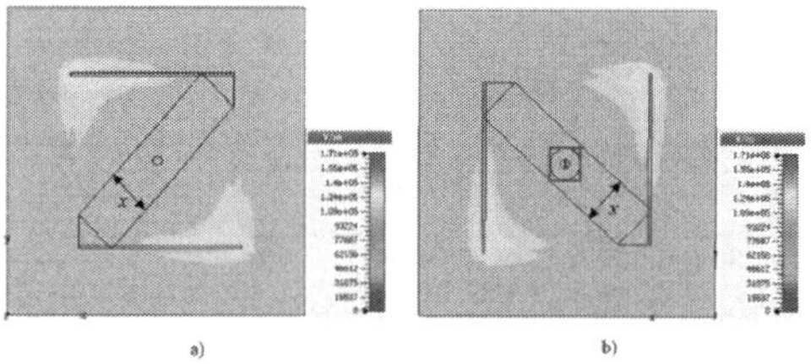

Figure 5. Electrical field distributions of the printed antenna: a) top view, b) bottom view.

perpendicular to the line of sight. Hence, the electric field received is proportional to the apparent length of the dipole. The electrical field is seen to be highly distributed along the two ends of the Alford-loop-structure. Due to the small thickness of the substrate, the electrical field generated by the "arm" on each side of the substrate cancels out each other and results in no electric field distribution along the "arm" (Lin et al. 2006).

4 ANTENNA FABRICATION

The parametric and simulation studies of the printed Alford-loop-structure antenna's designs indicate that it operates at 2.45 GHz with desirable performance characteristics. Hence, the designs were then materialised into a fully 3D printed Alford-loop-structure antenna, with both substrates and conductive antenna structures fabricated through additive manufacturing technologies. The final designs of the printed Alford-loop-structure antenna were directly printed onto a 1 mm thick ULTEM™ 9085 substrate with conductive silver nanoparticles ink using additive manufacturing technologies.

4.1 *Fabrication of the substrate*

The printed antenna's substrate was 3D printed using the Stratasys' Fortus 450mc fused deposition modelling (FDM) 3D printer, with *ULTEM™ 9085* and *ULTEM™ 9085 Support Material* as the modelling and support materials respectively. The high surface roughness of the printed substrate's surfaces was unfavourable for the deposition of silver nanoparticles inks onto its surfaces, due to poor wettability. Therefore, the substrate was then polished on both sides using a non-adhesive 200mm diameter abrasive disc of grit P800 with Struers RotoPol-25 Twin Table polisher, to decrease its surface roughness.

4.2 *Fabrications of the conductive antenna structures*

After the fabrication and post-processing of the 3D printed substrate, the patterns of the antenna design were printed using UT Dots Inc.'s conductive nanosilver ink, *UTDAg40TE*, with Optomec's Aerosol Jet 5x system (Chua & Leong 2017, Tan et al. 2019). This ink also requires a sintering temperature between 120 – 140 °C to achieve good electrical conductivity. The 3D printed substrate was first wiped clean using ethanol, then air dried, and lastly placed on the heated platen for 15 minutes to achieve thermal equilibrium before printing. The critical dimensions of the antenna's conductive patterns, that were to be printed on the substrate, were shown in Figure 6. The front side of the printed antenna was first printed with 40 layers of silver nanoparticles ink and followed by a thermal sintering process in

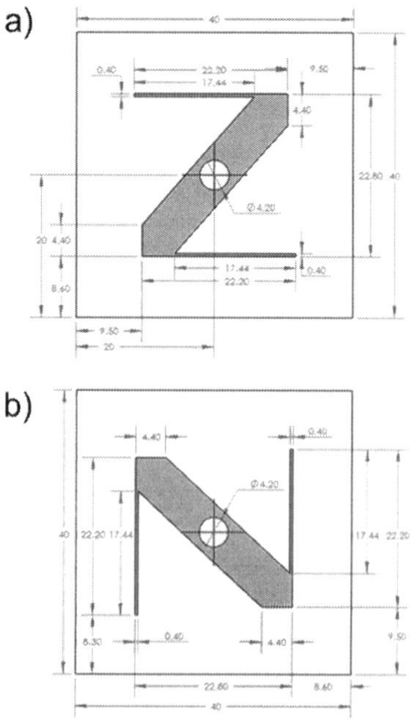

Figure 6. Critical dimensions of the printed antenna's conductive patterns: a) top view, b) bottom view.

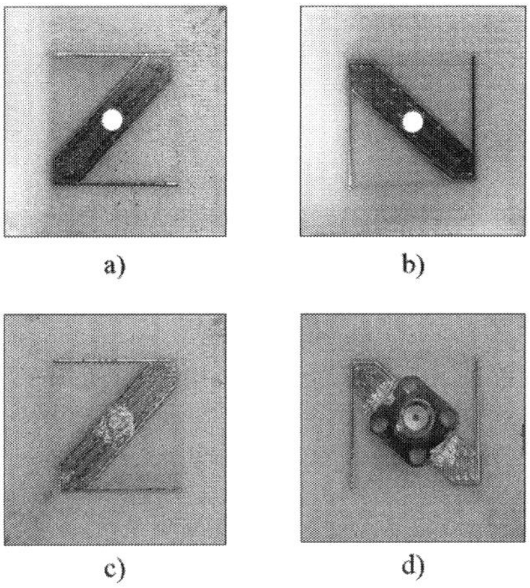

Figure 7. Conductive patterns of fully 3D printed Alford-loop-structure antenna on *ULTEM™* substrate: a) top view, b) bottom view, c) top view, mounted with coaxial RF connector, d) bottom view, mounted with coaxial RF connector.

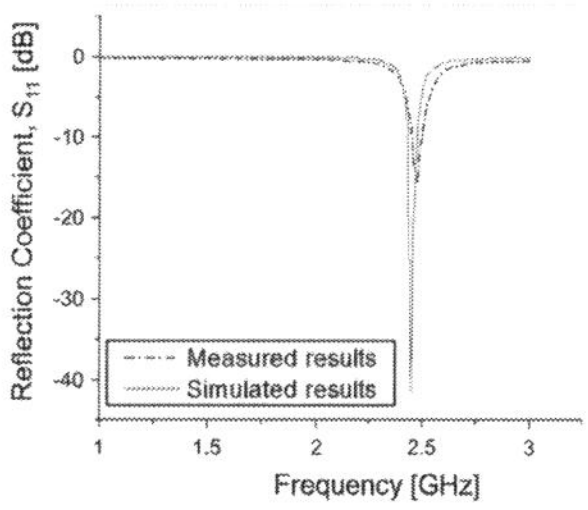

Figure 8. Simulated and measured reflection coefficient of a fully 3D printed Alford-loop-structure antenna.

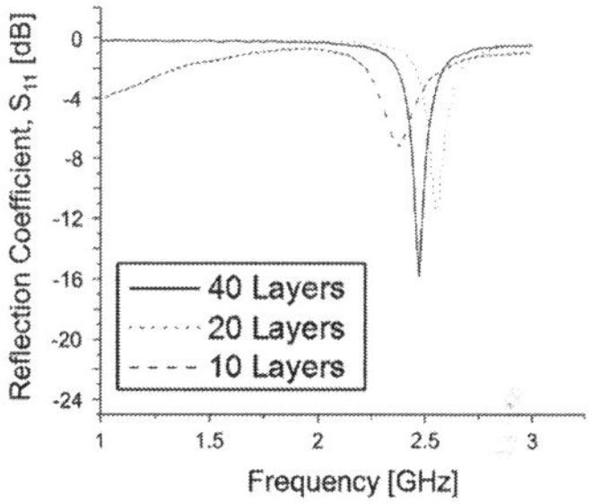

Figure 9. Measured reflection coefficient of a fully 3D printed horizontally polarised omnidirectional Alford-loop-structure antenna for various layers.

a convection oven at 150°C for 2 hours. These steps were repeated for the back side of the printed antenna again. The as-deposited printed patterns had poor electrical conductance due to the presence of electrically insulating organic additives encompassing each silver nanoparticle (Tan et al. 2020). Therefore, the thermal sintering process was required to decompose these organic additives in the as-deposited ink to enhance the electrical conductance of the printed patterns.

The completed conductive patterns of the printed antenna (top and bottom view) on the substrate were shown in Figure 7a and b. Lastly, a coaxial RF connector was mounted onto the printed antenna by inserting it into the feed point at the centre of the substrate. Silver epoxy was applied to various areas to connect the top and bottom conductive patterns to the coaxial RF connector (see Figure 7c and d). The silver epoxy was left overnight for it to cure.

5 MEASUREMENT RESULTS

Figure 8 compares the reflection coefficient of the fully 3D printed antenna with the simulated results. The simulated results showed that this antenna design can theoretically achieve a reflection coefficient of -41 dB at 2.45 GHz operating frequency. From the measured results, the gain of the fully 3D printed antenna is reduced significantly to -15.75 dB, where its operating frequency lies in the desired range of 2.45 – 2.5 GHz with a 50 MHz bandwidth.

Figure 9 also compares the reflection coefficient of the fully 3D printed antennas with a different number of conductive layers (for instance, 10, 20, and 40 layers). The collected data indicates that by increasing the number of printed conductive layers, which in return its electrical conductivity, can decrease its return loss. Therefore, it is believed that the thin thickness of the conductive layers, together with the uneven surfaces and porous structures of the FDM printed ULTEM™ substrate, contributed significantly to the gain reduction from the simulated results. At 40 layers, it was found that the efficiency of the printed antenna was at approximately 35%.

The radiation patterns of the fully 3D printed antenna were measured and characterised in an anechoic chamber at 2.45 GHz. The simulated and measured reflection coefficient gain were plotted in both E-plane and H-plane (see Figure 10). From the data, the fully 3D printed antenna was found to be horizontally polarised with the desired radiation pattern. The radiation pattern of the antenna in the E-plane is omnidirectional, while the radiation pattern in the H-plane is close to omnidirectional. The measured return loss gain is poorer to the simulated results, and this is highly attributed to the presence of air gaps, high surface roughness, and high porosity of the substrate. In addition, the thin thickness of the conductive layers also contributes to the poorer return loss gain. Moreover, the voltage standing wave ratio (VSWR) of this antenna design is approximately 1.03 when simulated and is less than 3 when measured from 2.45 GHz to 2.5 GHz.

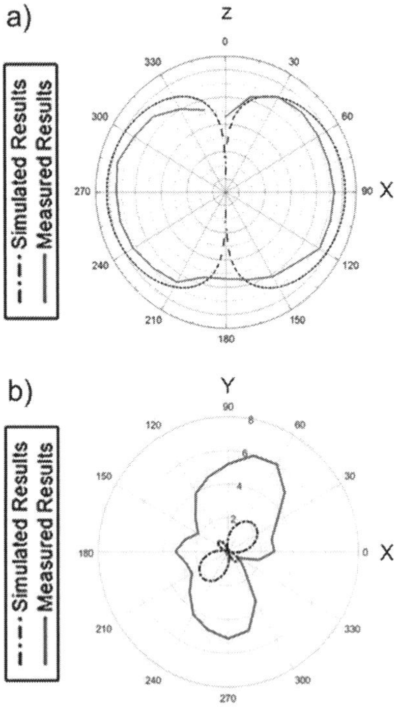

Figure 10. Simulated and measured radiation patterns of a fully 3D printed horizontally polarised omnidirectional Alford-loop-structure antenna: a) E-plane, b) H-plane.

6 CONCLUSION

A fully 3D printed antenna, which exhibits horizontally polarised omnidirectional properties, is designed and fabricated to radiate at the desired 2.45GHz. This printed antenna adopted an Alford-loop-structure design, and it demonstrated a VSWR of approximately 3. The radiation pattern of the antenna in the E-plane is omnidirectional, whereas the radiation pattern in the H-plane is close to omnidirectional. Future work includes the fabrications of high efficiency fully 3D printed antennas, in which conductive patterns with better electrical conductivity are printed on 3D printed substrates with lower porosity and surface roughness.

REFERENCES

Choong, Y. Y. C., Maleksaeedi, S., Eng, H., Wei, J., & Su, P.-C. (2017). 4D printing of high performance shape memory polymer using stereolithography. *Materials and Design*, *126*, 219–225.

Choong, Y. Y. C., Maleksaeedi, S., Eng, H., Yu, S., Wei, J., & Su, P. C. (2020). High speed 4D printing of shape memory polymers with nanosilica. *Applied Materials Today*, *18*, 100515.

Chua, C. K., & Leong, K. F. (2017). *3D Printing and Additive Manufacturing - Principles and Applications* (5th) ed.: World Scientific Publishing Co. Pte Ltd.

Deffenbaugh, P., Church, K., Goldfarb, J., & Chen, X. (2013). *Fully 3D printed 2.4 GHz bluetooth/wi-fi antenna*. Paper presented at the International Symposium on Microelectronics, Orlando, Florida, USA.

Ghazali, M. I. M., Gutierrez, E., Myers, J. C., Kaur, A., Wright, B., & Chahal, P. (2015). *Affordable 3D printed microwave antennas*. Paper presented at the 65th Electronic Components and Technology Conference (ECTC), San Diego, CA, USA.

Goh, G. L., Ma, J., Chua, K. L. F., Shweta, A., Yeong, W. Y., & Zhang, Y. P. (2016). Inkjet-printed patch antenna emitter for wireless communication application. *Virtual and Physical Prototyping*, *11*(4), 289–294.

Lin, C. C., Kuo, L. C., & Chuang, H. R. (2006). A horizontally polarized omnidirectional printed antenna for WLAN applications. *IEEE Transactions on Antennas and Propagation*, *54*(11), 3551–3556.

Mäntysalo, M., & Mansikkamäki, P. (2009). An inkjet-deposited antenna for 2.4 GHz applications. *AEU - International Journal of Electronics and Communications*, *63*(1), 31–35.

Ng, W. L., Chua, C. K. & Shen, Y. F. (2019) Print me an organ! Why we are not there yet. *Progress in Polymer Science*, *97*, 101145.

Pa, P., Larimore, Z., Parsons, P., & Mirotznik, M. (2015). Multi-material additive manufacturing of embedded low-profile antennas. *Electronics Letters*, *51*(20), 1561–1562.

Paulsen, J. A., Renn, M., Christenson, K., & Plourde, R. (2012). *Printing conformal electronics on 3D structures with Aerosol Jet technology*. Paper presented at the Future of Instrumentation International Workshop (FIIW) Proceedings.

Pozar, D. M. (2012). *Microwave Engineering* (4th) Edition ed.. United States of America: John Wiley and Sons, Inc.

Sayidmarie, K., & Yahya, L. (2013). Design and analysis of dual band crescent shape monopole antenna for WLAN applications. *International Journal of Electromagnetics and Applications*, *3*(4), 96–102.

Shemelya, C. M., Zemba, M., Liang, M., Espalin, D., Kief, C., Xin, H., & MacDonald, E. W. (2015). *3D printing multi-functionality: Embedded RF antennas and components*. Paper presented at the 9th European Conference on Antennas and Propagation (EuCAP), Lisbon, Portugal.

Tan, H. W., Tran, T., & Chua, C. K. (2016). A review of printed passive electronic components through fully additive manufacturing methods. *Virtual and Physical Prototyping*, *11*(4), 271–288.

Tan, H. W., An, J., Chua, C. K., & Tran, T. (2019). Metallic nanoparticle inks for 3D printing of electronics. *Advanced Electronic Materials*, *5*, 1800831.

Tan, H. W., Saengchairat, N., Goh, G. L., An, J., Chua, C. K., & Tran, T. (2020). Induction sintering of silver nanoparticle inks on polyimide substrates. *Advanced Materials Technologies*, *5*, 1900897.

Wei, K., Zhang, Z., & Feng, Z. (2012). Design of a wideband horizontally polarized omnidirectional printed loop antenna. *IEEE Antennas and Wireless Propagation Letters*, *11*, 49–52.

Whittow, W. G., Chauraya, A., Vardaxoglou, J. C., Li, Y., Torah, R., Yang, K., & Tudor, J. (2014). Inkjet-printed microstrip patch antennas realized on textile for wearable applications. *IEEE Antennas and Wireless Propagation Letters*, *13*, 71–74.

Industry 4.0 – Shaping The Future of The Digital World – da Silva Bartolo et al. (eds)
© 2021 Taylor & Francis Group, London, ISBN 978-0-367-42272-1

Paper in architecture: The role of additive manufacturing

T. Campos
School of Architecture, University of Minho, Guimarães, Portugal

P.J. Cruz & B. Figueiredo
Lab2PT, School of Architecture, University of Minho, Guimarães, Portugal

ABSTRACT: The introduction of Additive Manufacturing (AM) brought transformative approaches for the building industry by the development of new systems and the exploration of new components with complex geometries. This paper intends to demonstrate the different AM Techniques in Architecture by mixing different materials, such as cellulose and starch. The research focus on the ecological potentials and limitations of a completely biodegradable material, cellulose – polysaccharide present on plant walls – for architecture using 3D printing techniques. The sensitive nature of this project is to take advantage of the potential of cellulose, an abundant, recyclable and biodegradable material, while transforming into a pulp for application in 3D printing for generation of new models taking advantage of the potentialities of the techniques in AM.

1 INTRODUCTION

In the contemporary context, the work of the Japanese architect Shigeru Ban is paradigmatic in the integration of building components made of paper, where this material is present on the architectural expression of the building, revealing the potential of paper as building material (Latka 2017).

Nowadays the technology and the exploration of innovative materials, expand the way in which pulp can be integrated into the contemporary architecture. Cellulose is one of the most valuable materials and the main component of the plants used for the production of paper, the extraction of cellulose in its fibrous state is the basic process for the production of a pulp.

This paper describes a research which aims to analyse the use of mixtures based on cellulose in Additive Manufacturing (AM) processes, that is to infer their potential and constrains. Essentially, the research focuses on the mix of two raw materials, cellulose and clay, according to the organization of this article. The work initiated by the exploration of mixtures based on cellulose and starch (section 3), followed by studies of mixtures of cellulose and clay (section 4). This study allowed us to obtain conclusions by comparing the behaviour of cellulose pulp.

2 ADDITIVE MANUFACTURING CONTEXT

The AM technique used on this research was Liquid Deposition Modeling (LDM). This process consists in the continuous deposition, or extrusion, of layers of viscous materials (Rosenthal et al. 2017). The 3D printer used on the production of specimens and prototypes moves an extrusion nozzle and cartridge. The extrusion flow is controlled by a spindle, and the air pressure applied to the cartridge filled with the material. The CAD-CAM tool used to generate the G-Code was developed by our team in order to fully control the manufacturing process (Barbosa & Figueiredo, 2016). The application was implemented trough Grashopper® – an application programming interfaces of the 3D modelling CAD software Rhinoceros® that uses Visual Programming Language that allows to customise the extrusion path, speed, layer thickness, layer height and the creation of structures to support the object when printed (Figure 1).

Specimens, prototypes and tests produced in the research presented in this paper were made on the Advanced Ceramics R&D Lab (ACLab), located at the Institute Design of Guimarães, University of Minho, having taken advantage of the existing resources.

3 CELLULOSE, A RAW MATERIAL FOR AM

Cellulose is an abundant polymer on the walls of plants. To study these fibers it is necessary to take into account the basic principles of the anatomy of the wood, it's identification according to the type of species. Cellulose is a fiber with application in the most diverse areas of production, such as food, but also on the composition building materials. Cellulose is the main structural fiber of the plant Kingdom, being composed by a long chain polymer classified as polysaccharide or carbohydrate. In the words of Klemm et al. (2005), cellulose is the most common organic polymer and is considered as an almost

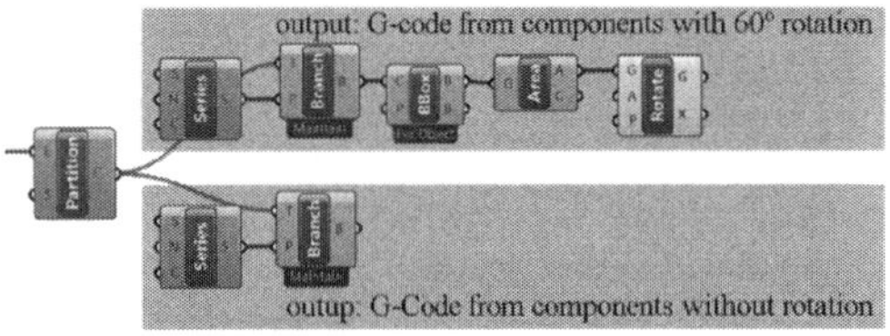

Figure 1. Grasshopper® interface illustrating the generation of the G-Code data.

Figure 2. Cellulose paste bleached in plates.

Figure 3. Cellulose paste bleached powder.

inexhaustible source of raw material for the increasing demand for environmentally friendly and biocompatible products (Latka 2017).

Two types of cellulose pulps used in the industry of paper and transformation were consider in this study: cellulose pulp in plates (Figure 2) and cellulose pulp bleached powder (Figure 3).

4 STARCH AND CELLULOSE

4.1 *Introduction*

As already mentioned, the main goal of this research was to infer material compositions of cellulose pulp with adequate behaviour for LDM process. The methodology adopted took three phases of work, the first phase was focused on the research of possible materials to combine with the cellulose that could reach the plasticity needed for the extrusion. Materials such as acetone, agar-agar, white glue and starch were used, but only starch presented workability for application in LDM. The main goal of this phase was to reach a pulp composition that presents a plasticity similar to clay mixtures with successful degree of viscosity for AM for extrusion. The second phase, after observing the behaviour of the materials used in the first phase, consisted in studying in detail the combination between cellulose and starch which generated a reason for the mixing. The starch needs to be mixed with water and then heated at low heat to form a gelatinous pulp. To create this mixture two cellulose pulps were analysed, one in plates, which needs to be previously mixed with water to obtain small grains of cellulose and a second one in powder. The third phase resulted in the application of pulp composed by cellulose and starch to 3D printing. Individual studies were carried out to each of the pulps with the aim of analysing parameters such as plasticity, validity, workability and strength for AM. The last stage consisted in the design and manufacture of a prototype wall composed by hexagonal blocks where different composite materials were tested.

Combining the materials and generating the pastes from the different materials, pulp of cellulose in plates, pulp of cellulose powder, starch and wood fibers, the main objective was the study of its properties, such as the plasticity of the pulp, the workability when applied in AM, the resistance in the production of prototypes and architectural components and their final appearance. The following sections of the paper describe in greater detail the different idealized mixtures and their behaviour when applied in AM.

4.2 *Starch*

Starch is a carbohydrate present in plants, when it comes into contact with water it dissolves and if subsequently boiled, it can transform into a gelatinous material, giving plasticity to the mixture when coupled to another material. It's the change from liquid to gelatinous state is its greatest advantage as a material for application in AM, because it confers a strong bond between the multiple cellulose particles. For the transformation of the material 50 grams of starch is required in 350 grams of hot or cold water. This preparation requires some care in its processing time so in order to preserve its characteristics, either in the weighing of the materials, so that the weight of attached starch is in conformity with the amount of water present.

4.3 *Mixture A*

The mixture A, composed by starch and cellulose pulp in plates was the first to be thought and the one that more fields of investigation potentiated for the accomplishment of the following ones. The cellulose pulp (Figure 2) is a raw material that needs to be processed before being mixed with other materials. For this transformation to occur, the pulp needs to be dissolved in water until the formation of small grains of cellulose. In order to generate the mixture A, it is necessary to follow only two steps:

1. The weighing of 350 grams of cellulose pulp;
2. Attach 150 grams of previously boiled starch, aggregation of both materials.

In order to start the study, a set of parameters was defined with the goal of studying and validating the mixture A. These parameters are divided into two phases, (1) the validation of the mixture for application in the AM and (2) its workability for the composition of architectural elements. These steps were applied throughout all of the defined mixtures. All the results obtained in the different tests and specimens produced were observed, being possible to infer some conclusions. In order to validate the use of the mixture in AM it was performed a simple test of extruding small circles. In the various small circles different parameters were tested, variation in the air pressure, differentiation of fluidity and diameters of the extrusion nozzles. In order to evaluate the workability of the mixture for application and production of architectural elements, different specimens with different wall curvatures were developed. The tests produced allowed to verify geometric constrains when the mixture was used in the LDM technique (Figure 4).

4.4 *Mixture B*

Composed of starch and cellulose pulp powder, the mixture B was the one that contributed the most for the development of this research. Similarly mixture A, to achieve mixture B, it is necessary to follow two steps:

1. The weighing of 60 grams of cellulose pulp;
2. Package of 300 grams of previously boiled starch and aggregation of both materials.

When the mixture A was studied, the major problems encountered during its use for AM were related to the final shape of the mixture, when the cellulose grains are saturated with water, the fluidity of pulp is excessive, if the grains are dry, they generated problems in the printing process. All problems previously encountered with the use of the processed cellulose plates were solved by simply attaching powder instead of grains to the mixture.

As described in the mixture A, some study parameters were defined for the validation of the mixture for AM.

The first test consisted on the manufacture of three single pieces, one cylinder and two truncated cones. After verification of the workability of the mixture B, a second test was carried out, such as in mixture A — cellulose in plates and starch printed different models with different wall inclinations. The third test consisted on the manufacture of simple and organic models, where the flexibility of curvature of the parts was tested.

Finally, a last test was performed with a higher degree of requirement than the previous ones. The Alveolus Block (Figure 5) is a regular base model and multiple conical openings where their union gives rise to a unique model.

These experiments were successful in verifying the ability of mixture B of being used in LDM to manufacture models in some degree of complexity.

The greatest disadvantage observed when using the mixture is in the drying phase, where the detachment and lamination of the walls of the objects produced is verified. The lamination is verified when the walls of the models lose water, through the drying, which causes the detachment of the surfaces.

Concluding, the greater the degree of complexity of the models manufactured, the more difficult is the response of the mixture, which consequently leads to the appearance of lamination, cracks and deformations in the model walls.

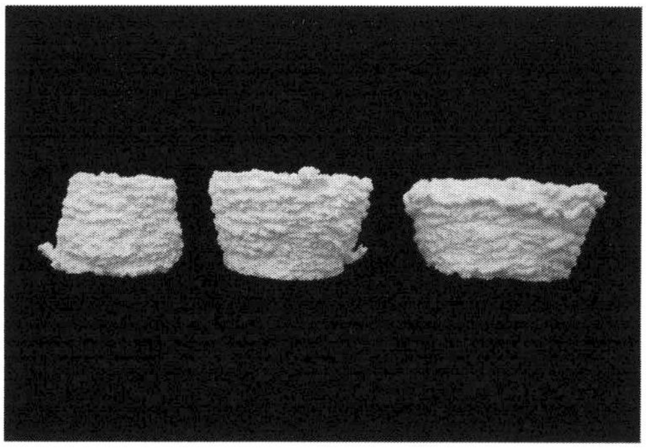

Figure 4. Models printed with mixture A, with different inclinations, to study the curvature limitation.

Figure 5. Alveolus block print with mixture B.

4.5 *Mixture B.1*

Composed by starch, cellulose pulp powder and wood fibers, the mixture B.1 is an improvement of the mixture B that came up with the purpose of solving the problems previously observed during the production and manufacture of components or models. With the same composition, with the exception of wood fibers, the main purpose when placing them in the mixture was to solve, in particular problems related to the retraction of the material. As observed in the individual study of mixture B, its workability exceeded the initially drawn targets, and the most fragile point observed was the constant lamination and detachment between the different layers that make up the manufactured model. It was thought that by placing sawdust, the retraction of the material wild be lower than previously observed, by not absorbing the water, eliminating the possibility of excessive saturations. The following steps were followed:

1. The weighing of 60 grams of cellulose pulp;
2. Attach 300 grams of previously boiled starch and aggregation of both materials.
3. Weighing of 30 grams of sawdust and subsequent aggregation of composite materials.

All tests performed on the study of mixture B were repeated. The mixture B.1 shows some improvements if compared to mixture B (Figures 6,7). The sawdust improves the fluidity of the mixture thus allowing a more careful, clean printing and more accurate surfaces. The higher the percentage of sawdust attached, the more accurate is the fabrication process. The retraction is also attenuated by the presence of the sawdust and it is verified a considerable decrease in the detachment and lamination produced by the evaporation of the water during the drying process. Similarly to what was observed in the mixture B, the retraction was mainly noticed in the thickness of the different layers (Figure 7).

The introduction of sawdust clearly change the appearance of the printed model. The greater the amount of sawdust is present in the mixture, the darker it gets (Figure 8). The mixture B.1 allowed to produce extruded surfaces with better finish than the ones observed in the models produced with the mixture B.

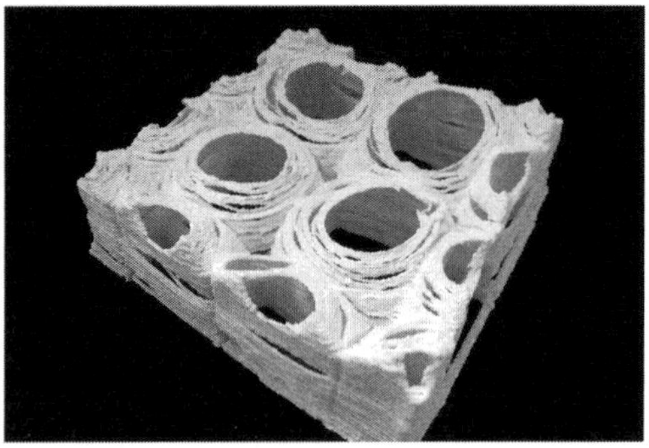

Figure 6. Lamination of the Alveolus Block walls after drying in greenhouse. Model produced with mixture B.

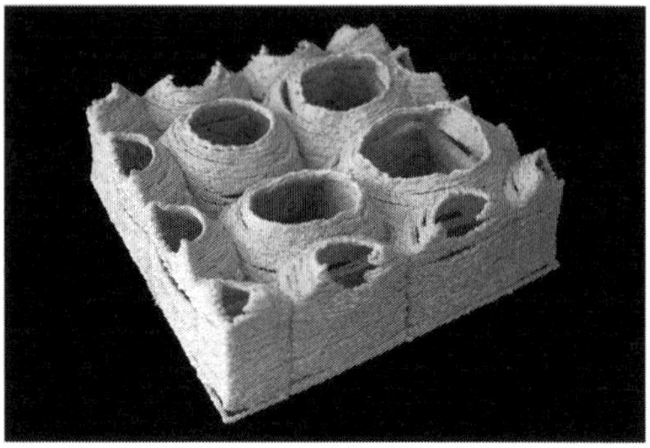

Figure 7. Reduction of the lamination observed on the walls of the Alveolus Block, when compared to the previous block, printed with the mixture B. Model produced with mixture B.1.

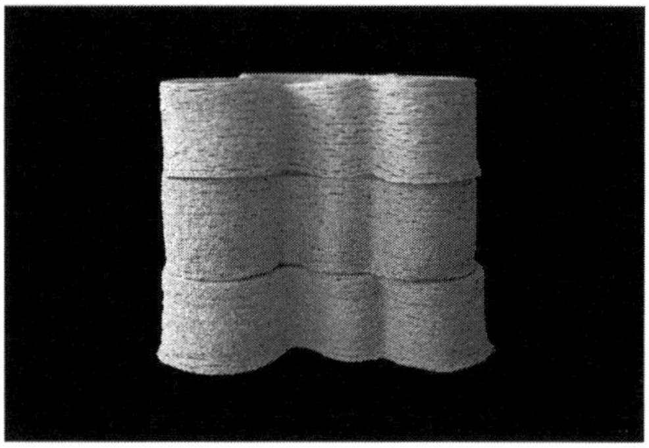

Figure 8. Final finishing of the AM Aalto pieces, printed with mixture B.1. Upper piece (15gr of sawdust), central piece (45gr of sawdust) and lower piece (30gr of sawdust).

5 CLAY AND CELLULOSE

5.1 *Introduction*

Clay is a mineral from sedimentary rocks that according to the place where it is found, can give rise to multiple types of clay bodies by the aggregation of other materials and minerals. Most clay bodies for architectural ceramics are earthenware and stoneware — both sedimentary clay types — as well as porcelain. These terms, used in common language to reference pottery, here designate technical expressions of the mixtures of clays and additives (Bechtold 2015).

The ceramic paste used on this study is a mixture of stoneware and water — GRES 130 MP, a ceramic paste produced and commercialized by Vicar, a Spanish company specialized in ceramic materials. With the right amount of moisture, this paste showed exceptional workability in LDM. The placement of the cellulose fibers in the paste intends to increase the potential of the mixture, evidencing a greater potential for application in architecture, through the resolution of different problems verified, when using only ceramic paste. These problems consist of:

1. Excessive retraction caused walls deformation;
2. Excessive retraction of the material taking into account the initially proposed dimensions;
3. Fissure between connections;
4. Decrease in weight of printed models.

To solve the above mentioned problems, it was believed that the placement of cellulose fibers would enhance the workability, behaviour and strength of the mixture. This research was divided into two mixtures:

Mixture C, composed by clay, cellulose pulp in plates and water; Mixture D, composed by clay, cellulose pulp powder and water.

5.2 *Mixture C*

Composed of cellulose pulp in plates and stoneware paste. According to the tests performed in ACLab, the adequate moisture content of this stoneware paste in order to obtain models with good print quality by LDM should be between 33 and 35% (Cruz et al. 2017). As described in mixture A, the cellulose pulp in plates, needs to suffer a previous transformation to be added to the clay. To achieve Mixture C, it's necessary to follow only two steps:

1. The weighing of 1000 grams of ceramic paste and subsequent placement of water missing;
2. The weighing of the quantity of transformed cellulose pulp and aggregation of both materials.

Given the possibility of placing several percentages of cellulose in the ceramic paste, a study was defined generating multiple mixtures until obtaining an interesting consistency for exploration:

1. 1Kg of ceramic paste and 25gr of cellulose pulp;
2. 1Kg of ceramic paste and 50gr of cellulose pulp;
3. 1Kg of ceramic paste and 75gr of cellulose pulp;
4. 1Kg of ceramic paste and 100gr of cellulose pulp;

Since ceramic paste is a widely used material in the manufacture of architectural components, using the techniques of AM, the study of the mixture C is based on the possibility of improvement of the clay paste. Precisely, were defined parameters of studies that consist in the workability and feasibility in the mixture when the cellulose is attached and its behaviour when applied in real manufacturing context. The first phase of work resulted in the analysis of the workability and feasibility of the mixture when attached to cellulose. In order to validate this phase, a first test consisted of the printing of a set of truncated cones, were the inclination of the walls is varied in order to observe the limitation of the curvature (Figure 9). A second test was carried out, as already stated in the study of the previously mixtures, which consisted in the manufactured of pieces inspired by geometry of the cups of the architect Alvar Aalto (Figure 10).

All tests performed with the mixture C showed better results than those observed with mixture A. The presence of cellulose fibers decreases all the problems previously mentioned, except the retraction of the pieces. The retraction index verified with the ceramic paste is similar to the one observed in the mixture C.

5.3 *Mixture D*

As mentioned, the Mixture D was composed of cellulose pulp powder and ceramic paste. Their preparation followed the same steps of the previous mixture, except for the cellulose pulp. Given the possibility of placing several percentages of cellulose in the ceramic paste, a study was defined generating multiple mixtures until obtaining an interesting consistency for exploration:

1. 1 Kg of ceramic paste and 10 gr of cellulose pulp;
2. 1 Kg of ceramic paste and 25 gr of cellulose pulp;
3. 1 Kg of ceramic paste and 50 gr of cellulose pulp;
4. 1 Kg of ceramic paste and 75 gr of cellulose pulp;

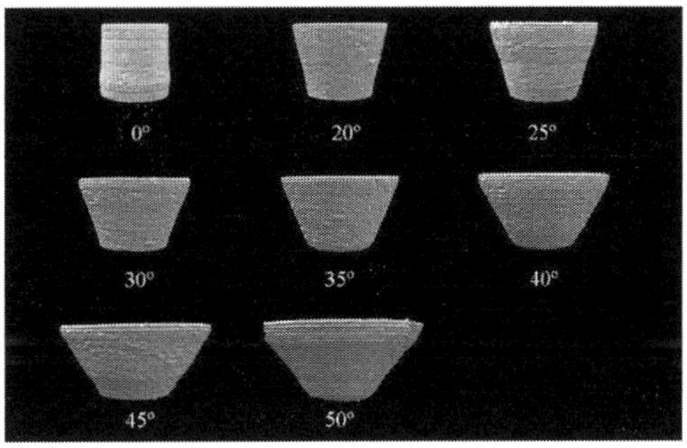

Figure 9. Final finishing of a set of truncated cones. Observation of limitation of the curvature of the mixture C.

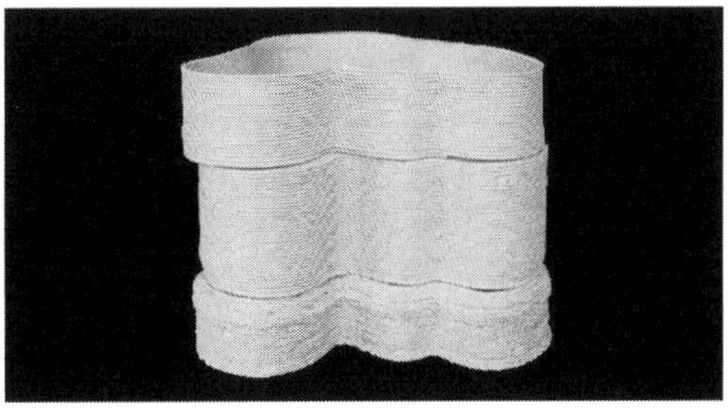

Figure 10. Top - ceramic paste, middle - ceramic paste and 25 gr of cellulose) and bottom - ceramic paste and 75 gr of cellulose.

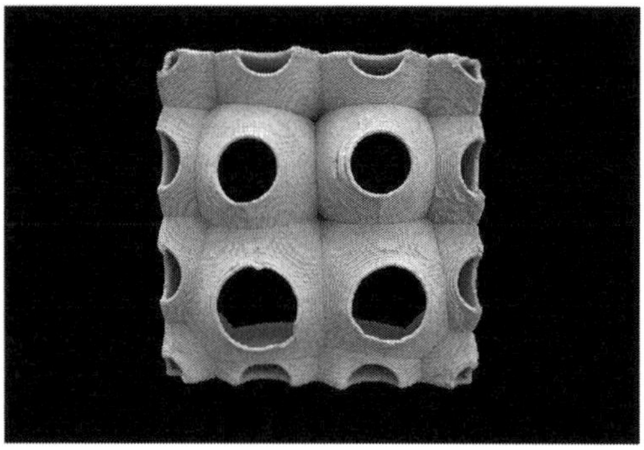

Figure 11. Alveolus Block produced with mixture D, composed by stoneware paste and 10gr of powdered cellulose.

When prepared the mixture D was necessary to considerate the preparation, since the powdered cellulose doesn't have moisture, when mixed with ceramic paste absorbs some of its moisture, considerably altering the expect moisture content, to ensure good workability. To control this factor, a parametric ratio was defined. For each 5 grams of powdered cellulose, was added 6 grams of water. As previously described, study parameters were defined. The first work phase, similar to described in previous subchapters, started by printing of the truncate cones.

Due to the excellent performative response of the mixture D in the first discriminated phase, the second working phase resulted in the production of the previously mentioned component – Alveolus Block.

Mixture D had a better behaviour than the observed throughout the individual study with mixture C. With the accomplishment of these studies, the retraction index observed, with or without cellulose is almost the same, varying the greater the percentage of cellulose attached. The deformations and cracks are attenuated by the presence of cellulose in mixture.

6 CONCLUSION

As demonstrated throughout this paper, cellulose is a material with great potential in AM. The application of composite mixtures (cellulose and starch) in 3D printing open multiple possibilities for the development of architectural components, the Alveolus Block can be understood as an example of this if applied on the assembly of a wall or facade. By analysing the cellulose mixtures tested in the research it's possible to conclude that the mixture with the greatest potential for AM is the mixture B.1. With the introduction of sawdust, some of the negative parameters previously observed, in the study of the mixture B, were attenuated, namely the decrease of the lamination and detachment of the surfaces of the models, generating cleaner surfaces, thus allowing a greater control on the extrusion and fluidity of the mixture. Mixture A, due to the granulometry of the transformed cellulose grains, originated some problems, namely the nozzle clogging and the production of models whose surfaces look careless. By analysing the clay mixture, the presence of the cellulose pulp in the mixture was able to attenuate the problems mentioned above.

ACKNOWLEDGMENTS

This work has the financial support of the Exploratory Research Project, with the reference MIT-EXPL/ISF/0006/2017, MIT Portugal-2017 Program, financed by National Funds, through FCT/MCTES.

We thank to RAIZ – Institute Research of Florest and Paper, for their support and partnership in this research, namely through the supply of cellulose. We are grateful to the Institute of Design of Guimarães for hosting and supporting the Advanced Ceramics R&D Lab on the use of their facilities and equipment.

REFERENCES

Barbosa, I. & B. Figueiredo (2017). Optimized Brick – Print Optimization. *Challenges for Technology Innovation: A agenda for the Future*, 201–210.

Betchold, M., A. Kane & N. King (2015). *Ceramic Material Systems: in architecture and interior design*. Basel, Switzerland: Bikhauser GmbH.

Cruz, P.J.S., U. Knaack, B. Figueiredo & D. Witte (2017). Ceramic 3D printing – The future of brick architecture. In *IASS 2017 – Interfaces: Architecture. Engineering. Science*.

Latka, J. (2017). *Architecture and the Built environment — Paper in Architecture*. Delft, Holland: Delft University of Technology, faculty of Architecture.

Rosenthal, M. et al. (2017). Liquid Deposition Modeling: a promising approach for 3D printing. *European Journal of Wood and Wood Products 76 (2)*, 797–799.

Industry 4.0 – Shaping The Future of The Digital World – da Silva Bartolo et al. (eds)
© 2021 Taylor & Francis Group, London, ISBN 978-0-367-42272-1

Towards an integrated sensor system for additive manufacturing

F. Morante & M. Palladino
Nuovo Pignone Tecnologie, Firenze, Italy

M. Lanzetta
University of Pisa, Pisa, Italy

ABSTRACT: Product defects are a major challenge for the understanding and application to new products of mass additive manufacturing. This work starts from an overview of typical defect categories, especially focusing on direct metal laser sintering (DMLS), their causes and the capability/reliability of in process sensors. Individual sensor applications are available in the literature. The proposed sensor system preliminary explores the qualitative and quantitative correlation and integration of an accelerometer mounted on the powder spreading recoater, with photodiodes, a visible and an IR camera. An online control strategy is also proposed for early detection and prevention of product defects.

1 INTRODUCTION

Additive manufacturing (AM) refers to a group of technologies used to produce components by material increase (i.e. layer by layer) starting from a three-dimensional mathematical (CAD) model (ISO/ASTM International 2013). AM, formerly rapid prototyping, dates back from the nineties and is rapidly growing, towards mass customization. Defects prevention, identification and correction are essential for competitiveness. AM machines will be increasingly complex and will require many sensors. Most current machines are only able to provide an alarm and manual control (Table 1). This work proposes a control system for the next generation machines. The ultimate goal of the proposed system is to achieve full integration from design to actual manufacturing, including hybrid processes and process chains (Rossi et al. 2020).

1.1 *State of art*

Foster et al. (2015) used a camera to identify defects on the parts after the laser exposure. They showed examples of 3D reconstructions of builds via image segmentation and slice edge detection. Jacobsmuhlen et al. (2015) used a camera to identify lack of powder on the bed and they created a feedback system for re-deposition.

Mathieu et al. (2017) used acoustic sensor to estimate the intensity of collisions between recoater and parts. Eschner et al. (2018) proposed the use of acoustic sensor to estimate the melt pool spatter emission.

They used artificial intelligence algorithms to find correlation between process sounds and parts defects.

Kleszczynski et al. (2014) used an accelerometer and a camera to study collision between blade and parts. They studied supports quality with different laser power. J. zur Jacobsmühlen et al. (2015) analyzed correlation between maximum acceleration and overhang area exposed with downskin strategy. They also found a correlation between acceleration intensity and dimension of not covered area. Reinartz (2016) registered a patent about sensors use for additive manufacturing monitoring. It claimed the metrological detection of coating perturbations (i.e. vibration detected with accelerometers), forces (i.e. forces acting against the travel direction), drive power, acoustic sound and deflections of coating slide.

Krauss et al. (2012) used an IR camera to study thermal deviation during the process. Schilp et al. (2014) proposed a defect indicator based on the evaluation of the temperature profile of each pixel belonging to the slice. The indicator consists of the total time the local temperature stayed above a certain temperature. They showed the generation of 2D and 3D spatial maps of this descriptor as a graphical means to detect the location of defects within each layer and within the entire part. Krauss et al. (2014) evaluated the local maximum temperature and the cool-down behavior in terms of a characteristic time as well.

Craeghs et al. (2011) analyzed melt pool behavior during the balling phenomena. They measured the melt pool intensity when laser do U sketch. Craeghs et al. (2012) analyzed melt pool intensity during support construction and when parts come out from the powder bed. Extended literature on spatter number, pool dimension and light emissions control can be transferred from (arc) welding control (Lanzetta et al. 2001).

Table 1. Correlation between process anomalies, sensors and commercial monitoring kit. © 2019 Baker Hughes, a GE company, LLC - All rights reserved.

	*Sensors**				*Kits***
Process anomalies	S1	S2	S3	S4	
Powder bed					2,3,4,6,7,8
Lack of powder	○	●	●	●	
Recoater collision	●	●	○	○	
Recoater vibration	●	○			
Particle drag		●	○	○	
Melting					1,3,4,5,6,9
MeltPool instability		○	○	●	
Spatter emission		○	●	○	
Hot/Cold spot		○	●	●	
Balling		○	○	●	
Gas flow					3,4,5,6
Instability		○	○	●	
Inhomogeneity		○	●		
Laser scanning					2,3,4,5,6,8
Geometric deformation		●	○		
Lack of power		○	●	●	
Thermal					3,4,5,6
Deformation	○	○	○	○	

* S1. Accelerometer; S2. Camera; S3. OT; S4. MeltPool
** 1. QM meltpool 3D (Concept Laser), 2. QM coating (Concept Laser), 3. EOSTATE MeltPool (EOS), 4. EOSTATE PowderBed (EOS), 5. EOSTATE OT (EOS), 6. PrintRite3D (B6 Sigma Inc.), 7. Melt Pool Monitoring (SLM Solutions), 8. (MPM) system, Layer Control System (SLM Solutions), 9. Sapphire System (Velo 3D).

1.2 *Market*

Different PBF system developers (EOS, Concept Laser, Arcam, Velo 3D, SLM Solutions, Additive Industries, Renishaw) and third part equipment developers (B6 Sigma, Inc) offer *in situ* monitoring modules and toolkits (2017). Most of them are mainly used to collect data and provide the user with some post-process data reporting. It's possible to generate alarms in some cases or to have an online feedback control strategy. Current commercial machines are usually closed and do not allow custom feedback.

Table 1 summarizes some of these commercial toolkits.

2 PROCESS DESCRIPTION

Powder Bed Fusion (PBF) as an AM process offers the opportunity to make prototype, customized components and small batch production of complex metal components. In the first step a thin layer of metal powder is spread onto the build platform. In a second step the powder is molten into solid material by moving a laser beam across the current cross-section of the part. After this, the building platform goes down and the two process stages are repeated until the solid metal part is fully produced. The produced components show very good mechanical properties, which are widely comparable to conventionally processed. Today, the main applications for PBF components are medical implants, automotive and aeronautical parts. The production of complex metal parts by PBF is an advantageous option in many cases but today is use only for few applications. Control and process monitoring could increase the process performance in order to increase economic efficiency and reproducibility of mechanical properties.

This work has been made using EOS® M280 and M290. These machines have a 400 W power laser. It works at a wavelength of 1064 *nm*. Nickel based alloys have been used.

3 SENSOR SYSTEM

Errors come from different sources. Of course, early detection of errors is preferable, e.g. design for additive manufacturing, however we will show that most defects can be detected and corrected in process as well. Once specific thresholds are set could be possible to use these instruments in production in order to assure the final parts quality. Using an intelligent control system could be possible to set up an online control system. Monitoring information could also be used in order to correct feed-forward model-based software as shown in Figure 1.

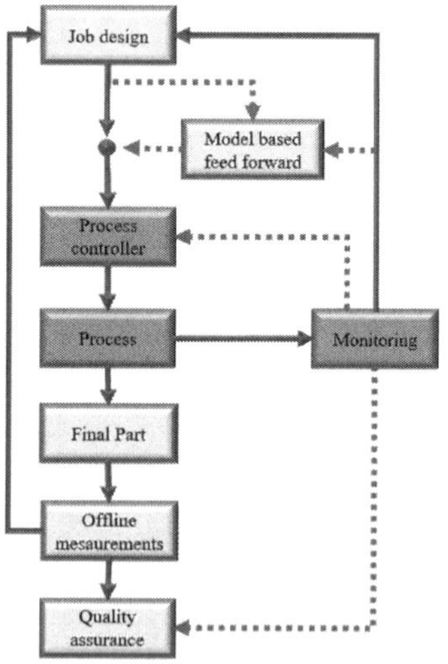

Figure 1. Monitoring sensor system. © 2019 Baker Hughes, a GE company, LLC - All rights reserved.

4 A NEW PROCESS SENSOR

The accelerometer is mounted on the recoating system (Figure 2). It helps to monitor the collision between the blade and the printed parts. It's possible to calculate the maximum acceleration for each layer and set threshold alarms in order to try to prevent the recoater jam. In Figure 3 it's possible to see the correlation between the sensor signal and the presence of parts that come out from the powder bed.

Figure 4 shows the maximum acceleration (every vertical line is a layer) over time before a job interruption. The lower horizontal line represents the 'contact threshold'. After this value we can see collision between recoater and part.

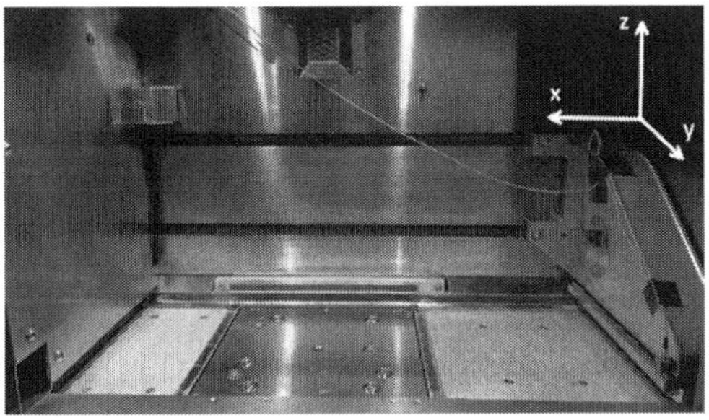

Figure 2. Accelerometer configuration in the building chamber. © 2019 Baker Hughes, a GE company, LLC - All rights reserved.

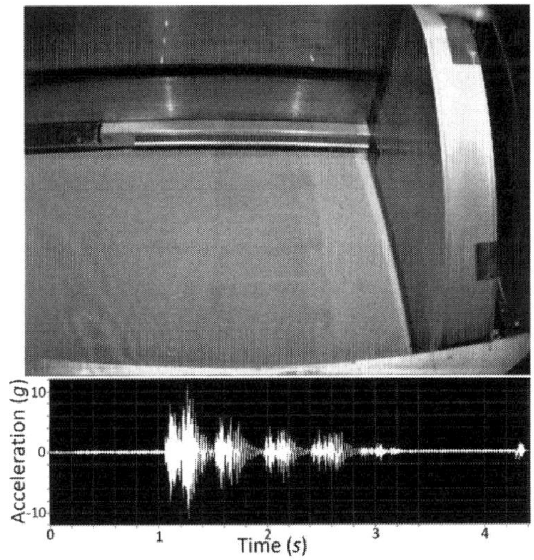

Figure 3. Correlation between overhang parts and x axis acceleration peak. © 2019 Baker Hughes, a GE company, LLC - All rights reserved.

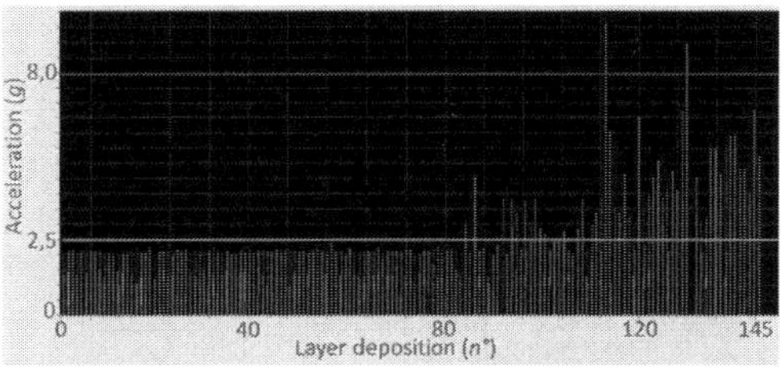

Figure 4. Acceleration maximum during job construction. © 2019 Baker Hughes, a GE company, LLC - All rights reserved.

The upper horizontal line represents the 'stop threshold'. Above this value the interruption probability in the following layers became very high. The lower threshold depends only on recoater speed, blade type and machine. The upper one is also job dependent.

5 DEFECT AND SENSOR OVERVIEW

Process's deviations by the desired conditions could cause defects on printed parts. *In situ* monitoring allows to observe the process and identify the problems that could induce defects. Process anomalies, part defects and anomalies sources are correlated according to the scheme in Figure 5. *In situ* monitoring isn't only an observer of the process but could actively interfere in order to correct anomalies. In this work four different sensors have been used. The accelerometer has been used to estimate recoater collision. The camera has been used to record and observe the construction phases.

The IR camera and the two photodiodes have been used to evaluate the melt pool intensity.

In Table 1 are shown process anomalies and sensors that could be used for their detection. The white dot indicates that the defect is only observable. The black one that is also quantifiable.

In Table 2 are shown the correlations between possible defect sources and process anomalies. In Table 3 are shown the correlations between process anomalies and defects that could be found on the parts.

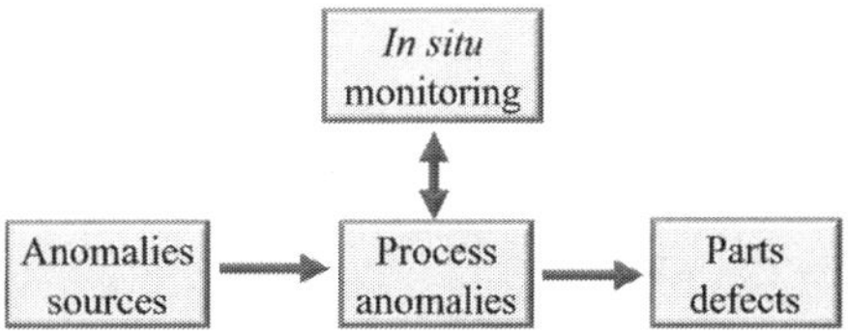

Figure 5. Process scheme © 2019 Baker Hughes, a GE company, LLC - All rights reserved.

Table 2. Process anomalies versus anomalies sources.

	Anomalies sources														
	Machine							Design							Powder
Process anomalies	Optical chain	Building platform	Machine calibration	Laser	Filtration system	Recoating system	Software	Layer thickness	Process parameters	Part orien-tation	Supports	Part shape	Powder bed parameters	Gas flow parameters	
Powder bed															
Lack of powder						•							•		
Recoater collision		•	•			•		•	•	•	•	•			
Recoater vibration		•				•		•	•		•	•	•		•
Particle drag		•				•		•	•		•	•			•
Melting															
Meltpool instability	•		•	•	•		•	•	•					•	•
Spatter emission			•	•	•		•	•	•					•	•
Hot spot			•	•	•		•		•		•	•		•	
Cold spot	•		•	•			•		•			•		•	
Balling			•	•			•	•	•						
Gas flow															
Instability					•									•	
Inhomogeneity					•									•	
Laser scanning															
Geometric deformation	•		•				•								
Lack of power	•		•	•	•		•								
Thermal															
Deformation		•			•			•	•	•	•	•		•	

Table 3. Process anomalies versus parts defects

		Parts defects				
Process anomalies	Incomplete parts	Geometric defects	Surface defects	Residual stress, cracks and delamination	Porosity	Microstructural inhomogen. and impurity
Powder bed						
Lack of powder	•	•	•	•	•	•
Recoater collision	•	•	•		•	•
Recoater vibration			•			
Particle drag		•	•		•	•
Melting						
MeltPool instability			•		•	•
Spatter emission			•		•	•
Hot/Cold spot			•	•	•	•
Balling		•	•		•	•
Gas flow						
Instability			•		•	•
Inhomogeneity			•	•	•	•
Laser scanning						
Geometric deformation		•				
Lack of power			•		•	
Thermal						•
Deformation	•	•		•		

6 SENSOR INTEGRATION

6.1 *Lack of powder*

Low powder density or low dosing factor can cause lack of powder (Figure 6). This issue is not acceptable if it last more than one or two layers. It could cause incomplete parts. The camera images showed the uncovered areas. The IR camera and the photodiodes shows the area without powder like a colder area. This is due to the higher heat exchange coefficient of the bulk material compared to the powder. Recoater vibrations on z axis are more intensive when a lack of powder happens.

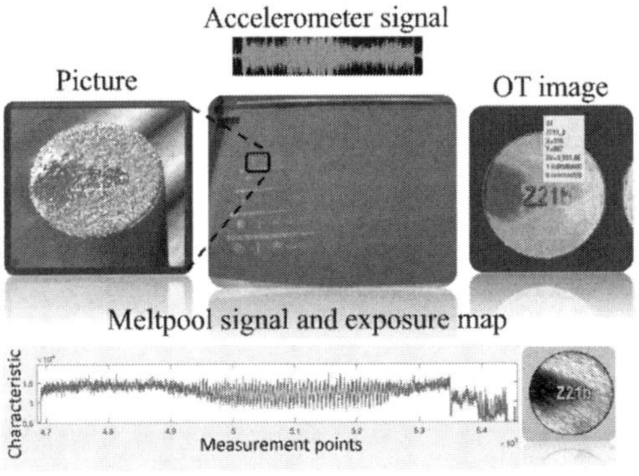

Figure 6. Lack of powder effect on part and monitoring signals (IR camera, photodiodes and accelerometer).

6.2 *Thermal deformation*

Thermal gradient can cause deformation on parts, especially on the thinner one. Because of the thermal deformation parts can appear from the powder bed and have collision with the recoater. After the collision, the powder bed can be damaged. This event could cause defects on the final part. An excess of powder causes hotter areas during the melting as show in Figure 7. This melting anomaly could affect the final part.

6.3 *Feedback control*

A feedback control (Figure 8) could use accelerometers and cameras to detect powder beds defects. The system can automatically re-spread the powder if some issues are detected. Towards cameras and photo-diodes it's possible to detect melt pool anomalies. The system can alert the users if some value overcame a pre-determined range.

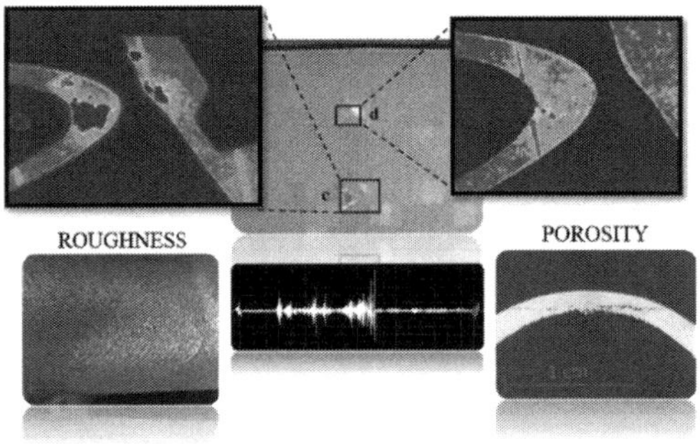

Figure 7. Sensor integration monitoring of a thermal deformation. Porosity and roughness on the final parts.

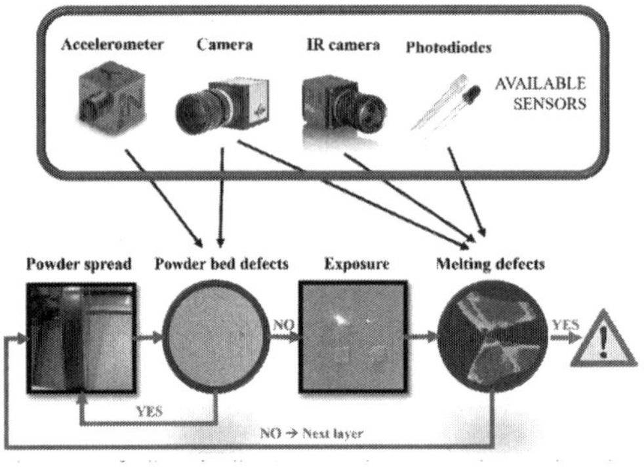

Figure 8. Online feedback control strategy leveraging data fusion from different sensors.

7 CONCLUSION

The two proposed cases have shown that an integrated sensor architecture is able to provide a promising detection of process errors and correction.

Further research is required to extend the number of detectable defects also to more integrated sensors. The main limit is the possibility to find universal correlation between defects in the process and defect on the final parts. In fact, it depends a lot from part shapes and process parameters. Also powder material and machine type could influence this correlation.

REFERENCES

Craeghs, T. Clijsters, S. Witt, G. Kruth, J.P. Bechmann, F. & Ebert, M.C. (2012). Detection of process failures in Layerwise Laser Melting with optical process monitoring. *Physics Procedia 39*: 753–759.

Craeghs, T. Clijsters, S. Yasa, E. & Kruth, J.P (2011) Online quality control of selective laser melting. *Proceedings of the Solid Freeform Fabrication Symposium*: 212–226.

Eschner, N. Weiser, L. Häfner, B. & Lanza, G. (2018). Development of an acoustic process monitoring system for selective laser melting (SLM). *Proceedings of the 29th Annual International Solid Freeform Fabrication Symposium* Austin, *TX*.

Foster, B. K. Reutzel, E. W. Nassar, A.R. Hall, B.T Brown, S.W. & Dickman, C.J. (2015). Optical, layerwise monitoring of powder bed fusion. *Solid Freedom Fabrication Symposium Proceedings*: 495–307.

Grasso, M. & Colosimo, B.M (2017). Process defects and in situ monitoring methods in metal powder bed fusion: a review. *Measurement Science and Technology* 28.

ISO/ASTM International (2013). cur. Standard Terminology for Additive Manufacturing-Coordinate Systems and Test Methodologies. 52921–1.

Jacobsmühlen, J. Kleszczynski, S. Witt, G. & Merhof, D. (2015). Elevated region area measurement for quantitative analysis of laser beam melting process stability. 26th International Solid Freeform Fabrication Symposium; Austin, *TX*.

Kleszczynski, S. zur Jacobsmühlen, J. Reinarz, B. Sehrt, J. T. Witt, G. & Merhof, D (2014). Improving Process Stability of Laser Beam Melting Systems. *Proceedings of the Fraunhofer Direct Digital Manufacturing Conference*.

Krauss, H. Eschey, C. & Zaeh, M. (2012). Thermography for monitoring the selective laser melting process. *Proceedings of the Solid Freeform Fabrication Symposium*: 999–1014.

Krauss, H. Zeugner, T. & Zaeh, M. (2014). Layerwise monitoring of the selective laser melting process by thermography. *Physics Procedia 56*: 64–71.

Lanzetta, M. Santochi, M. & Tanussi G. (2001). On-line control of robotized Gas Metal Arc Welding. *CIRP Annals*, *50 (1)*:13-16.

Mathieu, M. e Caelers, G. (2017). Study of in-situ monitoring methods to create a robust SLM process. Preventing collisions between recoater mechanism and part in a SLM machine. *Degree project in Mechanical Engineering. KTH Royal Institute of Technology*.

Reinarz, B. (2016). Verfahren und vorrichtung zur herstellung von bauteilen in einerstrahlschmelzanlage. *Patent EP 2 723 552 B1*.

Rossi, A. & Lanzetta, M. (2020). Integration of Hybrid Additive/Subtractive Manufacturing Planning and Scheduling by Metaheuristics. *Computers & Industrial Engineering*, 106428.

Schilp, J. Seidel, C. Krauss, H. & Weirather, J. (2014). Investigations on Temperature Fields during Laser Beam Melting by Means of Process Monitoring and Multiscale Process Modelling. *Advances in Mechanical Engineering*.

Industry 4.0 – Shaping The Future of The Digital World – da Silva Bartolo et al. (eds)

3D printing artefacts made of the ashes after the fire of the National Museum in Brazil

J.R.L. dos Santos
Ministério da Ciência, Tecnologia, Inovações e Comunicações - MCTIC
Instituto Nacional de Tecnologia – INT, Rio, Brazil
Pontifícia Universidade Católica do Rio de Janeiro – PUC Rio
Departamento de Artes e Design – DAD, Rio, Brazil

S.A.K. de Azevedo
Universidade Federal do Rio de Janeiro
Museu Nacional – MN
Laboratório de Processamento de Imagem Digital – LAPID, Rio, Brazil

C.E.F. da Costa
Pontifícia Universidade Católica do Rio de Janeiro – PUC Rio
Departamento de Artes e Design – DAD, Rio, Brazil

ABSTRACT: The National Museum, an Institution part of the Federal University of Rio de Janeiro, Brazil's oldest and most important historical and scientific museum was destroyed by fire on September 2018. 3D CAT scan files were used to generate 3D printed models by adding ashes from wooden coal obtained from the leftover burned remnants. Results has an emotional connection with the Museum history by combining the geometry of the lost artifact with the ashes of the museum itself.

1 INTRODUCTION

In the decree of creation of the Royal Museum, dated from June 1818, D. João VI, then King of Portugal, Brazil and Algarves, stated education, culture and diffusion of science would be the major objectives of the new institution. Since then, the so-called National Museum since 1830 organized and managed important scientific collections, remaining faithful to these goals throughout its 200 years of existence.

With the advent of the Republic, in 1892, the National Museum (Figure 1) occupied the Paço de São Cristovão at Quinta da Boa Vista in the city of Rio de Janeiro, formerly the Imperial Brazilian Palace, thus uniting important references in the history of Brazil. The residence of the monarchs of the Empire (D. João VI, D. Pedro I and D. Pedro II) and the seat of the first constituent of the Republic, besides representing one of the most significant architectural monuments in Brazil, began hosting the most important scientific institution National of the time, sheltering from then on, the history of politics, arts and science in Brazil. On January 16, 1946, as a national institution, the National Museum was incorporated into the University of Brazil, now the Federal University of Rio de Janeiro, thus intensifying the research and teaching activities (Azevedo, 2007).

Mariliasuchus amarali Carvalho & Bertini, 1999 represents a crocodyliform species (Figure 2) with reduced dependence on water in comparison to modern crocodiles, and specialized teeth suggesting complex feeding habits that are not currently well understood (Zaher et al, 2006). The fossilized remnants of these animals were found close to the city of Marília (São Paulo), in Upper Cretaceous layers on the Adamantine Formation, Bauru Basin (dated to around 80 million years). The specimens of this species are excellently preserved, with a good portion of partially unchanged cranial characteristics. This fact is rare in vertebrate Paleontology, where most well preserved bones found are post-cranial with almost no remnant of the cranial bones (Azevedo & Carvalho, 2008). A very well preserved specimen, MN6298-V (lost in the September 2 fire), was used in this research.

2 THE USE OF NON-INVASIVE IMAGE TECHNOLOGIES IN PALEONTOLOGY

According to Rahman, Virtual paleontology—computer-aided visualization of fossils—is becoming increasingly important in paleontological research. Computed tomography (CT) is the most widely used scanning method in paleontology (Anderson et al. 2003; Abel et al, 2012). This technique is applicable to a wide range of preservation types and specimen sizes, and is perhaps the most effective means of generating

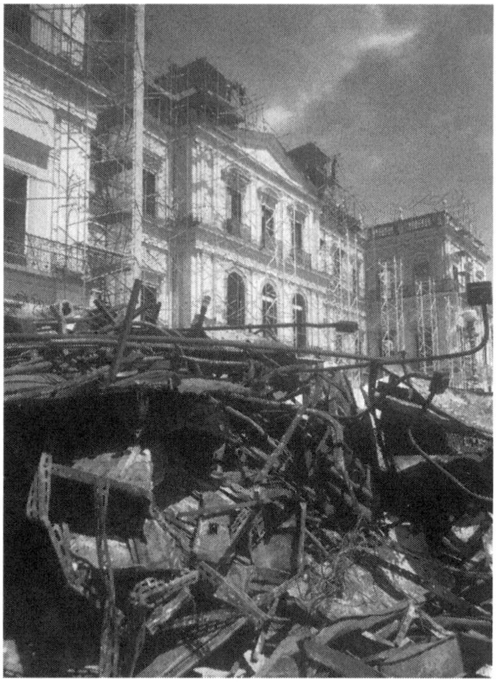

Figure 1. View of the National Museum after the fire.

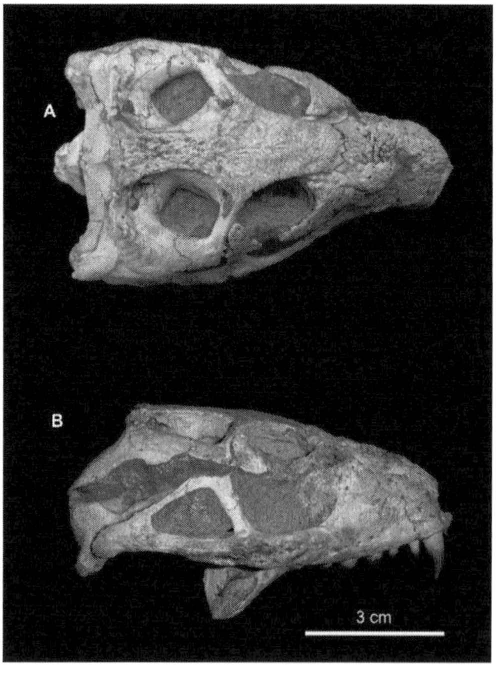

Figure 2. Skull of Mariliasuchus amarali – Museu Nacional – UFRJ.

data that can be used to reconstruct a virtual fossil (Figure 3).

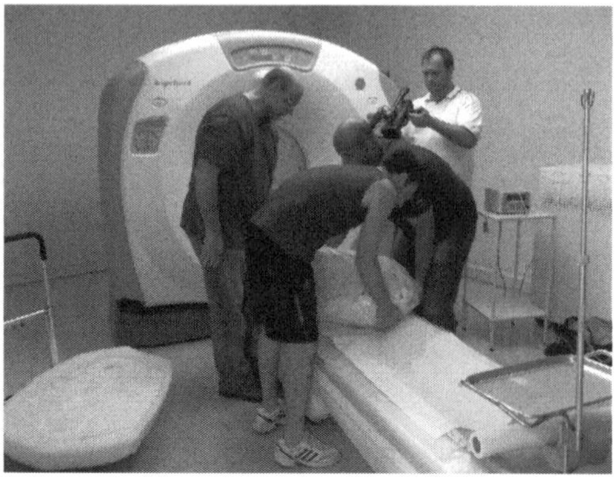

Figure 3. Researchers from the Museu Nacional – UFRJ positioning a large skull of an extinct vertebrate inside the CAT scanner equipment.

3 ADDITIVE MANUFACTURING TECHNOLOGIES

Being a deposition layering process, in additive technology, the fabrication process builds the part systematically by adding material instead of cutting it away, and a wider range of shapes can be achieved, including internal cavities or intricate shapes that would be difficult or even impossible by the use other technology. Additive technology makes it possible to materially translate into models and prototypes any virtual computerized three-dimensional drawing.

The additive method is based upon the successive overlapping of thin layers of specific material substances, according to the appropriate technical method, and is carried out by transforming the 3D files into a STL (Standard Triangulation Language) extension, which consists basically of X, Y and Z coordinates. Once the STL file is generated, the next step is the horizontal slicing of the whole 3D volumetric file, using a software appropriate to the specific hardware being used, and calculating the supporting structures when necessary. The building process starts with the sequential deposition of material layers, the layer width ranging from microns to fractions of millimeters, depending on the chosen technology.

This process is then followed by a post-processing stage, an essential procedure in all current Additive Manufacturing (AM) technologies, where the model has to be cleaned to remove the support material and/ or residues used during the building process. In the case of AM processes that work by using a laser beam to harden photosensitive materials, it is also necessary to position and expose the model inside an ultraviolet light camera in order to solidify the model completely.

Stereolithography was the selected technology for this project, which was the very first additive technology commercially available, introduced in 1988 in California, USA. It is a technique based on the polymerization of photosensitive resins by means of ultraviolet (UV) light emitted from a laser source and focused on a polymer containing ashes from wooden coal obtained from the leftover burned remnants (Figure 4).

When exposed to the laser beam, the photocurable resin changes from a liquid to a solid state, generating a physical slice. This procedure is then repeated sequentially until it reaches the final dimension of the physical model to be built. Today this technology is one of the most accurate, presenting in general a very high surface quality. This process also requires the addition of physical supports (Figure 5) during the construction of the prototype (which have to be manually removed after it has been built) as well as exposure to UV lamps (Figure 6) after the process in order to completely harden the model (Figure 7).

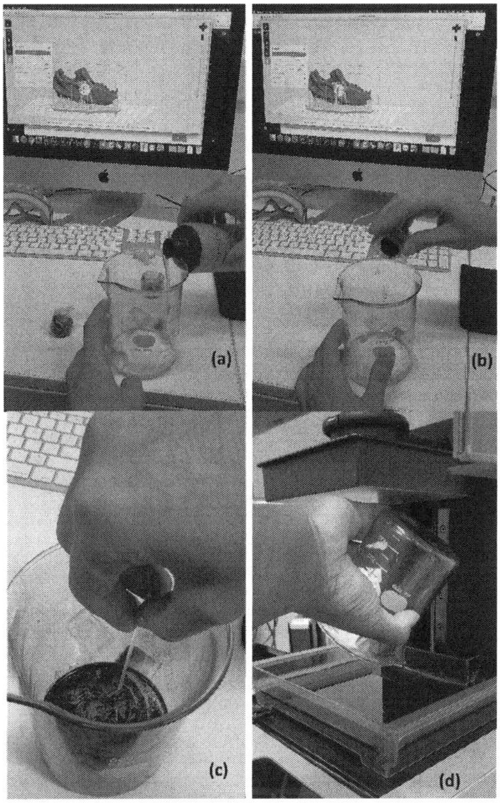

Figures 4. Sequence of the mixture of resin and coal ashes poured on the 3D printer.

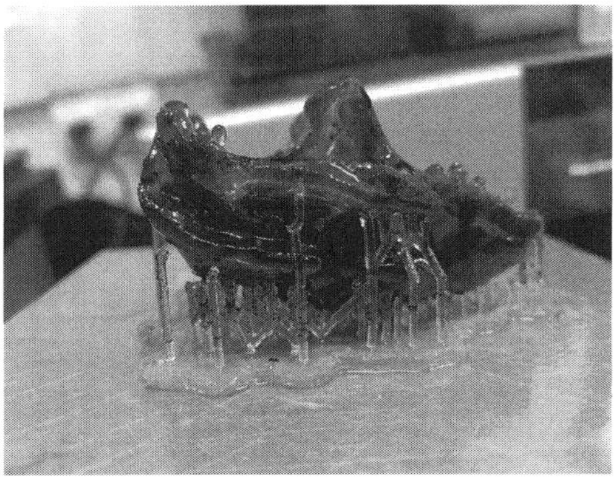

Figure 5. 3D printed model with physical supports of the skull of the *Mariliasuchus amarali*.

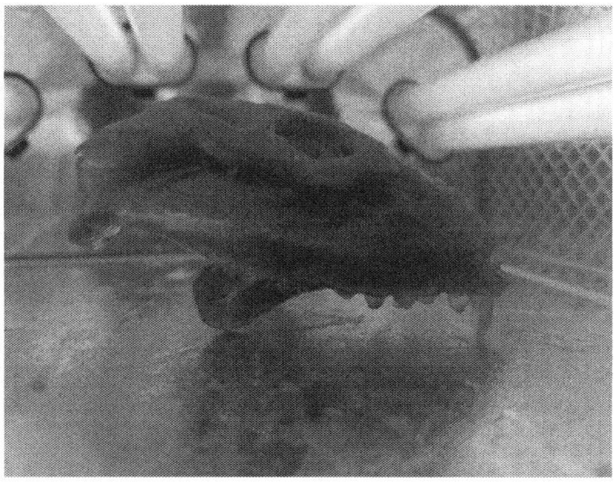

Figure 6. Hardening process through UV light.

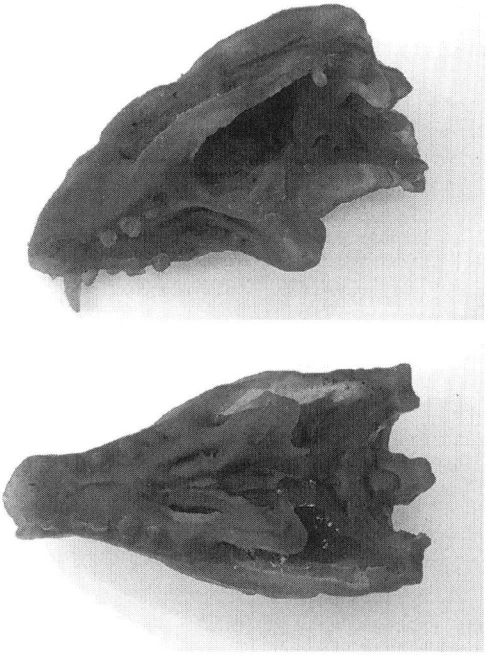

Figure 7. 3D printed model of the skull of the *Mariliasuchus* containing ashes from wooden coal obtained from the leftover burned remnants.

4 CONCLUSION

The burned remnants of Brazil National Museum, the oldest and most important historical and scientific museum, have been used to generate 3D printed models. This is a working progress project, several models were selected to be physically materialize through 3d printed technologies. Results has an emotional connection with the Museum history by combining the geometry of the lost artifact with the ashes of the museum itself.

REFERENCES

Azevedo, S. A., 2007. Museu Nacional. Banco Safra, São Paulo, 359p.

Azevedo, S. A. K. & CARVALHO, L. B., 2008. O uso da tomografia computadorizada no estudo de vertebrados fósseis no Museu Nacional/UFRJ – Use of computed tomography in the study of fossilized vertebrates in the National Museum/UFRJ. 1-32. In: Heron Werner Jr & Jorge Lopes, eds., Tecnologias 3D: paleontologia, arqueologia, fetotologia/Tecnologuies 3D: paleontology, archaeology, fetology. Ed. Revinter, Rio de Janeiro, 190p.

Carvalho, I.S. & Bertini, R.J., 1999. Mariliasuchus: um novo Crocodylomorpha (Notosuchia) do Cretáceo da Bacia Bauru. Geologia Colombiana, 24: 83–105.

Hopkinson, N., Hague, R., Dickens, P. (eds.) Rapid manufacturing – an industrial revolution for the digital age (London: John Wiley, 2005).

Rahman, I.A.; Adcock, A.; Garwood, R. J. Virtual Fossils: a New Resource for Science Communication in Paleontology (Springer Science+Business Media New York, 2012)

Sutton MD. Tomographic techniques for the study of exceptionally preserved fossils. Proc R Soc B. 2008;275 (1643):1587–93.

Zaher, H.; Pol, D.; Carvalho, A.B.; Riccomini, C.; Campos, D. & Nava, W., 2006. Redescription of cranial morphology of Marilisuchus amarali, and its phylogenetic affinities (Crocodyliformes, Notosuchia). American Museum Novitates, 3512, 40pp.

Industry 4.0 – Shaping The Future of The Digital World – da Silva Bartolo et al. (eds)
© 2021 Taylor & Francis Group, London, ISBN 978-0-367-42272-1

Slot-die simulations for 3D printing of Perovskite Solar Cells (PSCs)

A. Omar & P. Bartolo
Department of Mechanical, Aerospace and Civil Engineering, University of Manchester, Manchester

ABSTRACT: To use simple yet transformative capabilities of 3D printing is the way to enable the mass access of Solar energy, and to utilize materials which are also environmentally friendly, abundant, and of low cost. This is possible by adopting the emerging perovskite solar cells, which can be easily produces in an ultrathin film form factor. Extrusion-based 3D printers can use syringes which allow the deposition of thin layers of paste and liquid materials using the correct processing parameters. Furthermore, the use of the proposed Slot-nozzle accessory which can be 3D printed by using the same printer allow the deposition of a thin layer in one pass, reducing the time for printing, and cost of fabricating the cell. This also enables the fabricating of different sized and shape cells, minimizing any post processing, and maintaining the efficiency and possible the stability of the cell.

1 INTRODUCTION

The current highlighted efforts in developing renewable energy can be cited in various disciplines, each offering its own share of knowledge in order to solidify its capabilities and push then even further. The capability of solar energy can supply 5000 times that of the world consumption of energy. To manufacture devices which possess the photovoltaic effect should also be sustainable, as the heavily researched silicon solar cells are expensive, complex, and produce harmful greenhouse emissions during their processing and fabrication. The use of a novelty such as perovskites provide much anticipated solutions to the industry in the form of an excellent light absorber, tunable electronic properties, abundance, and low cost. Furthermore, as of this year perovskite solar cells (PSCs) have surpassed silicon solar cells in terms of power conversion efficiency by 3.8% to that of 23.3% (Rong et al. 2018). However, there are still strides to be made in PSC research in the form of enabling access to such technology through sustainable manufacturing methods which bring the best of this material in form of mass production and/or customization. The aim of this paper is to provide an efficient solution which can be accessed by any individual to satisfy the energy demand and scarcity of resources. The paper offers a solution in the form of an accessory which allows the printing of the PSC materials by using any extrusion-based 3D printer which can be equipped with a syringe, while reducing time, cost, and allowing customized printing of different sizes and shapes.

2 UNLOCKING THE FUTURE

The role in which manufacturing plays is the matching the capabilities of the technology with the expectations set for it to be commercialized and easily accessible. The main goal to be met is the levelized energy cost (LEC) of 9.0 cents/kWh for residential use by 2020, which will drop even further to 5.0 cents/kWh by 2030. Cai et al. (2018) used a three-process technique of estimating the cost of a perovskite solar cell to 107\$/m^2 with an uncertainty range of 87-140\$/m^2 where a silicon solar cell cost 112\$/m^2. Thus, the second goal is to manufacture a large PSC with a high efficiency, lifetime >20 years, stable, and while disposing of lead. The best efficiencies reported are for small area cells of 0.1 cm^2. Thus, developing techniques to translate solar panel developments from laboratory-scale to industrial-scale is essential to improve the possibility of mass production.

2.1 *Slot-die coating*

Slot die coating is a method which deposits a large homogeneous wet film at a high cross-directional uniformity and reproducibility. A solution seeps from a slot gap onto the substrate; afterwards, solidification takes place resulting in a dry film on the substrate. Depositing different thickness layers is achievable by controlling the flow rate, and the movement speed of the die head. Di Giacomo et al. (2018) used a slot die coating process to make a module with 25 cells connected resulting in an area of 168$/m^2 and a PCE of 11%. This is a promising result in the manufacturing of large-area PSCs and shows that other techniques can produce similar and improved results. Key parameters which are typically used and highlighted in fabricating thin films are the gap height between the meniscus guide and substrate and/or the height between head and substrate, coating speed, and pump rate in order to achieve the desired film thickness.

2.2 *Additive manufacturing*

Additive manufacturing involves the building of 3D structures on a layer-by-layer basis. The process is usually divided into three stages: (1) pre-processing: a 3D model in the form of a computer-aided design (CAD) file is tessellated forming a stereolithographic (STL) format; (2) processing: a specific additive manufacturing system deposits layers of material, thus creating a model; (3) post-processing: additional processing to achieve the desired quality of the final product (Zhakeyev et al. 2017). Additive manufacturing comprises different fabrication approaches and allows to reduce material waste, energy consumption and lead times, providing also freedom of design and the fabrication of parts without the need of special tools.

Based on the fabrication principle not all additive manufacturing technologies can be used to process perovskite. Vat polymerization process uses UV light to cure the material which damages the perovskite. Binder jetting, powder bed fusion, energy deposition and sheet lamination limit the materials to either a solid-state form where the perovskite is a liquid. Material jetting is suitable for printing liquid perovskite but not for the remaining materials, and finally extrusion which is adjustable for many high-quality materials as a solid or even liquid with the right nozzle. Therefore, building on the knowledge of the slot-die coating, it is possible to use a slot-nozzle which operates with the same knowledge, that is then coupled with an extrusion-based additive manufacturing system provides an approach to fabricate industrial-scale PSCs. Translating the success of slot die coating, an additive manufacturing system using a mini slot-die head fabricated a PSC with an area of 47.3cm^2 and efficiency of 11.6% (Vak et al. 2015). This is possible with the addition of a solution dispenser to the head to reduce cross-contamination and material waste. Controlling film thickness is possible through the 3D printer settings, as well as cutting the need for post-processing (e.g. annealing) afterwards. 3D Printing systems are usually controlled using a G-code.

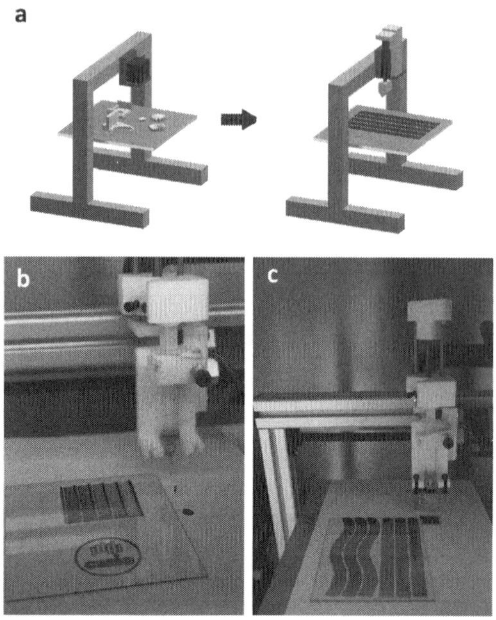

Figure 1. (a) The 3D printer used (b) Point slot nozzle (c) Slot nozzle (Vak et al. 2015).

Upon further analysis, the missing link which could enable the manufacture of a PSC is a head which can print materials which a conventional additive manufacturing system, in specific an extrusion-based printer. Understanding what makes a slot-die head special could unlock the potential of a degree of automation in additive manufacturing system to be employed for PSCs. The literature does not highlight the principle and design of Slot-die coating and lacks specifically in that of printing solar cells when looking from that scope. (Crone, 2016) presents knowledge which helps understand how (Vak et al. 2015) got his version of a slot nozzle. A slot die is first used to make thin layers, presents a range of printing speeds 0.10-1000m/min, range of ink viscosity 1-40,000 (mPa s), coating width of up to 4m, and most importantly layer thickness of 0.05-1000 um. However, it has a very small range of coating accuracy < 1%, and the conventional system is based on a continuous coating system, whereas the applicability of such tool on an extrusion-based printing allows the smart dosing and discrete release of material to print/coat specific areas as required. Considering a n-i-p PSC layers are all fabricated with slot-die coating with efficiencies of up to 7%, the process resembles a simple extrusion-based process, which allows the use of similar parameters used in such process in an extrusion-based printer in order to make a similar product, but with greater flexibility.

3 METHODOLOGY

This section will describe the process of design and simulation of a Slit-nozzle accessory which purpose is to spread the material into a sheet rather than being extruded as a filament.

The design of the slot die nozzle/head is to be designed to produce films of higher width, in order to speed up the process, print different material phases, and produce thin and uniform films. The benefit of such approach is that the thickness of the deposited layer could be controlled based on the flow rate of the material fed, and the coating speed. The dilemma at hand is the commonly used material to fabricate slot-die is very expensive, difficult, and time consuming. They would require a machine shop dedicated for the making of a die, and in the case one breaks or the need for a different size, this would be expensive. To stay in line with the scope of manufacturing a cheap PSCs, as much as possible using an additive manufacturing system, that would also be the preferred route in fabricating a slot die. Therefore, material selected to act as the structural material for the die is any polymer which could be 3D-printed instead of steel or aluminum as it is in line with the scope of the paper to adopt a fully additive manufacturing approach.

To design the slot die, analytical models are used in order to design the optimal head. The equation below demonstrates how the system controls the thickness (x) through coating width (B), coating speed (v), and pump flow rate $((\frac{dV_p}{dt}))$. (Crone, 2016)

$$\frac{dV_p}{dt} = bvx \rightarrow x = \frac{\frac{dV_p}{dt}}{Bv} \quad (1)$$

The work by (Whitaker et al. 2018) adds the gap height between the head and the substrate into consideration when printing the material. Gap height of 40-60um has been tested, and multiple heights which are relative to the printing thickness will be tested using constant parameters. Beeker et al. 2018 published an open-source parametric 3-D printed slot die system for the processing of polymer thin film semiconductors to nanoscale, as well as reported cost reduction of over 17,000%. The published open-source CAD model could possibly be used by altering a few parameters; however, it is designed for a specific type of printer. Therefore, a proposed model has been constructed using Solid works to fit the designated under the name of PAPS (Plasma-assisted printer system) as well as other printers as well which utilize syringes, by adding the suitable Luer lock model to the model. Figure 2 below demonstrates the proposed design of the slit accessory, which reduces the printing time down to 33%, as the piece will act as an accessory to the already installed heated syringe head in Figure 2.

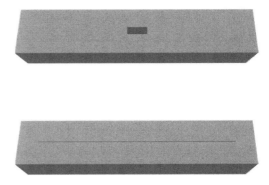

Figure 2. CAD model of slit accessory (Top) Top view, (Bottom) Bottom view.

The next step, is to test the the designed slit accessory, and this is deemed possible using Finite Element Analsysis (FEA) software packages. The software chosen to undergo the simulation is Ansys, specifically fluent. First, define the internal geometry as fluid, the inlet, and the outlet. Followed by applying a default mesh as a starting point, followed by the setup of the simulation through applying a pressure-inlet Boundary Condition (BC), and pressure-outlet (BC) for a value of 1 bar as that is deemed to be an approximate value for the pressure-assised extrusion 3D printer to be used. Finally, the extrudate was tested for the three different materials to be extruded out, Titania paste, Perovskite liquid, and Carbon paste. The solution initally found was sufficient but to insure quicker convergence on the solver, a mesh size of 9E-04 is used for all models, pressure of 1 bar, and this was tested for different accessory nozzle sizes (1, 2, and 4 cm) for all 3 materials. This yielded a total of 9 simulations and proving the concept design which is to be printer and used for the experiments.

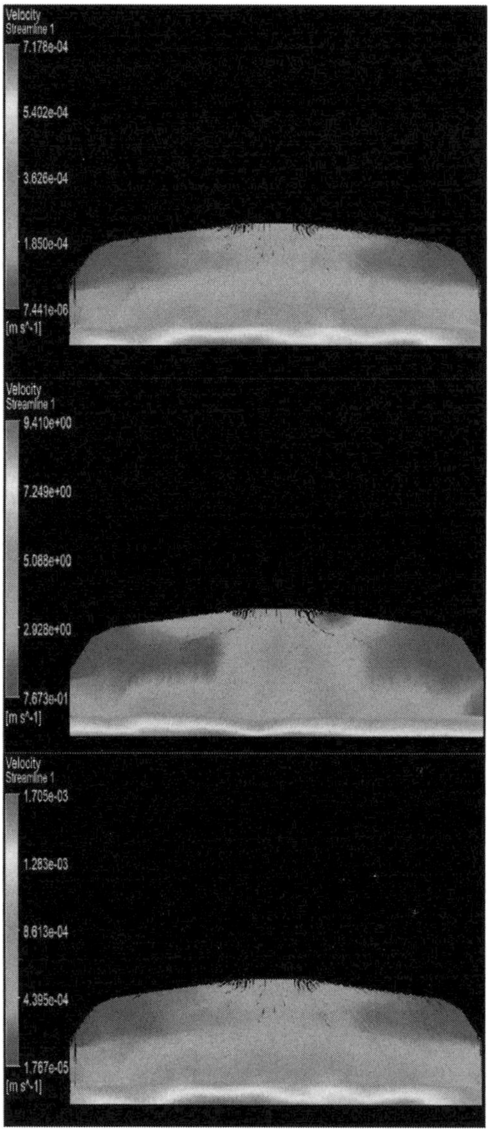

Figure 3. Top – 1cm Slit-accessory flow simulation of TiO2 paste. Middle - perovskite liquid precursor. Bottom - carbon black paste.

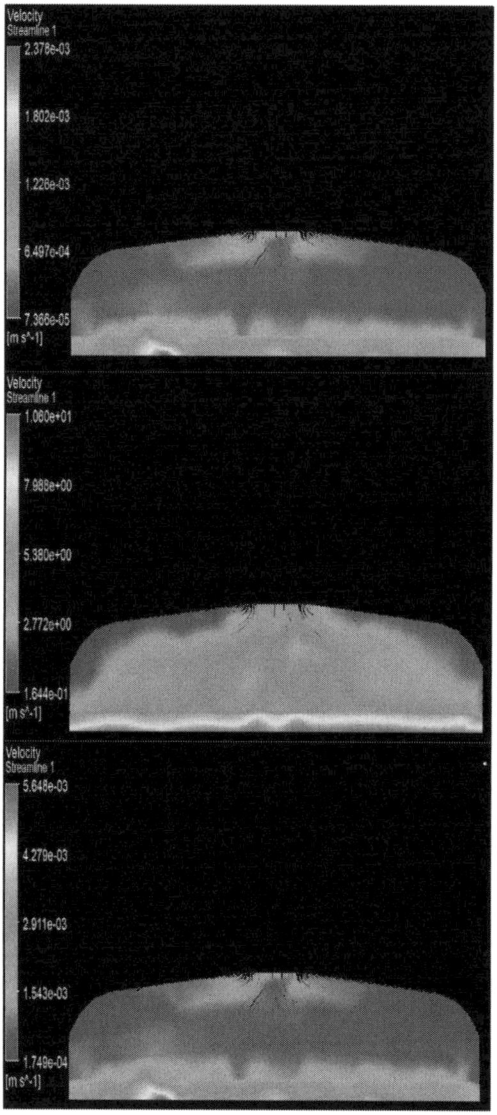

Figure 4. Top – 2cm slit-accessory flow simulation of TiO2 paste. Middle - perovskite liquid precursor. Bottom - carbon black paste.

The series of Figures illustrate the repsectful change in magnitude of velocity, and the uniformity of the velocity as a function of the nozzle size. All three designs will be in theory able to deposit the materials which are required to construct the PSC. Moreover, the simulations allow the further experimentation with parameters such as pressure applied as will be in the experiment to control the flow of the material if required. Where the current values are appropreiate to the rate of required of the material to be deposited onto the substrate, and the only parameter which is not considered is the speed of printing "Extruder head speed". Nevertheless, the 4 cm wide accessory nozzle demonstrates not just a higher coverage, but also uniform velocities with the exception of the perovskite test where there is a small spot which velocity is concentrated.

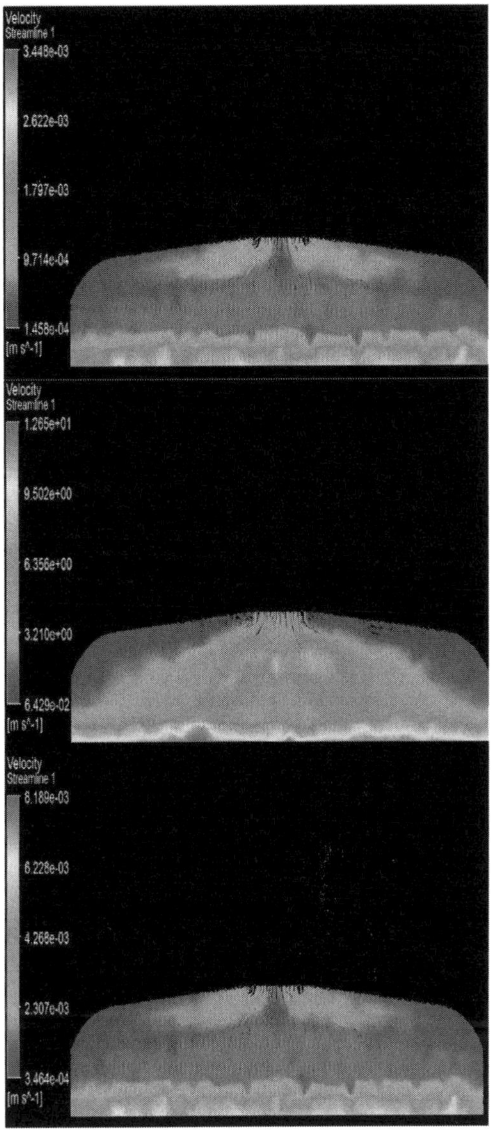

Figure 5. Top – 4cm slit-accessory flow simulation of TiO2 paste. Middle - perovskite liquid precursor. Bottom - carbon black paste.

4 CONCLUSION

This paper reviews general concepts related to perovskite solar cells, discusses the PSC operation mechanism, different architectures suitable to handle charge transport, and the materials used are detailed. Finally, different manufacturing methods used to produce small, intermediate, and large area PSC. It is also discussed the use of additive manufacturing to mass produce and upscale PSCs while maximizing its capabilities. The research project will carry on with the work to manufacture a planar PSC (FTO/TiO_2 paste/$CH_3NH_3PbI_3$/Carbon black paste) with an area over 25cm^2 and an efficiency of over 10% using an extrusion-based 3D printer through the use of the nozzle accessory which was simulated using ANSYS in order to confirm the concept as well as the appropriate size, and internal geometry. Further optimization will also be applied to the design to ensure a uniform deposition across the manifold and avoid any defects.

REFERENCES

Beeker, L., Pringle, A. and Pearce, J. (2018). Open-source parametric 3-D printed slot die system for thin film semiconductor processing. *Additive Manufacturing*, 20, pp.90–100.

Burkitt, D., Searle, J. and Watson, T., 2018. Perovskite solar cells in N-I-P structure with four slot-die-coated layers. *Royal Society Open Science*, 5(5), p.172158.

Cai, M., Wu, Y., Chen, H., Yang, X., Qiang, Y. and Han, L. (2016). Cost-Performance Analysis of Perovskite Solar Modules. *Advanced Science*, 4(1), p.1600269.

Crone, D. (2016). *Slot-Die coating principle and application.*

Di Giacomo, F., Shanmugam, S., Fledderus, H., Bruijnaers, B., Verhees, W., Dorenkamper, M., Veenstra, S., Qiu, W., Gehlhaar, R., Merckx, T., Aernouts, T., Andriessen, R. and Galagan, Y. (2018). Up-scalable sheet-to-sheet production of high efficiency perovskite module and solar cells on 6-in. substrate using slot die coating. *Solar Energy Materials and Solar Cells*, [online] 181(Thin film solar cells and applications), pp.53–59. Available at: https://www.sciencedirect.com/science/article/pii/S0927024817306190.

Rong, Y., Hu, Y., Mei, A., Tan, H., Saidaminov, M., Seok, S., McGehee, M., Sargent, E. and Han, H. (2018). Challenges for commercializing perovskite solar cells. *Science*, 361(6408), p.eaat8235.

Vak, D., Hwang, K., Faulks, A., Jung, Y., Clark, N., Kim, D., Wilson, G. and Watkins, S. (2015). Solar Cells: 3D Printer Based Slot-Die Coater as a Lab-to-Fab Translation Tool for Solution-Processed Solar Cells (Adv. Energy Mater. 4/2015). *Advanced Energy Materials*, [online] 5(4). Available at: https://onlinelibrary.wiley.com/doi/full/10.1002/aenm.201401539.

Whitaker, J., Kim, D., Larson, B., Zhang, F., Berry, J., van Hest, M. and Zhu, K. (2018). Scalable slot-die coating of high performance perovskite solar cells. *Sustainable Energy & Fuels*, 2(11), pp.2442–2449.

Zhakeyev, A., Wang, P., Zhang, L., Shu, W., Wang, H. and Xuan, J. (2017). Additive Manufacturing: Unlocking the Evolution of Energy Materials. *Advanced Science*, 4 (10), p.1700187.

Industry 4.0 – Shaping The Future of The Digital World – da Silva Bartolo et al. (eds)
 ISBN 978-0-367-42272-1

Remote monitoring and controlling of an additive manufacturing machine

A. Alhijaily & P. Bartolo
University of Manchester, Manchester, UK

T. Alhijaily
IT Department of Control and Investigation Board, Madinah, Saudi Arabia

ABSTRACT: Industry 4.0 strives to integrate the physical environment with the digital world. With the increasing popularity of additive manufacturing (AM) in the recent years, it is time to remote monitoring and controlling such technology. This aids the production of personalized products. However, there are few attempts to remotely control or monitor AM systems in the literature. This paper discusses the development of a method to remotely control and monitor an AM machine by utilizing Industry 4.0 components such as the internet of things and cloud computing. Several software were developed to achieve this aim, such as custom web server and a mobile application. The mobile successfully communicated with the machine from the server and sent various commands, such as G-code, requesting motors data and viewing the printing process. This opens up new dimensions for additive manufacturing and mass personalization.

1 INTRODUCTION

The fourth industrial revolution, Industry 4.0, aims to integrate the digital world with the physical objects. Such interaction enables remote monitoring and controlling of machines and components. This concept is a result of information technologies integrated with physical devices such as sensors and actuators. It provides ease of access and interaction with machines without requiring the presence of operators.

Additive Manufacturing (AM), also known as 3D printing, is a manufacturing process in which a part is made by adding material layer by layer instead of removing material as in conventional machining. Moreover, AM is also the ideal technology for mass personalization which aims at satisfying each individual needs and requirements (Tseng et al., 2010). Remote monitoring and controlling of such technology aid the production of personalized products.

Cloud computing is another component of Industry 4.0 which is an information technology model where computations and storage are delivered to users over the internet (Marston et al., 2011). It is an emerging concept in computer science which eliminates the need for building standalone servers and networking infrastructures. Another component of Industry 4.0 is the Internet of Things (IoT) which is defined by Oztemel & Gursev (2018) as a technological paradigm that connects physical objects, devices, and structures which are embedded by digital chips and sensors and allows them to transfer and process data. These two components enable the concept of remote monitoring and controlling.

Remote monitoring and controlling has been done before for various machines and using different technologies and approaches. Peng et al. (2019) developed a system which remotely controls different parameters in low carbon emission air conditioning system by utilizing various sensors and data acquisition systems. Moreover, Rastogi et al. (2019) presented a remote handling control system for a nuclear fusion machine but used virtual reality to enrich the visual aspect of the operation. As for remote monitoring and controlling in maintenance, Campos & Marín (2011) proposed a control-monitor -simulation architecture to use for maintenance off stone cutting machines and proved that this model was able to produce efficient fault analysis. Also, Mourtzis et al. (2017) applied Augmented Reality technology to support remote maintenance and allow communications between a beginner operator and an experienced one during maintenance which reduced the required maintenance resources.

There is a lack of remote monitoring and controlling regarding additive manufacturing machines in the literature. This research addresses this limitation through the development of a computational application to remotely control and monitor an additive manufacturing machine, using cloud computing and IoT. The focus is on sending G-code and motors positions to the machine as well as monitoring the process by extracting motors data and getting views from a camera that is attached to the machine. There are two devices for the process, a mobile as a controlling unit and a computer to interact with the machine and the sensors. Also, a web server is developed to allow communication between the mobile application and the computer software.

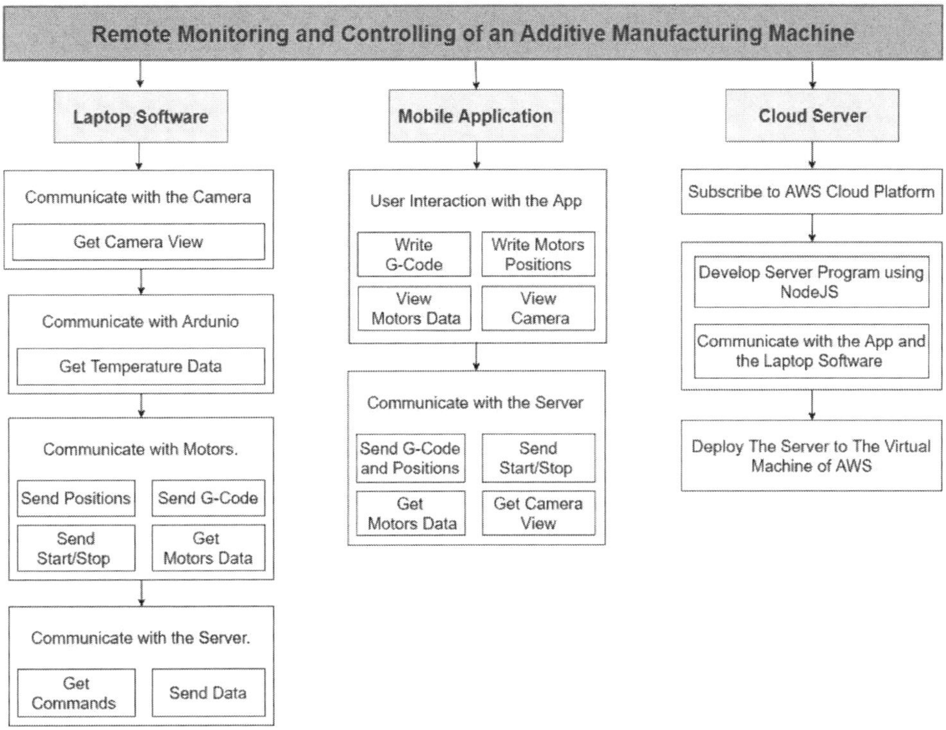

Figure 1. A diagram that shows the methodology of the research.

2 METHODOLOGY

The steps taken to achieve the aim of this research consists of configuring the needed hardware and developing some software. Figure 2 lists the required hardware as well as how they are used in the software. The hardware of this research are as follows:

- Additive Manufacturing Machine: the machine that is the focus of this research is a Plasma Assisted Bio-extrusion System (PABS) which was developed by Prof. Bartolo's group at the University of Manchester (Figure 3). PABS consists of pressure assisted and screw-assisted extrusion heads as well as plasma jets. PABS allow the fabrication of hybrid tissues or organ-like structures that have compositional variations based on the region (Liu et al., 2018). The machine's motors are controlled by using Pewin32PRO2 software.
- Web Camera: a normal camera that is attached to the machine to stream the process of printing to the user.
- Laptop: the laptop is connected to the machine to control the G-code and to the camera to receive the camera view. It is connected to a custom-built server to send the information to.
- Mobile: the controlling and monitoring unit that sends and receives information from the server.

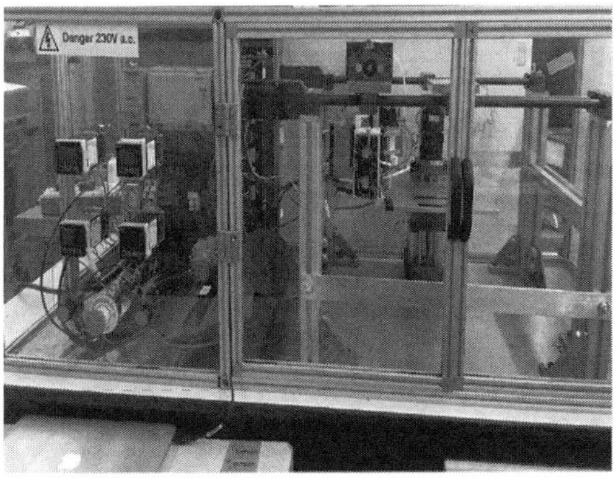

Figure 2. The plasma assisted bioextrusion system.

For the software part, there are three main programs that are developed to allow the process:

- Web Server: a custom server that is built from scratch using Node.js which is an open source JavaScript environment used to develop both back-end API and websites. The server is uploaded to Amazon Web Services which is a cloud computing platform provided by Amazon.

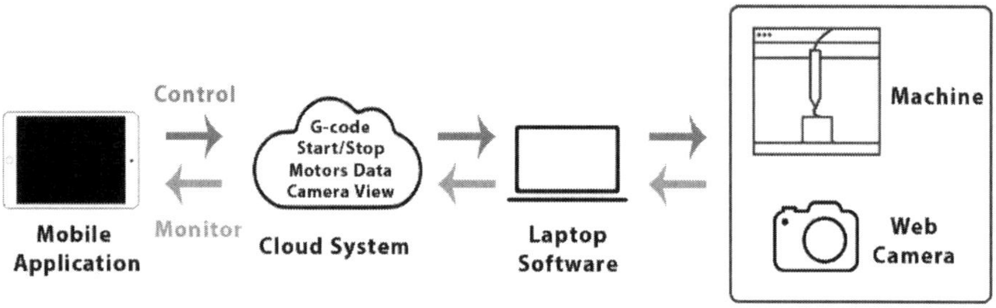

Figure 3. The architecture of the system.

This web server allows communications between the laptop and the mobile.

- Laptop Software: this software is developed using C#. It listens to the user commands coming from the server which then launches different AutoHotKey scripts that are responsible for all the communications between Pewin32PRO2 and the laptop. Also, the camera streams the views to this software. All the data received from the camera and Pewin32PRO2 are then sent to the server.
- Mobile Application: the interface that allows the user to control and monitor the process. It is developed for Android operating system using Android Studio and Java. There are three menus on the application: G-code menu, motors menu, and camera views menu. G-code menu allows the user to send G-code, start and stop printing, and reset axes positions Motors menu shows positions, velocities, and following error of the motors as well as buttons for moving the motors to specified positions. Finally, the camera menu shows the view of the camera which allows the user to follow the process of printing and to stop the process in case of an emergency.

When the previously mentioned components are connected, the process of accessing the machine remotely is done. Figure 4 shows the architecture of the system and how each component is placed in the system.

3 RESULTS

After installing the required software on the laptop and connecting it to the machine and the camera, some tests were conducted to check whether the system is working as intended or not. When everything was checked, various results were generated from testing the full system.

Since the connection between the two ends of the project (the mobile application and the laptop software) is the most important part, it is checked first. When the software development for the laptop finished, the software was installed on the laptop. After installation, the connection of the software to the server was checked, and the server responded back that a connection was registered. For the mobile application (Figure 5), the files were exported to the mobile using Android Studio. Then, the server confirmed that there is a connection from this mobile. The connection of the application and the software to the server is the first step of remote monitoring and controlling the machine.

One of the objectives of this research is to send user commands from the mobile application to the laptop software. There are two types of commands for the user: commands without responses and the other is commands with responses from the other end. First, direct commands are tested which are present in both types. A simple command was sent first from the mobile to the server which fired an event in the software to display that the command has arrived. This simple command served as a test of the connectivity from the mobile to the laptop through the server which proved to be successful. The other way around is similar; the laptop sends commands to the server which fires an event that calls functions in the mobile and again the operation was successful. Testing command and response was done by using double commands method. In this case, the first end sends a command which calls a function in the other end that sends another command back to the first end.

After configuring the hardware and the software and checking that everything is sending and receiving data, it is time to control the printing process. There are three controlling commands in the application: sending G-code, starting/stopping the printing process, and moving the motors to specific positions for testing purposes. When the user writes the G-code and presses the send G-code button, the software takes the fed G-code and opens a new file then download the G-code to the machine. The next step is pressing the start printing button which sends a command line to the terminal to start printing the downloaded G-code. Then the user can stop printing by pressing the stop button. The previously mentioned steps worked successfully, and the machine performed the input G-code. Also, the motors can be moved to specific positions by inputting the required position values in the application.

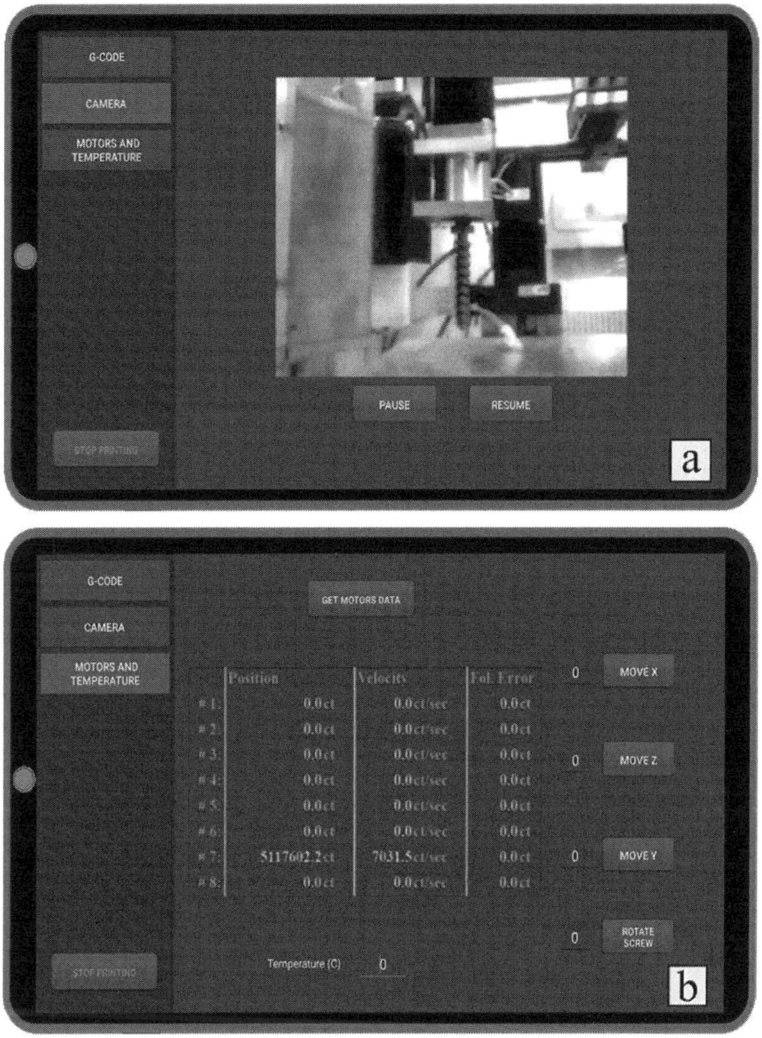

Figure 4. The user interface of the application. a) Camera menu showing the process. b) Motors and temperature menu.

Monitoring is the next step after successfully controlling the machine. The two monitoring functions in the system are: following the printing process using a camera and receiving data about the position, velocity and following error of each motor. The camera sent the view back to the application through the software and the server. The second function is monitoring motors data; this is done when the user requests the data from the software. After requesting the data, the software responds with the data fetched and sends it back to the mobile application.

4 DISCUSSION

The system introduced in the previous chapters serves various goals and bypasses different limitations of normal monitoring and controlling. Integrating this system with AM enhances the ability of producing parts and components with low supervision.

Any machine cannot be fully utilized if it only works with the presence of workers near the machine. This disadvantage reduces the productivity of the workshop. Hence, the developed remote monitoring and controlling system bypasses this by allowing the access of the machine from anywhere in the world at any time. The availability zones for AWS according to (Amazon Web Services, Inc., 2019), include 61 zones within 20 regions across the world. This spread of services allows for low delay communications between the two ends of the system.

Utilizing Industry 4.0 opens up new improvements in the industry such as mass personalization.

When the developed system is fully utilized, it can be used for improving the capabilities of Industry 4.0 to achieve mass personalization. For example, some parts of the system can be modified to allow external users to use the machine and manufacture a personalized part without being physically near the machine.

The aspect of monitoring in the system streamlines the process of detecting failures or deviations from the required task. For example, with the introduction of a camera in the system, visual feedback can be received which can help in detecting off-bed printing. Also, the camera can be used to stop printing when catastrophic events occur (e.g., collision between the printing head and parts). Another monitoring aspect in the system is getting each motor data from the software, which can help in detecting problems in the motors such as when one motor is not moving when a specific command is sent. Moreover, it can be used to monitor how fast each motor is moving and check if that match the provided G-code. This system can return the following error of each motor which is the difference between the actual position of the motor and the command of the motor controller. Another plus for this system is to examine problems within the machine by utilizing the controlling and monitoring aspect of it. The user can send data about where the motors should locate to and check both the motors data and the camera to detect the problems.

5 CONCLUSION

Industry 4.0 is trying to digitize physical environment to achieve various goals such as mass personalization. Internet of Things and Cloud Computing are two of the components of Industry 4.0. When these two are combined and utilized, remote monitoring and controlling over the internet can be achieved. This concept has been discussed before in the literature and used for various machines. However, additive manufacturing did not receive the deserved attention in this area. In this paper, the concept of remote monitoring and controlling was applied to PABS machine which is an additive manufacturing machine. For this reason, three software were developed: a Node.js web server, a C# windows software, and an android mobile application. These software were connected together to convert user inputs on the mobile to commands that communicate with the software that controls the machine's motors. Various commands can be sent to the machine, such as G-code related commands and motors commands. Also, to monitor the machine, a camera was used to stream the printing process to the user as well as allowing the user to request motors data. Applying this concept to the machine allows controlling and monitoring the machine from any location in the world and at any time which results in high utilization of the machine. Also, the developed system opens new opportunities for mass personalization.

Further work will be done for this system to improve it, such as:

- Introducing more parameters in the system such as the temperature of extrusion or the heated bed.
- Allowing users to schedule printing for future prints.
- Implementing notifications system to send messages to the user when an event occurs, such as when a print finishes printing.

REFERENCES

Amazon Web Services, Inc. (2019). Global Cloud Infrastructure | Regions & Availability Zones | AWS. [online] Available at: https://aws.amazon.com/about-aws/global-infrastructure/[Accessed 24 Feb. 2019].

Campos, J. and Marín Martín, R. (2011). Remote maintenance and fault analysis system for custom-made machine. IFAC Proceedings Volumes, 44(1), pp.14976–14981.

Liu, F., Wang, W., Mirihanage, W., Hinduja, S. and Bartolo, P. (2018). A plasma-assisted bioextrusion system for tissue engineering. CIRP Annals, 67(1), pp.229–232.

Marston, S., Li, Z., Bandyopadhyay, S., Zhang, J. and Ghalsasi, A. (2011). Cloud computing — The business perspective. Decision Support Systems, 51(1), pp.176–189.

Mourtzis, D., Zogopoulos, V. and Vlachou, E. (2017). Augmented Reality Application to Support Remote Maintenance as a Service in the Robotics Industry. Procedia CIRP, 63, pp.46–51.

Oztemel, E. and Gursev, S. (2018). Literature review of Industry 4.0 and related technologies. Journal of Intelligent Manufacturing, (Online First), pp.1–56.

Peng, W., Su, D. and Higginson, M. (2019). A novel remote control system for air conditioning in low carbon emission buildings using sensor fusion and mobile communication technologies. Building and Environment, 148, pp.701–713.

Rastogi, N. and Kumar Srivastava, A. (2019). Control system design for tokamak remote maintenance operations using assisted virtual reality and haptic feedback. Fusion Engineering and Design, 139, pp.47–54.

Tseng, M., Jiao, R. and Wang, C. (2010). Design for mass personalization. CIRP Annals, 59(1), pp.175–178.

Industry 4.0 – Shaping The Future of The Digital World – da Silva Bartolo et al. (eds)
© 2021 Taylor & Francis Group, London, ISBN 978-0-367-42272-1

Virtual prototyping for fashion 4.0

A.S.M. Sayem
Manchester Fashion Institute, Manchester Metropolitan University, Manchester, UK

ABSTRACT: Virtual prototyping shows enormous potential to make the fashion manufacturing industry greener and leaner. Yet the technology is not well embraced by the industry. No matter what the reason is, this is the only way towards e-manufacturing for fashion 4.0. This paper provides an overview of the technology and its features.

1 INTRODUCTION

Creative and technical designs are two integral parts of the fashion-product-development-process (FPDP) practised in the industry today (Glock & Kunz, 2000). The aspect of creative design covers the process of fashion design resulting from designers' imaginations coupled with market research information. On the other hand, the aspect of technical design includes pattern drafting based on the anthropometric information of the target market to facilitate geometric fabric cutting prior to clothing manufacture. Both manual and computer-aided techniques exit for both creative and technical fashion designs. However, the new generations of designers are naturally fond of computer-aided techniques following the millennial trend. A notable number of computer-aided design (CAD) systems is available today for use in the fashion industry. They offer efficiency and timesaving solutions to many complex and complicated tasks and facilitate smooth communication over the worldwide web platform between one place and almost any geographically remote corner of the world.

2 2D CAD FOR FASHION INDUSTRY

General graphics design software packages such as Illustrator® from Adobe Systems Incorporated (USA) and CorelDRAW® from Corel Corporation (Canada); or more customised software systems for the fashion industry such as Kaledo® Style Lectra (France), Vision® fashion studio from Gerber Technology (USA), Tex-Design™ from Koppermann Computersysteme GmbH (Germany) are extensively used around the world for fashion drawing and illustration (Sayem et al. 2010). Specialised CAD software packages for drafting and grading flat patterns of garments were introduced into the industry in the 1980s (Burke, 2006) and they have become very popular within the fashion industry. The commonly known 2D CAD software packages that offer options for drafting flat pattern pieces using the body measurements information of the target consumers are: cad.assyst from Assyst (Germany), Modaris from Lectra (France), Accumark PDS from Gerber Technology (USA), PAD Pattern Design from PAD System Technologies Inc. (Canada), TUKAcad from Tukatech (USA), GRAFIS from the company Dr. K. Friedrich (Germany) and Audaces Apparel Patterns from Audaces (Brazil) (Sayem et al. 2010). In addition to pattern drafting, they support importation of existing block patterns with the help of a "digitiser". Usually, they support automatic pattern grading with the help of pre-developed grade rule tables. Figure 1 illustrates a pattern piece of a shirt panel drafted in a commercial CAD system.

3 3D CAD FOR FASHION INDUSTRY

Several researchers have demonstrated effective techniques of draping digital pattern pieces on virtual mannequins and simulation of virtual clothing prototype (Fozzard & Rawling, 1991, 1992, Okabe et al. 1992, Volino et al. 1996, Kang & Kim, 2000b, 2000c, Chiricota, 2003, Fuhrmann et al. 2003, Thalmann & Valino, 2005, Luo & Yuen, 2005). This technology has already reached a stage of sufficient maturity to be implemented in the industry (Sayem et al. 2010, 2011). Commercial 3D CAD systems that include virtual avatars and support simulation of 2D pattern pieces on them started to appear on the market since 2001 (Goldstein, 2009). They include the software packages like Vstitcher from Browzwear (Singapore), Accumark 3D from Gerber (USA), Modaris 3D from Lectra (France), TukaCAD 3D from Tukatech (USA), PDS 3D from EFI OptiTex (Israel) and Vidya from Assyst (Germany).

3.1 *Virtual human models*

3D clothing CAD systems usually come with a library of built-in virtual avatars, which can be customised with the variation of different

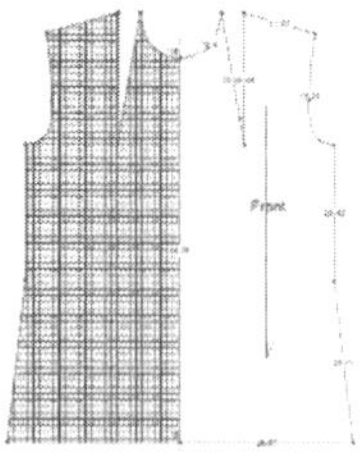

Figure 1. Front panel of a ladies' shirt drafted in a commercial 2D CAD system.

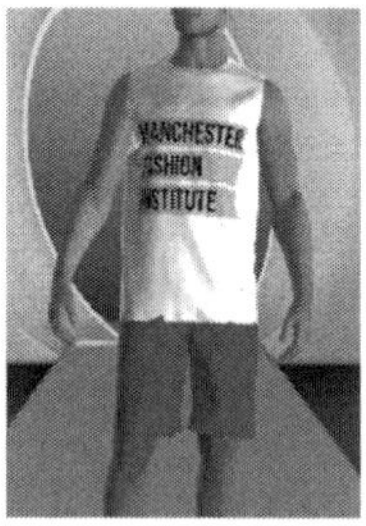

Figure 2. An example of a virtual fashion prototype produced in a commercial 3D CAD system.

anthropometric and demographic information such as age, gender, measurements, body posture, skin tone, hair style and colour, and even the pregnancy stages (Sayem et al. 2010) They work as a platform for 3D clothing generation. The systems allow to position 2D digital pattern pieces on the avatars and to stich them virtually to facilitate virtual simulation of drape. With the help of 3D bodyscanning technology, it is now possible to capture a full human body very quickly as a 3D scan and to extract measurements out of it. This helps to morph virtual avatar with required size and shape to produce a virtual twin of a human model. Figure 3 illustrates the morphed avatars in the different available 3D CAD systems. However, the commercial systems differ in the number of parameters they utilise to morph virtual avatars (see Figure 4 presents the number of morphing parameters utilised in a few available 3D CAD systems.

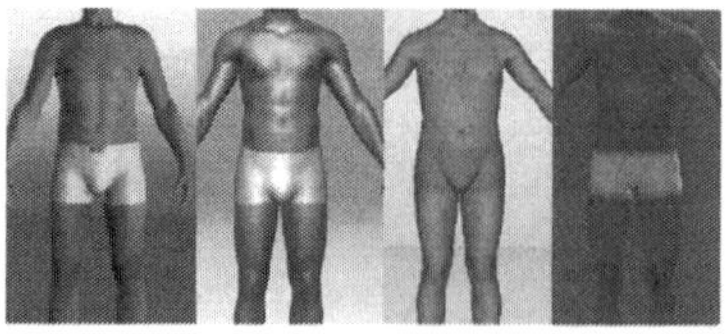

Figure 3. Morphed male avatars in different CAD systems.

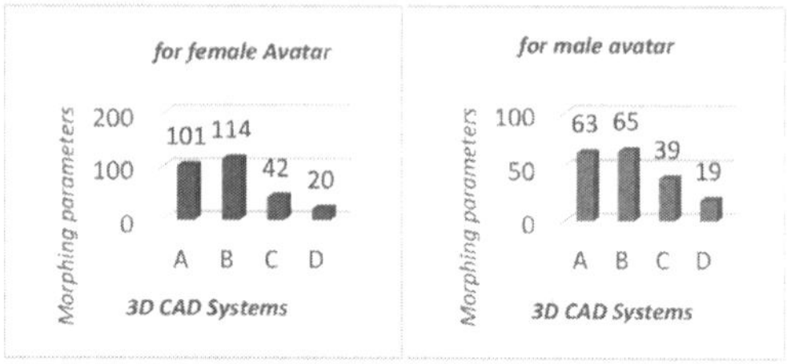

Figure 4. Number of morphing parameters utilised in commonly available CAD systems.

3.2 *Fabric library and virtual drape simulation*

The 3D fashion CAD systems include a built-in library of fabrics and other related materials together with their physical and mechanical characteristics Once digital pattern pieces are positioned on avatars, drape simulation can be performed applying relevant material properties (Figure 5). This can be done by selecting appropriate fabric from built-in fabric library or by adjusting the properties a close one. In a few cases, it is also possible to input new fabric properties taken from an objective fabric measurement system such as KES-f and FAST to view differential drape. Very recently few software companies have brought their own fabric testing instruments on the market for use in acquisition of materials parameters of drape modelling and do not include the KES-f and FAST data converters anymore. However, these newly introduced fabric testing systems need go through the processes of standardisation (Power, 2013), approval, and accreditation from the international standardisation bodies for use in the industry.

3.3 *Virtual fit analysis*

Available 3D CAD systems allow their users to check garment fit in various fabrics and sizes and facilitate the virtual review of prototype garments. Colours, motifs and trims etc. can be easily and quickly changed on virtual garments to help decision making, as presented in the Figure 6.

However, it is apparent from the works of Kim (2009) and Kim & LaBat (2013) that only a visual check of virtual prototype does not provide any conclusive clue for decision making on

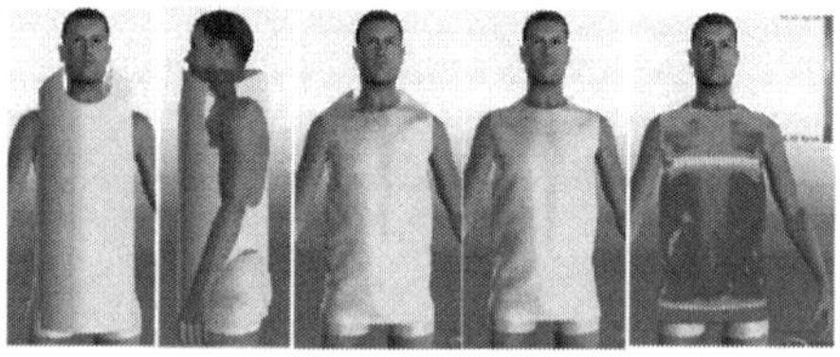

Figure 5. Steps in virtual simulation of clothing.

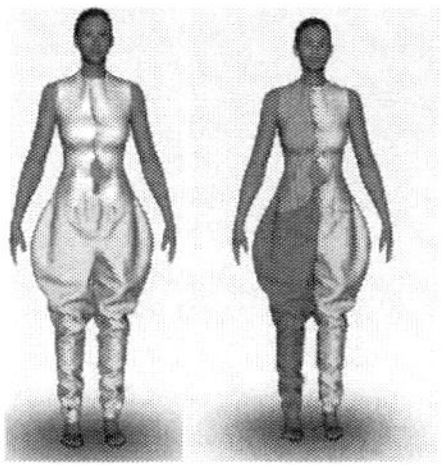

Figure 6. Examples of same virtual garment in different colours.

the state of virtual fit. They simulated a pair of woven trousers based on fabric properties and found that the appearance of virtual fit significantly differed from that of real fit of garments. Particularly wrinkles that existed on physical prototypes they made were not accurately reproduced on virtual prototypes. The work by Lim (2009) include a virtual comparison of a women's wear simulated two different systems utilising identical material properties. He found that the visual appearances of the simulated garments looked different in two systems he used. Power et al. (2011) also highlighted similar limitation of visual assessment of virtual fit. They virtually simulated two different fabrics having vastly different characteristics and found a very similar appearance of the virtual simulations resulted from those two fabrics. This supports the findings of Kim (2009), Lim (2009) and Kim & LaBat (2013). This indicated the existing subjective evaluation is not sufficient to evaluated virtual fit of clothing and asks for an objective approach in addition to this.

3D CAD systems offer several technical tools for virtual fit analysis such as tension, pressure, stretch and ease mapping tools, as can be seen in the Figure 7. These tools offer the opportunities of both subjective and objective fit evaluations in addition of visual check of the simulated drape and fit on virtual avatars (Sayem et al. 2010, 2012, Lim & Istook, 2011). This provides an opportunity to review and forecast the clothing fit at the pre-manufacture stage and to take decision on the correctness of drafted pattern pieces (Sayem, 2017, Sayem et al. 2017).

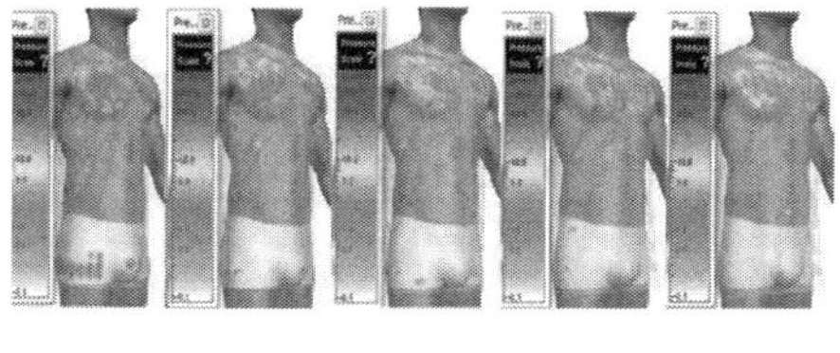

Figure 7. Examples of Tension maps of virtual garments.

4 PRESENT BUSINESS SCENARIO AND VIRTUAL PROTOTYPING

In these days of globalisation, offshore sourcing has become a key business strategy for many fashion retailers (Christerson & Appelbaum, 1995, Firoz & Ammaturo, 2002; Gereffi & Memedovic, 2003). To take the advantages of labour market differentials in many lower labour cost countries, most of the world's clothing production today takes place in the far eastern counties. However, most of the activities of fashion design and product development are done and still controlled from the retailers' head offices in Europe and America. The geographical distance between manufacturers and retailers causes a considerable long period of time for making prototypes at manufactures' factories according to designers' specifications and transporting those prototypes to the retailers. Often this process needs a repetition in order to rectify any problems related to assembly or fit or to incorporate any change in design. Together with the distance involved, it inevitably increases the development lead-time and cost even further. Therefore, there is always pressure from the retailers' side to curtail the development lead-time and to minimise the cost involvement in physical prototyping in order to cope with rapid fashion changes and business competition. Virtual prototyping offers a solution to this problem. It is now possible to drape 2D digital pattern pieces on a virtual avatar to simulate the 3D appearance of clothing close to an acceptable reality. This gives the opportunity of identifying flaws in 2D patterns through a process of virtual fit checking to rectify initial problems with 2D pattern pieces without any need for a physical prototype with real materials. As the process of virtual fit assessment can also be communicated over the worldwide web platform between suppliers and retailers from any corner of the world, the product-development lead-time can be significantly shortened and the dependency on physical prototypes can be reduced (Ernst, 2009, Sayem & Bednall, 2017).

5 CONCLUSION

The suppliers claim several advantages of using these 3D systems such as improved communication of design data throughout the supply chain and product life cycle and lowering of product development time and costs (Ernst 2009, Sayem et al. 2012, 2014). However, the fashion designers and industry are yet to get any true benefit from the prevailing virtual fashion technology, despite such software systems being available on the market for more than a decade now (Sayem, 2017, Sayem et al. 2014). Little information is available on how and at what extent such technology is being used in the industry around the world and what practical benefits of using them are being experienced. No academic literature reports the reasons behind the slow and sluggish adoption of virtual simulation technology within the fashion industry (Poterfield & Lamar, 2017). It is important to strengthen the research and development (R&D) from industry and academia to address these current issues.

REFERENCES

Burke, S. 2006. FASHION COMPUTING - Design Techniques and CAD. UK: Burke Publishing.

Chiricota, Y. 2001. Geometrical modelling of garments. Intern. Journal of Clothing Science & Technology, 13 (1), 38–52.

Christerson, B. & Appelbaum, R., 1995. Global and local subcontracting: space, ethnicity, and the organization of Apparel Production, World Development, 23 (8), 1363–1374.

Firoz, N.M. & Ammaturo, C.R., 2002. Sweatshop labour practices: The bottom line to bring change to the new millennium case of the apparel industry, Humanomics, 18 (1/2), 29–45.

Fozzard, G.J.W. & Rawling, A.J., 1991. Simulation of dressing and drape for garment CAD, Proceedings from the 6th International Forum on CAD, 157–62.

Fozzard, G.J.W. & Rawling, A.J., 1992. CAD for garment design – Effective use of the third dimension, Proceedings of the Eighth National Conference on Manufacturing Research, 183-189.

Fuhrmann, A.; Gross, C.; Luckas, V. & Weber, A., 2003. Interaction-free dressing of virtual humans. Computers & Graphics, 27, pp. 71–82.

Glock, R. E. & Kunz, G. I., 2000. Apparel Manufacturing Sewi ng Product Analysis. 3rd ed. New Jersey: Prentice Hall.

Goldstein, Y., 2009. Virtual prototyping: from concept to 3D design and prototyping in hours", in Walter, L., Kartsounis, G.-A. & Carosio, S. (Eds), Transforming Clothing Production into a Demand-Driven, Knowledge-Based, High-Tech Industry, Springer, London.

Gereffi, G. & Memedovic, O., 2003. The global apparel value chain: What prospects for upgrading by developing countries? [online]. United Nations Industrial Development Organization (UNIDO), Vienna. Available from: https://www.unido.org/sites/default/files/2009-12/Global_apparel_value_chain_0.pdf (Accessed 31 Jan 2019)

Kang, T.J.& Kim, S. M., 2000a. Development of three dimensional apparel CAD system Part 1: flat garment pattern drafting system. International Journal of Clothing Science & Technology, 12 (1), 26–38.

Kang, T.J. & Kim, S. M, 2000b. Development of three dimensional apparel CAD system, Part II: prediction of garment drape shape. International Journal of Clothing Science & Technology, 12 (1), 39–49.

Kim, D., 2009. Apparel Fit Based on Viewing of 3D Virtual Models and Live Models (Unpublished doctoral Thesis). The University of Minnesota.

Kim & LaBat 2013. An exploratory study of users' evaluations of the accuracy and fidelity of a three-dimensional garment simulation, Textile Research Journal, 83 (2), 171–184.

Luo, G. Z & Yuen, M.M.F. 2005. Reactive 2D/3D garment pattern design modification, Computer-Aided Design, 37, 623–630.

Lim, H.S. 2009. Three Dimensional Virtual Try-on Technologies in the Achievement and Testing of Fit for Mass Customization (Unpublished doctoral Thesis). North Carolina State University.

Okabe, H.; Imaoka, H., Tomih, T. & Niwaya, H. 1992. Three dimensional apparel CAD system, Computer Graphics, 26 (2), 105–110.

Poterfield, A. & Lamar, A.M.T. 2017, Examining the effectiveness of virtual fitting with 3D garment simulation, International Journal of Fashion Design, Technology & Education, 10 (3), 320–330.

Power, J. 2013. Fabric objective measurements for commercial 3D virtual garment simulation. International Journal of Clothing Science & Technology. 25(6), 423–439.

Sayem, ASM; Kennon, R. & Clarke, N. 2010. 3D CAD systems for the clothing industry. International Journal of Fashion Design, Technology & Education, 3 (2), 45–53, URL: http://dx.doi.org/10.1080/17543261003689888.

Sayem, ASM; Kennon, R. & Clarke, N. 2012. Resizable trouser template for virtual design and pattern flattening, International Journal of Fashion Design, Technology & Education, 5:1, 55–65, DOI: 10.1080/17543266.2011.614963.

Sayem, ASM; Kennon, R. & Clarke, N. 2014. 3D Grading and Pattern Unwrapping Technique for Loose-fitting Shirt, Part 1: Resizable Design Template. Journal of Textile and Apparel, Technology & Management. 8(4), http://ojs.cnr.ncsu.edu/index.php/JTATM/article/view/5145/2850.

Sayem, ASM. 2017. Objective analysis of the drape behaviour of virtual shirt, part 1: avatar morphing and virtual stitching. International Journal of Fashion Design, Technology & Education. 10(2), 158–169.

Sayem, ASM. & Bednal, A. 2017, A Novel Approach to Fit Analysis of Virtual Fashion Clothing, IFFTI annual Conference 2017, The Amsterdam Fashion Institute (AMFI), Amsterdam, 28/3/2017.

Thalmann, N. M. & Volino, P. 2005. From early draping to haute couture models: 20 years of research. Visual Computer, 21, 506–519.

Volino, P.; Thalmann, N.M.; Shen J. & Thalmann, D. 1996. An evolving system for simulating clothes on virtual actors, IEEE Computer Graphics & Applications, 16 (5), 42–51.

Industry 4.0 – Shaping The Future of The Digital World – da Silva Bartolo et al. (eds)
© 2021 Taylor & Francis Group, London, ISBN 978-0-367-42272-1

Could the food market pull 3D printing appetites further?

S. Killi & A. Morrison
The Oslo School of Architecture and Design, Oslo, Norway

ABSTRACT: We situate the 3D printing of food in a 'culinary turn' of uptake in domestic settings. We suggest a shift from a pull-by market technology approach to 3D printing as a technology to an interest in creative, participative domestic experimentation with food and 3DP. There is potential for new appetites maybe 'pulled' by the lure of play and uptake of 3D printing in the cultural making and appetites shaped in the wider context of nutrition, sustainability and a circular economy.

1 INTRODUCTION AND BACKGROUND

1.1 *3D printing food*

The first 3D printer designed for food would probably be fablab@home in 2005 (Lipton et al. 2009), though several attempts had been made with regular printers prior to this. Most of the technology available today is in the format of Fused Deposition Modeling or FDM. The different technologies will be discussed more in depth in section two. Interestingly these techniques, squeezing something edible onto either a cake (frosting) or sandwich (mayonnaise), are akin to something almost all of us have done at some point in our lives. It is quite easy to relate to the technique, but also to the material challenges; if the frosting is too runny it's hard to make a precise pattern. Then again, how many of us actually need a precise frosting tool or an artistic pattern for the mayonnaise on our sandwich?

In many ways we see the same challenges as with printing plastic things: what is our daily need for small plastic knick-knacks? Yet, we do actually have to eat every day, often in places and situations that are challenging. Also, don't we very often customize our food? This involves seasoning and mixing, experimenting and testing. We suggest that there are more people who design and prepare their food than there are people designing and producing any other things.

However, we need to locate these aspects in relation to cultures of making and consuming food as well as to the looking critically at recent promotion of 3D food printers. For example, the online TechRepublic, suggesting the democratic reach of emerging technologies, has the following text in a recent article: '3D-printed food offers new possibilities such as intricate designs, automated cooking, mass manufacturing, and personalised meals. But will it ever replace the traditional methods we know today?' (Yang et al. 2017). The issue here, and address below, is just how such technologies become not only popularized, but whether we are on the cusp of a much more substantial democratization, and indeed domestication, of this Additive Manufacturing technology through emergent and situated use. We see this as more than a matter of replacing methods: it is one of a transformation in the ecology of food production and consumption in an age of actual and potential climate, economic and market change.

Is there not only potential for domestic 'appetites' for 3D technology to be furthered, as part of an uncritical market driven consumerism but also a role for how 'appetite' for food may itself be implicated in what we call 'a culinary turn' for culturally framed 3D printing? Might new shaping and experience of taste, tasting and indeed appetites - in an age of healthier and more sustainable eating - work to co-construct the situated uses and application of this much hyped technology?

We address these issue below through discussion of two potential and related aspects: technology and possible market need. We present a selection of projects, emergent uses and potential developments in order to arrive at a discussion of how not just how food trends, such as pulled pork, may be realized through 3D food printing but a new interplay between the sourcing, making and satisfying consumption of food. This is printing that centres on crafting enticing, desirable and satisfying sustenance not only a model of more mass produced fast food of questionable nutritional quality in unhealthy packaging conveyed at great distance for a maximum number of consumers. As our title indicates, the question is whether there is an opportunity for 3D food printing to effect a different market pull.

1.2 *Technology push and market pull*

The term technology pull has long been used about Additive Manufacturing, or more conveniently, 3D printing. Such extended use has led to people asking whether claims that 3D printers are central to a post-industrial revolution have support (Lipton et al.

2009). When it comes to the projection of 3D printers as core home appliances, it's hard to see their actual need.

What is missing we suggest is market pull. 3D printing is indeed one of the more inspirational emergent technologies of the 21st century. It has been strongly hyped and circulated in the media as a technology of user and transformation (Lipton 2016; Garner 2012). Print and electronic media were filled with children and scientists showing off all sorts of plastic pieces they had made at home; these included small figures of Yoda, a referee whistle, maybe a smart phone casing.

According to these mediated promotional and technologically determinist views of 3D printing, the future looked bright. This still continues, but 3D printers used today are probably mostly used in schools, higher education, design offices and hobbyists. In our design university, students can play with 50 or so 3D printers, with most use centered on the development of models and prototypes for conventionally manufactured products.

We have tried to innovate with this digital-physical production technology in our product design pedagogy (Kempton et al. 2017). Our aim has been to reach beyond the clear engineering and material focus of much research and teaching so as to support interest in a more 'plastic', non-linear and rapid, iterative mode of design exploration. We have researched into related matters such as bespoke design and making (Killi et al. 2016).

However, we still see there as being untapped potential for 3D printing beyond the design school and beyond the marketing of likely of consumer use. With all the projection of 3D printing and the allure of alternative means of production and use, is there perhaps a matter of appetite that has gone largely unnoticed? Could food be the market pull that might lift 3D printer somewhere near its anticipated potential? How might a 3D food printer be applicable in the home?

In the next section we will take a look at what has been done regarding printing food so far, and what might be possible in the near and distant future. The third section will discuss the market potential, is it a marked pull out there? In part, our work is akin to the connected multi-category approach advanced by Council and Petch (2015).

2 TECHNOLOGY FOR PRINTING FOOD

2.1 *What exist on the market today?*

In their article "An Overview of 3D printing Technologies For Food Fabrication", Sun et al. (2015) provide a comprehensive overview of existing technologies. Their findings cover the main categories of standard 3D printing used for polymers, adapted to printing food. In the coming sections we discuss the main technology categories.

2.2 *Fused Deposition Modeling (FDM)*

Fused Deposition Modeling (FDM) is one of the three original additive manufacturing methods. It was patented in 1989 and is by far the technology most people know. When the patent ran out in 2007, we saw an explosion in low price printers, using the fusing principle wrapped into different machine design.

Many of these printers are long gone but some, like Ultimaker, are still around. Very soon after the introduction of these low-priced 3D printers, pictures emerged of them printing food.

Earlier though, serious thoughts about printing food had arisen, with the first patent issued to Nanotech Industries in 2000. There clearly have been attempts to utilize FDM technology for food. In short, this method could be described as an automation of decorating a cake or spreading mayonnaise on a sandwich etc.

On googling 3D printing food, what strikes one is that most images show what would be more correctly described as mimicking a chef's hand movements. Importantly, food as a material behaves differently than a polymer. It is this matter that has been a challenge for those wanting to print food.

When in 2016 William Kempton made what was probably the first 3D printed gingerbread house - the traditional medieval Norwegian wooden beam supported churches still standing to day and the tradition of gingerbread biscuit making as part of Christmas celebrations - he had to change the recipe for the gingerbread dough quite drastically. This was done in order to be able to build the structure in one go without building it in flat sections with subsequent mounting. This same technique is used with stencils and with 3D printers.

Looking more closely at the picture in Figure 1 one sees the layers and how the dough collapsed during the build process. The examples demonstrate that, as a material, food needs to be adapted to 3D printing techniques and processes.

Figure 1. Gingerbread "Stav Church". Design, production and photo William Kempton.

2.3 *SLS and 3DP*

The second key Additive Manufacturing technology on the market is selective laser sintering (SLS). This is less known and is also harder to see being applicable for food manufacturing. As far as ascertained, we were the first to test this out at the Oslo School of Architecture and Design (AHO) back in 2007.

In our SLS machine we replaced the nylon polymer powder with fine flour sugar, designed some fun shapes and tried to sinter candy. The material behaved quite differently from the Nylon polymer. The result was a proof of concept, showing that it is possible to laser sinter flour sugar, but with some technical challenges. Figure 2 illustrates one of the objects made: a chain of melted sugar with a sugar drizzle from the un-melted sugar in the powder cake.

3DP technology is a slightly simplified version of powder sintering. Instead of a laser, heaters and inert atmosphere 3DP uses powder and print heads that draw patterns using a liquid binder. 3D Systems launched their Chefjet and Chefjet Pro in 2014. Colorful candy could be printed out relatively easy.

2.4 *Other additive manufacturing solutions and food*

Other manufacturing options do exist, such as technology using the inkjet method, exemplified by Foodjet and Foodini. Droplets of food are deposited via nozzles onto, for instance, a pizza or a cake. More futuristic methods are ones that adapt the digital material technology developed by Connex.

This entails exchanging different polymer materials with edible materials.

The idea behind the digital material is to customize material values on a micro level, leading to a heterogynous material that, for example, could change gradually from soft to stiff. Each layer would be different from the previous one, on almost a molecular level.

In theory one could have containers with meat (or meat surrogate), fat (fat surrogate), salt and pepper. A burger with your preferred distribution and measures of all ingredients could be designed and produced, as is shown in Figure 3.

Figure 2. Selective laser sintered flour and sugar necklace. Photo Anders August Kittelsen.

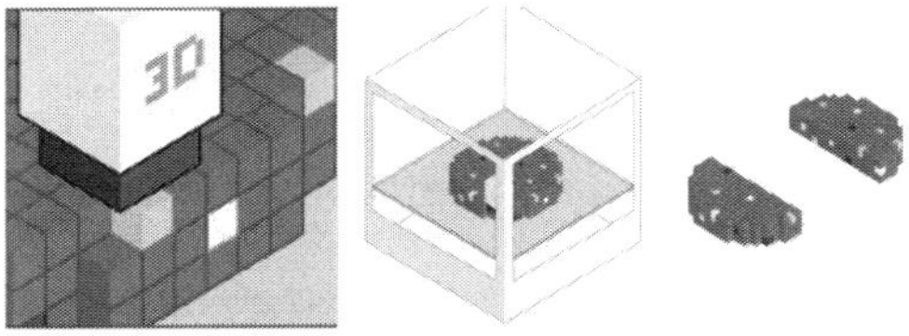

Figure 3. Digital burger, meat (red), fat (yellow), salt (white) and pepper (black); distributed on a micro level. Photos Raoul Koreman.

There is of course an important ethical discussion when printing of meat comes into the picture. This will not be addressed in this paper. As we indicate later, food needs to be understood in its cultural frames and 'tastes'. The academic publisher Bloomsbury indicates this in a six volume series on the cultural histories of food.

2.5 *Some conclusions on technological aspects for printing food*

We now summarize some of the main uses of 3D printing technologies and the making of food. Printing of food uses the same technology developed for polymers, the big different lies in the materials. Physical, mechanical and chemical values for food do differ quite substantially from polymers. Several of the problems with polymer printers occur with food printers, such as speed, post-processing and print environment. Most technology available today emphasizes shape, not taste. The possibility to generate artistic, edibles shapes seems to be the main use. Except for candy, it is necessary for a secondary process to be enacted, such as heat treatment, and this seems to be hard to combine in the printing process. The price of food printers seems to be higher than for similar polymer printers. This might be partially due to a much lower volume of machines produced. Finally, printing food opens up a whole new range of challenges due to health, machine and materials hygiene protocols.

Having summarized these technical features we next turn to the second part of the paper that a focuses on unpacking the notion and potential of market pull when it comes to 3D printing and food. This shifts the focus form the technological to the notion and activities of the market as a cultural and experiential force and entity.

3 MARKET PULL AND 3D FOOD PRINTERS

3.1 *What might the market be?*

In the context of food, food sourcing, food types, nutrition and appetites and gastronomic experimentation, we now turn to a wider discussion of some of the emerging practices and situated implications of the notion of market pull and 3 D printing of food.

Typically, market pull is used to refer to the way the market provides an opportunity or had a given need for a new product, service, solution or approach. Often this view is seen as providing a solution to a problem. In a more experimental and experiential view of innovation and research by design, the 'stance' may be one of posing new problems or potential pathways, less determinist solutions. Users and customers, however, may also make a pull-by-market approach apparent through requests for improvements to products or services.

In the case of 3D printing of food, the market itself may not be well defined or known. We suggest that there are opportunities to use design based inquiry, between making and analysis (e.g. Morrison & Sevaldson 2010; Lury 2018), to discuss and investigate ways to shift from earlier push of the technology, with its DIY/democratic and market claims (Gershenfeld 2008) and imaginaries (e.g. Stein 2017), to ones that are more pulled by the lure of culinary play and participatory experimentation (Wilde & Bertran 2019), alongside domestic use and engagement with 3D food printing as part of a wider gastronomic and cultural shaping of appetite, nutrition and culinary creativity. This is short is a mode of cultural market shaping by socio-technical pull.

A quick Google search for the 3D printing of food launches a substantial volume of pictures showing different kind of edible objects being "printed" out. As mentioned previously, a large part of these are artistic shapes and futuristic fantasies of Star Trek replicators. That said, however, the first permanent 3D printed food restaurant was launched in 2018 in the Netherlands (Watkin 2018). Pop-up venues, though, have been around at least since 3D systems launched their Chefjet series in 2014 (3D Systems, 2014).

In the following sub-section we look at 3D printing as a home appliance. This is what the regular 3D printer was earlier presented as likely to become.

3.2 *What are the potentials for a domestic 3D food printer?*

Are there any similar products with which we could compare the 3D food printer? First, there are vast numbers of kitchen gadgets available. While new items constantly enter the market, one product, the Sodastream, serves as a useful comparison. This is a machine that produces carbonized water with a selection of flavors. The brand has been available since the 1970s and seems to have a good grip on the market. It has been noted in the media (Zalik 2010) that in 2010, 20% of the households in Sweden owned a Sodastream machine. If a 3D food printer could move even close to such an impact this would be a major shift in perception and use. In considering this, we turn to four themes that could emphasize the potential for this transformation.

We suggest that our technological conceptualizations of 3D printing and food needs to be extended and even enhanced by attention to social and cultural perspectives, including patterns of making and eating, cultural views on the journey between materials and taste, types and sources of food (from e.g. organically grown to lab produced etc.), and where patterns of social making and consumption are considered. This appoints 3D printing of food and its nutritional and stylistic qualities to studies and framings that have shown that food is always located in socio-technical contexts and uses. Importantly, uptake and market pull in 3D printed food is as much a matter of taste as technology, including play, creative experimentation and the shaping of new and future culinary forms, styles and appetites (Warriner & Sweetapple 2017).

3.3 *Sustainability*

The reduction of plastics is now a major part of changes in the food sector. Today standard 3D printers produce an enormous amount of what could easily be labeled plastic garbage. That includes parts that serve as successful models and prototypes, but there is probably even more waste plastic from failed prints, for example due to flawed design, technical problems, user errors etc.

Several solutions are emerging (Woem et al. 2018) but there will remain huge challenges to reduce plastic waste for 3D printers at home. One could only imagine the extent of this waste if just 20% of the homes printed out plastic parts on a weekly or even monthly basis.

With food rather than plastics as material, there is one potentially huge advantage: outputs could be consumed! Even flawed results could be eaten; we usually eat what we make unless is severely damaged. Waste food matter may be composted or reused. Ideas of such using 3d printing leftovers to make new meals are in their infancy.

3.4 *Actual need*

If one looks back over the last weeks, or even months, how often have we had the need to print out a plastic part? Maybe there was something broken you could have fixed if you had a 3D printer, or maybe you were having a party and it would be fun to make something novel? Perhaps you printed a new toy for your children?

In reality, most of us would seldom use a 3d printer for these purposes. To compare the 3D food printer with the Sodastream, you could easily be using it every day as we typically eat several times a day. The potential is considerable, if likely to be suited to specific interests and tastes, times of day and food source material, seasonally or in volume.

3.5 *Skills*

Owning a saw, a hammer, nails and planks does not make you a carpenter. Having the tools to create 3 dimensional objects does not necessarily mean you will be able to design something useful or appealing. On the other hand, most of us are able to cook, out of bare necessity, and we practice a lot. There is daily room for improvement one could say.

If a regular 3D printer were to become a household appliance it would probably not be a wild guess to say that people would print out parts not designed by them, but downloaded from some source. This could of course also be the case for food, but the threshold to actually design the food oneself would be much lower.

3.6 *Custom-made or bespoke food*

One of the main benefits stated from the 3D printing community is customization. Of course this is a very valuable possibility embedded in this emerging technology. However, many products have no need for customization of for having artifacts tailor made to each user; many products work better in predefined shapes and sizes. To customize packaging, for instance, would provide few benefits in many cases and could often be solved when we pick and wrap for instance fruit or vegetables in a plastic/paper bag at the grocery store. With food, however, we could see a different diversity due to taste. The same edible product could have different mixes of for instance spices or other ingredients. Even more important, special nutrition needs due to allergy or dietary reasons could be a lifesaving benefit.

4 DISCUSSION AND CONCLUSION

The 3D printing industry is growing: according to Gartner's Hype cycles (Gartner 2018) it has moved past the "slope of enlightenment" and ideas for how and where to use this technology are in no way scarce (Ventola 2016). As a home appliance however, the 3D printer seems to have lost its momentum. This is not that surprising when we consider how the technology has evolved, and the extended skills needed to utilize it (Killi 2017). The voice-controlled replicator from Star Trek seems to be something for the future. The obvious future use will of course be on long journey to Mars and other places in space, but even though several companies explore the concept of space tourism (Molitch-Hou 2017), it's not likely to be a near future product.

However, if we would imagine some sort of local/domestic production it seems like there is potential for the market to pull the technology forwards into a different domestic use domain, namely food. This mode of double consumption - both hardware/software and edible substance - moves the 3D printer into a more sensory domain of creative use centered on taste, pleasure and sociality. This is in contrast to the projection of home printing that has had limited reach despite the technology having been pushed for some years but where need has not materialized.

Granted, new attention to printing 3D food may also fall foul of contexts of uptake, experimentation and effective use. Yet, there is potential for the engagement of users in different processes of making, shaping, sharing and consuming the outputs of the technology, connecting it to sustenance and potentially wider design driven drives and use needs centering on sustainable product and good production (Huang et al. 2013). 3D food printers may join a bevy of other domesticized food making tools and techniques that have been popularized into the home through cooking channels, YouTube video tutorials, and urban dwellers' fatigue with fast food and related disposable cultures of consumption.

We suggest that in time, perhaps following more public use, 3D food printing may join the microwave and the rice steamer, juice maker and the smart refrigerator and interfaces of ceramic hobs as part of how the making and enjoyment of food play out in affluent societies and domestically (Truninger 2013). Kitchens are infused with devices and activities and are venues for not only preparation and conformity but also expression and enjoyment in acts of both making and devouring food. Changing practices of making via printing and seen use in professional and commercial food contexts may well provide market pull towards domestic use.

Within households 3D printers could also be used to generate a diversity of products from single or blended materials. All in all, there will be potential for the food market to also be a different supplier of materials, both fresh and pre-prepared. Such relations have yet to be taken up in new innovation models that address matters of developing the dynamics of the circular and experience economy (e.g. Bocken et al. 2014; Pine & Gilmore 1998).

However, there is also potential for 3D food printers it be deployed to print in diverse contexts of need, such as potentially even emergency situations, whether with solar power or with limited materials. These contexts may indeed feed one another, as it were, so that current site-based and distribution-oriented food production and sale may be augmented or even altered. This suggests we may need to reconsider notions of technologically framed automation, and think through ones that involve a mesh of humans and technologies at work together to produce food. This may be whether for storage in a home freezer (as opposed to supermarket ones) or customized for daily mealtimes or special dinners. This is in contrast to the burgeoning of fast food delivery services from restaurants via providers such as Deliveroo and Foodora.

Experimentation with ways of 'printing food' and examples of novel, compelling and interesting use is beginning to emerge. These 'proto-print' food examples may help create further conceptualization and promotion of new appetites. Maximizing profit from homogenic products in which nutrition and appeal are not necessarily primary culinary values or from the app-based restaurant-service delivery systems may not drive such shifts. Digitally ordered fast food, delivered by bicycle and scooter (despite critiques of its gig-economy employee exploitation) could even be transformed into a print on demand at home through a subscription arrangement through the delivery of materials and even software.

Making food this way make take more time that opening a pizza carton or removing the plastic cover from a microwave meal. Again we need to heed responses from research that shows that people are not always keen on food that is printed (e.g. Lupton, 2019), indicating a need to consider user perceptions (Gayler et al. 2018) a mode of making not yet widely experienced or domesticated.

Yet, it is possible for such food preparation to also become part of a wider routine of the making of meals that may have a mix of techniques, activities and materials, from frying and steaming to baking. We do need to still be reminded that food is a cultural form as is its consumption. Lupton and Turner (2018) they found, for example, that embedding insects in 3D printed food was not always well received.

Appetites and desire still need to be primed; traditional forms of food such as the Korean dumpling called Ori-mandu (Lee et al. 2017) may be totally altered in 3D printing and the essence of its identity and cultural lineage may need to be followed through into new machine printed forms and interfaces (Gayler 2017) and new connotations in culture that may return such prints back into hand forming (Vannucci et al. 2018).

In our view, the 3D printing of food is not a matter of merely functions, shape or prototyping; it is about taste, convenience, making life better and more sustainable. One could imagine there might be a demand for a device that could address this in our homes, at public events, in poignant cultural contexts and for on-site responses to special need and situated demand.

Not pulled pork but 'pulled three dimensional printed food' that is motivated by connections and making 'meals' with hardware and software, materials and techniques, appetites and tastes, nutrition and sustainable consumption.

REFERENCES

Bertling, J. & Rommel, S. 2016. A critical view of 3d printing regarding industrial mass customization versus individual desktop fabrication. In P. Ferdinand, U. Petschow & S. Dickel (eds), *The decentralized and networked future of value creation*: 75–105. Cham: Springer.

Bertran, F., Jhaveri, S., Lutz, R., Isbister, K. & Wilde, D. 2018. Visualizing the landscape of human-food interaction research. *Designing Interactive Systems (DIS '18 Companion), Hong Kong*, 9-13 June 2018. New York: ACM. 243–248.

Bertran, F. & Wilde, D. 2018. Playing with food: Reconfiguring the gastronomic experience through play. In R. Carrilho Bonacho (ed.), *Experiencing food: Designing dialogues*: 3–6. London: CRC Press.

Bocken, N., Short, P., Rana P. & Evans, S. 2014. A literature and practice review to develop sustainable business model archetypes. *Journal of Cleaner Production* 65(15): 42–56.

Council, A. & Petch, M. 2015. *Future food: How cutting edge technology & 3D printing will change the way you eat*. Tumwater: GYGES3DCOM.

Gartner. 2012. *Gartner's hype cycle special report for 2012*. Available: http://www.gartner.com/technology/research/hype-cycles/.

Gershenfeld, N. 2008. *Fab: The coming revolution on your desktop–from personal computers to personal fabrication*. New York: Basic Books.

Gruijters, K. 2016. *Food design*. Houten: Lannoo Publishers.

Gartner. 2018. *Hype cycles for emerging technologies*, 2018. Available: https://www.gartner.com/en/documents/3885468.

Gayler. T. 2017. Towards edible interfaces: Designing interactions with food. In *Proceedings of the 19th ACM International Conference on Multimodal Interaction (ICMI '17) Glasgow*, 13-17 November 2017. New York: ACM. 623–627.

Gayler, T., Sas, C. & and Kalnikaitē, V. 2018. User perceptions of 3d food printing technologies. In *Extended Abstracts of the 2018 CHI Conference on Human Factors in Computing* Systems *(CHI* EA *'18), Montreal*, 21-26 April 2018. New York: ACM. Paper LBW621, 1–6. Available: https://doi.org/10.1145/3170427.3188529.

Huang, S. H., Liu, P., & Mokasdar, A. 2013. Additive manufacturing and its societal impact: A literature review. *The International Journal of Advanced Manufacturing Technology* 67(5-8): 1191–1203.

Kempton, W., Killi, S. & Morrison A. 2017. Meeting learning challenges in product design education with and through additive manufacturing. *International Journal of Systemics, Cybernetics and Informatics* 15(6): 119–129.

Killi, S. 2019 (ed.), *Additive manufacturing: Design, methods and processes*. Singapore: Pan Stanford.

Killi, S., Kempton W. & Morrison A. 2015. Design issues and orientations in additive manufacturing. *International Journal of Rapid Manufacturing* 5(3-4): 289–307.

Lee, B., Hong, J., Surh, J. & Saakes, D. 2017. Ori-mandu: Korean dumpling into whatever shape you want. In *Proceedings of the 2017 CHI Conference Extended Abstracts on Human Factors in Computing* Systems *(CHI* EA *'17), Denver*, May 201.7 New York: ACM. 456–456.

Lipton, J., Cohen, D. & Heinz, M. et al. 2009. Fab@Home Model 2: Towards ubiquitous personal fabrication devices. In *Solid Freeform Fabrication Proceedings 2009, Austin*, 3-5 August 2009. Austin: University of Texas at Austin. 70–81.

Lupton, D. 2019. "I can't get past the fact that it is printed": Consumer attitudes to 3D printed food. *Food, Culture & Society* 21(3): 408–418.

Lupton, D. & Turner, B. 2016. Both fascinating and disturbing: Consumer responses to 3D food printing and implications for food activism. In T. Schneider, K. Eli, C. Dolan & S, Ulijaszek (eds), *Digital food activism*. 169–185. London: CRC Press.

Lupton, D. & Turner, B. 2018. Food of the future? Consumer responses to the idea of 3d-printed meat and insect-based foods. *Food and Foodways* 26(4): 269–289.

Lury C., Fenshan, R., Heller-Nicholas, A. et al. 2018. *Routledge handbook of interdisciplinary research methods.* London: Routledge.

Molitch-Hou, M. 2017. The future of building and 3d printing in space. Available: https://www.engineering.com/3DPrinting/3DPrintingArticles/ArticleID/14481/The-Future-of-Building-and-3D-Printing-in-Space.aspx.

Morrison, A. & Sevaldson, B. 2010. *Research by design.* Special Issue, *FORM Akademisk* 3 (1).Available: http://www.formakademisk.org.

Pine, B. & Gilmore, J. 1998. Welcome to the experience economy. *Harvard Business Review* July-August: online. Available: https://hbr.org/1998/07/welcome-to-the-experience-economy.

Southerland, D., Walters, P. & Huson, D. 2011. Edible 3d printing. In *Proceeding of NIP & digital fabrication conference, Vol. 2, Minneapolis*, 2-6 October 2011. Springfield: Society for Imaging Science and Technology. 819–822.

Stein, J. 2017. The political imaginaries of 3D printing: Prompting mainstream awareness of design and making, *Design and Culture* 9(1): 3–27.

Sun. J., Zhou, W., Huang, D., et al. 2015. An overview of 3D printing tehcnologies for food fabrication. *Food and Bioprocess Technology* 8(8): 1605–1615.

Truninger, M. 2013. The historical development of industrial and domestic food technologies. In A. Murcott, W. Belasco & P. Jackson (eds), *The handbook of food research*: 82–108. London: Bloomsbury.

Vannucci, E., Bertran, F., Marshall, J. & Wilde, D. 2018. Handmaking food ideals: Crafting the design of future food-related technologies. In *ACM* Conference Companion Publication on Designing Interactive Systems (DIS '18 Companion), Hong Kong, 9-13 June 2018. New York: ACM. 419–422.

Ventola, L. 2016. Medical application for 3d printing: Current and projected uses. *Pharmacy & Therapeutics* 39 (10): 704–711.

Warriner, G. & Sweetapple, K. 2017. *Food futures: Sensory explorations in food design.* Barcelona: Promopress.

Watkin, H. 2018. First permanent 3d printed food restaurant premieres in the Netherlands. *All3DP.* 10 April. Available: https://all3dp.com/first-permanent-3d-printed-food-restaurant-premieres-netherlands/.

Wilde, D. & Bertran, F. 2019. Participatory research through gastronomy design: A designerly move towards more playful gastronomy. *International Journal of Food Design* 4(1): 3–37.

Woem, A., McCaslin J. & Pringle A. et al. 2018. ReRapable Recyclebot: Open source 3-d printable extruder for converting plastic to 3-d printing filament. *Hardware* X: 4 (e00026): 1–58.

Yang, F., Zhang, M. & Bandari, B. 2017. Recent development in 3d food printing. *Critical Reviews in Food Science and Nutrition* 57(14): 3145–3153.

Zalik N. 2010. Sodastream to float stock on Wall street. *Haaretz*. 20 October. Available: https://www.haaretz.com/1.5128093.

Industry 4.0 – Shaping The Future of The Digital World – da Silva Bartolo et al. (eds)
© 2021 Taylor & Francis Group, London, ISBN 978-0-367-42272-1

Configuration of fashion design in the contemporaneity: Innovation through additive manufacturing

D. Nogueira da Silva & E. Pereira das Neves
São Paulo State University (UNESP), FAAC, Bauru, Brazil
CIAUD, Lisbon School of Architecture, Universidade de Lisboa, Lisbon, Portugal

M. Santos Menezes
São Paulo State University (UNESP), FAAC, Bauru, Brazil

F. Moreira da Silva
CIAUD, Lisbon School of Architecture, Universidade de Lisboa, Lisbon, Portugal

ABSTRACT: Additive manufacturing has emerged in recent years as a possibility of innovation in artefacts creation, being used in several areas of the industrial production, as an alternative to contemporary yearning. In fashion design, it can bring to market the differential that the user wants to find and, although there is too much to be developed, it has presented the possibility of popularization soon. So, through a literature review, the present paper reflects about the contemporary culture observing in the development of the activities in fashion design, investigating its consuming configuration.

1 INTRODUCTION

Additive manufacture or three-dimensional printing has emerged in recent years as a possibility of innovation in the manufacture of artifacts, being used in several areas of industrial production.

In fashion design, it has been no different, although the use in the fashion industry is still in its early stages, the first collections that make use of additive manufacturing techniques emerge and create expectations in the market.

Just as additive manufacturing presents challenges for fashion use, such as a whole fresh way of thinking about the product, it ensures true innovation by allowing creations and geometries that cannot be reproduced in traditional manufacturing.

As the aim of placing "Fashion Design in the face of the demands of contemporary society and proposing" the use of additive manufacturing in fashion to meet these needs, the present article makes use of a narrative literature review about the subjects involved: consumption in the contemporary, the design and application of additive manufacturing in fashion design. So, our principal aim is reflecting on this theme and presenting the work of pioneering designers in using three-dimensional printing to create garments.

2 CONSUMPTION IN THE CONTEMPORARY

2.1 *About the contemporary*

The contemporary age began in the year 1789, the time of the French revolution, and extends to the present day. Within this period several historical facts are inserted, shaping society and determined their habits, which changed and shaped the Design concept. Bonsiepe (1992) states that the transformations of society, culture, and technology are reflected in the changes in the concept of design itself and the strategies of project discourse. The contemporary project must take into account the present time in the same way that analyses the past that propitiated the moment in which it is, incorporating observation, study, and research, being always in motion, accompanying, updating and reflecting on the issues present in the society and the cultural dynamics surrounding individuals. Since it focuses on the human being, it is necessary that the time in which he lives and the issues with which he relates are considered and is on this premise that the contemporary design can be established.

Complex changes occur in the way of living and being in the world, involved in information and communication systems, new and different relationships among people, things and objects, spaces and environments, and other thoughts and ideas that contribute in the people' life and generate new ways, attitudes, and habits in the quotidian, emerge daily (Moura, 2014).

Analyzing what drives habits and lifestyles is a matter of the contemporary design, once all these changes reflects in the process of Design. There is a need for interaction and dialogue between different fields of knowledge and languages, such as sociology, philosophy, history, communication, psychology, among others, making the contemporary design transdisciplinary. In the same way, the frontiers between dialogical areas such as art, engineering, architecture, and fashion are broken, and they can work together with areas that at first seem more distant like medicine, physics, chemistry, biology. The project is thinking more comprehensively, incorporating knowledge, attitudes, and challenges.

We notice the way contemporary design works are done through collective systems, where groups of creation and development of projects and products relate uncommon knowledge from unfamiliar areas of training and performance of its members. In these groups, there is no hierarchy, which confirms the interdisciplinary action and the transdisciplinary possibilities present in the project's action and attests to the decline of the concept of authorship, another striking feature in the contemporary. According to Burdek (2016), at present, objects and products gain new meanings and are used in uncommon life situations. Therefore, the complex contemporary world cannot be mastered by the designer.

The relationship of contemporary human being with objects and the election process of them has changed. In the search for personalization, for the individualization and the search for its identity, the human being looked for pleasure and emotions through the objects.

The design project begins to be developed from the premise that its result impacts the individual, society and the environment, being subject to match with requirements such as sustainability, universality of use, social inclusion, incorporation and development of innovation, without neglecting the ergonomic issues of usability and comfort, guaranteeing the satisfaction of desires, a good use experience and construction of meanings for the user (Neimeyer, 2014). Thus, we see that design must incorporate political and social attitudes and challenges in search of a broader and consistent design thinking, making its configuration diverse, with innumerable possibilities of relations and associations.

However, to a better comprehension of the contemporary design and its needs, it is necessary to analyze the history of consumption, the scenario in which we insert design and the means of production. Here, the next section provides a brief scenario about the evolution of consumption and the evolution of the fashion field and its production technology.

2.2 *Consumption history*

Lipovestsky (2007) divides the history of consumption into three major cycles to understand the way of consuming in what the author calls hyper-modernity: the first cycle goes from the mid-1880s to the Second World War; Second begins around 1950 and runs through the late 1970s; The third begins in the 1980s and is becoming more and more present until today.

The first cycle is characterized by the expansion of large-scale production, pursuiting profit by the drop in sales prices, the popularization of the consumption of non-durable and durable goods, and the beginning of the great magazines that opened space for the emergence of marketing and advertising. In the second occurs "the hyper-consumption society", where there was an increase in the population's quality of life and access to products that were previously restricted to the elite.

The growth and improvement of living conditions during cycle two became the project and goal of western societies. From this point, the consumption began to take a new direction: the cult of well-being began to reduce the logic of consumption with a view to the social distinction of classes to promote a more individualistic consumption model, focused on satisfying particular needs (Andrade & Hatadani, 2010). This model evolved and took on much larger proportions reaching the third cycle which has its apogee in the present.

Today, consumption is intrinsically liked to live in the contemporary to be unlikely to disassociate it from the human being, establishing a culture based on consumption.

Since the organization of society was established around consumption, it underwent changes giving way to a more complex model, shifting the focus from the collective to the individual plane. The capitalism of the industrial era of the past has found its decline in the face of the new strategies employed by the companies, which contribute to building a new economic model.

Unlike previous Ford regulations, the production model has become less focused on mass-produced standardized products and more on innovative strategies such as differentiation of products and services, the proliferation of variety, acceleration of the pace of new product launches and the exploitation of consumers' emotional expectations. Production centered capitalism has been replaced by seductive capitalism focused on the pleasures of consumers through images and dreams, forms and stories. (Lipovestky & Serroy, 2011).

From this premise, the competitiveness of firms is no longer based on cost reduction, in the exploitation of economies of scale or permanent productivity gains, but in more qualitative, immaterial or symbolic competitive advantages. Contemporary companies focus on new sources of value creation, through strategies focused on the aesthetic-affective tastes of consumers which, according to Lipovestky & Serroy (2015), forge the so-called post-Fordian or post-industrial model of a liberal economy. The place of the consumer moved before the user served the consumption and was little heard, now the review consumption is reviewed, and the user is part of the creation process having a voice and its desires being considered.

This new stage of capitalism, coined by the authors as hyper-consumption, has very specific characteristics, being one of them the amplification of the existing mentality of consumption. However, this reaches places that until then were not considered mercantile - such as family, school, ethics. We are dealing with a fresh subjective, emotional, experiential consumption more concerned with self-satisfaction than with social display and status-seeking. In this way, the hyper-consumers goal is to make materialistic existence more qualitative and more balanced, without giving up the advantages of the modern world. (Lipovetsky, 2007)

A responsible manner of consumption has emerged in the refusal of consumerism without consciousness. The so-called "conscious consumption" means consuming better, with more quality and in a more responsible way towards the environment, representing a "form of suspicion concerning the noble institutions, to the reflexivity of individual behaviors, to qualitative searches". (Lipovetsky, 2007). Thus, we can conclude, as Kim et al. (2014) state that consumer experience has become a value orientation that is changing the way the consumer thinks, gains information and decides about consumption activities.

3 THE FASHION IN CONTEMPORARY

For fashion, this can be an excellent example of contemporaneity. Agabem (2009) affirms that it is the fashion that defines the discontinuity in time, marking it according to its actuality, being or not in fashion. The spirit of ephemerality and obsolescence, which seek for the new and the different, are also other important characteristics that justify fashion as an example of contemporaneity.

Benarush (2008) proposes that we understand clothing as an artifact of material culture, this being the discipline that aims to understand the objects we produce and consume and how they fit into symbolic and ideological systems. From this point of view, we can see how the pieces of clothing can be a testimony of their time possessing cultural, social and national identity, being one of the main reflections of the time to which it is contemporary, they materialize a time, reflecting the ideological notion of this and representing the society that created them.

When we analyse a piece of clothing we can infer details of its construction, the fabric, the used techniques, and technologies, in this way it becomes a document from which one can build knowledge about a certain time, the technology that surrounds it, aspects symbolic and ideological in which its production is inserted. Nowadays it is possible to find fashionable features that define contemporary design and its various aspects. The same groups that reflect the current design, also find these characteristics clothing production.

Another attribute of fashion lies in the fact that, to a certain extent, it encourages perceptual and programmed obsolescence, which encourages the continuous development of new products, often with revised, similar or recycled ideas and projects. These participate in the creation and feed a series of segments and collections in the dynamics of the consumer market, avid for the new, differentiated, personalized, customized. (Moura & Coelho, 2014)

The relationship between tradition and innovation is another feature present in a contemporary design which is reflected in fashion. Even today we use the principles of the first techniques used by the human being that has been modified and perfected with the emergence of new technologies, just as there is a rescue of handcrafted techniques such as crochet, knitting, and embroidery indicating how tradition and technology can coexist in the current world.

Faced with globalization, which in the beginning presupposes mass production, against this system, we find a valuation of manual techniques of making objects process that stimulates new actions in Fashion Design aiming at goods production with identity, culture and attractive aspects to the market. In this way, designers maintain in their projects modern traits and establish relationships with the contemporary through differentiated technologies or sustainable issues present in their products.

Contemporary design also establishes a relationship between design and technology when it seeks innovation linked to the differentials of the market in industrial production, this same aspect is also present in fashion design, not only in the contemporary but throughout in all its history. The industrial revolution itself, for example, began with the weaving of cotton, with the first steam engines that worked in the manufacture of fabrics, and later the use of synthetic fibers for clothing made the discoveries in the area popularized and integrated daily life.

Today, an important technology emerges as a possibility of innovation in the artefacts production: digital manufacture through additive manufacture. This technology meets the needs of the consumer, in line with various requirements present in today's consumer ideology, such as collaborative creation and interdisciplinarity among unfamiliar areas of knowledge. Using this mode of production for fashion designers can bring to market the differential that the user wants to find, presenting a possibility of popularization, being the bet of some designers in creation and experimentation providing a high level of innovation and significant differentiation to the fashionable product.

The first collections that already use the technique emerge and create an expectation in the market. With this in view, we studied the technique and know the possibilities of its application in fashion.

4 ADDITIVE MANUFACTURING IN FASHION

Also known as 3D printing, additive manufacturing had its idealization in the 1980s, with the initial goal of creating prototypes for products developed in the industry. In 1981, Hideo Kodama, a member of the Nagoya Municipal Research Institute, created two additive methods for making three-dimensional plastic models with thermosensitive polymers (Wohlers & Gornet, 2014), but the first registration of the technology made in 1986 under the name of Stereolithographi Apparatus (SLA), by Charles Hull.

Charles Hull was also the co-founder of the 3D System Corporation, one of the largest in the 3D printing technology industry. In the 1990s the initial prototype process went through technological variations and other systems besides the SLA were developed, such as Selective Laser Sintering (SLS) and Fused Deposition Modeling (FDM), the latter was registered by the company Stratasys that holds more prominence in the Printing sector (Hull, 2015).

Throughout the years 2000, technology underwent several changes. In 2010 there was a decline in costs, marked with the introduction in the market of a device with a value less than one thousand dollars and from 2012 alternative processes of 3D printing were open to the market, culminating in the dissemination and popularization of technology (Kuhn & Minuzzi, 2015). From then on, a series of researches and experiments were initiated and soon the technology began to be incorporated into fashion, allowing the transgression of the existing limits and bringing relevant contributions to the innovation in the area.

Black Drap Dress was the first clothing manufactured by additive manufacturing. It was developed in 2000 by Dutch industrial designer Jiri Evenhuis in collaboration with Finnish designer Janne Kyttanen (Kuhn & Minuzzi, 2015). the dress is part of the MoMA (Museum of Modern Art) collection in New York City. The garment was produced using Nylon as its raw material and printed by the SLS (Selective Laser Sintering) process, where lasers are used to melt the dust and build 3D shapes.

The designer Janne Kyttanen started a group of designers and technology researchers, the Freedom of Creation, where she developed White Drape Dress, her second piece of clothing in 3D (3D Systems, 2019). Since then the group has developed different fabrics using this technology. From what we observe, the fabrics are composed of additive manufacturing structures that are grouped in a weave which visually resembles the techniques used in the conventional fabrication of the knit.

On the runways, Dutch fashion designer Iris Van Herper created the first collection to feature a three-dimensional printed piece in 2010. The fashion show took place at Amsterdam Fashion Week, where the designer presented her collection entitled Crystallization. Since then this fashion designer has been presenting pieces created from 3D printing in all of her collections, experimenting with different compositions and raw materials. In her Paris Haute Couture parade in July 2011, the Skeleton dress was the highlight. In January of the following year, the designer presented the collection Micro; among the pieces there was the dress entitled Cathedral, a piece printed with sculptural details in three dimensions built from the technology SLS, having as its raw material the polyamide (Iris Van Herpen, 2019).

Another highlight was the collection Hybrid Holism which made use of a technique called 'Mammoth Stereo lithography', a rapid prototyping method where a part is printed from the bottom up in a container with stiffening polymers when hit by the laser. The technique allowed for a transparent piece with liquid appearance. Another innovation came in the hands of the designer in 2013: Voltage collection presented the first flexible printed dress. The piece, conceived in collaboration with MIT Media Lab, counted on the development of a differentiated texture, that incorporated a hard material and another soft one granting the piece greater softness and elasticity (Iris Van Herpen, 2019).

Besides aesthetic innovation, 3D printing has also been used in studies to create the so-called intelligent parts that can, for example, respond to changes in the environment or in the body that uses it, such as that developed by designer Anouk Wipprecht, where a mechatronic system printed in 3D, with sensors that respond to the presence and movements around the wearer. The piece called Spider Dress 2.0, is inspired by the instincts of a spider found in the Amazon Rainforest (Wipprecht, 2016).

Using additive clothing makes great expectations and promises. Customized products according to preferences and tailor-made to users' bodies counteract the culture of mass consumption and fast fashion, which can be a revolution in the way we produce and consume fashion (Yap & Yeong, 2014).

When we deal with jewelry and accessories, the transition from production mode to 3D printing takes place more easily, since these relate differently to the body and most cases with a smaller area of it. Comfort and trim is best solved and we can see that 3D printing already has a great application in the area, either for the production of molds or for getting the ultimate product.

In using three-dimensional printing technology for manufacturing clothing, we have the substitution of the textile material, or a new way of conceiving it different from the traditional and more used, that is given by the interweaving of the yarns of the weave and the warp. In apparel projects that use digital manufacturing, the relationship between the material and the body needs to be rethought, making up one barrier and the substantial challenge for the advancement and popularization of the use of technology in the area.

However, even though it still seems a tool far from the productive reality of fashion apparel, the 3D printer has already been used by designers and fashion brands to build innovative clothes with interesting aesthetic results. Besides the previous example of the pioneering stylist in the area, Iris van Herpen, we can also mention the brand Chanel that presented in 2015 a version of its classic tailleur made in a 3D printer; and the brands Versace and Ohne Tite that blend 3D printing with traditional sewing and embroidery processes into their creations.

There are also other professionals developing such experiments as the English designer Catherine of Wales who exhibited the collection entitled "DNA Project", a result of her master's degree in Digital Fashion, where 3D printing formatted the production base to explore the creation of accessories inspired by the visual structure of human chromosomes (Wales, 2019).

The creations of such designers and their experiments affect products that focus on the relevance of a new perspective on the use of digital production and its implications in fashion. Such constructions motivate reflection on the advantages in the process of prototypes and even end products ready for use. It also raises issues such as the threat to artisanal production and intellectual property.

As a tool that proposes a revolution in industrial production modes, three-dimensional printing also suggests the "do it yourself" that raises questions about copyright and makes it necessary for the answers and solutions to be studied. But while production from digital files is not popular enough to trigger copyright control measures, stylists and designers conduct trials and experiments in the area (Yap & Yeong, 2014).

Another aspect addressed by 3D printing is the creation of digital media. With this form of creation, it is possible to perceive the project almost instantly while it is created through software, which accelerates the development process and allows the designer to digitally visualize the created object in real-time, allowing tests, experiments, and change in shapes. Thus, a new way of creating fashion is also configured from technology.

This form of production allows the designer freedom of creation, where new shapes, thicknesses, and textures of materials can be experimented. These projects already have added value because they are innovative and exclusive conceptions. The use of 3D printing also guarantees independence for the designer in the production of his project, since there is no need for a productive chain until the final product is obtained.

The configuration of the textile material is another relevant question in this type of construction: thought in the three axes of Cartesian space, where the thickness is relevant, already used in the construction of technical fabrics, such as the three-dimensional mesh used in flexible heating systems. In the design of clothing and footwear, although different pieces have already been printed, it is possible to observe that many of these are more like samples than for daily use (Melnikova et al. 2014).

For textile surfaces to be produced commercially from 3D printing, the used materials need larger development to improve the comfort and flexibility of these fabrics to produce clothes for daily use. Currently, 3D printing materials are still relatively expensive and limited, which prevents many designers and stylists from using technology. However, the development of new printers and materials has been constant in recent years, different materials are being inserted into the possibilities of raw material for production. Liquid polymers such as latex, silicone, polyurethane, and Teflon, as well as textile fibers such as cotton, viscose, and polyamide, are being tested by research companies (Yap & Yeong, 2014).

For clothing manufacturing, a new way of thinking textile materials is needed. From prehistory, mankind produces fabrics through the interlacing of knit and warp yarns and for good garment designs to be made from rapid prototyping this formula must be practically forgotten. In the traditional mode of clothing production, a flat surface is transformed through modeling and begins to wear a three-dimensional surface, the body. Digital manufacturing allows design and modeling to be performed in a virtual environment so that there is a revolution in the creation process as well.

The solutions currently found by designers using rapid prototyping transform and adapt the interlacing structure of the yarns of the traditional textile fabrication process. The use of elastic and flexible materials and the movement of the structures allowed in this type of construction brings the physical comfort as well as the visual to the structure. Even with the progress of this technology in product manufacturing, some problems make it difficult to produce and sell commercial parts, such as being a slow process and the cost of raw materials being high when compared to traditional methods (Hopkinson & Dikens, 2001). Designers, scientists, and companies develop projects in the area, creating solutions, products and even researching new raw materials that can be used in three-dimensional printing.

5 FINAL CONSIDERATIONS

By reflecting on contemporary design, investigating the motivations of consumer choices in today's society, and understanding how these are reflected in fashion, it is possible to understand how three-dimensional printing can meet contemporary trends and signify the necessary aesthetics and functional innovation in fashion. In this way, the research in the area becomes important and presents a great potential of development, adding value and desirable characteristics to the product of fashion.

The use of new technologies is an important issue when we refer to innovative projects, combined with studies and creativity it can provide great changes, meet the wishes of the consumer, and introduce new techniques and materials in our day to day and popularize them. The fashion, which is observed throughout its history, has always been attentive to new technologies and new materials and never took time to incorporate these into its universe, which should not be different from additive manufacturing.

Three-dimensional printing can mean new applications and possibilities for clothing, attributing characteristics such as greater flexibility, movement, tactile or visual differential of the final piece. In the future, the popularization of technology can meet the wishes of the contemporary user: with the fall in price and the popularization of the technique the consumer can intervene in the creation process or even create and produce his pieces, contributing to his aesthetic vision, or even personalized products.

As we have seen, since the first wearable piece created from the technology, Black Drap Dress, different creations have been tried out with special attention to the designer Iris Van Herper, the partnerships established with the companies responsible for the technology allow for greater experimentation resulting in the insertion the use of different raw materials, different shapes and varied aesthetic results. In this way, three-dimensional printing technology is growing, as well as growing its applications to Fashion Design. For its diffusion, studies and experiments that make use of the technique are desirable.

REFERENCES

3D Systems. 3d Systems Logo. FOC Textiles in permanent collection at MOMA. Available in: <https://www.3dsystems.com/blog/foc/foc-textiles-to-permanent-collection-at-moma?utm_source=freedomofcreation.com%26utm_medium=301>. Access in: 15 ago. 2019.

Agabem, G.O., 2009. *Que É Contemporâneo? E Outros Ensaios*. Santa Catarina: Argos.

Andrade, R.R. & Hatadani, P.S., 2010. O Hiperconsumo e a Personalização de Produtos de Moda. *Iara: Revista De Moda, Cultura E Arte*, V. 3, pp. 1–16

Benarush, M.K., 2008. A Memória das Roupas. In: *Dobra-[S]*, V. 3, 2008.

Bonsiepe, G., 1992. *Teoria e Prática do Design Industrial*. Lisboa.

Burdek, B.E., 2016. *Design: História, Teoria, e Prática do Design de Produtos*. Ed. Blucher.

Featherstone, M., 1995. *Cultura de Consumo e Pós-modernismo*. São Paulo: Studio Nobel.

Herpen, I.V., 2016. *Iris Van Herpen Womenswear*. Paris.

Hopkinson, Hague & Dickens, 2006. *Rapid Manufacturing: An Industrial Revolution for the Digital Age*. J. Wiley & Sons.

Hull, C. W. 2015. "The Birth of 3D Printing." *Research-Technology Management* 58 (6): 25–29.

Kuhn, R. & Minuzzi, R., 2015. Panorama da Impressão 3D no Design de Moda. In: *Moda Documenta*.

Lipovetsky, G. & Serroy, J., 2015. *A Estetização do Mundo: Viver na Era do Capitalismo Artista*. São Paulo: Companhia das Letras.

Lipovetsky, G. & Serroy, J., 2011. *A Cultura-mundo Resposta a uma Sociedade Desorientada*. Ed. Companhia sas Letras: São Paulo.

Lipovetsky, G., 2007. *A Felicidade Paradoxal: Ensaio Sobre a Sociedade do Hiperconsumo*. São Paulo: Companhia.

Melnikova, R., Ehrmann A. & Finsterbusch, K., 2014. 3D Printing of Textile-Based Structures by Fused Deposition Modelling (Fdm) With Different Polymer Materials. Iop *Conference Series: Materials Science and Engineering*, [S.L.], V. 62, P.012018–012025.

Moura, M. & Coelho, L., 2014. Design Contemporâneo Brasileiro: Moda, Pluralidades e Singularidades. In: *2nd Internacional Fashion And Design Congress*. Milão.

Moura, M., 2014. *Design Brasileiro Contemporâneo: Reflexões*. São Paulo: Estação das Letras e Cores.

Niemeyer, L., 2014. Design Contemporâneo no Brasil. In: *Design Brasileiro Contemporâneo: Reflexões*. São Paulo: Estação das Letras e Cores.

WALES, Catherine. About ProjectDNA. Disponível em: <http://catherinewales.eu/read-this/>. 2019.

Wipprecht, A., 2016. *Spider Dress 2.0*. São Francisco, EUA.

Wohlers T, Gornet T (2014) History of additive manufacturing. WohlersReport. http://wohlersassociates.com/history2014.pdf

Yap, Y.L. & Yeong, W.Y., 2014. Additive Manufacture of Fashion and Jewellery Products: A Mini Review. *Virtual and Physical Prototyping*, [S.L.], V. 9, N. 3, P.195–201.

Industry 4.0 – Shaping The Future of The Digital World – da Silva Bartolo et al. (eds)
© 2021 Taylor & Francis Group, London, ISBN 978-0-367-42272-1

Multimodal virtual reality based maintenance training system for Industry 4.0

M.H. Abidi
Raytheon Chair for Systems Engineering, Advanced Manufacturing Institute, King Saud University, Riyadh, Saudi Arabia

H. Alkhalefah
Advanced Manufacturing Institute, King Saud University, Riyadh, Saudi Arabia

A. Al-Ahmari, E.S. Abouelnasr & A.M. El-Tamimi
Department of Industrial Engineering, College of Engineering, King Saud University, Riyadh, Saudi Arabia

ABSTRACT: Computer-based training systems created to simulate machine assembly maintenance are normally operated by means of ordinary human–computer interfaces (keyboard, mouse, etc.), but this usually results in systems that are far from the real procedures, and therefore not effective in terms of training. In this study, we show that an enhanced solution may come from a multimodal virtual reality based maintenance system. In this regard, a research work is presented that is dealing with development of a haptic assisted maintenance training system. The proposed system applies an interaction environment in which each of the vital maintenance tasks can be simulated by the operator using a stylus-based commercial haptic device, run by means of explicit haptic-rendering methods to provide accurate feedbacks during part handling. A brief user based study is included to help evaluating the developed system and haptic based approach.

1 INTRODUCTION

Virtual Reality (VR) is now an established technology and extended its wings of applications in numerous fields. In the field of manufacturing, it is applied into design, assembly, layout planning, and maintenance (Abidi et al. 2016). Maintenance is the combination of three main activities, disassembly, fault diagnosis, and assembly. Therefore, a lot of researchers explored these activities separately using various tools and techniques (Abate et al. 2009).

When sophisticated systems are not designed and manufactured by keeping in mind future maintenance procedures, they imply high costs and time taking process even for a simple replacement. Due to these important factors, maintainability has to be considered an additional value (design wise) and a competitive feature (DiGironimo et al. 2005). The term maintainability represents the combination of various features that are to be established at the design stage. Actually, at this stage the vital parts on which maintenance tasks carried out can be realized, the instruments required for fault diagnosis can be recorded and specific devices for maintenance can be developed. Thus, it is plausible that the design for maintainability has an affirmative effect on the operational costs that the user will face using the system (Ivory et al. 2001). At initial stages it looks like that the cost is increasing due to this, however, the complete lifecycle cost is reduced (Serger 1983). Performing simulations of maintenance operations within a VR environment gives a person the facility to directly interact with 3D virtual models for maintenance purposes. Engineers can apply it to assess aspects of human-centric design for maintainability (approachability, reach, tool usage, part mount/dismount ability). Popular and tested methods based on the use of digital manikins are the first-person method provides a more direct, instinctive control over the interaction task, thus accelerates the maintenance checks, along with the chance to realize improved design solutions from the maintainability point of view. Moreover, maintenance operators (mechanics, technicians) can be trained within a highly interactive, intuitive VR environment, thus coalescing benefits of a risk free training environment with the value of the "learning by doing".

Since, assembly/disassembly is a vital component of maintenance, therefore from several decades researchers are focusing on these areas. In the 1980s, the idea of computer-aided assembly process planning was introduced for solving real-world assembly restrictions in many industrial realms including complex systems. Computer based methods chiefly covers four implementations as follows according to various principles. Knowledge-based implementation transfers personal experience into rules or library, and offers dependable references for action (Tonshoff et al. 1992, Zha et al. 1998, Chang et al. 1988), while this execution needs more proficient knowledge for a better use. Interaction-based implementation could guide the process step by step while the tiresome work appears to be ineffectual in practice (Manli et al. 2007, Baldwin et al.

1991). Inverse operation-based implementation focus more on combining both assembly and disassembly, and makes the structure more practicable (Wilson et al. 1994, Homen de Mello et al. 1991) while it might be incongruous if the inverse operation was unclear.

Virtual maintenance (VM) technology, as an extension of VR technology in the maintainability domain, provides a creative, intuitive, and complete way to perform maintenance task simulation and analysis by a virtual prototype in a virtual maintenance environment (Li et al. 2003, Pouliquen et al. 2007, Gartner et al. 2010). VM can intuitively represent the entire maintenance process, including human manipulation, cooperation, and tool operation (Amos et al. 2008, Elzendoom et al. 2009, Ieronutti et al. 2007, Liu et al. 2010). In the recent years, the predominance of VM allowed more comprehensive applications, such as maintenance training (Lina et al. 2002, Kara et al. 2011), product design (Pouliquen et al. 2007), process simulation, and analysis (Amos et al. 2008, Elzendoom et al. 2009). Consequently, VM also provides a reasonable basis for maintenance process analysis, and flaws can be exposed during the early stage of product design to avoid the hysteresis of conventional ways. Referring to the application of VM in spacecraft domain, United Space Alliance is leading the pack on the basis of powerful technology and software/hardware support, in which virtual reality technology was deployed comprehensively and systematically, by using a specific series of hardware, on both spacecraft assembly and ground operations design and training, and brought both visual and ergonomic experiment results for training and safe operations. These applications provided a feasible reference for us to solve aerospace design and training problems based on VR. Abidi et al. (2015), presented a haptic assisted low cost virtual assembly platform. A case study of complex assembly of blower pump was successfully developed. Geng et al. (2017) developed a virtual maintenance system for huge systems that is satellite maintenance and troubleshooting.

This study presents the integration of the haptic feedback into VR based maintenance solutions to offer a user force feedback to enhance the handling of the virtual components of product involved. Since, most of the haptic based VM systems are still complex and faces several issues (such as lag in response, synchronization between haptics and physics engine frequency etc.). Therefore, the developed system in this research work tackles these issues and a step to move forward in this field is presented.

2 SYSTEM DEVELOPMENT

This section summarizes the methodology that was used to develop the system.

2.1 *Structural design of the system*

Figure 1 illustrates the relationships between the three modules.

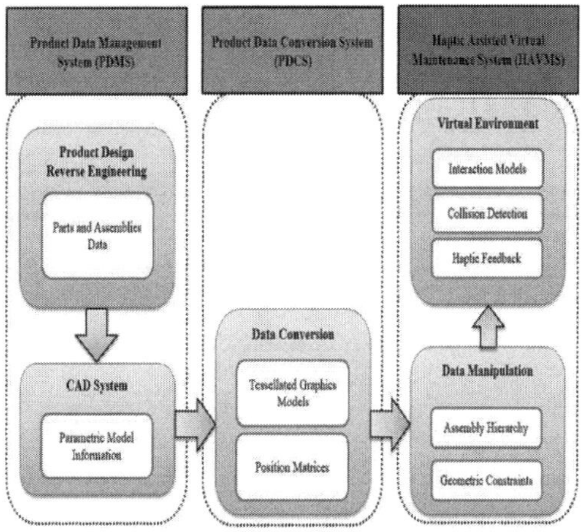

Figure 1. Relationship between three modules.

The proposed system is based on three main segments: product data management system (PDMS), product data conversion system (PDCS), and Haptic Assisted Virtual Maintenance System (HAVMS). The PDMS segment creates CAD files, which serve as an input to the PDCS module. The output of the PDCS module is a VR software file (i.e. a *.VRNative file) that is used for developing the VR based simulation platform.

PDMS module. The purpose of this segment is to create product information (in eXtensible Markup Language (XML)) using a specification taxonomy for data coming either from design efforts or via reverse engineering. The product information is then used to create a CAD model that is exported in IGES (initial graphics exchange specification) format, STL (standard tessellation language) format, and different CAD software format. A nomenclature for identifying the suitable level of detail for parts, sub- assemblies, and assemblies were established. The nomenclature use the XML file format to represent (1) product dimensional data; (2) positional info with respect to a global reference system; (3) spatial connections between different product parts, subassemblies, and assemblies; and (4) inter-part relationships. The nomenclature include (1) product data in the form of low-level geometric entities, such as lines, planes, and circles, and (2) the associations between the product geometric info. The XML nomenclatures for the selected case studies was used to create 3D object models in a CAD system. Figure 2 illustrates the structure and operation of the PDMS module.

PDCS module. The objective of this module is to translate the geometrical format derived from a CAD system to the VR software file format. In this activity, design of experiments techniques were used in combination with the ProductView adapter to translate CAD files to a VR software format. There are various

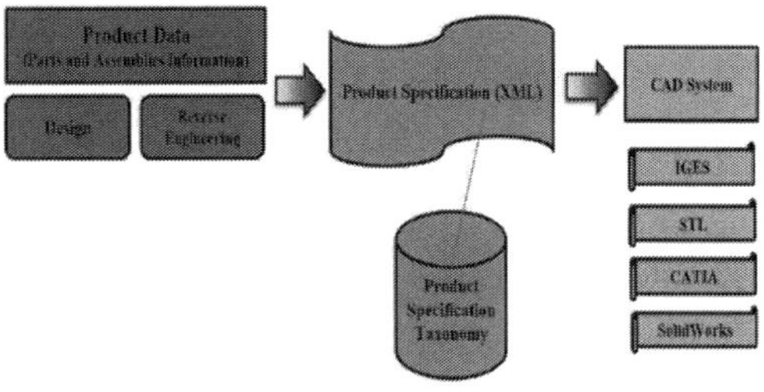

Figure 2. Structure and operation of the PDMS module.

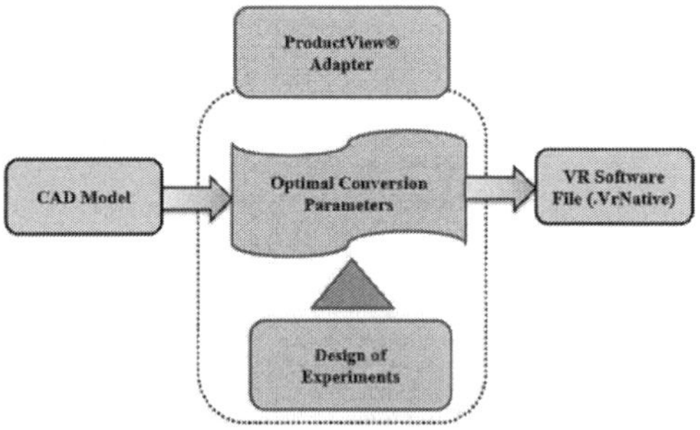

Figure 3. Structure and operation of the PDCS module.

factors linked with the translation procedure, such as the tolerance, chord length, edge length, and number of levels of detail. Sequential experimentation was applied to find out the optimized parameter settings, as measured by the perceived quality of the subsequent VR models. Figure 3 demonstrates the structure and operation of the PDCS module.

HAVMS module. Visual subsystem of the HAVMS is based on a handy open-source scene graph library, whose core part, the Simple Scene Graph (SSG), is, in turn, based on OpenGL for real-time 3D visualization. A number of advanced visual/simulation methods have been employed in HAVMS with the objective of providing lifelike highly interactive VR environments. VR environment includes many features such as stereoscopic rendering, advanced lighting, reflections, hard and soft shadowing, to name a few. The HAVMS has several sub-systems, such as (1) an input manager to accomplish communication with different input devices such as the keyboard, mouse, and haptic; (2) a VR world manager that handles task interaction and graphical scene updates; (3) a visual manager that updates a graphical 3D display; (4) an audio manager that takes care of 3D surround sound system; (5) a maintenance/disassembly simulation event manager that handles inter-part constraints and senses inter-part collisions in real time; and (6) haptic rendering engine. Figure 4 illustrates the structure and operation of the HAVMS module.

A generalizing method to obatin realistic entity behaviour depend on the simulation of the dynamics

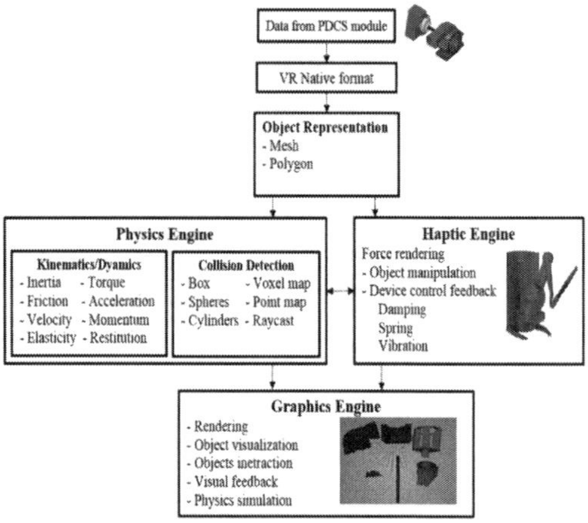

Figure 4. Structure and operation of the HAVMS module.

of entities perceived as rigid bodies. This kind of methods estimates the motion of these bodies in line with classic physics laws. Each body has allocated mass properties and shape and is affected by forces such as gravity, contacts (collisions between bodies), user-defined and different types of constrains. Contact forces are also treated as constraint reactions instigated by the collision of two bodies. The collision detection phase, where contacts occurs, updates the world of simulated bodies with contact constraints and the following simulation deciphers them in corresponding repulsion forces. The Open Dynamics Engine (Open DE, a.k.a. ODE) open-source library and AGEIA PhysXTM has been incorporated with the SSG in HAVMS for adding rigid-body dynamics simulation features. For optimized computing efficacy on multiprocessor computers, the dynamics simulation is realized as a discrete thread which shares the scene graph data structure with the visualization thread. The overall system architecture is shown in Figure 5.

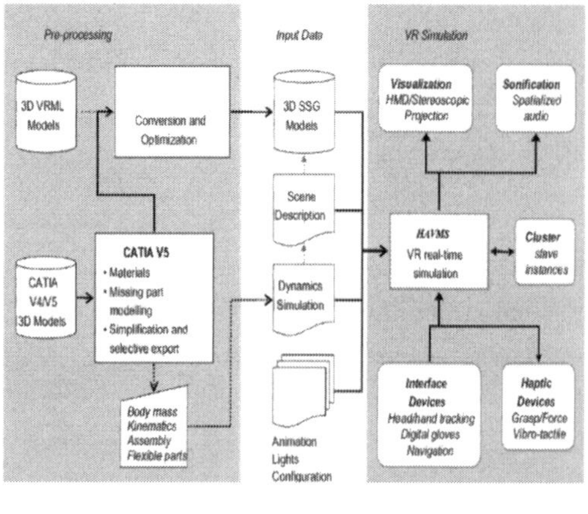

Figure 5. System architecture of HAVMS.

3 CASE STUDY

A case study has been selected to assess the functionality of the system. Actually maintenance includes three activities, i.e. disassembly, fault diagnosis, and assembly. In this work, a radial engine assembly was selected as a case study.

3.1 *Equipment used*

A Hp Elite PC with Intel i7 processor with NVidia Quadro P5000 graphics card was used in this work. For haptic feedback 3D Systems Touch X (formerly called as Phantom Desktop) was used. A virtual maintenance operation for a radial engine was selected as a case study. Radial Engine components were developed in SolidWorks, and exported to the HAVMS system via STL file.

3.2 *System description*

Once the appropriate model and configuration files were loaded, the application was then started using three separate threads graphics, physics, and haptics. The OpenHaptics toolkit started a discrete high priority, high frequency haptic loop, which is responsible for interacting with the haptic device. The coding of the system had been done using C++ as the programming language and Microsoft Visual Studio 2013 as the development environment. Object oriented programming concepts were applied to create a linkage among different libraries to get the desired functionality. The system was implemented using the following modules:

Graphic Engine. For rendering graphics, processing of images, and visualization, the OpenGL/GLUT libraries were used. To display the graphics for components embodied by geometric primitives GLUT was used while triangle meshes were showed using the GLM loader.

Haptics Interface. OpenHaptics toolkit by 3D systems, that is responsible for the device control for 3D Systems Touch X (formerly called as Phantom Desktop), and supports polygonal objects, material properties, and force feedback. The toolkit allows both lower level and higher-level programming access to the haptic device.

Physics Engine. Physics engine includes an integrated solver for fluids, particles, cloth and rigid bodies. AGEIA PhysX^{TM} physics engine was used. It a steady open source physics engine, which provides the physically based modelling of components. Convincing physical behavior is simulated by creating hard contacts or non-penetration constraints whenever bodies collide with one another. Figure 6 these modules.

When the program has started, the user can pick, select and handle components in the virtual environment. The user guides the locus of the virtual haptic handle, denoted by a cone, using the haptic device handgrip. An entity can be selected while intersecting the virtual cone by pressing the lone switch available on the stylus of haptic device. Once an entity is selected in virtual environment, the haptic stylus and digital art are connected together using the virtual spring coupling method. Forces are then sent back and forth between the haptic device and the engaged component to simulate the part's interaction with the virtual environment. Figure 7 shows the HVAMS on the desktop environment.

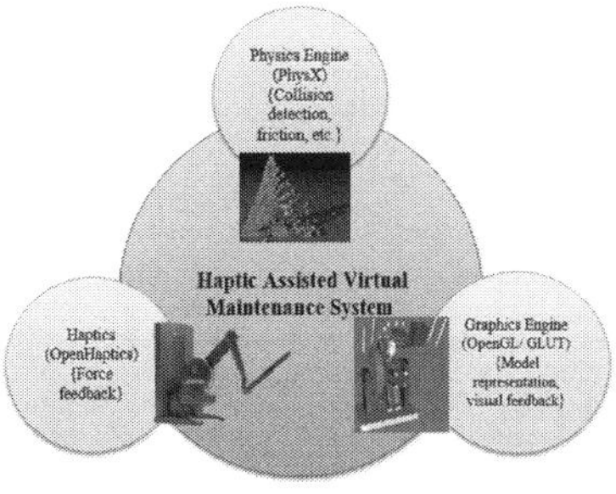

Figure 6. Three modules of HAVMS.

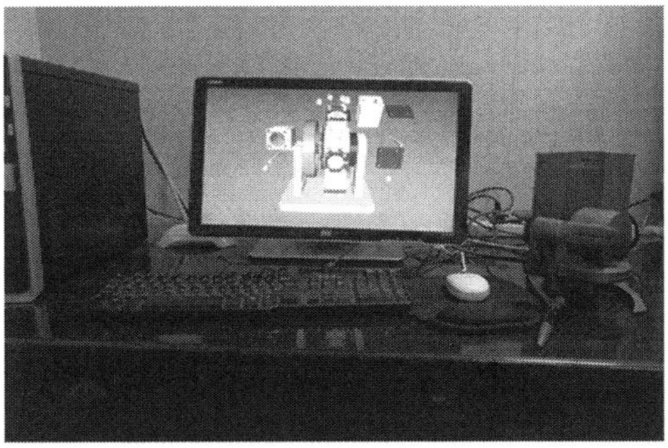

Figure 7. HAVMS on desktop environment.

3.3 *Participants*

Thirty persons participated in the case study to assess the functionality of the HVAMS. Several questions were asked from them to record their feedback about the system. Questions were marked by them on a Likert scale (1-5).

3.4 *Results*

Most of the participants gave positive comments about the HVAMS. Participants were excited with the force feedback mechanism. It gave them a sense of presence in the virtual environment. Figure 8 shows the user's feedback on the subjective measures.

Based on Figure 8, it can be said that participants gave positive feedback about the HVAMS. Only measure that was the below perceived unsatisfactory mark

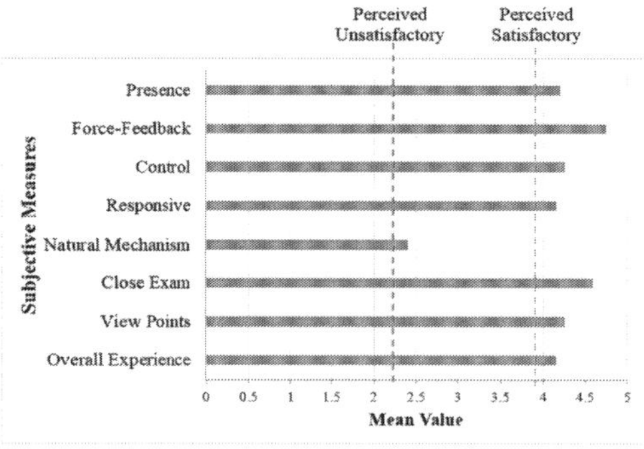

Figure 8. Graphical representation of subjective measures.

(2.5) was natural mechanism, because the force feedback was coming through a stylus type device, which is not a natural mode as in real world. However, for other measures like viewpoints, presence, responsiveness, etc., the HVAMS performed above satisfactory mark.

4 DISCUSSION AND CONCLUSIONS

Commonly available CAD systems only simulates the visuals of any product or component. To make the VR more intuitive it is important to add the other senses, and force-feedback is one of the senses which is very important and crucial to make VR close to natural. A plethora of VR based systems are developed and available, however most of them relies on visual feedbacks only. Some of them that are integrated with haptics are equipped with tactile feedback rather than force feedback. Only few of the VR based systems are integrated with force feedback.

In this work, a haptic based multi-modal system for maintenance training is developed. The system successfully performed the radial engine assembly. This implies that HVAMS is capable to handle large parts and products with several parts. A small-scale user based study was also performed to get their feedback about the developed system. Subjective measures were recorded based on the feedback received from the users. The results indicated that the developed system was up to the mark for all the subjective measures except one that was natural mechanism.

Future works include to assess the system with some objective measures also. Moreover, the system will be run on the semi-immersive environment to make it closer to the real world.

ACKNOWLEDGEMENT

The authors are grateful to the Raytheon Chair for Systems Engineering for funding.

REFERENCES

Abate A.F., Guida, M., Leoncini, P., Nappi, M., & Ricciardi S. 2009. A haptic-based approach to virtual training for aerospace industry. *Journal of Visual Languages and Computing* 20: 318–325.

Abidi et al., 2015. Haptics Assisted Virtual Assembly. *IFAC Papers Online* 48(3): 100–105.

Abidi et al., 2016. Development and Evaluation of the Virtual Prototype of the First Saudi Arabian-Designed Car. *Computers* 5(4): 26–49.

Amos H.C. Ng, Adolfsson, J., Sundberg, M., DeVin, L.J. 2008. Virtual manufacturing for press line monitoring and diagnostics, *International Journal of Machine Tool and Manufacture* 48: 565–575.

Baldwin, D.F., Abell, T.E., Lui, M.M., DeFazio T.L., & Whitney D.E. 1991. An integrated computer aid for generating and evaluating assembly sequences for mechanical products, *IEEE Trans. Robot. Autom.* 7 (1): 78–94.

Chang, K.K., & Wee, G. 1988. A knowledge-based planning system for mechanical assembly using robots. *IEEE Expert* 3(1): 18–30.

DiGironimo, G., & Leoncini, P. 2005. Realistic interaction for maintainability tests in virtual reality, in: Proceedings of the *Virtual Concept 2005, Biarritz, France, November 8–10, 2005.*

Elzendoorn, B., deBaar, M., & Chavan, R., Goodman, T., & Heemskerk, C. 2009. Analysis of the ITER ECH Upper Port Launcher remote maintenance using virtual reality, *Fusion Engineering Design* 84: 733–735.

Gartner, M., Seidel, I., Froschauer, J., & Berger, H. 2010. The formation of virtual organizations by means of electronic institutions in a 3D e-Tourism environment, *Information Sciences* 180(17): 3157–3169.

Geng et al., 2017. A virtual maintenance-based approach for satellite assembling and troubleshooting assessment. Acta Astronautica 138: 434–453.

Homen de Mello, L.S. & Sanderson, A.C. 1991. A correct and complete algorithm for the generation of mechanical assembly sequences, IEEE Transactions of Robotic Automation 7(2): 228–240.

Ieronutti L., & Chittaro, L. 2007. Employing virtual humans for education and training in X3D/VRML worlds, *Computers in Education* 49: 93–109.

Ivory, C.J., Thwaites, A. & Vaughan, R. 2001. Design for maintainability: the innovation process in long term engineering projects, in: Proceedings of the Conference "*The Future of Innovation Studies,*" Eindhoven, Netherlands.

Kara, A., Ozbek, M.E., Cagiltay, N.E., Elif Aydin, Maintenance, sustainability and extendibility in virtual and remote laboratories, *Procedia Soc. Behav. Sci.* 28: 722–728.

Li, J.R., Khoo, L.P., & Tor, S.B. 2003. Desktop virtual reality for maintenance training: an object oriented prototype system (V-REALISM), *Computers in Industry* 52: 109–125.

Lina, F., Ye, L., Duffy, V.G., & Su, C. 2002. Developing virtual environments for industrial training, *Information Sciences* 140 (1–2): 153–170.

Liu, X., Peng, G., Liu, X., & Hou, Y. 2010. Development of a collaborative virtual maintenance environment with agent technology, *Journal of Manufacturing Systems* 29: 173–181.

Manli, L., Shaoguang Y., Weibing D., Gang, S., Yugang, L. 2007. The assembly visualization prototype system, *Spacecraft Environment Engineering* 24(2): 113–115.
Pouliquen, M., Bernard, A., Marsot, J., & Chodorge, L. 2007. Virtual hands and virtual reality multimodal platform to design safer industrial systems, *Computers in Industry* 58: 46–56.
Serger, J.K. 1983. Reliability investment and life-cycle cost, *IEEE Transactions on Reliability* 32: 259–263.
Tonshoff H.K., Menzel E., & Park H.S., 1992. A knowledge-based system for automated assembly planning, *Ann. Int. Inst. Prod. Eng. Res*: 41(1): 19–24.
Wilson, R.H., & Latombe, J.C. 1994. Geometric reasoning about mechanical assembly, *Artificial Intelligence* 71(2): 371–396.
Zha, X.F., Lim, S.Y.E., & Fok, S.C. 1998. Integrated knowledge-based assembly sequence planning, *Int. J. Adv. Manuf. Technol.* 14(1): 50–64.

Industry 4.0 – Shaping The Future of The Digital World – da Silva Bartolo et al. (eds)
© 2021 Taylor & Francis Group, London, ISBN 978-0-367-42272-1

Knowledge-driven holistic decision making supporting multi-objective innovative design

A. Khudhair & H. Li
Cardiff University, Cardiff, Wales, UK

ABSTRACT: The lack of collaboration between different disciplines due to the isolation of information has caused low productivity in the construction industry. The separation of data can be related to the existence of proprietary data formats in different types of software tools, which has caused a loss of data and set back the decision-making process. This paper proposes a multi-objective decision-making framework to orchestrate different functions holistically through automatic application models conversion and supported by a data-exchanging knowledge base. The framework leverages the developed common data as a central layer and also understands the specific data that is needed. The proposed structure aims at developing a new method to implement holistic decision-making, which will further enhance the current BIM and Industry 4.0 revolution in the construction industry. Future work will include the study of human decision perception and convert that into a knowledge model to manage the collaboration process better.

1 INTRODUCTION

The construction industry has an enormous environment of information that can be important and sensitive and may need to be exchanged among different disciplines. The isolation and fragmentation of data (Juan & Zheng 2014) are increasing the gap for better collaboration and integration among disciplines instead of placing them more closely. The need for more efficient information exchange and a new generation knowledge-based holistic decision-making framework is becoming crucial.

BIM plays an essential role in providing a productive data environment (Juan & Zheng 2014), which allows a collaborative work atmosphere. However, the isolation of information due to the lack of interoperability between disciplines (Juan & Zheng 2014), since several construction disciplines require various sets of information and models to be shared, is limiting the achievement of the full potential of BIM. Interoperability is an issue that cannot be solved immediately. It more closely resembles a lifetime process that should be maintained and updated as new technologies and concepts become available in the industry. A better understanding of how different models can work together by identifying the common and specific data that come from different models and data sources can help in building a new federated concept.

Although many modelling analyses tools have been developed to close the gap between the architectural and structural domains and are influential in generating and analysing a given model, they do not define a comprehensive multi-objective decision-making framework. The need for more useful information exchange between architectural, structural, and other domains is becoming crucial.

The main objective of this study is to develop a new method to implement a multi-objective decision-making framework to orchestrate different functions, which will further enhance the current BIM and Industry 4.0 revolution in the construction industry. This paper is organised as follows: Section 2 reviews the most-related work in the automation between architectural and structural models. Section 3 describes and explains the multi-objective framework and is followed by a conclusion.

2 RELATED WORK

2.1 *Improving interoperability within the BIM environment through IFC standards*

Having a proprietary data format in each application that is available in the construction industry can obstruct the data exchange process. An open data exchange format is required. The Industry Foundation Classes (IFC) is an open standard that is developed by the Alliance for Interoperability (IAI) to allow interoperability between software vendors by using a standard file format (Grilo & Jardim-Goncalves 2010). IFC can describe and relate all the data exchanged throughout the building lifecycle without being associated with a particular project (Juan & Zheng 2014). Although it is a richly expressive standard, it lacks formal definitions of its entities, attributes and relationships (Venugopal et al. 2012).

A good understanding of the semantic mapping from IFC schema to the targeted area is necessary since, in the construction industry, much information exists, and some of this information can be related while some cannot. For engineers, it is essential to understand the changes in elements and features within the IFC schema, but they are less interested in understanding how this schema was formed. In order to realise the definitions stated in the IFC schema and how they can be related to the domain requirements, three main points should be considered. Firstly, are the definitions in different domains already included in the available version of IFC schema or is there a need to add new definitions? Secondly, do common definitions exist between different areas in the IFC schema that can be related and can reduce the duplication of information and efforts? Thirdly, are there similar definitions that do the same work but are stated within different names in the IFC schema (Wan et al. 2004)?

Although many software tools based on IFC are available in the construction industry now, the development process can take several years to achieve complete coverage of all information and all domains.

2.2 *Automatic model conversion between architectural and structural models using IFC*

The architecture and structural models are the two models under study in this paper since they are the most closely connected. In order to investigate the research activity from a structural perspective, six articles were retrieved and analysed by discussing the main applications and challenges shown in these articles. The period for the collected research is between 2003 and 2018, and are based in three countries: China, UK and Singapore, including different universities. The main schema used in all these researches is the IFC.

Wan et al. (2004) assessed the capabilities of how well the IFCs can support structural analysis by analysing the information requirements of SAP2000 and other software tools from a software point of view. It was verified that the existing gaps are no longer limited to one software. It is becoming more common to have missing information in other structural analysis software as well. They stated that it is not necessary to add new classes to the IFC schema, except in some cases where information is not covered within the schema (Wan et al. 2004). It is preferable to avoid new classes since amending the schema can be inconvenient for the people working with it. It is more advisable to use property sets, attributes or manage these small problems with programming (Wan et al. 2004). This study was based on IFC2, which is an old version of the IFC schema. Hence, it is necessary to reconsider such review to see how far the new versions of IFC have developed over the years. This research was continued with more team members (Chen et al. 2005) by developing an IFC-based web for collaborative building design between architects and structural engineers based on IFC standards, Web and XML technology. One of the reasons behind using XML technology is that it has an excellent structure to store and maintain information and it also allows this information, which is stored in an XML file, to be extracted and transformed into other formats in a text document form (Qin et al. 2011). The developed server was able to obtain the information needed effectively from the architectural model to create the structural model.

Deng & Chang (2006) conducted another study showing the extraction of information from an IFC architectural model to construct the structural model for various structural analysis software packages since these software tools input data formats are different and may not work when imported in different. By finding the relationships and differences of data between these two models, they were capable of handling a unidirectional conversion. Moreover, Qin et al. (2011) proposed a centralised building information modelling framework to process the data exchange between the IFC architectural and structural models based on IFC and XML format. This research was a developed version of the previous study but with better improvements. However, the unified system did not cover the exchange of meshes, load cases boundary conditions and analysis results. Further integration of other building lifecycle phases can be done by applying their proposed system.

Liu et al. (2010) created an integration tool to exchange information from the architectural model to PKPM software since PKPM software, which is widely used in China, does not support information exchange with the IFC-compliant structural model. The IFC data model, PKPM, and some language programming software such as C++, and FORTRAN were used to draw the structure of the integration tool. The final implementation was able to read the IFC format file and extract all the information needed for the structural analysis. Even though the tool was successful, it still does not support the information exchange of loads and support conditions of building elements, and it is also a one-direction process.

Moreover, Hu et al. (2016) proposed an approach to improve the interoperability between the architectural and structural models on one side and among several structural analysis software on the other side. The method was based on combining different algorithms with the IFC unified information model. It was implemented in both client and browser servers to allow central and remote collaboration between different users. The study addressed many challenges, such as 3D model displaying and the optimisation of model transmission. The developed platform was tested to verify its capability of performing a bidirectional conversion. The limitations in this proposal were: It cannot handle the conversion of complex shapes such as variable arches with solid elements. It does not support the transformation of finite element meshes, analysis results, and classification. Load combination and restraint conditions are not addressed yet.

It shows that the possibility of converting one model into another is achievable. The construction industry is facing a very intricate world, especially with BIM's ability to model more complex buildings and infrastructure than before. Having separate models such as cost model, structural model, and other models that has a diverse perspective of one object, which is "The Building", raises the following question: can we think of a holistic approach to analyse a building overall and not individually, model by model? The aim is to understand how different models can work together by identifying the common and specific data that come from different models and various data sources. As an initial attempt, models such as architectural and structural models are considered with the prospect of studying additional models such as energy, safety, cost and others in the outlook.

3 A MULTI-OBJECTIVE DECISION MAKING FRAMEWORK

The design stage is one of the essential stages in setting the right path for a project (Ham et al. 2008), especially that the volume of information grows as the construction project becomes complex. Eadie et al. (2013) stated that the design stages are more engaged with BIM than the latter stages, and this might be due to the discontinuity between the lifecycle stages of a building since using BIM alone will not be able to move the project from the office to the field. BIM needs to be supported by other technologies to achieve that. In this study, we are focusing mainly on the information required in the design stage, with the prospect of covering up additional models from different stages in the future.

3.1 *Architecture and methodology*

The building design process cannot be divided into several specific design functions. For instance, if designing specifically for safety, another perspective such as sustainability might be neglected. Many available software tools are designed to work in specific design areas. However, using separated individual models is not the best way to work with as the project becomes more complex, especially as each model has its input and output. This study concentrates on how different functions can be linked collectively, which will result in establishing a federated model.

The proposed framework, illustrated in Figure 1, is a multi-objective framework that can trigger various functions by leveraging numerous models collectively to make a better decision. For instance, the building has architectural, structural, safety, cost

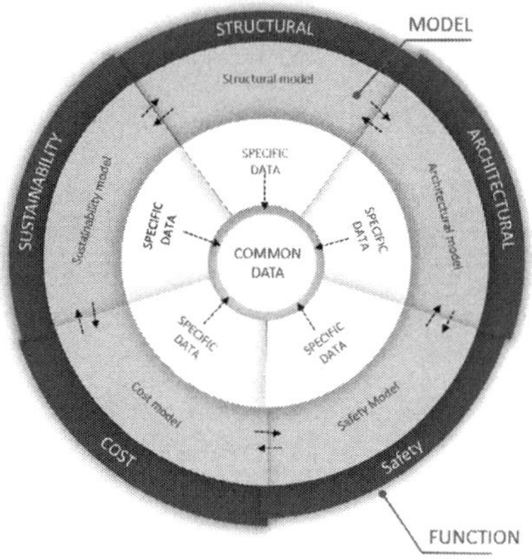

Figure 1. Illustration of the conceptual architecture of the proposed framework for buildings.

perspectives and others. The more models are included, the more advanced the system can be. The suggested framework will be able to summon these models separately, but also link them jointly by sharing the common data among them since they are all the sub-functions of one object, which is "The Building." The framework should be able to leverage these common data as a mutual layer and at the same time understand the specific data since each model has its data. The common layer symbolises the fundamental baseline to achieve a holistic decision since it considers several models at the same time. The system should be capable of converting models as well as extracting the necessary information from these models by channelling a data flow from a different model and different sources. A unidirectional data exchange is also required.

By focusing on buildings as our primary target, two sets need to be identified, which can form input data for one function in this framework. Firstly, identifying the common information shared between all these sub-functions/models. Secondly, determining the specific data for each of those sub-functions/models. For instance, the aim is to understand: What is required for the architectural model? What is mandatory for the structural analysis model? Moreover, what are the common data shared in both models in the design stage?

Defining the required information is divided into three parts, as shown in Figure 2. The first part: software tools by spotting the data requirements when forming the model, the data required when transferring the model, and the missing information, including the reasons behind that. The second part: Manuals and standards evaluations by finding the requirements needed from numerous manuals, and that can involve manuals from different countries or regions. The final part is to examine previous projects and case studies to detect and link the data identified in this part to the first two parts. These exchange requirements need to

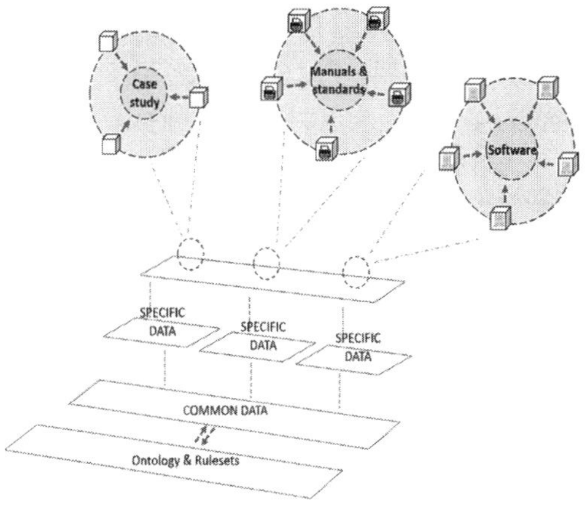

Figure 2. The process followed in defining the required information.

be collected, stored, visualised, interpreted and validated. The purpose of data collection is to utilise this data to establish a hierarchy that can be used to build a knowledge base (ontology) and rulesets that can form the core of the proposed platform.

3.2 *Decision making process - two aspects (human perception/regulations and standards)*

Decision-making is an extremely challenging task that necessitates the development of a systematic business intelligence framework. Researchers continuously mention how technologies and algorithms can be utilised. However, they miss a critical aspect which is human perception. Design decisions are influenced by architects, engineers, contractors, and all the stakeholders who are engaged in the design process of a building.

To accomplish an improved holistic decision making, two aspects of the process should be explored - first, standards, regulation and collaboration frameworks which are utilised to assist the holistic decision making. However, different persons or groups on the same project using the same standards and regulations can make different decisions and have varying levels of efficiency, which raises the critical question of why humans behave differently even though they are using the same methods and regulations. Secondly, how humans make decisions. The focus will not be on examining the human itself, but instead focusing on how to standardise a framework to fit the human decision-making process by infusing human perceptions, since humans are one of the critical components that can assist BIM to innovate.

3.3 *Data requirements for architectural and structural models*

The architectural and structural models share much information. However, the design perspective from the architectural side is different from that of structural engineers. An architect is concerned when designing a building on identifying the spatial arrangements of various architectural components. On the other hand, for a structural engineer, the mechanical properties are the most vital part of the design process. In order to do a structural analysis, the information required can be classified into five categories: geometrical information, material information, load information, member section information, and other information (Chen et al. 2004).

In addition to that, a reduction in dimensions is necessary when a 3D geometrical object from an architectural model is transferred to a structural model since these two models can recognise the same element, but for a different purpose. For example, the information of a section stored in a structural element can only contain the cross-section shape and specific size, while in the architectural element it provides not only the cross-section shapes and size but also the space position and rotation (Wang et al. 2015).

In the approach by Venugopal et al. (2015), the components were split into two classes: primary components and element components. Firstly, the primary components which exist in the spatial structure of the project. For example, beam, wall, column, door, others. Secondly, the element component that is used as an add-on to the primary component to achieve a given role. The element component itself is categorised into three more groups: element parts such as covering, stair flights, others; element accessories which are the entities attached to or embedded in a primary building component; element modifiers that are used to modify the shape of the primary building component. The classification architecture used in (Venugopal et al. 2015) can support in the development of a hierarchy structure of building components by showing the primary and secondary component that is required in a building design stage.

3.4 *A brief case study*

To be able to define the required information, certain questions need to be answered. For instance, is this information required by the structural model, the architectural model or any other model? Moreover, questions such as: are the width and thickness of this column needed in this model? Or are both required in addition to the length? Many other questions need answers. To convert one model into another, we need to investigate each model by hand and independently. Consequently, we will be able to transfer them automatically.

This part of the study represents the initial step that is considered in the software tools analysis. Three software are selected: ArchiCAD representing the architectural model; ETABS and SAP2000 representing the structural model, as shown in Figure 3. The reason behind choosing these three software tools is that all are compatible with IFC format, and

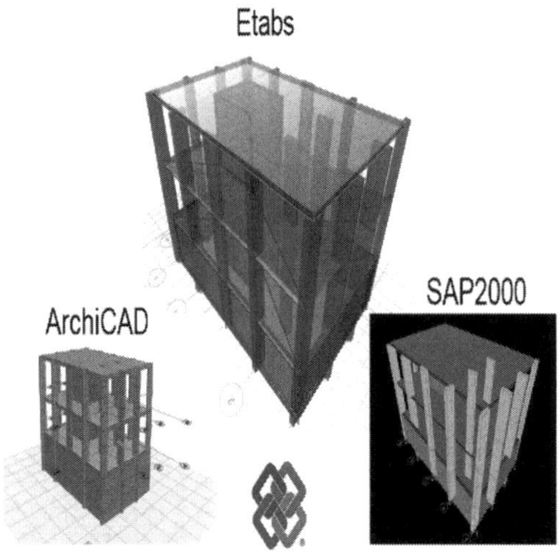

Figure 3. Illustration of a building created in more than one software tool.

they are the most popular used software tools in the construction industry. Three models were created.

In order to define the requirements from the created models, three elements were selected for this study, which is the column, slab, and wall. The information identified from the architectural model, as shown in Table 1, was focusing on project details (unit system and project information), grid system (spacing and storey height), connectivity, geometry (type, shape and dimensions of the element),

Table 1. Required information for various models.

Architectural model	Structural model	Analysis model
Project details (unit system and project information)	Project details (unit system and project information)	Project details (unit system and project information)
Grid system (spacing and storey height)	Grid system (spacing and storey height)	Grid system (spacing and storey height)
Connectivity	Connectivity	Connectivity
Geometry (type, shape and dimensions of the element)	Geometry (type, shape and dimensions of the element)	Geometry (type, shape and dimensions of the element)
Location of elements	Location of elements	Location of elements
Material Properties	Material Properties	Material Properties
-	-	Support Conditions
-	-	Load conditions
-	-	Load combinations

Location of elements in addition to Material Properties. These results might be because ArchiCAD is an architectural software, and it is built according to the architectural perspective to do a specific task.

On the other hand, the two other models were created using structural software tools. The utilisation of two software was to pinpoint not only the unified information between architectural and structural models but also between the structural software. The information that were shown, Table 1, were project details, grid system, and geometry and properties definition such as material and section properties. Moreover, information such as boundary conditions, meshing, load conditions and load combinations, which play an essential part in the structural analysis, were also needed. This case study validates the pervious result which showed that the structural analysis model requires more attributes than the architectural model and that these attributes are mainly related to the mechanical properties of the building, which are not required by the architect. However, additional structural information needs to be added manually in order to have a complete structural analysis model. These results are initial results that are still under research.

4 CONCLUSIONS

The lack of interoperability and knowledge of human resources caused an increase in the BIM budget, which made it less accepted by many companies. A digitalised direct exchange and sharing of information are necessary for the current construction industry to improve collaboration and integration. However, the presence of different concepts and data platforms between different disciplines is delaying this achievement. The proposed platform is a multi-objective decision-making framework to orchestrate different functions holistically through automatic models' conversion and is supported by a data exchanging knowledge base. For different objectives, we need to consider different models. To achieve five goals, we need to create five different models. However, these models share some commonalities. This study focused on how different functions can be collectively placed together, which will result in forming a federated model. The architecture and methodology of the proposed framework were presented that were accompanied by identifying the data requirements of the architectural and structural models, supported by a case study.

Further work will include identifying and verifying the two sets of the data needed, which are the common and specific data that come from different models and different data sources in order to provide an innovative data engine that suits holistic decision-making. Furthermore, an ontological approach will be considered to construct a knowledge base that will be supported by rulesets related to structural design. Identifying these exchange requirements and mapping them to the IFC schema can help in achieving better interoperability. How this can improve productivity within a project, and how this will affect the decision-making process are essential questions that need to be investigated in the future.

REFERENCES

Chen, P. H. et al. 2005. Implementation of IFC-based web server for collaborative building design between architects and structural engineers, *Automation in Construction*, 14 (2005), 115–128.

Deng, X. Y. & Chang, T.-Y. P. 2006. Creating structural model from IFC-based architectural model, *Joint International Conference on Computing and Decision Making in Civil and Building Engineering*, 3687–3695.

Eadie, R. et al. 2013. BIM implementation throughout the UK construction project lifecycle: An analysis, *Automation in Construction*. Elsevier B.V., 36, 145–151.

Grilo, A. & Jardim-Goncalves, R. 2010. Automation in Construction Value proposition on interoperability of BIM and collaborative working environments, *Automation in Construction*. Elsevier B.V., 19(5), pp. 522–530.

Ham, N. H. et al. 2008. A study on application of BIM (Building Information Modeling) to pre-design in construction project, *Proceedings - 3rd International Conference on Convergence and Hybrid Information Technology*, ICCIT 2008, 1, 42–49.

Hu, Z. Z. et al. 2016. Improving interoperability between architectural and structural design models: An industry foundation classes-based approach with web-based tools, *Automation in Construction*. Elsevier B.V., 66, 29–42.

Juan, D. & Zheng, Q. 2014. Cloud and Open BIM-Based Building Information Interoperability Research, *Journal of Service Science and Management*, (April), 47–56.

Liu, Z.-Q. et al. 2010. IFC-based integration tool for supporting information exchange from architectural model to structural model, *J. Cent. South Univ. Technol*, 17, 1344–1350.

Qin, L. et al. 2011. Industry foundation classes-based integration of architectural design and structural analysis, *Journal of Shanghai Jiaotong University (Science)*, 16 (1), 83–90.

Venugopal, M. et al. 2012. Semantics of model views for information exchanges using the industry foundation class schema, *Advanced Engineering Informatics*, 26, 411–428.

Venugopal, M. et al. 2015. An ontology-based analysis of the industry foundation class schema for building information model exchanges, *Advanced Engineering Informatics*. Elsevier Ltd, 29(4), 940–957.

Wan, C. et al. 2004. Assessment of IFC for structural analysis domain, *ITcon*, 9 (May), 75–95.

Wang, X. et al. 2015. Research of the IFC-based Transformation Methods of Geometry Information for Structural Elements, *Journal of Intelligent and Robotic Systems: Theory and Applications*, 79(3–4), 465–473.

Industry 4.0 – Shaping The Future of The Digital World – da Silva Bartolo et al. (eds)
© 2021 Taylor & Francis Group, London, ISBN 978-0-367-42272-1

Portable eye tracking and the study of vision

N. Alão
Faculty of Architecture, University of Lisbon, Lisbon, Portugal

ABSTRACT: This article intends to fulfill two objectives: on one hand, to present the technology of 'portable eye tracking' as an effectihve way of studying vision analytically and especially with regard to three-dimensional immersive practice (which other eye tracker models do not allow); on the other hand, taking as basis a study (Alão 2017) carried out in Lisbon between 2014 and 2016, it pretends to show the limitations of this equipment, pointing out the only way that seems trustworthy to work with it, in order to obtain results with scientific value. It's also points towards future resolutions. 120 analyses of three-dimensional architectural space were performed by 30 observers in 4 different architectural locations, with the use of portable eye tracking technology, featuring 120 samples. Thereafter, 60,000 video frames were analyzed to better understand the types of elements we have to read visual information.

1 INTRODUCTION

In order to understand eye tracking technology, it is necessary first to look at vision as a whole, and for its mechanics, in particular, to understand its basic concepts and understand "what we see and the way we see". For this, it is essential to give to the function of 'attention' the importance it holds in this whole process. Vision is a complex phenomenon that encompasses an enormous series of functions in all its process of construction of the visual image. However, this process is fed by information that comes to the brain through the result produced in the retina, by the incidence of focused light. These incidence, or projections, are called 'retinal projections': the brain does not see them, but uses them to produce what we call visual images. They are the result of fixations, generated in the process of eye movements. At this point, it is common knowledge that, given the characteristics of the focused light touched and the conception of the retina, the eye needs to be in constant movement to acquire information (Yarbus 1967, MacKay 1999, Viosca 2018). These constant but not random movements are untidy. Regarding to eye movements, these are grouped according to their increasing amplitudes: Fixations, Micro-saccadic Movements, Saccadic Movements and Macro-saccadic Movements. The last ones involve not only the eye movements but also other wider movements of the human body. These are the events, within the process of vision, detected and discriminated by the eye tracker, that allows for the development of more detailed studies in this field.

2 THE EYE AND THE IMAGE

The eye is the organ of the body that allows us to detect light, with the aim of perceiving information from the outside world. It transforms this luminous information in nervous electric impulses that feed the mind with the necessary elements to construct a coherent and complete visual image. Retina is then the inner surface of the eye, the place where retinal images are projected. Covered with photosensitive cells of two types - rods and cones - allows two different types of vision: scotopic vision, with monochromatic features solved by the rods - cells extremely sensitive to light, unable to distinguish different wavelengths, giving only monochromatic vision; Photopic vision, with color features solved by the cones - cells not very sensitive to light, forcing the reception of large quantities of it to enter into operation but enable to distinguish different wavelengths, giving us chromatic vision. These cells are in the minority in the retina and its distribution is not uniform.

If we observe an horizontal cut through the center of the eyeball, it is usually indicated that the image projected on the retina, through the center of the eye where it undergoes an inversion, is projected on its spheroidal surface. Usually, in this diagrams of the humen eye, a special emphasis, in the small area of the macula, is evidenced. This area embodies in fovea which defines here the central view of greater definition and detail with amplitude between 1° to 2° degrees. Outside this area, the projected image on the retina is less defined, decreasing this away from the fovea. Thus, being the fovea of very small size, it is easy to understand that the eye sees by focal points, and in order to perform a complete sweep of the global visual landscape it needs to be in continuous movement, passing over and over through all the visual elements, at least on those that we consider of greater importance.

3 EYE MOVEMENTS AND VISUAL ATTENTION

These constant eye movements, uncertain but not random, are defined by constant jumps of different amplitudes, presenting as movements of the type Brownian, of two distinct components: Pulse and Step (MacKay 1999).

Moving through the visual landscape, when the eye finds a sign that attracts the attention of the mind, stops its movement and fixates on the vision element, from which it wishes to absorb information - visual information. This part of the movement, fixation, has duration of the order of the hundredths of a second to the tenths of a second, and is shorter or longer, depending on the attention given to a particular observation. Successive fixations can overlap, connected by micro-saccadic eye movements, to obtain a longer and wider fixation. This insistence on a number of micro-sacádicos is evidence of the importance of visual attention. Visual attention usually also reduces the field of attention on a small particular detail of the visual object observed against the rest of all other visual elements near by. After this acquisition of visual information it is necessary for the mind to find another point of interest to collect. Then the eye goes back into rapid movement with another jump - a pulse - repositioning the gaze over another different point of visual interest, creating another fixation - a step. This jump, between two fixations, is a sudden movement, extremely fast, in the order of milliseconds, and during which the eye has no visual perception, returning to see only when fully stabilized in the next fixation. According to Ibbotson & Krakelberg (2010), there is a saccadic suppression of 100 ms always before and after each saccade.

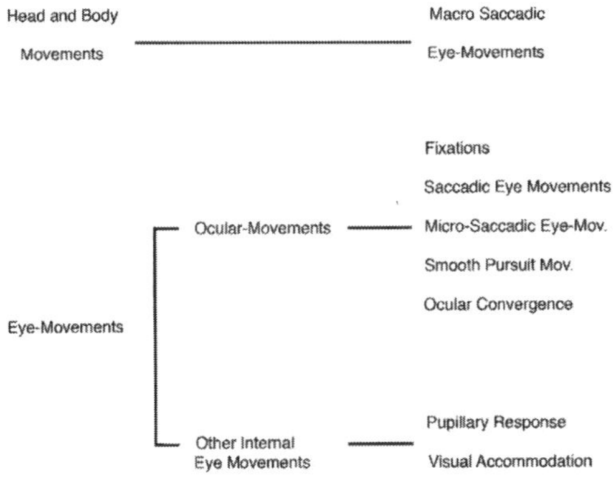

Figure 1. Summary diagram of eye movements during vision.

These are the two main components of eye movement in the vision process - Pulse/Step, Saccadic/Fixation. These phases of eye movements can also be classified according to their amplitudes and movements: fixations, micro-saccades, saccades and macro-saccades.

4 THE EYE TRACKING

These types of eye movement, as well as stop times in each fixation, can be accurately detected and measured by an eye tracker. Eye Tracking is a methodology which allows to know the track of the gaze during vision. There are two types of eye trackers: fixed (table-mounted eye trackers) and portable (head-mounted eye trackers). The most commonly used are the desktop ones, whose methodology uses a fixed bidimensional target on a support, as well as a fixed observer in front of it. In the most trustworthy models, the user has to keep the head fixed by supporting it on its own stand using 2 points of support, chin and forehead, and possibly even two other lateral points, to avoid any movement of head, which is not very comfortable. A portable eye tracker consists of a pair of glasses with a mounted eye tracker device which the observer puts on the face and can be used with total freedom of movement. The device is a micro-video camera able to detect two kinds of information: video information of the visual landscape background and coordinates of gaze positioning, overlapped by the software. The user of a portable eye tracker is allowed to move freely during data collection, and this technology should detect macro-saccades. However, portable eye trackers typically cannot distinguish saccadic from macro-saccadic. In these cases recognition must be done manually.

5 HOW DOES NA EYE TRACKER WORK

The Eye Tracker works as follows: the light that enters in the eye, will project on your inner surface, the retina, giving rise to the creation of the retinal images and, consequently, to the beginning to the process of vision. However, part of this light focuses on the retina, which is not absorbed in the process of vision, is reflected out of the eye through the pupil. As the retina is full of small blood vessel, that light is red. That's why this occurs in flash photography, when people are photographed with red eyes. This is the same light that the Eye Tracker leverages to identify the direction of the gaze. Some devices use only the infrared light for this purpose because it does not interfere with the light that you see. This beam of light leaving from inside the eye, and that matches the visual cone primary, is so the element essential, that the eye tracker uses to identify the direction of the gaze.

The eye tracker offers a first series of raw data, in an excel sheet, consisting since the simple positioning coordinates of the facial gaze to the coordinates provided by the system, to the image of the visual field observed, as well as the time of coordinates of each movement, de radius of the pupil in each event, the amplitudes of the pulses, and some other elements. This raw data is not easy to understand separately but is very useful in understanding the final video reading, and allows to do more research in the future.

6 THE PORTABLE EYE TRACKER

In works and studies, requiring the use of a portable eye tracking, the determination of the events behind the vision - fixations and saccadics - is not complete, so this is the main issue to be addressed in this article.

Portable eye tracker is used, mainly, in field development works and non in-laboratory essays. Especially when we want to study vision in a 3D immersion environment, ie when studying architectural environment. In the case of a study of this kind, using fixed eye trackers is not acceptable, the 3D space cannot be replaced by a two-dimensional image - photography or another - so the user should wear a portable eye tracker. If in the cases of the use of the fixed eye trackers, especially when the head is fixed, the only detected eye movements are the saccadic movements and micro-saccadics, distinguished due to the size of their amplitudes (macro-saccadics do not exist because the head does not move) in the case of vision in immersive 3D space, the movements of the head and trunk are present and that's why we have to consider also the macro-saccades.

With the movement of the head and trunk, the eye tracker also moves in relation to the visual world. Both coordinates systems do not overlap so it is not possible to obtain reliable information on the location of the direction of the gaze in three-dimensional space vision. This location must be performed manually, frame by frame, for all the images resulting from the data collection. On the other hand, the purpose of the movement of the body and the head, the brain has achieved a system of eye repositioning that allows an individual on the move can fix an external target and keep his orientation. This automatic mechanism is named Vestibulo- Ocular Reflex - VOR.

7 THE VOR – VESTIBULO-OCULAR REFLEX

A question then is the ocular compensation during the movements of the head, in current vision. This compensation is meant to, during a move, keep the gaze about a fixed point of the visual world. So, eye movement VOR is an automatic movement, contrary to the movement of the body and the head but equivalent to them. This movement is not detected by portable eye trackers, unless we analyze all the frames, one by one, every film made in each observation. This is the only way to identify all the positions of the fixations due to VOR movements. Note that the coordinate system of the portable eye tracker is located on the glasses, coupled to the head, and so every time your head move also moves the coordinate system in the space 3D. In the Figure 2, can be noted three consecutive different times, with 3 points of passage of the visual beam always incident on the same element observed, but where the eye tracker does not identify a fixation.

In this study, used as the basis for this article, it was found that the percentage of this type of VOR fixings is too significant to be neglected. 120 films were analyzed 20s each with 25im/s, which meant a total of 40m of film and 60,000 frames observed, having determined 8003 eye fixations of various types. On the graphic of Figure 3, can be seen the

Figure 2. Three consecutive images, of a VOR fixation not detected by the eye tracker.

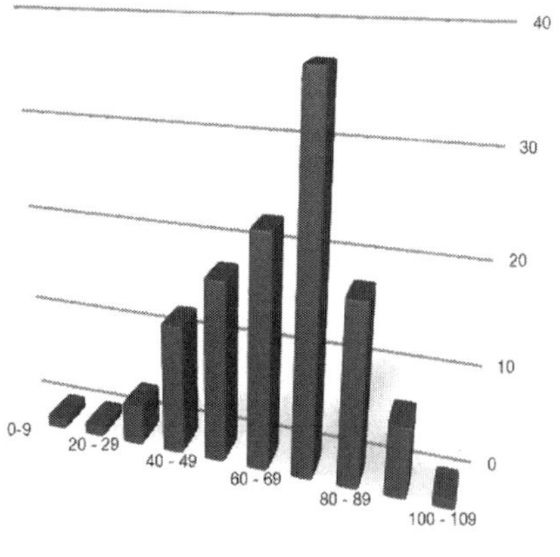

Figure 3. Graphic of distribution of the samples, according to the number of the fixations within the intervals.

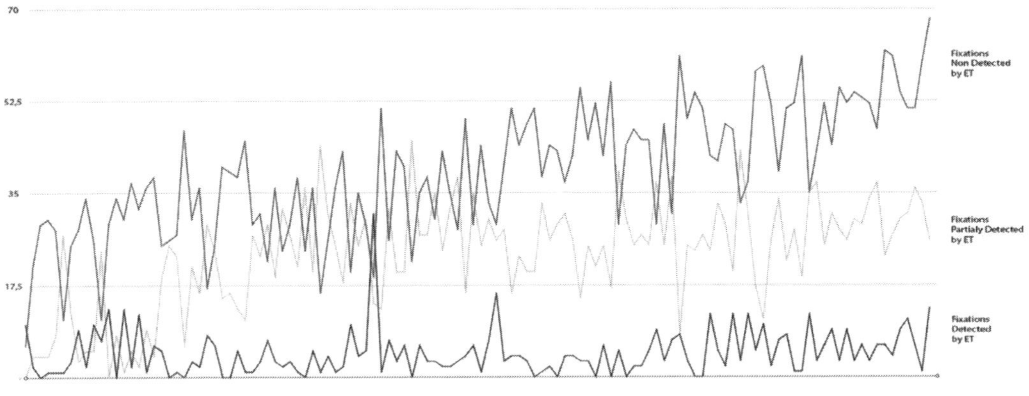

Figure 4. Distribution of fixation type according the samples.

distribution of the samples, acording its number of fixations, within the intervals.

Of these 8003 fixations, 548 are single eye fixations that can be normally determined by the eye tracker with for more than 200ms, in a percentage of 6.84% of fixations, 4673 are VOR fixations that the eye tracker is not able to detect, in the percentage of 58.39% of the total result. The remaining 2,782 fixations are combinations of overlapping between normal fixations of >200ms and VOR-type fixations, which means that even these combinations are deficiently determined regard to its time of duration and their visual attention, comprising a total of 34.76% of the total fixations. On Figure 5, there are some

Elementos Notáveis				L 01 - C.C.Alvaláxia		
Nº Fix		1ª Fixação		1ª Leitura		For
Fix>200ms	131	Central	4	Esquerda	8	Hor
Fix ROV	962	1º Q	2	Errática	7	Diag
Fix Mult.	753	2º Q	4	Vertical	0	Vert
		3º Q	9	Direita	15	Circ
		4º Q	11			

Elementos Notáveis				L 02 - Ig. São Domingos		
Nº Fix		1ª Fixação		1ª Leitura		For
Fix>200ms	144	Central	0	Esquerda	10	Hor
Fix ROV	1150	1º Q	11	Errática	8	Diag
Fix Mult.	741	2º Q	7	Vertical	8	Vert
		3º Q	5	Direita	7	Circ
		4º Q	7			Elip

Elementos Notáveis				L 03 - Lg. São Miguel		
Nº Fix		1ª Fixação		1ª Leitura		For
Fix>200ms	99	Central	0	Esquerda	8	Hor
Fix ROV	1232	1º Q	18	Errática	7	Diag
Fix Mult.	604	2º Q	4	Vertical	5	Vert
		3º Q	6	Direita	10	Circ
		4º Q	2			Elip

Elementos Notáveis				L 04 - Terreiro do Paço		
Nº Fix		1ª Fixação		1ª Leitura		For
Fix>200ms	174	Central	2	Esquerda	10	Hor
Fix ROV	1339	1º Q	5	Errática	4	Diag
Fix Mult.	684	2º Q	15	Vertical	9	Vert
		3º Q	1	Direita	7	Circ
		4º Q	7			Elip

Figure 5. Some visual events detected in the samples according the locals. The 1st left group of data are the fixations by types.

visual events described by cases of study, and we can read the amount of each fixation type, in each case of study and observe the huge difference between isolated fixations bigger than 200 ms and the fixations of VOR type, only identified in manual work. In Figure 4 we have a comparison between all fixation types in the following sequence of colors: > 200 ms isolated fixation in black bottom line; VOR type fixation in grey upper line; Multiple fixation in light grey midle line. With few exceptions it is easy to confirm that the fixations dependent on VOR movements are superior to simple isolated fixations, bigger than 200 ms. The latter are easily detected by a portable eye tracker, but not the former.

8 CONCLUSION

The Eye Tracking research methodology is very useful in the scientific study of vision, in the field of visual perception, leading to easy and clear analysis of eye movements, and allowing clear and scientific results. The variety and quantity of information about, vision and eye movements, that Eye Tracking allows you to obtain, is huge and with tendency to develop, according to the research needs. In addition to the coordinates x, y, of the location of the gaze, the eye tracking allows you to still know precisely that position in a given specific moment of time, as well as its duration - we could say it introduces the time coordinates in a three-dimensional space.

Of the variety of models of eye tracker, which serve a variety of purposes and research, the portable features facilitate and allow the use in differentiated sites and even outdoors, allowing an immersive 3D approach in the field of visual perception. The equipment which is referred in this article, demonstrates a great capacity for use both in the diversity as in the depth of analysis it allows.

However, in the case of portable eye trackers, given the presence of the head and torso movements and consequently macro-saccadic movements, and because proper software does not exist yet, for a full and reliable use of the same equipment, the analysis of the data requires a handmade labor of all images (frames) resulting from the collection of eye tracking, one by one, separately and comparatively.

In completion, it should be noted that any analysis that passes only by the media and software provided by the eye tracker is totally inadequate. It will lead to the annulment of the results obtained as scientific. It is to alert for the need to develop an appropriate software to these abilities which does not yet exist.

REFERENCES

Alão, N. (2017). Reconstituição Geométrica da Visão. FAUL Lisboa.

Caldas, A. (2008). Viagem ao Cérebro e algumas das suas com-Petências. Univ.CatólicaEd. Lisboa

Dutchowski, A. (2007). Eye Tracking Methodology-Theory and Practice. Springer-Verlag. London

Gregory, R. (1972). Olho e Cérebro-Psicologia da Visão. Zahar Editores. Rio de Janeiro

Hecht, E. (2002). Óptica. Fundação Calouste Gulbenkian. Lisboa

Ibbotson, M. & Krekelberg, B. (2011). Visual Perception and Saccade Eye Movements. Elsevier. Bethesda

MacKay, W. (2011). Neurofisiologia sem lágrimas. Ed. Fund.Calouste Gulbenkian. Lisboa.

Yarbus, A. (1967). Eye movements and vision. Plenum Press.New York

Materials for healthcare applications and circular economy

Industry 4.0 – Shaping The Future of The Digital World – da Silva Bartolo et al. (eds)

Self-assembly of a bioinspired, photoactive-centered nanocontainer for drug delivery

S. Sonkaria & V. Khare
Institute of Advanced Machinery and Design, Seoul National University, Seoul, South Korea

V. Delorme
Institut Pasteur Korea, Tuberculosis Research Laboratory, Gyeonggi-do, South Korea

P. Purwar & J. Lee
Seoul National University, Mechanical and Aerospace Engineering, NMSL, Seoul, South Korea

ABSTRACT: The process of compartmentalization has unique evolutionary importance in biology by facilitating structural order, through the delivery of self-assembled building blocks which are both intrinsically and materially diverse in nature. For example, bacterial micro-compartments formed under biogenic control comprise of semi-permeable protein shells that permit in specific cases, the assembly of photo-sensitive quantum confined materials. Functionalised nanocontainers with porous cavities formed by crystalline photoactive 'quantum sized' centers within a shape sensitive supramolecular polymer shell can offer tunable surface properties to immobilize, transport and release therapeutic agents at target sites via externally controlled triggers. Using the physical properties of shape, size, geometry and texture of the nanocontainer, we demonstrate the drug loading capacity of polymer-trapped TiO_2 core structures and their potential for the development of drug-based technologies via controlled triggered release mechanisms.

1 INTRODUCTION

1.1 *Drug discovery and drug delivery: On a part*

While the discovery of 'new and promising' chemical entities' (NCEs) is pivotal to advancing pharmaceuticals in healthcare, strategies to deliver drugs are gaining primary importance as strong determining factors for the success of therapeutic agents. Indeed, the therapeutic pipeline of promising drugs is currently experiencing a slow pace for clinical approval due to stability and poor solubility (Brus. 1991) issues in the vast majority of cases. A barrier to the availability of an increasing number of effective drug formulations is likely due to their lipophilic nature, inevitably compromising their bioavailability and the desired efficacy. There is an on-going technical challenge to solubilize and improve the therapeutic outcome of disease with 'better' drugs, also reiterating the opinion that conventional approaches to delivering drugs show behavioral characteristics that are often deemed undesirable. Nonetheless, increasing the complexity of drug formulation is also introducing new challenges to ensure the marketable growth of a 'new wave' of drug types. A current concern is the burden of cost increase on patient drug prescription prices (Figure 1). The approval of more cost effective drugs requires the development of better standardized formulation methods or the significant improvement of existing ones.

Novel technologies are being sought that directly address concerns related to drug safety and efficacy with minimal impact to the patient. From a pharmaceutical perspective, this means that unwanted side effects are significantly reduced without affecting the performance of the drug itself. Hence, the most attracttive strategies are those providing control over the localization of a drug and its subsequent release specifically at the target site, both at a rate and time-dependent manner. For example, a pulsatile drug release system (Deepika et al. 2011) designed to deliver drug regimens at specific sites has obvious advantages. Such devices however are technology challenging due to complex drug-release patterns for diseased sites that are small in area and largely inaccessible using systems controlled outside of the body.

1.2 *The potential role of 'nano' and technology in drug delivery*

In view of these challenges in therapeutics development and drug formulation, it seems attractive to start forging partnership between nanotechnology and biology, at scales relevant to bioprocesses. Indeed, nature operates on multi-dimensional scales in extremely compact environments, posing problems of fundamental importance in drug delivery. The practicality of miniaturizing sophisticated drugs

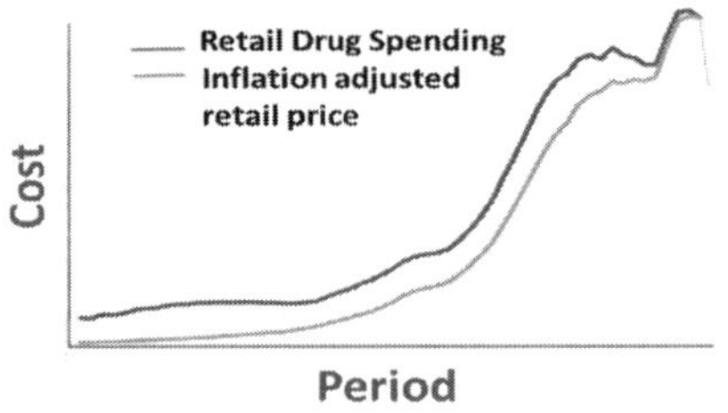

Figure 1. Normal and inflation-adjusted per capita spending on retail prescription drugs, 1960-2017 (Peterson-Kaiser Health System Tracker).

to navigate complex biological surroundings and locate target sites requires not only sophisticated approaches but also sophisticated materials. Such materials should be biocompatible with the human host and biodegradable in biological fluids within a reasonable time, releasing only armless components upon degradation. Meanwhile, biochemical function should be permitted at the vicinity of the materials, allowing interactions at the biological scale. Biomimetic materials are likely to be the best candidates for such a task, but they need to display fully tunable properties like size, shape, geometry, material and composition, which in turns define properties and functions of the materials. In essence, synthetic surfaces may be fabricated from engineered chemical building blocks to perform defined biomedical functions, in manners similar to that of biological building blocks, like template-induced polyclonal antibody catalysts (Sonkaria et al. 2004; Gul et al. 2003; Sonkaria 2016) or shape-driven aptamers for target recognition (Cai et al. 2018), for example.

On the physical front, in the first instance materials should be adaptable to a particular function or indeed multifunctional as most biological entities are. A leading consideration here in morphological terms is *shape*. Material assembly from building blocks into 'nanocaged' structures has unpresented utility in drug cargo loading and transport. The nanocontainment-like binding patterns (if knowledge allows) could be specifically used to design diverse structures to suit different drug types. Since biological environments are essentially heterogeneous in nature, we envisage that biological function is not only confined to shape and size as a nanocontainer but also the intricate surface properties, such as texture. Soft/hard interfaces might intrinsically change and adapt to environmental signals such as chemicals and solvents, light, pH, heat, electromagnetic fields, voltage-current inputs or energy particle interactions as with photosensitive materials. It is possible also that signal regulation may be achieved at the bio-synthetic material interface.

Similarly to bio-driven nano vehicles (Liu et al. 2015a), synthetic nanocages and hollow polymer particles (Ramli. 2017) have the potential to engage in host-guest interactions and release their guest structures through an appropriately triggered response mechanism (Liu et al. 2015). Hybrid materials combining polymers with photonic materials (Karewicz et al. 2018) for example, can function both as nano drug carriers and stimuli-responsive materials and formulated into nanomedicines (Shen et al. 2017). Porous synthetic materials mimicking the intrinsic properties of biopolymers and configured as drug vehicles to respond to external stimuli, will likely prove to be important advancement in this direction. For example, exploiting the intrinsic phase change properties of cargo loaded liposomes responsively tuned to near infrared (NIR) wave signals using plasmonic gold nanoparticles provides sufficient heat energy (Li et al. 2016; Ali et al. 2017) to destroy localized diseased cell populations. Hence, exploiting the property enhancement of hybrid bio-synthetic materials that use light to trigger the release of drugs by disruption of lipid carriers (Li et al. 2017) can provide controlled release for drug delivery applications (Sahandi Zangabad et al. 2017).

In this preliminary study, we report a dual functional semiconductor copolymer sensitive to both temperature and UV irradiation. We show that a self-assembled synthetic photonic polymer material with a TiO_2 quantum-sized semiconductor (NC-QDTiO_2) core is capable of retaining and releasing magnetic nanoparticles (MNP) tagged with a drug conjugate upon activation under a UV source.

2 EXPERIMENTAL

Carboxylate functionalized magnetic nanoparticles (MNPs) were synthesized in-house. All reagents were purchased from Sigma-Aldrich unless specified elsewhere. Synthesis of polymer nanocaged TiO_2 quantum dots was achieved by a one pot reduction process similar to the method previously reported by Khare *et al* (Aye et al. 2010). Surface coupling of carboxylate MNPs with the drug (Ampicillin) was performed using a standard coupling reagents (N-hydroxysuccinimide). Coupling and surface modification was assessed by absorption spectroscopy (UV-vis), Fourier-transform infrared (FTIR) spectroscopy and High-resolution transmission electron microscopy (HRTEM).

3 RESULTS AND DISCCUSION

In this preliminary investigation, we explored the host-guest potential of a biocompatible and biodegradable polymer nanocaged quantum dot semiconductor material (NC-QDTiO_2). The drug loading ability of NC-QDTiO_2 was assessed with a magnetically tagged drug conjugate by drug adsorption onto the surface of NC-QDTiO_2.

Group activation of the anionic ligand (carboxylate) coated MNP surface provided adequate charge functionalised MNPs by agitation provided the driving force for the coupling reaction for covalent be formation to occur. Nanometer sized MNP-drug conjugated molecules were magnetically cleared from free drug particles after incubating the reaction overnight at RT. The conjugate was incubated with fluorescent dye molecules specific for binding hydrophobic alkyl chains localized between the MNP surface and the drug. The estimated size of the immobilized drug MNP conjugate was approx. 0.4 μm MNP-drug conjugated samples (see Figure 2).

Table 1 shows spectrophotometric data of the MNP conjugated drug molecules caged within the polymer TiO_2 colloid particles in the presence of dye. After removal of unbound drug and dye constituents from the conjugate, significant differences were observed in the optical properties of the MNP-drug-dye colloid and free particles. Loss of the absorbance peaks in the conjugate observed for the drug, MNP and dye alone.

Enhancement of the signal of the drug in the complex occurred in the UV range strongly overlapping

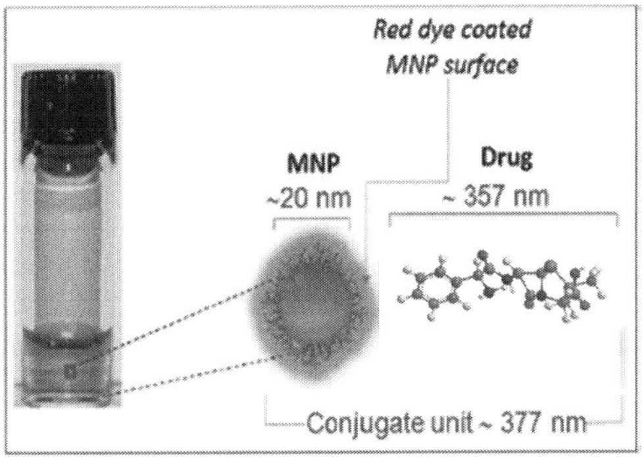

Figure 2. Polymer caged MNP-drug conjugate assembled on TiO_2 photoactive semiconductor quantum dots.

Table 1. Comparison of the optical peak signal of the conjugated drug with free drug, dye and MNP.

Constituent	Max Absorbance	Observation
Drug	322	Drug peak only
MNP	376	MNP peak only
Dye	541	Dye peak only
MNP with colloid (No UV exposure)	No Peaks	Drug-dye conjugated MNPs
MNP with colloid (UV exposure)	No Peaks	31% absorbance decrease

the MNP region. This complexity indicated a strong association of drug molecules with MNP surfaces by at 322, 376 (half peak) and 541 nm, respectively, revealed significant alteration in the organization of the drug and dye with colloid in the conjugated structure. conjugation chemistry, but with little change in wavelength absorption. The expected enhancement of light absorption of the drug was influenced by the conjugated pi structure of the dye attached to the alkyl carbon chains also located towards the inner MNP surface. The π -π* transitions in the dye however shifted from a low energy range to a higher wavelength, reflecting the complexity of binding in the conjugated environment in the colloid and dominance of the stronger UV absorbance range of TiO_2 QDs (300 – 380 nm).

The porous nature of NC-QDTiO_2 and their possible use for remote activation of therapeutic agents after drug loading as MNP-drug-dye colloid conjugates was explored under UV irradiation after purification. Visual evidence for the physical entrapment of drug conjugated particles was initially explored by Transmission Electron Microscopy (TEM). Non-covalent incorporation of the drug complex partitioned within cage-like structures of interior or exterior compartmentalized pockets is predicted from computational modelling of TiO_2 QDs locked within the polymer network (results not shown). Here, Figure 3 shows Bright field TEM micrographs of densely packed MNP-drug-dye conjugates in NC-QDTiO_2 in the absence of UV light. Dark contrasted white encircled area shown in Figure 3(a) can be attributed to the absorption of complex formed with nanocages. We have previously shown that TiO_2 embedded in nano sized polymer cages are of the order 0.8 – 2.0 nm (Choi et al. 2019). Aggregates of 20 nm MNPs are also visible in surrounding areas as would be expected with the high concentration of drug deposits. TEM images of the MNP-drug-dye conjugate after UV light exposure is shown in Figure 3 (b). Disintegration of the drug-bound complex is clearly visible from the polymeric caged QDs indicating photo-activated release of drug triggered by UV radiation. Furthermore, the morphology after UV exposure is very similar to the polymeric caged QDs. However, the presence of regional dark contrast and MNPs are indicative of the fact that release has not approached completion.

Fluorescence intensity measurements of the released conjugated drug correlates well with the UV of the observed TEM data. Figure 4 shows a comparison between UV light-exposed supernatant (less aggregated MNP fraction) and sediment fractions (more aggregated MNP fraction) of the MNP-drug-dye conjugate after thorough washing. In contrast, the encapsulated non-UV light exposed samples shows substantially low fluorescence signal indicating that drug release is the result of TiO_2 excitation at local sites.

We predict a dual functional responsiveness that begins with encapsulation of the nano sized drug

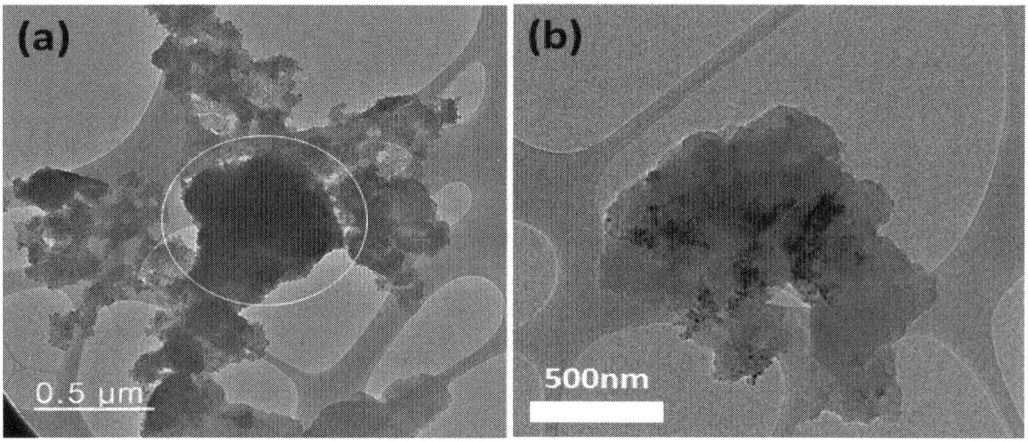

Figure 3. TEM micrographs of (a) MNP-drug entrapped in polymer TiO_2 nanocages (b) MNP-drug entrapped in polymer TiO_2 nanocages after UV light exposure. The crystalline contrast is clearly visible showing magnetic particles encircled by the red border of average size around 16 – 20 nm.

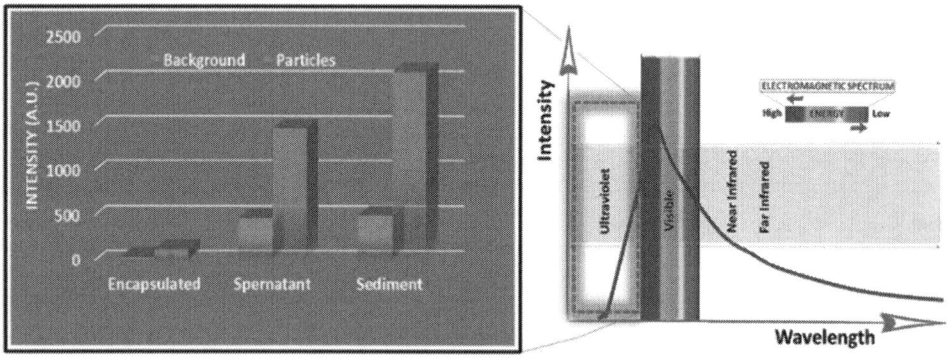

Figure 4. Fluorescence intensity measurements of UV exposed MNP-drug-dye conjugate samples (supernatant and sediment fraction) compared with the non-UV exposed fraction (encapsulated). The wavelength-intensity profile indicates drug release from photons.

conjugate inside polymer cages with dimensions suitable for the entrapment of small drugs. The second relates to local thermal energy generated regions at or around the site of the drug. The opening of nanocages that allows the release of drugs may function by heat generated expansion. This mechanism need to be explored in more detail.

4 CONCLUDING COMMENTS

In this conference paper, we have analyzed the host-guest behavioral properties of polymer confined TiO_2 quantum dots. Drug filled nanosized polymer cages, self-assembled around quantum confined TiO_2 by a simple adsorption process, can be photo-actively triggered to release the drug on demand to the surrounding environment. Such stimuli-responsive materials may show promising potential as drug-carriers and drug releasing agents. This preliminary study opens the way to protein-like miniature cages that can be tailored to perform task specific processes at the biological scale. Here, drug release from polymer assembled TiO_2 nanocages were examined under UV irradiation. However, the use of semiconductor colloids with drug-release excitation properties in the energy range closer to biological tissue absorption (*i.e.* UV-vis, NIR and far NIR range) will be explored.

ACKNOWLEDGEMENTS

This work was supported by the National Research Foundation of Korea (NRF), funded by the Korean Ministry of Science (MSIT), under the grant numbers 2012R1A2A2A01047189, 2017R1A2B4008801 and 2017M3A9G6068246 and the INDO-KOREA JNC program of the National Research Foundation of Korea Grant No. 2017K1A3A1A68. VD was supported by the French ministry of Foreign Affairs.

REFERENCES

Ali, M.R.K., Rahman, M.A., Wu, Y., Han, T., Peng, X., Mackey, M.A., Wang, D., Shin, H.J., Chen, Z.G., Xiao, H., Wu, R., Tang, Y., Shin, D.M. and El-Sayed, M.A. 2017. Efficacy, long-term toxicity, and mechanistic studies of gold nanorods photothermal therapy of cancer in xenograft mice. *Proceedings of the National Academy of Sciences*. 114(15), pp.E3110–E3118.

Ayi, A.A., Khare, V., Strauch, P., Girard, J., Fromm, K.M. and Taubert, A. 2010. On the chemical synthesis of titanium nanoparticles from ionic liquids. *Monatshefte für Chemie - Chemical Monthly*. 141(12), pp.1273–1278.

Cai, S., Yan, J., Xiong, H., Liu, Y., Peng, D. and Liu, Z. 2018. Investigations on the interface of nucleic acid aptamers and binding targets. *Analyst*. 143(22), pp.5317–5338.

Choi, D., Sonkaria, S., Fox, S.J., Poudel, S., Kim, S.-y., Kang, S., Kim, S., Verma, C., Ahn, S.H., Lee, C.S. and Khare, V. 2019. Quantum scale biomimicry of low dimensional growth: An unusual complex amorphous precursor route to TiO2 band confinement by shape adaptive biopolymer-like flexibility for energy applications. *Scientific Reports*. 9 (1), p18721.

Formulating Poorly Water Soluble Drug. Robert O. Williams III, Alan B. WattsDave A. Miller. (2012) Springer, New York, NY.

Gul, S., Sonkaria, S., Pinitglang, S., Florez-Alvarez, J., Hussain, S., Thomas, E.W., Ostler, E.L., Gallacher, G., Resmini, M. and Brocklehurst, K. 2003. Improvement in hydrolytic antibody activity by change in haptenic structure from phosphate to phosphonate with retention of a common leaving-group determinant: evidence for the 'flexibility' hypothesis. *The Biochemical journal*. 376(Pt 3), pp.813–821.

Jain, D., Raturi, R., Jain, V., Bansal, P. and Singh, R. 2011. Recent technologies in pulsatile drug delivery systems. *Biomatter*. 1(1), pp.57–65.

Karewicz, A., Lachowicz, D. and Pietraszek, A. 2018. Photonics in Drug Delivery. In: Van Hoorick, J., et al. eds. *Polymer and Photonic Materials Towards Biomedical Breakthroughs*. Cham: Springer International Publishing, pp.131–151.

Li, H., Yang, X., Zhou, Z., Wang, K., Li, C., Qiao, H., Oupicky, D. and Sun, M. 2017. Near-infrared light-triggered drug release from a multiple lipid carrier complex using an all-in-one strategy. *Journal of Controlled Release*. 261, pp.126–137.

Li, Z., Huang, H., Tang, S., Li, Y., Yu, X.-F., Wang, H., Li, P., Sun, Z., Zhang, H., Liu, C. and Chu, P.K. 2016. Small gold nanorods laden macrophages for enhanced tumor coverage in photothermal therapy. *Biomaterials*. 74, pp.144–154.

Liu, J., Detrembleur, C., Mornet, S., Jérôme, C. and Duguet, E. 2015a. Design of hybrid nanovehicles for remotely triggered drug release: an overview. *Journal of materials chemistry B*. 3(30), pp.6117–6147.

Liu, J., Detrembleur, C., Mornet, S., Jérôme, C. and Duguet, E. 2015b. Design of hybrid nanovehicles for remotely triggered drug release: an overview. *Journal of Materials Chemistry B*. **3**(30), pp.6117–6147.

Ramli, R.A. 2017. Hollow polymer particles: a review. *RSC Advances*. **7**(83), pp.52632–52650.

Sahandi Zangabad, P., Karimi, M., Mehdizadeh, F., Malekzad, H., Ghasemi, A., Bahrami, S., Zare, H., Moghoofei, M., Hekmatmanesh, A. and Hamblin, M.R. 2017. Nanocaged platforms: modification, drug delivery and nanotoxicity. Opening synthetic cages to release the tiger. *Nanoscale*. 9(4), pp.1356–1392.

Shen, S., Wu, Y., Liu, Y. and Wu, D. 2017. High drug-loading nanomedicines: progress, current status, and prospects. *International journal of nanomedicine*. 12, pp.4085–4109.

Sonkaria, S., Boucher, G., Flórez-Olvarez, J., Said, B., Hussain, S., Ostler, E.L., Gul, S., Thomas, E.W., Resmini, M., Gallacher, G. and Brocklehurst, K. 2004. Evidence for 'lock and key' character in an anti-phosphonate hydrolytic antibody catalytic site augmented by non-reaction centre recognition: variation in substrate selectivity between an anti-phosphonate antibody, an anti-phosphate antibody and two hydrolytic enzymes. *The Biochemical journal*. 381(Pt 1), pp.125–130.

Sonkaria, S., Kim, H.-T., Kim, S.-Y., Kumari, N., Kim, Y. G., Khare, V. and Ahn, S.-H. 2016. Ionic liquid-induced synthesis of a graphene intercalated ferrocene nanocatalyst and its environmental application. *Applied Catalysis B: Environmental*. 182, pp.326–335.

Industry 4.0 – Shaping The Future of The Digital World – da Silva Bartolo et al. (eds)
© 2021 Taylor & Francis Group, London, ISBN 978-0-367-42272-1

Physical and biological properties of electrospun PCL meshes as a function of polymer concentration and voltage

E. Aslan, E. Daskalakis, C. Vyas & P. JDS. Bartolo
Mechanical, Aerospace and Civil Engineering, University of Manchester, Manchester, UK

ABSTRACT: This paper investigates the use of solution electrospinning to produce polycaprolactone (PCL) wound dressings and explores the role of polymer concentration and voltage on the morphological, mechanical, and biological properties of a polycaprolactone electrospun mesh. Polymer solutions were prepared with acetic acid at different concentrations (16% (w/v), 20% (w/v), and 24% (w/v)). Electrospun meshes were produced using different voltages (10 kV and 14 kV) with a constant flow rate (0.5 mL/h). Morphological observations show that the fibre diameter decreases by increasing the applied voltage, and increases by increasing PCL concentration, which leads to improved mechanical properties. Electrospun meshes were biologically assessed using human adipose-derived mesenchymal stem cells. Results showed that cells were viable on all meshes and able to proliferate.

1 INTRODUCTION

Skin is the largest tissue in our body and protects the internal organs against significant external problems such as burns, sores, and trauma (Dias et al. 2016, Timmons, 2006). Normally, skin has considerable self-healing capacity in healthy people, however, in diabetes and infected wounds the skin heals slower or becomes a chronic wound (Dias et al. 2016, Pereira & Bártolo, 2016, Timmons, 2006). A clinical solution to these challenges is a key aim within tissue engineering. In order to overcome these chronic skin conditions different products have been developed with a key focus on the use of electrospun wound dressings to treat or regenerate the skin.

Electrospinning provides the possibility to produce fibres in a wide scale from the nano- to micrometre with versatility and simplicity. The nanoscale fibres of a electrospun mesh resemble the morphological structure of the fibrous extracellular matrix (ECM) in the body (Law et al. 2017). Thus, electrospinning usage in tissue engineering area is quite common in areas such as skin regeneration, controlled drug release, wound dressings, neural tissue repair, and ligament regeneration (Deldar et al. 2018, Góra et al. 2011, Law et al. 2017, Miguel et al. 2017).

Polycaprolactone (PCL) is a commonly used synthetic polymer in tissue engineering due to its biocompatibility, biodegradability and ease of processing (Woodruff & Hutmacher, 2010). PCL has been used to fabricate three-dimensional (3D) scaffolds for bone, cartilage, and nerve regeneration using additive manufacturing and electrospun for a wide range of applications.

This work investigates the use of electrospinning PCL wound dressings for skin applications. In this preliminary study, the effect of voltage and polymer solution concentration on the morphological, mechanical, and biological properties of the PCL meshes is investigated. Few studies have systematically investigated the role processing parameters have on mesh properties thus will provide an insight into mesh design for tissue engineers (Anindyajati et al. 2018, Cipitria et al. 2011; Doustgani et al. 2012).

2 MATERIALS AND METHODS

2.1 *Materials*

PCL in a pellet form was supplied by Perstorp Caprolactones (Cheshire, UK). Acetic acid (AA) glacial was supplied by Fisher Scientific (Loughborough, UK).

2.2 *Fabrication of PCL nanofibre meshes*

Electrospun meshes were produced using solution electrospinning (Spraybase, Ireland) (Figure 1). Three different polymeric solutions (16% (w/v), 20% (w/v), and 24% (w/v)) were prepared, the flow rate was kept constant at 0.5 mL/h, and two different voltages (10 and 14 kV) were applied to the needle and collector. A circular metallic collector (diameter =15 mm) was used and the distance

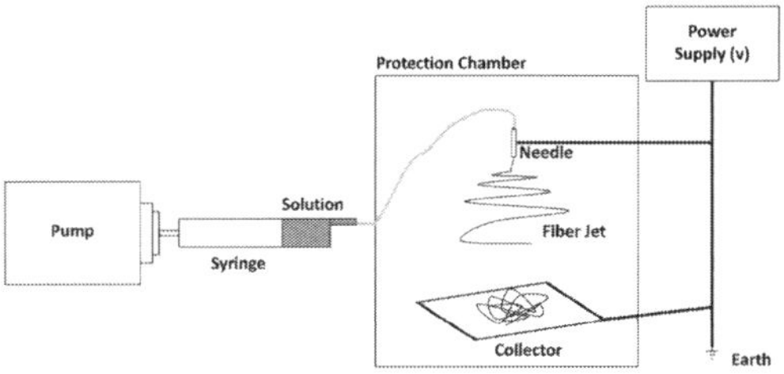

Figure 1. Electrospinning experimental setup.

between the needle and the collector was kept constant at 10 cm.

2.3 *Morphological characterisation*

The HITACHI S-3000N (Hitachi, Japan) scanning electron microscopy (SEM) was used to assess mesh morphology. Meshes were platinum coated using the EMITECH K550X sputter coater (Quorum Technologies, UK) and imaged at an acceleration voltage of 10 kV. The images were analysed using Fiji software with the DiameterJ plugin to assess fibre diameter and porosity (Hotaling et al. 2015). SEM images were taken from at least three random points from each mesh.

2.4 *Mechanical characterisation*

Tensile tests were conducted using the Instron 3344 (Norwood, USA). Meshes were spun onto a metallic rectangular platform (12 mm x 50 mm x 3 mm) covered with aluminium foil. Meshes were carefully removed from the foil and placed in the machine exactly at the centre and stretched in a dry state at a rate of 5 mm/min using a 10N load. Tests were performed five times for each mesh type.

2.5 *Biological characterisation*

Human adipose derived stem cells (hADSCs) (STEMPRO, Invitrogen, USA) were used to investigate the cytotoxicity of the meshes. MesenPRO RS^{TM} basal media, 2% (v/v) growth supplement, 1% (v/v) glutamine and 1% (v/v) penicillin/streptomycin (Invitrogen, USA) was used for cell culture. Cells were cultured in an incubator (37°C, 5% CO_2, and 95% humidity) until an appropriate cell density was achieved and harvested at passage 4 through trypsinisation.

Meshes were sterilised using 80% ethanol for 2 h, rinsed in phosphate buffered saline (PBS) solution three times and air-dried overnight in a sterile laminar flow cabinet. The meshes were transferred into 24-well plates and kept in the incubator prior to cell seeding. A 75 μl cell suspension containing 25,000 cells was added to each mesh (n=4) at day 0. A tissue culture plastic (TCP) control with the same number of cells was used in order to evaluate the cell seeding efficiency after day 1.

At day 1, all meshes were transferred to new plates, and fresh media added. Cell proliferation was measured using the Alamar Blue assay (Sigma-Aldrich, UK). Through this assay a fluorescent signal is produced by the reduction of resazurin to resorufim intracellularly by mitochondrial enzyme activity based on their cellular metabolic activity and can be detected by a microplate reader (Infinite 200, Tecan, Hammedorf, Germany). The Alamar blue assay was used at day 1, 3, 7 and 14.

At each time point, a 10% by volume (50 μL) of resazurin solution (0.01% (v/v)) was added to each sample and incubated for 4 hours. After incubation, 150 μL of each sample was transferred to a 96-well plate and the fluorescence intensity was measured (540 nm excitation/590 nm emission wavelength) with a plate reader. Samples were washed twice in sterile PBS to remove the resazurin solution before the addition of fresh media. Cell culture media was changed every 3 days.

3 RESULTS & DISCUSSION

3.1 *Morphological characterisation*

SEM images of the electrospun meshes show the effect of both voltage and polymer concentration on the morphology of the meshes (Figure 2). The results

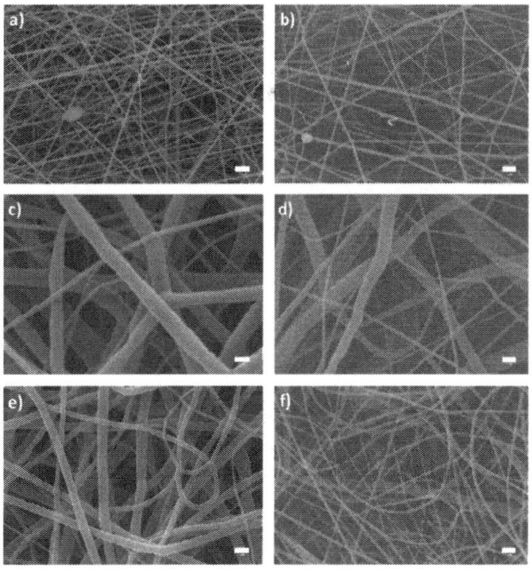

Figure 2. Representative SEM images of 16% PCL concentration meshes obtained at a) 10kV, b) 14kV; 20% PCL concentration meshes obtained at c) 10kV, d) 14kV and 24% PCL concentration meshes obtained at e) 10kV, f) 14kV. (Scale: 1 μm).

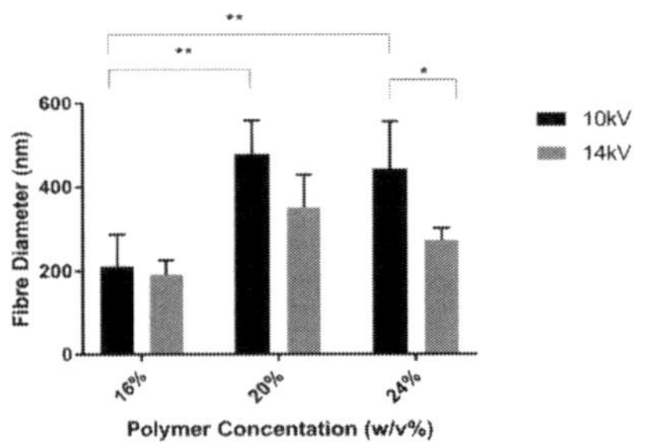

Figure 3. Average fibre diameter at different conditions.

show that the fibre diameter ranges from 190 to 470 nm, values suitable for tissue engineering applications, providing enough surface area for cell attachment.

As observed the fibre diameter is dependent on both voltage and polymer concentration, being possible to observe that the fibre diameter slightly decreases by increasing the applied voltage and increases by increasing the polymer concentration (Figure 3). In the case of meshes produced with a solution containing 24% of PCL at 14 kV, the dominant effect of the applied voltage reduces the fibre diameter.

The porosity distribution and percentage for the meshes show that the porosity decreases by increasing the PCL concentration in the solution (higher viscosity), related to increasing fibre diameter (Figure 4). The higher voltage led to an increase in mesh porosity as an increase in voltage leads to a reduction in fibre diameter.

The fibre distribution in the meshes show a more uniform distribution in the meshes produced using 16% and 24% of PCL (Figure 5). The fibre diameter ranges from 190 to 470 nm.

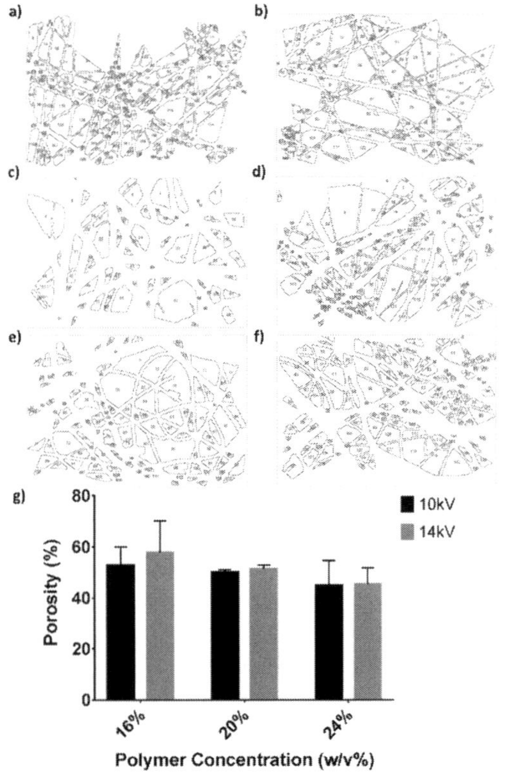

Figure 4. Porosity distribution in the 16% PCL concentration meshes obtained at a) 10kV, (Porosity: 53%), b) 14kV, (Porosity: 58%); 20% PCL concentration meshes obtained at c) 10kV, (Porosity: 50%), d) 14kV, (Porosity: 52%) and 24% PCL concentration meshes obtained at e) 10kV, (Porosity:44%), f) 14kV, (Porosity:45%) & g) percentage of porosity.

3.2 *Mechanical characterisation*

The Young's modulus of the meshes show that the mechanical properties increase with increasing voltage and constant PCL concentration, which is associated with a decrease in the fibre diameter (Figure 6). Reducing the fibre diameter increases the density of the fibres within the same area, which leads to increasing Young's modulus as the mechanical forces are distributed across a greater number of fibres. When the voltage is kept constant, the Young's modulus increases with increasing PCL concentration. In this case, the concentration of the polymer shows a dominant character for the mechanical properties when compared to the voltage effect, as the force required to deform a thicker fibre is higher than the forces needed to deform a thinner fibre.

The highest average Young's modulus, 11 MPa, belongs to meshes produced with 24% PCL at 14 kV. Furthermore, meshes produced from 16% PCL concentration at 10kV present the lowest mechanical strength with 3 MPa. According to these mechanical results, the Young's modulus in all cases are suitable for skin tissue applications (Jacquemoud et al. 2007).

3.3 *Biological analysis*

Cell proliferation results observed up to day 14 of cell seeding is presented in Figure 7. The results indicate that the fluorescence intensity increases in all samples until day 14 except for 24% PCL meshes at 10 kV. As observed the effect of polymer concentration on cell proliferation is more significant than the effect of the applied voltage. The observed result can be partially explained by the effect of both voltage and polymer concentration on both fibre diameter and porosity.

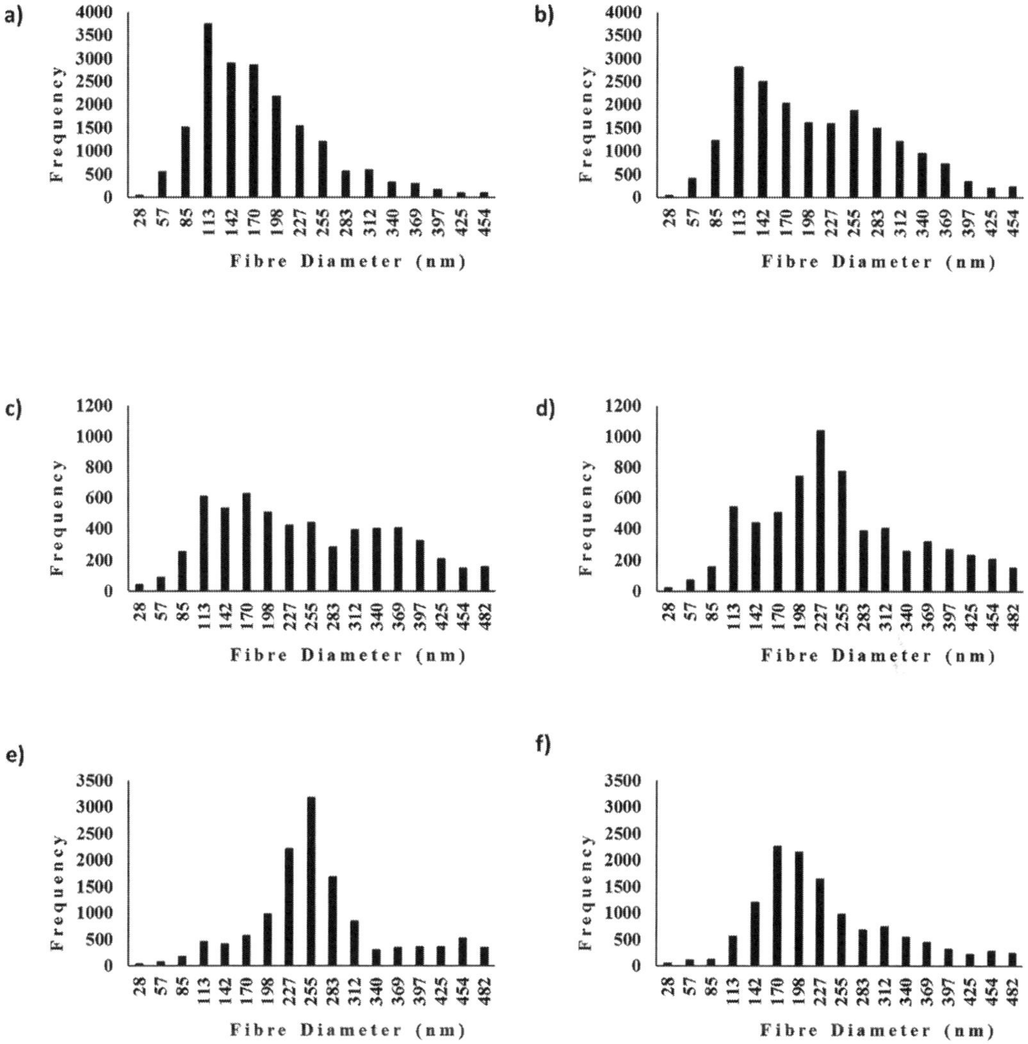

Figure 5. Fibre distribution 16% PCL concentration a) 10kV, b) 14kV; 20% PCL concentration c) 10kV, d) 14kV and 24% PCL concentration e) 10kV, f) 14kV.

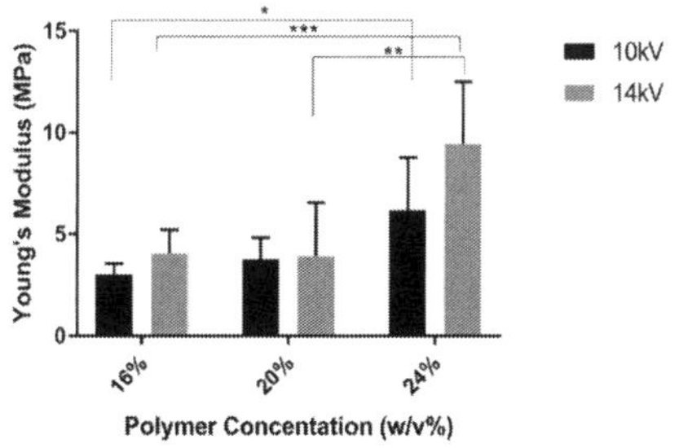

Figure 6. Young's modulus of the PCL meshes as a function of voltage and concentration.

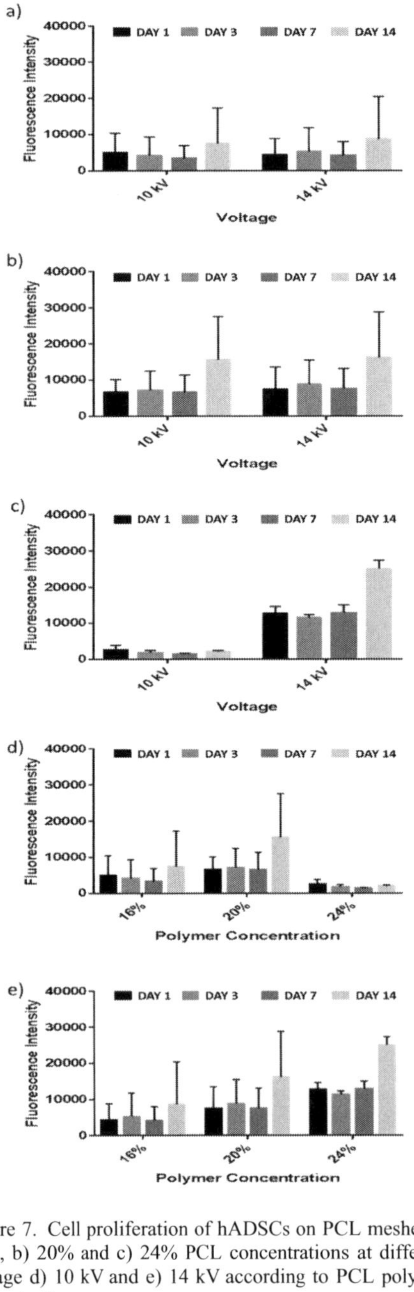

Figure 7. Cell proliferation of hADSCs on PCL meshes a) 16%, b) 20% and c) 24% PCL concentrations at different voltage d) 10 kV and e) 14 kV according to PCL polymer concentrations.

4 CONCLUSIONS

The concentration of PCL in the solution has a significant impact on the produced fibres. As observed, when the concentration of PCL increases the viscosity of the solution increases leading to the increase in fibre diameter. The surface tension also increases leading to the increase of the field strengths and the increase of porosity. The applied voltage has also a high impact on the morphology of the produced fibres. Results show that by increasing the applied voltage, the electrostatic forces increase leading to an increase in the fibre stretching mechanism, which reduces the fibre diameter and increases the mean porous area.

From the mechanical analysis it appears that the concentration of PCL has a major impact on the mechanical properties.

Biological tests with hADSC show that all the produced meshes can support cell attachment and proliferation. Overall, high cell proliferation values were observed in 24% PCL meshes produced at 14kV, however, meshes produced from 20% PCL concentration presented better and more uniform cell proliferation at both voltages.

This preliminary investigation will aid in the design of an electrospun mesh with suitable morphological, mechanical, and biological properties for skin tissue engineering. Further studies are required to systematically elucidate the relationship between processing parameters and physical and biological properties.

REFERENCES

Anindyajati, A., Boughton, P. & Ruys, A. J. 2018. Modelling and Optimization of Polycaprolactone Ultrafine-Fibres Electrospinning Process Using Response Surface Methodology. *Materials (Basel, Switzerland)*, *11*(3).

Cipitria, A., Skelton, A., Dargaville, T. R., Dalton, P. D. & Hutmacher, D. W. 2011. Design, fabrication and characterization of PCL electrospun scaffolds—a review. *Journal of Materials Chemistry*, *21*(26), 9419.

Deldar, Y., Pilehvar-Soltanahmadi, Y., Dadashpour, M., Montazer Saheb, S., Rahmati-Yamchi, M. & Zarghami, N. 2018. An in vitro examination of the antioxidant, cytoprotective and anti-inflammatory properties of chrysin-loaded nanofibrous mats for potential wound healing applications. *Artificial Cells, Nanomedicine and Biotechnology*, *46*(4), 706–716. 2.

Dias, J. R., Granja, P. L. & Bártolo, P. J. 2016. Advances in electrospun skin substitutes. *Progress in Materials Science*, *84*, 314–334.

Doustgani, A., Vasheghani-Farahani, E., Soleimani, M. & Hashemi-Najafabadi, S. 2012. Optimizing the mechanical properties of electrospun polycaprolactone and nanohydroxyapatite composite nanofibers. *Composites Part B: Engineering*, *43*(4), 1830–1836.

Góra, A., Sahay, R., Thavasi, V. & Ramakrishna, S. 2011. Melt-electrospun fibers for advances in biomedical engineering, clean energy, filtration, and separation. *Polymer Reviews*, *51*(3), 265–287.

Hotaling, N. A., Bharti, K., Kriel, H. & Simon, C. G. 2015. DiameterJ: A validated open source nanofiber diameter measurement tool. *Biomaterials*, *61*, 327–338.

Jacquemoud, C., Bruyere-Garnier, K. & Coret, M. 2007. Methodology to determine failure characteristics of planar soft tissues using a dynamic tensile test. *Journal of Biomechanics*, *40*(2), 468–475.

Law, J. X., Liau, L. L., Saim, A., Yang, Y. & Idrus, R. 2017. Electrospun Collagen Nanofibers and Their Applications in Skin Tissue Engineering. *Tissue Engineering and Regenerative Medicine*, *14*(6), 699–718.

Miguel, S. P., Ribeiro, M. P., Coutinho, P. & Correia, I. J. 2017. Electrospun polycaprolactone/Aloe Vera_chitosan nanofibrous asymmetric membranes aimed for wound healing applications. *Polymers*, *9*(5), 183.

Pereira, R. F. & Bártolo, P. J. 2016. Traditional Therapies for Skin Wound Healing. *Advances in Wound Care*, *5* (5), 208–229.

Timmons, J. 2006. Skin function and wound healing physiology. *Wound Essentials*, *1*, 8–17.

Woodruff, M. A. & Hutmacher, D. W. 2010. The return of a forgotten polymer—Polycaprolactone in the 21st century. *Progress in Polymer Science*, *35*(10), 1217–1256.

Industry 4.0 – Shaping The Future of The Digital World – da Silva Bartolo et al. (eds)

Microcrystalline cellulose as filler in polycaprolactone matrices

M.E. Alemán-Domínguez, Z. Ortega, A.N. Benítez & M. Monzón
University of Las Palmas de Gran Canaria, Las Palmas de Gran Canaria, Spain

L. Wang
University Xi'an Jiaotong, Xi'an, China

M. Tamaddon & C. Liu
University College London, London, UK

ABSTRACT: Polycaprolactone is a biomaterial widely used for tissue engineering applications. However, its hydrophobicity hinders cell attachment and proliferation on its surface. In this study microcrystalline cellulose has been proposed as a functional filler for polycaprolactone matrices expected to improve these properties. Composite material samples containing 0, 2, 5, 10 and 20% w/w of microcrystalline cellulose have been manufactured by compression molding and evaluated in terms of their mechanical properties, swelling behavior, water contact angle values and sheep mesenchymal cells viability. The results confirm that the presence of the additive is able to increase the swelling ability of the material (the samples containing 20% w/w of additive are able to absorb an amount of water 6 times higher than the value for polycaprolactone ones), the Young's modulus (from 224±14 MPa for polycaprolactone to 388±30 MPa for the composites containing 20% of microcrystalline cellulose) and the bioaffinity of polycaprolactone based composite materials.

1 INTRODUCTION

Polycaprolactone has been one of the most explored materials to be used as scaffolding in bone regeneration. However, its lack of hydrophilicity hinders the cell attachment and proliferation (Mondal et al. 2016, Mirhosseini et al. 2016, Mehr et al. 2015). Different alternatives have been explored in order to improve the hydrophilicity of this material. Filling the polycaprolactone matrix with additives has been one of them. Ceramic fillers have been the most explored options, especially bioactive ceramics (Park 2011, Domingos et al. 2017, Nyberg 2017, Lin et al. 2018, Tsai et al. 2017). However, less attention has been paid to organic additives, although some of them could be interesting options in order to improve simultaneously the mechanical properties of the matrix and its water affinity. The authors of this paper have previously proposed the introduction of carboxymethylcellulose (CMC) (Alemán-Domínguez et al. 2018a). CMC was confirmed as a suitable additive in order to modify the hydrophilicity of the polycaprolactone matrix. However, the cell affinity did not show a similar trend, as no significant modification was observed for the composite materials compared to pure polycaprolactone. In this study, another organic additive has been proposed as a functional filler of polycaprolactone matrices: microcrystalline cellulose (MCC).

Microcrystalline cellulose has been used as a reinforcement additive in thermoplastic matrices for industrial applications (Izzati et al. 2015, Haafiz et al. 2013). However, the suitability of this additive for biomaterials aimed to be used in Tissue Engineering has been barely explored. The authors of this paper (Alemán-Domínguez et al. 2018b) have already shown the potential of the composites based on polycaprolactone:microcrystalline cellulose blends to obtain 3D printed scaffolds. However, the behavior of those 3D printed scaffolds can be affected not only by the material properties, but also by the quality of the 3D printing process. The influence of the 3D printing parameters on the characteristics of the structures was demonstrated by Drummer et al. (Drummer et al. 2012), who concluded that the processing temperature can change the crystallinity of polylactid acid 3D printed structures and, as a consequence, their mechanical properties.

Therefore, the evaluation of the bulk properties of the composites obtained by filling this additive in polycaprolactone matrices is highly relevant in order to understand the relevance of the materials properties on the improvement of the matrix to be used as a support biomaterial in Tissue Engineering applications. Herein, we present such an evaluation with bulk samples of polycaprolactone:microcrystalline cellulose composite materials, so the results of this study are complementary to the previous work and, moreover, they could be used as essential information to extrapolate the data for exploring other manufacturing techniques.

2 MATERIALS AND METHODS

2.1 *Materials*

Polycaprolactone (PCL) Capa ® 6800 with mean molecular weight 80000 Da was kindly supplied by Perstorp, UK. Microcrystalline cellulose (MCC) was purchased from Sigma Aldrich. The following reagents were used for cell culture: DMEM low glucose (Sigma-Aldrich, UK), 100 units/ml penicillin-streptomycin (P/S Gibco, UK), PBS, phosphate buffered saline (Life Technologies), fetal calf serum-columbia (First Direct, First Link, UK), trypsin-EDTA (0.5%) (Thermo Fisher Scientific).

2.2 *Preparation of composite materials*

The polycaprolactone pellets were milled at 8000 rpm in an Ultra Centrifugal Mill ZM 200 Retsch device. The powder obtained from the milling process was mixed with the appropriate amount of microcrystalline cellulose to obtain the following mixtures: PCL:MCC 98:2, PCL:MCC 95:5, PCL: MCC 90:10 and PCL:MCC 80:20. These blends were homogenized and, afterwards, they were subjected to a compression molding cycle in a Collin P 200 P/M press at 85°C and a pressure of 10 bar.

2.3 *In vitro swelling*

The in vitro swelling of the composite samples was evaluated both in phosphate buffered saline and water. For both liquids, the samples were maintained overnight in a desiccator and weighed in this state before immersion (dry mass, m_d). The immersion times were 1h, 2h, 24h and 48h. These periods were chosen in order to evaluate how long it takes to reach the equilibrium swelling. After these periods, the samples were removed and dried with filter paper in order to remove the excess of liquid. Once dried, they were weighed (wet mass, mw). This values allow determining water and PBS uptake according the equation (1):

$$\% \text{ Uptake} = 100 \cdot (m_w - m_d)/m_d \quad (1)$$

After determining the equilibrium swelling time following this gravimetric procedure, the samples were subjected to a more precise evaluation of the amount of water they are able to store by thermogravimetric analysis. Three replicas of each type of composite material were immersed in water for 48 hours. Afterwards, they were dried with filter paper and subjected to a thermal cycle consisting of a heating stage from 25°C to 200°C at 20°C/min. Afterwards, the temperature was steady at 200°C for 2 minutes. During the cycle an air flow of 10 ml/min was provided and alumina crucibles were used. This evaluation was carried out in a TGA/DSC 1 Mettler Toledo device. The mass loss values at 150°C were identified as the water released from the samples. Therefore, it was possible to correlate the water uptake from the samples with these values.

2.4 *Water contact angle*

The water contact angle was measured with a contact angle meter OCA15 Dataphysics at room temperature and the data were analysed with the SCA20 software. The liquids used were water and diiodomethane. This two substances were chosen because their difference of polarity to evaluate the surface energy by the application of the Owens, Wendt, Rabel and Kaelble method (Annamalai et al. 2016).

2.5 *Mechanical testing*

Four replicas of each material consisting of sheets of 60x20 mm2 were obtained by compression molding and subjected to tensile testing. This testing was carried out in a Zwick Roell Z0.5 device in displacement control mode at a crosshead speed of 10 mm/min.

2.6 *Shore D hardness*

The hardness of polycaprolactone and its composites was measured with a PCE-DD-D durometer (resolution 0.5 and accuracy ±2). Two sheets of material were piled up and 10 readings at different points of the surface were obtained.

2.7 *X-Ray diffraction*

X-Ray diffraction (XRD) analysis was performed with an X-ray diffractometer (Bruker D8 Advance A25, with a Cu target, λ=0.154 nm). The samples were scanned from 3 to 90 °. The system was operated at 40 kV and a current of 20 mA.

2.8 *Cell culture*

To analyze the effect of the additive in terms of cell attachment and proliferation, sheep bone marrow mesenchymal stem cells were cultured on the surface of PCL and composite samples.

Four 10 mm diameter discs of each blend of material (PCL, PCL:MCC 98:2, PCL:MCC 95:5, PCL:MCC 90:10 and PCL:MCC 80:20) were cut and sterilized by immersion in 70% ethanol for 30 minutes. Afterwards, they were washed three times with PBS and three times with media to remove any trace of ethanol before carrying out the cell culture.

Regarding the seeding procedure, 50 μl of media containing 30000 cells were seeded on each sample. The cells were allowed to attach to the surface for 1 hour in an incubator at 37°C and 5% CO_2. After this time, 750 μl of media were added and the samples were incubated for 8 days. The viability of the cells was analyzed after 1, 3 and 8 days of cell culture through the resazurin-based Presto Blue ® assay.

3 RESULTS AND DISCUSSION

3.1 *In vitro swelling*

The in vitro swelling has been pointed out as an indicator of the modification of the hydrophilicity of a material to be used for tissue engineering applications (Jonnalagadda et al. 2015). The results obtained during the present study can be observed in Figures 1 and 2.

These in vitro swelling data show that MCC increases the hydrophilicity of the samples and this increase is related to the content of the additive in the composite material, even for short periods of time. For example, during the first hour of immersion in PBS, PCL samples are able to uptake only a 0.19% from their initial mass value, while PCL:MCC 80:20 ones are able to uptake a 1.14% (6 times the referred amount for PCL). This behavior was previously observed for the loading with carboxymethylcellulose (Alemán-Domínguez et al. 2018a) and it entails a clear advantage when compared to other additives, like phosphate glass fiber. According to Liu et al. (Liu et al. 2014), filling polycaprolactone with phosphate glass fiber allows to increase the water uptake of the material, but periods of time larger than one day were required. However, when compared with polycaprolactone based composites obtained by carboxymethylcellulose (CMC) filling (Alemán-Domínguez et al. 2018a), it is possible to observe that MCC composites are able to absorb lower amounts of water and PBS (at equilibrium water swelling time -48 hours- the value for PCL:CMC 80:20 was 6% and for the equivalent group of sample of PCL:MCC the value is 3%).

Regarding the in vitro swelling equilibrium time, there is not significant difference ($p>0.05$) between the uptake values at 24 h and 48 h. Therefore, it is possible to confirm that at 24 h the material has reached the equilibrium of water and PBS uptake.

Regarding the more accurate data from the TGA analysis (Figure 3), it is possible to observe a correlation between the amount of water released during the thermal cycle and the concentration of the additive in the composite material. This correlation is a linear expression, so the relationship between the presence of the MCC in the composite materials and the increase in their hydrophilicity gets confirmed.

The higher hydrophilicity of CMC-based composites is confirmed even clearer by the comparison of the equations from the thermogravimetric analysis. The equation that correlates water uptake to the concentration is a quadratic one when CMC is used as additive, but a linear one when the additive is MCC.

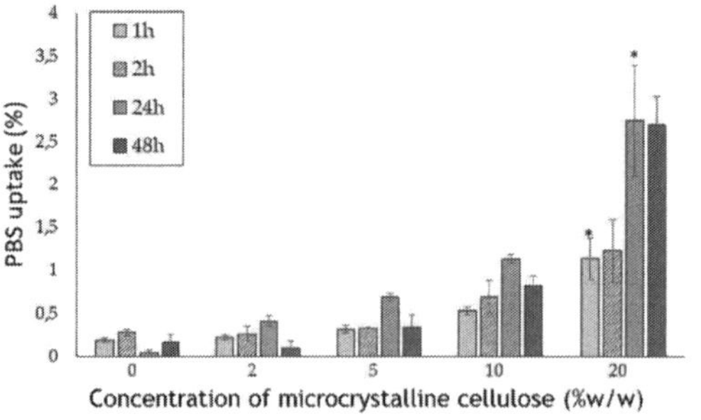

Figure 1. PBS uptake (%) for 1 h, 2h, 24h and 48h of immersion (**$p<0.05$ compared to the group of pure PCL samples).

3.2 *Water contact angle*

In Table 1 it is possible to observe the water and diiodomethane contact angles. It is possible to observe a decrease of the water contact angle when a 2% and a 5% wt:wt of MCC is introduced in the matrix. However, for the samples containing a higher amount of additive, the water contact angle increases up to 95.2±2.7° for PCL:MCC 80:20. On the other hand, a decrease in the diiodomethane water contact angle indicates that the surface

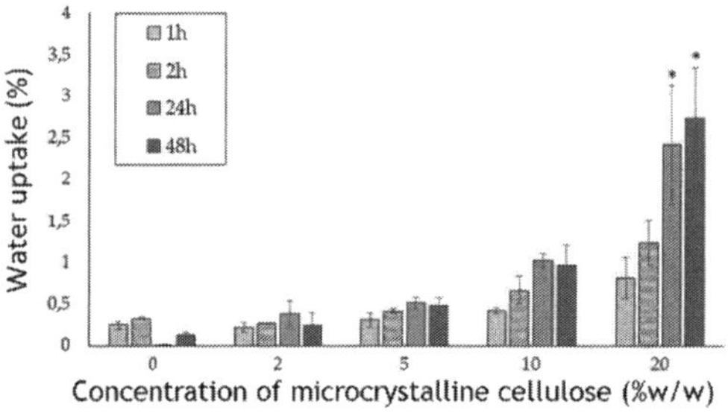

Figure 2. Water uptake (%) for 1h, 2h, 24h and 48h of immersion (**$p<0.05$ compared to the group of pure PCL samples).

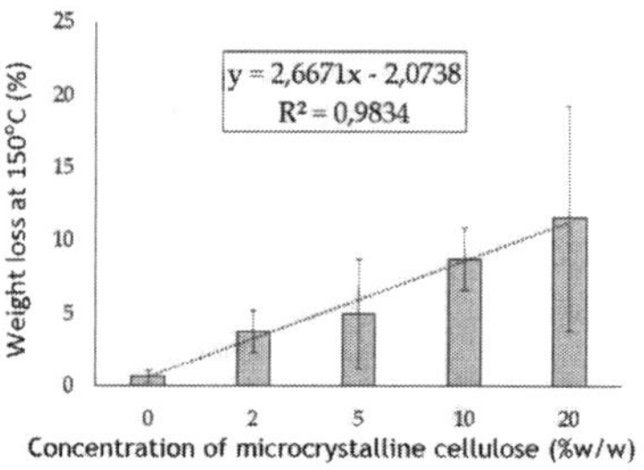

Figure 3. Correlation of weight loss at 150□C and the concentration of microcrystalline cellulose loaded in PCL matrices.

Table 1. Wettability data for PCL and its composites (*p<0.05; **p<0.01 compared to PCL values).

	Water contact angle	CH_2I_2 contact angle	Surface energy
	degrees	degrees	mN/m
PCL	87.1±3.8	39.9±2.0	39.20
PCL:MCC 98:2	80.1±0.3**	44.2±10.3	41.50
PCL:MCC 95:5	74.8±0.8*	41.1±10.2	42.03
PCL:MCC 90:10	85.3±2.7	48.4±1.1**	35.67
PCL:MCC 80.20	95.2±2.7**	37.0±2.6*	38.75

becomes more hydrophobic (Can-Herrera et al. 2016). In agreement with the data for water contact angle, the diiodomethane contact angle gets higher for the samples containing low concentrations of additive but it decreases when the concentration of MCC reaches the 20% wt:wt. This trend confirms an increase of the hydrophilicity of the samples containing a low concentration of the additive. The results for the surface energy confirm this conclusion, as this parameter shows a maximum value for the group PCL:MCC 95:5.

3.3 *Mechanical testing*

The values of the tensile modulus of the composite materials increase with the amount of MCC introduced in the blend, although this difference is significant only for those samples containing 20% w/w MCC (Figure 4). It is important to notice that there is a linear trend between the values of the tensile modulus and the concentration of the additive in the blend.

However, this increment of the elastic modulus implies a more brittle behavior of the blend with the increase of the additive. As a consequence, the stress at yield point decreases, as it is possible to observe in Figure 5. This change of the thermoplastic matrix from the ductile behavior to the brittle or quasibrittle one is a common trend in composite materials where the filler is not a thermoplastic (Bazhenov 2011) because the filler particles act as discontinuity point in the matrix.

The increase on the Young's modulus of the composite material has been reported for other fillers such as hydroxyapatite (Ródenas-Rochina et al. 2013, Rezai & Mohammadi 2013) or bioglass (Ródenas-Rochina et al. 2013, Ji et al. 2015). The possibility of modifying the stiffness of the material with the concentration of the additive is highly relevant, as it is possible to find studies that state the relationship between the stiffness of the support material and the differentiation and migration processes of stem cells (Bracaglia et al. 2017, Engler et al. 2006).

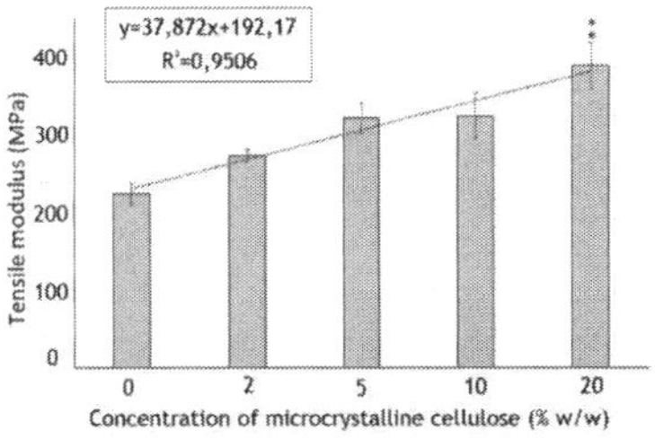

Figure 4. Tensile modulus of composite materials and polycaprolactone (**p<0.01 compared to pure PCL samples).

3.4 *Shore D hardness*

The hardness of the materials slightly increases with the introduction of the additive (Table 2). The Shore D of pure polycaprolactone is in accordance to the value from its datasheet. Hardness and stiffness are related parameters (Meththananda et al. 2009), so these results confirm the trend described by the mechanical characterization (Figures 4 & 5) about the increase in the stiffness of the composite material with the concentration of microcrystalline cellulose.

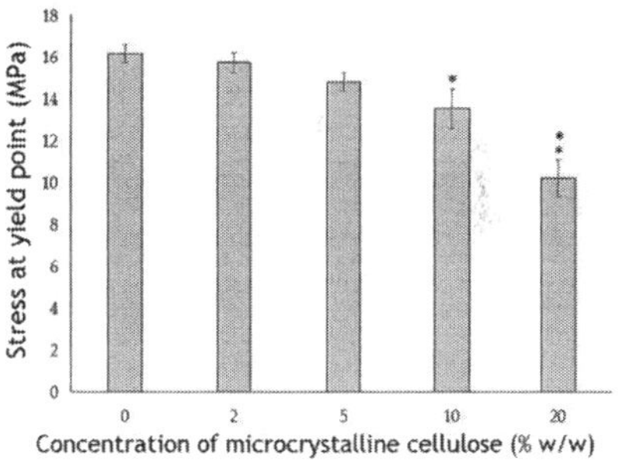

Figure 5. Mean values of the stress at yield point of composite materials and polycaprolactone (**p<0.01 and * p<0.05 compared to the group of pure PCL samples).

Table 2. Mean values of shore D hardness of composite materials and polycaprolactone (**p<0.01).

	Shore D hardness
PCL	50±2
PCL:MCC 98:2	50±2
PCL:MCC 95:5	52±2
PCL:MCC 90:10	52±2
PCL:MCC 80:20	55±3**

3.5 *X-Ray diffraction*

Figure 6 shows the XRD pattern of PCL and their composites samples. All the diffractograms show the characteristic peaks of PCL at 21.3° and 23.6° which correspond to the (110) and (200) crystallographic planes of this semicrystalline polymer (Mirhosseini et al. 2016, Martins-Franchetti et al. 2010). Table 3 shows the values of the area of these two peaks for the different materials evaluated. The area of both peaks increases with the introduction of the additive with a maximum located at a concentration of 5% wt/wt of microcrystalline cellulose.

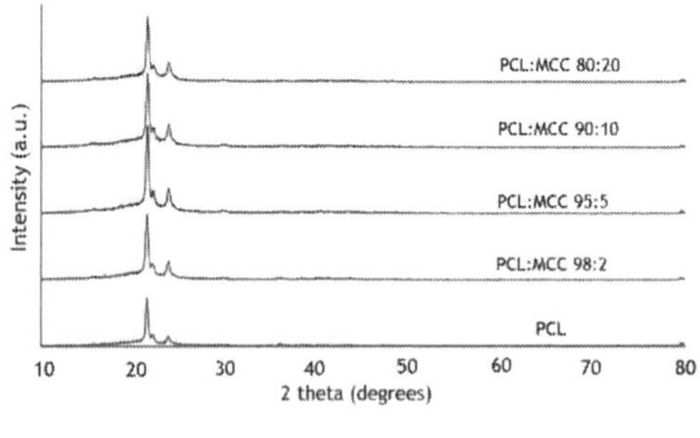

Figure 6. XRD patterns of PCL and its composites.

The sharper shape of the diffraction peaks at 21.4° and 23.6° and their increased area in the XRD patterns (Figure 6 and Table 3) can be related to an increment on the crystallinity of the matrix due to the presence of the filler particles. Therefore, the XRD data allow to confirm that the microcrystalline cellulose particles are able to act as nucleating points during the crystallization process. The increase in the crystallinity of the matrix explains the increase in the stiffness of the material confirmed by the tensile testing of the materials (Figures 4 & 5) and the related increase in shore D hardness (Table 2).

3.6 *Cell viability*

The results of the viability assay are shown in Figure 7. From Day 3 to Day 8, the population increases on

Table 3. Area of the main XRD diffraction peaks.

	Area of 21.3° peak	Area of 21.3° peak
	a.u.	a.u.
PCL	19238	3866
PCL:MCC 98:2	28059	7537
PCL:MCC 95:5	35072	10135
PCL:MCC 90:10	31256	9611
PCL:MCC 80:20	26360	6953

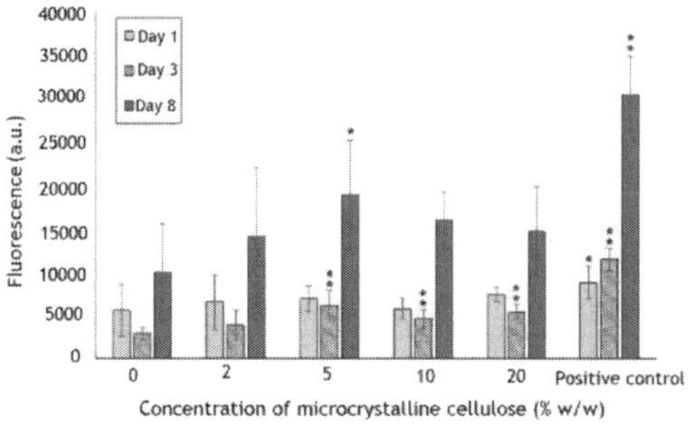

Figure 7. Results of viability tests for PCL:MCC composites using Presto Blue ® assay (* $p<0.05$; ** $p<0.01$ compared to the group of pure PCL samples).

all the groups of samples analyzed. This fact confirms the biocompatibility of the different composite materials in terms of cell viability. The fluorescence at the end of the experiment (Day 8) is higher for the samples with MCC than for those without additive.

It has been possible to confirm that the introduction of the microcrystalline cellulose has a concentration-dependent effect: the fluorescence at Day 8 is increased for the samples containing 2 and 5% w/w of microcrystalline cellulose. For the samples containing 10 and 20% w/w, the values are still higher than the PCL one, but it is possible to observe a diminishment compared to the values obtained for the samples containing 5% w/w of microcrystalline cellulose (Figure 7).

This concentration-dependent behaviour has been previously reported for other composite biomaterials, such as polylactic acid filled with Bioglass ® (Verrier et al. 2004) or polycaprolactone filled with hydroxyapatite(Ródenas-Rochina et al. 2013). The maximum fluorescence, related to the maximum cell affinity, is therefore observed for the group of samples containing 5% microcrystalline cellulose. This fact could be explained by the increase of surface wettability (Table 1). As previously described, the surface energy shows a maximum for the PCL:MCC 95:5 group, so this characteristic could have direct implications on the cell affinity of the samples.

However, when compared these data to the previously reported for PCL:MCC 3D printed structures,(Alemán-Domínguez et al. 2018b) it is possible to observe a modification of the position of the cell viability maximum. In that case, it was located at 2% wt/wt content of MCC. This change could be explained by the effect of the presence of the additive on the morphology of the surface of the 3D printed samples. The modifications of the morphology of the filaments have a definite effect on cell attachment and proliferation. Therefore, these results confirm the positive influence of the filling with microcrystalline cellulose and how the manufacturing process can modify the biological performance of the composite materials.

Besides, when these data are compared to the data from PCL:CMC composites (Alemán-Domínguez et al. 2018a),it is possible to confirm that microcrystalline cellulose provides an improvement on cell affinity that carboxymethylcellulose cannot offer. It is possible that the water affinity of the composite materials plays a role in this behavior. As stated above, for polycaprolactone: carboxymethylcellulose composites, water and PBS uptake values were higher. It can be stated that these values were high enough to hinder the adsorption of adhesion-signaling extracellular proteins (Alemán-Domínguez et al. 2018a, Chen et al. 2018). For PCL:MCC composites obtained in the present study, however, the hydrophilicity is kept within the values able to provide an improvement on cell proliferation.

Nevertheless, the processes that rule the attachment and proliferation of cells on biomaterials are complex and several parameters are involved, such as wettability, topography or surface charge (Chen et al. 2018). It would be interesting for future work to deeper analyze the influence of these parameters on the biological performance of polycaprolactone:microcrystalline cellulose composites.

4 CONCLUSIONS

The introduction of microcrystalline cellulose has a significant influence in the properties of polycaprolactone based composites. The results obtained in this study show the reinforcement effect that can be obtained by filling PCL with this additive. Besides, the composite materials have a higher bioaffinity for sheepbone marrow mesenchymal cells than pure polycaprolactone.

ACKNOWLEDGMENTS

The authors thank H2020-MSCA-RISE program through the BAMOS project, funded from the European Union's Horizon 2020 under grant agreement Nº 734156. They would like to thank also the project DPI2017-88465-R funded by the Science, Innovation and Universities Spanish Ministry.

M.E. Alemán-Domínguez would like to express her gratitude for the funding through the PhD Grant (PIFULPGC-2014-ING-ARQU-2).

REFERENCES

Alemán-Domínguez ME, Ortega Z, Benítez AN, Vilariño-Feltrer G, Gómez-Tejedor JA, Vallés-Lluch A. 2018a J Appl Polym Sci. 135 46134.

Alemán-Domínguez ME, Giusto E, Ortega Z, Tamaddon M, Benítez AN, Liu C. Three-dimensional printed polycaprolactone-microcrystalline cellulose scaffolds. 2018b J Biomed Mater Res B.

Annamalai M, Gopinadhan K, Han SA, Saha S, Park HJ, Cho EB, et al. 2016 Nanoscale 8 5764.

Bazhenov S. 2011 in Cuppoletti J (ed)Metal, Ceramic and Polymeric Composites for Various Uses (InTech) p171.

Bracaglia LG, Smith BT, Watson E, Arumugasaamy N, Mikos AG, Fisher JP. 2017Acta Biomater.56 3.

Can-Herrera LA, Ávila-Ortega A, de la Rosa-García S, Oliva AI, Cauich-Rodríguez JV, Cervantes-Uc JM. 2016 Eur Polym J 84 502.

Chen S, Guo Y, Liu R, Wu S, Fang J, Huang B, et al. 2018Colloid Surface B.164 58.

Domingos M, Gloria A, Coelho J, Bartolo P, Ciurana J. 2017Proceedings of the Institution of Mechanical Engineers, Part H: Journal of Engineering in Medicine. 231 555.

Drummer D, Cifuentes-Cuéllar S, Rietzel D. 2012 Rapid Prototyping J.18500.

Engler AJ, Sen S, Sweeney HL, Discher DE. 2006Cell126677.

Haafiz MKM, Hassan A, Zakaria Z, Inuwa IM, Islam MS, Jawaid M. 2013Carbohyd Polym. 98 139.

Izzati Zulkifli N, Samat N, Anuar H, Zainuddin N. 2015Materials and Design 69 114.

Ji L, Wang W, Jin D, Zhou S, Song X. 2015Mat Sci Eng C-Mater.46 1.

Jonnalagadda JB, Rivero IV, Warzywoda J. 2015J Biomater App30 472.

Lin YH, Chiu YC, Shen YF, Wu YHA, Shie MY. 2018J Mater Sci-Mater M.29.

Liu X, Hasan MS, Grant DM, Harper LT, Parsons AJ, Palmer G, et al. 2014 J Biomater App29675.

Martins-Franchetti SM, Egerton TA, White JR. 2010J Polym Environ18 79.

Mehr NG, Li X, Chen G, Favis BD, Hoemann CD. 2015J Biomed Mater Res A.103 2449.

Meththananda IM, Parker S, Patel MP, Braden M. 2009Dental Materials25956.

Mirhosseini MM, Haddadi-Asl V, Zargarian SS. 2016J Appl Polym Sci.13343345.

Mondal D, Griffith M, Venkatraman SS. 2016Int J Polym Mater Po.65 255.

Nyberg E, Rindone A, Dorafshar A, Grayson WL. 2017Tissue Engineering - Part A.23 503.

Park SA, Lee SH, Kim WD. 2011Bioprocess and Biosystems Engineering.34 505.

Rezaei A, Mohammadi MR. 2013Mat Sci Eng C-Mater.33390.

Ródenas-Rochina J, Ribelles JLG, Lebourg M. 2013J Mater Sci-Mater M. 241293.

Verrier S, Blaker JJ, Maquet V, Hench LL, Boccaccini AR. 2004 Biomaterials25 3013.

Industry 4.0 – Shaping The Future of The Digital World – da Silva Bartolo et al. (eds)
© 2021 Taylor & Francis Group, London, ISBN 978-0-367-42272-1

Towards a circular economy - recycling of polymeric waste from end-of-life vehicles, electrical and electronic equipment

C. Abeykoon
North West Composites Centre, School of Materials, University of Manchester, Manchester, UK

P. Chongcharoenthaweesuk
Axion Polymers, Salford, UK

P. Xu
School of Materials, University of Manchester, Manchester, UK

C.H. Dasanayaka
Business School, University of Cumbria, Lancaster, UK

ABSTRACT: Polymeric materials-based waste has been accumulating over the years and now these can be found not only on the land but also in the waters across the globe, and has become a serious global crisis in many ways. Among used plastics, end-of-life vehicles (EOLV) and electrical-and-electronic-equipment (EEE) are two major sources of polymeric-waste. These wastes can have variety of combinations of polymers, and the properties of such blends are not well-understood yet. In this work, the effects of six different process parameters were tested with three types of recycled polymers (derived from EOLV and EEE waste) under various processing conditions. The results confirmed that the properties and processing behavior of recycled materials can be highly variable and hence difficult to predict. Furthermore, findings highlighted the importance of knowing their actual constituents and also the selection of appropriate process settings to achieve the desired product properties while saving resources.

1 INTRODUCTION

Plastics are one of the popular and commonly used materials in the modern world. The global production of plastics exceeded approx. 335 million tonnes in 2016 and has forecasted a four-fold increase in production tonnage by 2050 (Crawford and Quinn, 2017). Obviously, such a massive production capacity of plastics also causes to generate tons of waste each year and hence demand for recycling also increases rapidly. At present, the accumulation of polymeric wastes has become a global issue and both manufactures and consumers have been forced to rethink the current take-make-waste extractive industrial model for reusing these materials. To tackle this evolving waste problem, a wide range of recycling techniques, under three main categories: mechanical, thermal and chemical, is available for polymers/composites depending on the types of polymers and the manufacturing techniques used (Yang et al. 2012). Of these three types, mechanical breakdown is the dominant and involves separation, re-melting and remoulding of materials. After separation, plastic wastes are sorted, cleaned and then used to make products directly. However, separation of materials remains one of the main issues and industrially common density separation is not good in performance. Optical, spectroscopic and electrostatic separations are accurate and commonly used in research settings (Balakrishnan and Sreekala, 2016). However, these consume a great deal of time and effort, and hence are not attractive for industry. Thermal recycling follows three main routes: combustion for energy, fluidised bed or pyrolysis processes (Yang et al. 2012). Ideally, thermosetting plastics can be burned as a source of energy as they usually have a high calorific value (Skrifvars & Nyström, 2001). The fluidised bed and pyrolysis processes can be used to recycle materials in the forms of fibres and small molecules but they are more complex and energy-consuming in principle (Pickering, 2004). Generally, chemical recycling involves using a chemical solution to dissolve the matrix (Skrifvars & Nyström, 2001). Given the nature of the solution/s and constituents of the products to be recycled, this may be applicable for some types of polymers only. Also, this might have limitations due to environmental, health/safety issues as well (Tukker et al. 1999).

At present, there are a few challenges or barriers for the evolution of the polymer recycling industry such as lack of adequate markets for secondary/recycled materials, high capital cost, lack of recycling technologies, and tight environmental/industry regulations (e.g., EU-directive for end-of-life vehicles (European Parliament and Council, 2000)) etc. Also, the variation and deterioration of mechanical properties over the life-cycle are other issues with recycled materials (Grigore, 2017). According to Scheirs (1998), a reduction in mechanical properties is not tolerable if higher than 10% after reprocessing. One of the solutions for this would be that mixing virgin polymers with recycled polymers to improve the properties, but even the same type of recycled and virgin materials may not form strong bonding and may lead to stress concentrations across mechanical bonds (Bream and Hornsby, 2001).

Clearly, there is an essential and urgent need for improving recycling of polymeric waste for the betterment of globe community and economy. This can only be achieved through a joint effort of all related parties: consumers, manufactures, authorities and so forth. Numerous previous researches have also focused on different aspects of polymeric waste recycling: end-of-life vehicles (Wong et al. 2018; Karaeen et al. 2017), electronic waste (Turner and Filella, 2017), recycling techniques (Ragaert et al, 2017), effects of contaminants on recycling and products (Eriksen et al. 2018), composites recycling (Shuaib, 2017; Yang et al. 2012), and so forth.

This work aims to investigate the processing behavior and properties of recycled polymers derived from end-of-life vehicles (EOLV) and electrical and electronic equipment (EEE) towards a circular economy. As was reported by International Organization of Motor Vehicle Manufacturers, 95 million cars and commercial vehicles were manufactured in 2016. Undoubtedly, these two industries contribute massively in producing polymeric waste globally and expected to be increased further in coming years.

2 EXPERIMENTL METHODS & MATERIALS

2.1 *Experimental materials*

For this study, three types of recycled polymers were used: MEP54, S1 and S2. Axplas® MEP54 is a Mix Engineering Polymers from materials fraction that heavier than a density of 1.12 g.cm^{-3} and also is used as a control sample. Samples named S1 and S2 were manufactured and supplied by Axion Polymers (UK) to investigate their nature and behavior.

2.2 *Separation and identification*

Three samples were prepared via laboratory density separation techniques. A beaker was placed on the magnetic stirrer plate and the solution was stirred by a magnetic bar. Magnesium sulphate was added as a solvent until solution's density reached to 1.3 g.cm^{-3} and was measured by hydrometers. This method was used to separate sink fractions included, PVC and other impurities from target polymers in the float fraction. 500 chips each of type of materials' samples were tested to identify its polymer compositions before and after the density separation.

2.3 *Compression moulding and machining*

A small scale LOSCA compression molder was used to prepare test samples of 8×8 cm as shown in Figure 1. The moulder was heated to the set moulding temperature before materials were placed in the cavity. The press was slowly closed applying a low pressure for a few minutes and then it was opened to release the possible volatile gases and closed again.

After the sample consolidated, the mould was cooled down until the cavity temperature reached below 100 °C. Finally, the mould was opened and the sample was removed. Test samples were prepared by considering six control parameters in moulding process: molding time, molding temperature, molding pressure, chip size, sample thickness, and non-plastic content. In each set of experiments, one parameter remained varied while the other five parameters were fixed.

2.4 *Testing techniques*

For all the tests, at least five specimens were tested and the average values were reported. All the measurements were performed at 23 ± 2 °C and 24 hours after moulding. First, 8×8 cm moulded samples were cut into rectangular specimens of 8×1 cm. Then they were machined to prepare dog-bone specimens for tensile testing.

2.4.1 *Tensile tests*

Tensile specimens were tested using an Instron 3365 universal testing machine with a crosshead speed of 1 mm.min^{-1} (by following ISO 527-1:2012) All specimens were 72 mm long, 12 mm wide and 0.8-3 mm thick with a 36 mm gauge length. A stress–strain diagram was obtained from each test, and the both Yong's modulus and tensile strength were obtained.

2.4.2 *Flexural test*

Bending specimens were tested using an Instron 4301 (USA) universal testing machine with a crosshead speed of 1 mm.min^{-1} (ISO 178:2010 +A1:2013). All specimens were 80 mm long, 10 mm wide and 0.8-3 mm thick. A stress–strain diagram was obtained from each test and the both flexural modulus and strength were obtained.

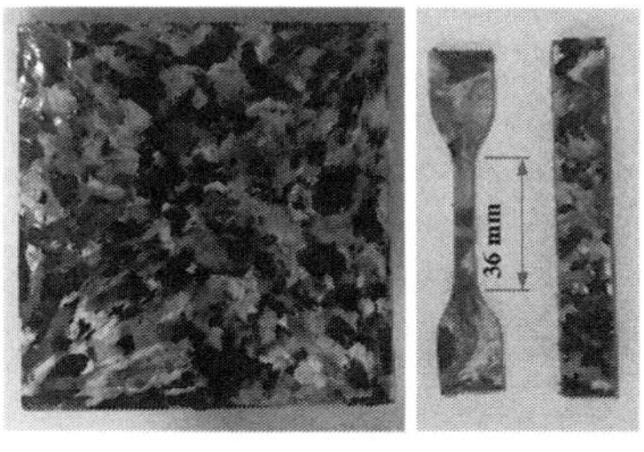

Figure 1. Photographs of a compression molded 8×8 cm sample and tensile and bending test specimens.

2.4.3 *Rheology tests*

The rheological properties of MEP54, S1 and S2 were determined by using a flat plate rheometer. Samples were solid disks of 25 mm in diameter and roughly 1 mm in thickness. Frequency sweep tests were conducted in the frequency range of 0.1–100 rad.s^{-1}, with 1% strain amplitude. It was aimed to explore the relationship of viscosity and elastic/loss modulus of samples verse increasing frequency. Temperature ramp tests were also conducted in a temperature range of 50-210 °C, with 5 °C.min^{-1} heating rate which simulated the heating rate in the compression molder. It was expected to explore a possible relationship of viscosity and elastic/loss modulus with temperature.

2.4.4 *Differential Scanning Calorimetry (DSC)*

The melting and crystallization behaviors of samples were measured under a liquid nitrogen atmosphere via a DSC (model QS100) with 3-10 mg samples in an aluminum pan. It was expected to determine T_g of all the constituents of recycled materials. The 1st step was increasing the temperature from 25 °C to 220 °C and then in the 2nd step this was reversed from 220 °C to 25 °C. Finally, in the 3rd step, the temperature was increased to 220 °C again. The aim of first heating and cooling was to erase the thermal history. Heating rate was at 5 °C.min^{-1} and it simulated the heating rate in the compression molder.

3 RESULTS & DISCUSSION

3.1 *Identification of composition*

A Fourier-transform infrared spectroscopy (FTIR) was performed to figure out the compositions and proportions of MEP54 (before and after density separation), S1 and S2. The results showed that the main constituents of these recyclates (derived from EOLV and EEE) are ABS, HIPS, PP, PC/ABS and PA (Nylon). Additionally, small amounts of PVC, PEI and PE can also be seen. Hence, only a portion of recyclates can be used directly in moulding. If a density of 1.3 g.cm^{-3} is taken as the minimum possible, only 51.30%, 92.13%, and 87.69% proportions of MEP54, S1 and S2 batches can be used respectively. Components like PVC and PA are not suitable to be moulded by compression and should be removed.

3.2 *Tensile properties*

Young's modulus and tensile strength were obtained for MEP54 via tensile tests in terms of the six control parameters mentioned in Section 2.3. As shown in Figure 2, Young's modulus and tensile strength are clearly influenced by the moulding time where tensile strength decreases as molding time increases. Modulus decreases until time reaches 5 mins and then increases again. Meantime, modulus also

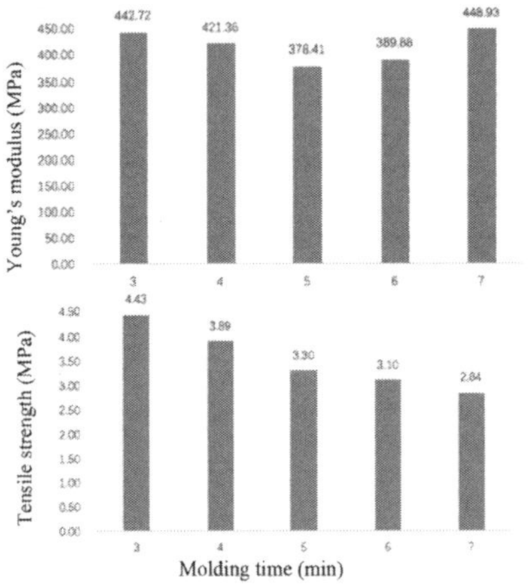

Figure 2. Young's modulus and tensile strength vs moulding time.

decreases as molding temperature increases until 190°C and then rises again (Figure 3). Tensile strength gradually decreases with the increase of both moulding temperature and time. The results show that the best operating temperature for compression moulding of MEP54 would be at 170 °C which will provide the best Young's modulus and tensile strength. This is also favorable in terms of energy savings as less heat is needed.

As shown in Figure 4, both modulus and tensile strength changes slightly with molding pressure. Based on this information, it can be concluded that 5 MPa would be sufficient for compression moulding of MEP54. Higher pressures above 5 MPa will neither degrade compression samples nor improve them but only wastes energy.

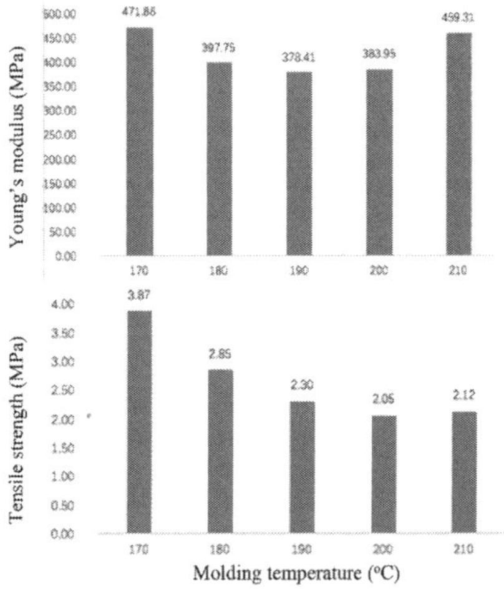

Figure 3. Young's modulus and tensile strength vs moulding temperature.

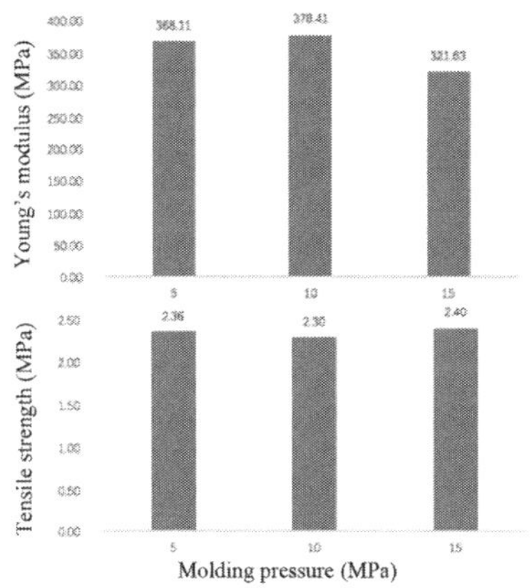

Figure 4. Young's modulus and tensile strength vs moulding pressure.

Size of the chips also has clear effect on the strength (see Figure 5) where decreasing chip size causes in strength reductions. This opposes the assumption that the smaller chip size leads to more connecting bonds within the samples which should make mechanical properties better. This result implies that although smaller chip size does bring more connecting bonds to the sample, these bonds themselves are weaker than original bonds that were already in chips, becoming defects in samples. This assumption can also be confirmed because almost every specimen was failed in one of the connecting sections (different chips are in different colours, making it easy to recognize by naked eyes). These defects lead to stress concentrations easily so both Young's modulus and tensile strength decrease sharply as the chip size becomes smaller. Overall

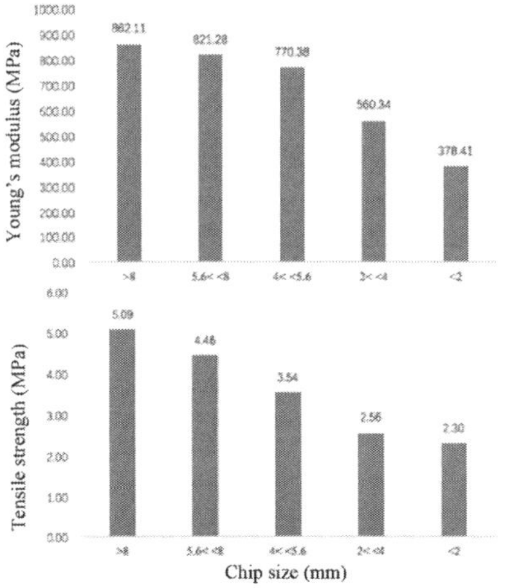

Figure 5. Young's modulus and tensile strength vs chip size.

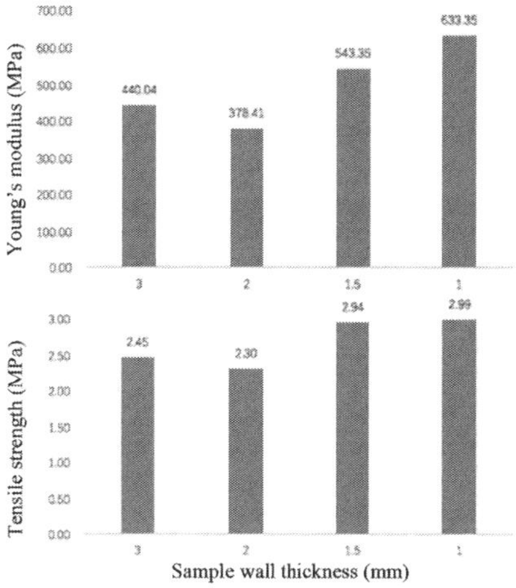

Figure 6. Young's modulus and tensile strength vs wall thickness.

these findings supports the fact that the original bonds that were generated in materials during the previous forming stage are stronger than the newly-formed connecting bonds and this agrees with the previous findings [9-11] which claimed deterioration of material properties across the life cycle.

Young's modulus and tensile strength were also affected by the wall thickness as was expected but by showing an opposing trend than that of presumed (i.e., the strength increases as wall thickness decreases). It seems that the amount of energy absorbed by materials (per unit weight or volume) in thin samples is greater than that in thicker ones (if they were moulded at the same set conditions).

3.3 *Flexural properties*

As shown by Figure 7, flexural modulus or strength was not affected much by the increase of molding time and the biggest variation of these are approximately 1.5% and 5%, respectively for five tested conditions. Figure 8 shows that both flexural modulus and strength increase with the molding temperature. The change of set mould temperature from 170 to 180 °C marked the sharpest increase in both parameters which may be explained by the typical melting temperature of PP which is close to 170°C. According to the composition tests, MEP54 contains of around 29% of PP by weight, hence a considerable portion of the material samples remain partially molten at 170°C and this fact was also confirmed by the data of DSC results as well.

Moulding pressure did not show much influence on both flexural modulus and strength (Figure 9). Flexural modulus marked a very slight decrease of around 2.2% (1419.87 to 1387.33 MPa) while the strength increased by 2.2%, (17.02 to 17.41 MPa) as moulding pressure changes from 5 to 15 MPa.

Hence, this again confirms that a moulding pressure of 5 MPa is adequate for forming MEP54 by compression.

As it is clear from Figure 10, flexural modulus and strength decreased by almost 16.57% and 30.46% respectively as chip size changed from 8 to 2 mm. The same trend was observed with Young's modulus and tensile strength as well, and the same explanation can be made here as well.

By following the same trend observed with Young's modulus and tensile strength, both flexural modulus and strength also increase quite significantly as sample thickness decreases (Figure 11). In particular, flexural modulus has increased by 14.4%, (1399.14 to 1600.64 MPa) while the strength has increased by 35.62% (17.07 to 23.15 MPa) as the sample thickness reduced from 3 to 1 mm.

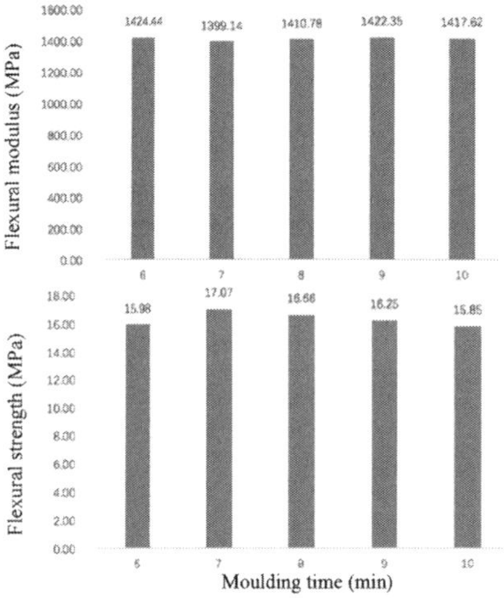

Figure 7. Flexural modulus and strength vs molding time.

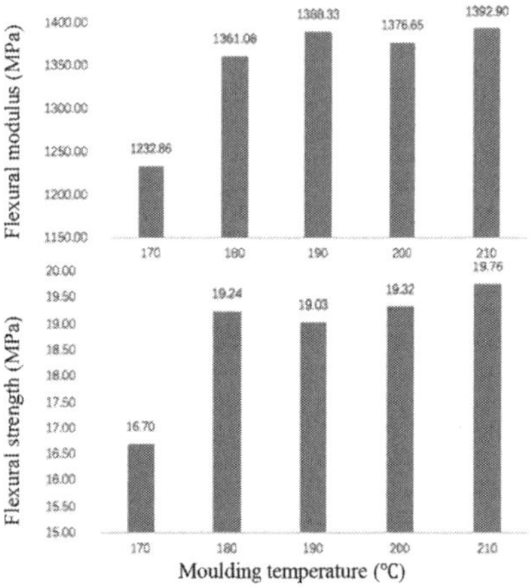

Figure 8. Flexural modulus and strength vs molding temperature.

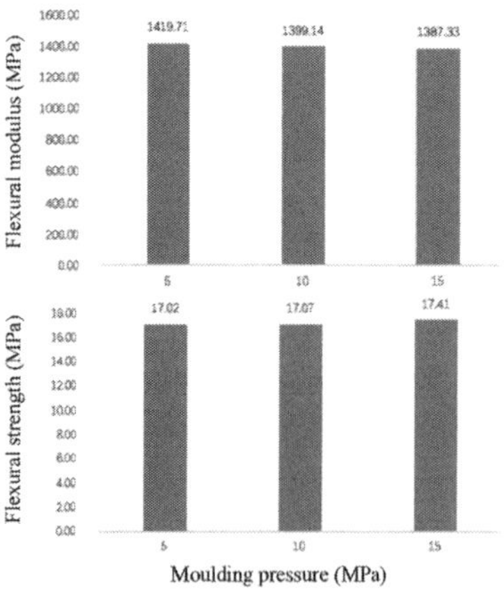

Figure 9. Flexural modulus and strength vs molding pressure.

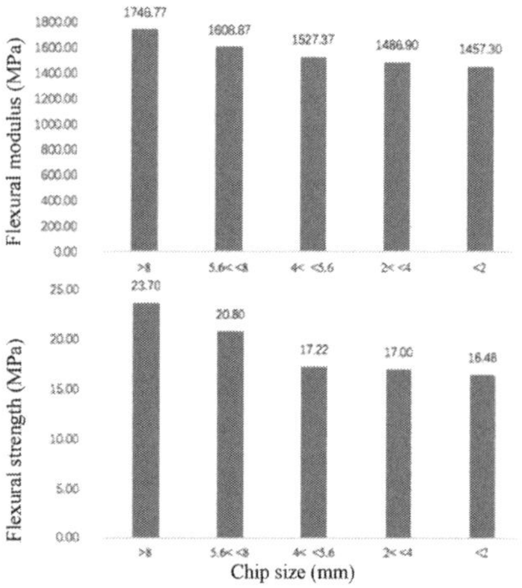

Figure 10. Flexural modulus and strength vs chip size.

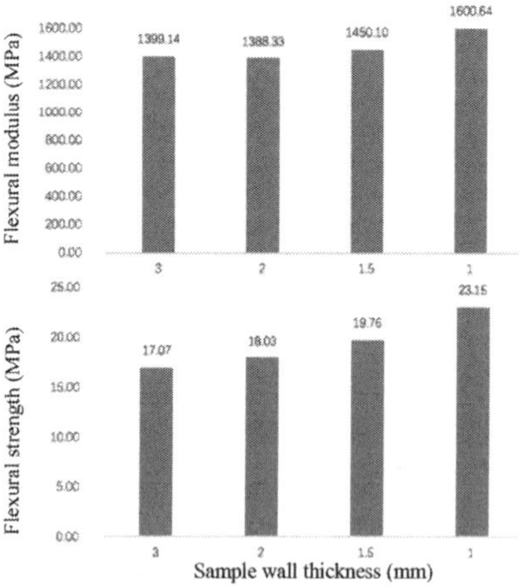

Figure 11. Flexural modulus and strength vs sample thickness.

3.4 *Rheological properties*

Typically, rheological properties of the polymers are quite complex in nature and depends on shear and viscoelasticity. In this study, two types' rheological tests were performed. Figure 12 shows the effect of increasing frequency of shear on the viscosity and storage/loss modulus of MEP54. The complex viscosity shows a decrease as frequency increases exhibiting a nature of a pseudoplastic fluid (shear thinning) and this can be explained as increasing shear force reduces the molecular chain entanglement for all constituents contained in MEP54, leading to a better fluidity of the blend.

The results of the temperature ramp test for MEP54 are shown in Figure 13 and it represents the variations of viscosity, modulus and tangent δ with the temperature. Clearly, two viscosity reduction points can be noticed for MEP54 at about 100 °C and 165 °C, which should be attributed to the melting of ABS (T_g of ABS $\approx$ 102 °C) and PP (T_m of PP $\approx$166 °C). Also, there is a significant reduction in δ at about 110 °C and this suggests a lower gain in the viscoelastic behavior, compared with the gain in the storage modulus. When the temperature is higher than 160 °C, two identical values of modulus are noticeable as δ is close to 1. Of the complex viscosity results, storage modulus (G') and loss modulus (G") for S1 and S2, G' is higher than G" in the low frequency region and it was reflected in the curves of δ as well. Moreover, the temperature ramp test of S1 was different from MEP54 and S2 as it kept reducing continuously with increasing temperature.

3.5 *Thermal properties*

Glass transition temperatures (T_g) of the six main constituents included within the three recycled materials used in this study (S1, S2 and MEP54) are given Table 1.

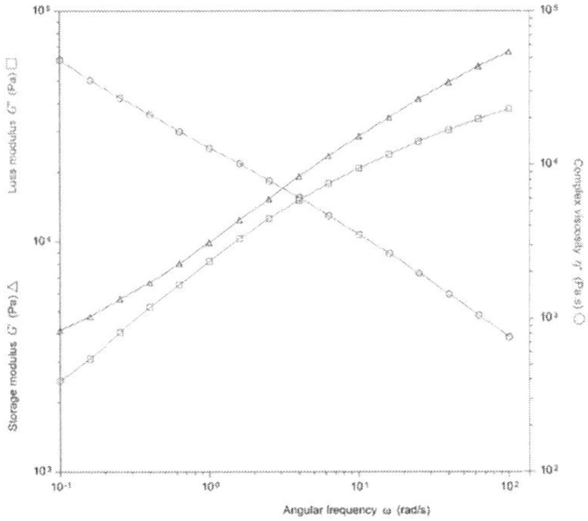

Figure 12. Modulus and viscosity vs angle frequency for MEP54.

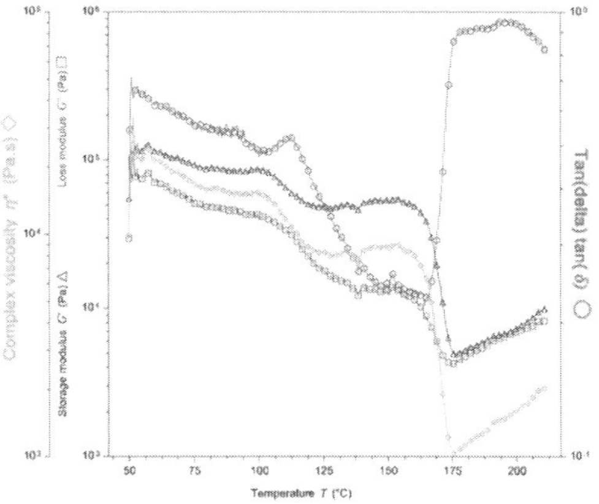

Figure 13. Modulus, viscosity and tangent (δ) vs angle frequency for MEP54.

Table 1. T_g of the constituents of 3 materials from DCS test.

	Sample No	T_g (°C)		Average T_g (°C)	
ABS	1	97.97		102.13	
	2	103.05			
	3	103.45			
	4	102.98			
	5	103.20			
HIPS	1	97.12		96.86	
	2	96.10			
	3	96.89			
	4	97.19			
	5	97.00			
Nylon	1	111.21		111.28	
	2	110.98			
	3	/			
	4	111.67			
	5	/			
PC/ABS	1	89.23	/	89.23	144.74
	2	84.68	/		
	3	/	143.50		
	4	87.67	/		
	5	87.82	145.98		
PP(with 20% Talc)	1	165.69		165.69	
	2	166.34			
	3	166.78			
	4	165.86			
	5	166.12			
PP-C	1	168.18		166.94	
	2	165.67			
	3	165.98			
	4	167.98			
	5	166.90			

For this also, all recycled materials from the same batch of recyclates were used and then at least five chips from each constituent were tested and results were averaged. Although the chips tested may not be recycled from one product, they showed only a little difference during DSC tests. There are two different T_g values for PC/ABS. Also, for samples No.3 and No.5 of PC/ABS, it is clear that there is another T_g around 145 °C. PC/ABS is a blend of PC and ABS and the T_g of PC is about 145 °C. Hence, if two polymer types are not mingled well in the blend, a T_g of around 145 °C for a PC/ABS blend is justifiable. As is clear, the amorphous polymers (ABS, HIPS, Nylon, and PC/ABS) marked T_g values lower than 100 °C. However, crystalline polymers (PP with 20% Talc and PP-C) cannot be melted temperatures below 170 °C. This result explains the sharp rise observed in tensile and flexural properties from 170 to 180°C as these materials may not be fully molten if the temperature is lesser than 170 °C.

4 CONCLUSIONS & FUTURE WORK

This study was aimed to study the nature and processing behavior of three different polymers under varying processing conditions. Test samples prepared via compression moulding were tested for mechanical, rheological and thermal properties followed by a SEM analysis as well. Mechanical testing showed that molding temperature, sample thickness and chip size were key parameters for tensile and flexural properties. Of all the samples tested, samples made from big chips (> 8 mm) moulded at 210 °C with 1 mm thickness exhibited the best mechanical performance. Moreover the results show that it is better to reduce the rubber content and impurities prior to moulding/processing, as the higher the rubber content the lower the mechanical performance of the samples. Meantime, rheological studies exhibited the behaviour of modulus, viscosity and tan δ in the range of different shear frequencies and temperatures where all the materials (MEP54, S1 and S2) exhibited the typical motion characteristics of a pseudo-plastic fluid. A DSC revealed the T_g values of different constituents of recycled polymer mix and it was noticeable that T_g of dominant constituents should be considered for selecting an appropriate set moulding temperature for better performance. Furthermore, SEM observations showed more nodules and craters on the surface of samples moulded in low set temperature conditions. Overall, the results highlighted the complexity of processing behavior and also the unpredictable nature of the properties of the recycled materials.

In the future, better separation methods should be used to separate materials. Adding the different constituents one by one to the mixture should help to discover the effect of different polymers on the properties of final material mixture. Also, it would be good to investigate the impact strength of recycled materials. Based on the observed tensile and flexural properties, the impact strength performance of these materials expected to be much more severe according to the relevant previous research but should be tested. The results from a DSC test can only show a T_g of single kind of polymer/constituent, not the T_g of a polymer blend, because the size required for DSC is too small. Thus the T_g of the blend should be obtained. Besides, it is better to investigate how some compatibilizers can improve the properties of these types of blends.

REFERENCES

Balakrishnan P. and Sreekala M. S., "Recycling of Plastics," *Recycl. Polym. Methods, Charact. Appl.*, pp. 115–139, 2016.

Bream C. E. and Hornsby P. R., "Comminuted thermoset recyclate as a reinforcing filler for thermoplastics: Part I characterisation of recyclate feedstocks," *J. Mater. Sci.*, vol. 36, no. 12, pp. 2965–2975, 2001.

Crawford C. B. and Quinn B., "Plastic production, waste and legislation," *Microplastic Pollut.*, vol. 306, pp. 39–56, 2017.

Eriksen M. K., Pivnenko K., Olsson M. E., and Astrup T. F., "Contamination in plastic recycling: Influence of metals on the quality of reprocessed plastic," *Waste Manage.*, 79, 595–606, 2018.

European Parliament and Council, "Directive 2000/53/EC on end-of-life vehicles," *Off. J. Eur. Union*, vol. L, no. 269, pp. 34–42, 2000.

Grigore M., "Methods of recycling, properties and applications of recycled thermoplastic polymers," *Recycling*, vol. 2, no. 4, p. 24, 2017.

Karaeen M., Hanieha A. A., AbdElall S., Sughayyer M., and Hasan A., "Concept model for the second life cycle of vehicles in Palestine", *Procedia Manufacturing*, 8, 707–714, 2017.

Pickering S. J., "Recycling technologies for thermoset composite materials," *Adv. Polym. Compos. Struct. Appl. Constr. ACIC 2004*, vol. 37, pp. 392–399, 2004.

Ragaert K., Delva L., and Geem K. V., "Mechanical and chemical recycling of solid plastic waste," Waste Management, 69, 24–58, 2017.

Scheirs J., Polymer recycling - science technology and applications. London: Wiley; 1998.

Shuaib N. A., "Energy efficient fibre reinforced composite recycling," *PhD Thesis*, The University of Manchester, 2016.

Skrifvars M. and Nyström B., "Material and energy recovery of composite waste by incineration," vol. 46, no. 0, 2001.

Tukker A., De Groot H., Simons I., and Wiegersma S., "Chemical Recycling of Plastics Waste (PVC and other resins)," *Delft, Netherland. TNO-*, pp. 7–107, 1999.

Turner A. and Filella M., "Bromine in plastic consumer products – Evidence for the widespread recycling of electronic waste," *Sci. Total Environ.*, vol 601–602, 374–379, 2017.

Wong Y. C., K. Al-Obaidi M., and Mahyuddin N., "Recycling of end-of-life vehicles (ELVs) for building products: Concept of processing framework from automotive to construction industries in Malaysia," *J. Cleaner Prod.*, 190, 285–302, 2018.

Yang Y., Boom R., Irion B., Van-Heerden D. J., Kuiper P., and de-Wit H., "Recycling of composite materials," *Chem. Eng. Process.*, vol. 51, pp. 53–68, 2012.

Industry 4.0 – Shaping The Future of The Digital World – da Silva Bartolo et al. (eds)

Ligno project: Development of composite waste material from wood and MDF applied to the design

G.R. Corrêa & M.L.A.C. de Castro
Federal University of Minas Gerais, Belo Horizonte, Brazil

J.C. Braga
CIAUD, Federal University of Uberlândia, Uberlândia, Brazil

F. Moreira da Silva
CIAUD, Lisbon School of Architecture, University of Lisbon, Lisbon, Portugal

ABSTRACT: This article presents the results of a research on the technological development of flexible material, from waste generated by the furniture industry. In a primary moment, a diagnosis was made on the generation of waste in 39 companies located in the metropolitan area of Belo Horizonte (MG, Brazil). Then, an innovative material was developed, which was tested and applied in the manufacture of products, incorporating design solutions. Procedures at this stage included: waste selection; development of the composite material; development of alternatives for the use of waste; design and monitoring of mold manufacturing; prototyping; manufacturing of finished products. As a result, it was possible to select a suitable sample for the material, systematize its production dynamics, verify its productive viability and generate alternative applications. The technology developed emerges as a sustainable possibility for a correct destination and, besides everything, profitability of these residues.

1 INTRODUCTION

The main residues generated by the wood production chain, through carpenters and furniture manufactures, are a direct consequence of the transformation of solid wood or reconstituted wood sheets: Medium Density Fiberboard (MDF), Medium Density Particleboard (MDP), Oriented Strand Board (OSB), plywood or agglomerated. The production of furniture in Brazil generates a large volume of wood residues and their by-products due to the cutting process itself and, mainly, by manufacturing waste.

At the furniture center of Belo Horizonte, there are 84 furniture manufacturing industries associated with the Furniture Industry Trade Union (Sindimov/MG). Each of these companies generates, on average, 300 tons of waste per month composed mainly of chips and powder of MDF and solid wood sawdust. Only 16% of the waste is correctly disposed of, that is, is incinerated, reprocessed or destined for licensed industrial landfills.

Although the National Solid Waste Policy (PNRS), which provides for the integrated management and management of solid waste, has been instituted in Brazil, most of the furniture industries still use waste in an irregular way. Residues resulting from the processing of reconstituted wood sheets may present some danger when they are not adequately disposed because of the presence of urea-formaldehyde in the composition and the high flammability of the material.

Fiksel (2009) considers the reduction of material consumption as a dematerialization strategy while Lovins et al. (2008) calls it as a resource productivity strategy. For Lovins, this strategy offers three significant advantages: it slows resource depletion, reduces pollution, and creates new job growth prospects. Therefore, the designer must rethink production systems so that less resources are used and the need for extraction is reduced, contributing to the minimization of the use of virgin raw materials (Leonard, 2011).

This paper presents the partial results of the Ligno Project, from which a composite material was developed as an alternative for the reuse of solid wood residues and MDF for the manufacturing of new products. The results have showed that, with design support, the composite is an alternative to reuse wood waste, to manufacture new products with complex shapes, to minimize the irregular disposal of waste, to generate economic gains for micro and small furniture industries, increase job generation and local community income.

1.1 *Overview of the furniture industry in Brazil*

The production chain of the furniture industry in Brazil is historically specialized in the production of articles made from wood, because of the abundance of raw materials of forest origin (Galinari et al. 2013). However, because of the environmental and legal restrictions to obtain solid wood, the manufacture of reconstituted wood panels, mainly of MDF produced from the reforestation wood (pine and eucalyptus), was expanded in Brazil.

There are 18 panel producing factories in Brazil and the country ranks 8th in the world ranking of the largest producers of reconstituted wood panels. According to the Brazilian Tree Industry (Ibá), in 2016 the production of reconstituted wood panels in Brazil was 7.3 million m^3, of which 86% was destined for the domestic market (Ibá, 2017).

The wood furniture manufacturers in Brazil correspond to approximately 17,500 industries (Galinari et al. 2013, Iemi 2016, Leão 2009). Most of these industries (74%) are micro-enterprises (up to 9 employees), and 21% of them are small enterprises (10 to 49 employees). The micro and small furniture industries employ low technology and produce high quality furniture, mainly residential (living room furniture, bedroom, kitchen, etc.), custom made and by ordering, using semi-handcrafted production processes (Abimóvel 2006).

1.1.1 *The problem of waste generated by industries*

According to the Ministry of Development, Industry and Foreign Trade, the solid waste in the furniture industry consists mainly of leftover materials used in the production of furniture (lumps, shavings and sawdust of solid wood and reconstituted wood). (Ministério do Desenvolvimento, Indústria e Comércio Exterior, 2002).

The diagnosis made by the Ligno Project on waste from furniture production, carried out in 39 carpenters in the city of Belo Horizonte, State of Minas Gerais, Brazil, pointed out that each industry generates, on average, 1,665 tons per year mainly composed of MDF (34%) and solid wood (17%). The study also found that more than half of these residues (53%) are incorrectly destined for vacant lots, for the lining of farms, for burning in ceramic kilns or burnt in the open (Corrêa et al. 2016).

Similar results were also found in the Triângulo Mineiro region (state of Minas Gerais, Brazil) where there are 800 formal and informal furniture companies. The diagnosis of the CACO project investigating 50 furniture industries in 2017 showed a high level of waste production: an average of 7.2 tons per year, per company. These companies also generated significant environmental liabilities, mainly because of the irregular disposal of waste on the peripheral areas of the cities (Braga et al. 2017, Moreira da Silva & Braga, 2018).

According to ABNT Standard NBR 10004:2004, furniture waste is considered to be Class II A waste - non-hazardous and non-inert, having characteristics of biodegradability, combustibility or water solubility (Associação Brasileira de Normas Técnicas 2004). However, waste from the furniture industry "may present some danger when they are not adequately disposed [...] especially if they are improperly stocked, close to industrial facilities or urban agglomerations" (Nahuz 2005, p.9).

Regarding the irregular open burning practice of MDF and MDP, Araújo (2012) reports that in this process there is formation of dioxins and furans, harmful to human health. Incineration, which is a more controlled process, can also generate toxic emissions if the process is not carried out with great control, requiring expensive equipment and filters. It is noteworthy that the urea-formaldehyde resin used for the manufacture of reconstituted wood panels is on the list of materials with carcinogenic potential by the International Agency for Research on Cancer (IARC).

The law 12,305 of August 2, 2010 establishes the National Solid Waste Policy (PNRS), which presents guidelines related to integrated management and solid waste management in Brazil. Meanwhile, as Nahuz argues, with rare exceptions, the furniture industry in Brazil does not have or apply permanent environmental conservation programs or integrated waste management plans (Nahuz 2005).

2 REVIEW OF LITERATURE

2.1 *Existing alternatives for the proper use and disposal of wood waste*

Among the principles of the National Solid Waste Law are the shared responsibility for the product life cycle and the recognition of reusable and recyclable solid reusable as an economic good and of social value, generator of work and income and promoter of citizenship. The change (Fiksel 2009; Jeswiet & Hauschild 2005) in industry behavior, encouraged by sustainable product design which can be very economically advantageous, is a strong connection between environmental and financial progress.

In this sense, Nahuz (2005) underlines three viable alternatives for the residues generated by the furniture industry. The first one would be the controlled burning of waste from wood and by-products, with the objective of generating electricity and steam, which demands the application of high technology and high cost. A second alternative involves chemical processes for processing wood residues into other products, which may include organic compounds for agricultural use (provided that it is solid wood without chemical treatment) and the development of composites to be used as raw material in the same furniture industry. The third alternative would be the implementation of a waste management program among its participants (Nahuz 2005). Currently there are initiatives and studies that cover some of these alternatives in an attempt to reintroduce the furniture residues in the production chain.

Hexion (2010), in patent document US7803855B2 (United States of America) presents the development of a composite made from the blend of phenol formaldehyde (PF), polyvinyl ester (PVA), animal or vegetable protein and wood. The aim of the technology is to improve the properties of the material with respect to the stability and strength of internal bonds, to lower the levels of the phenol formaldehyde (PF) release in the production of the plaques and to withdraw the urea in the confection processes. The processing of this composite involves the following steps: a) mixing the resin formed by phenol formaldehyde (PF), polyvinyl ester (PVA) and soy protein (vegetable origin); b) application of the resin to the wood, by spraying and mixing the elements; and c) the pieces of wood coated with the resin are released in-to their matrix-containing matrices and are then pressed at a temperature of 140° to 170° C.

Closer to the objectives of this research, Ciatec (2013) developed, as shown in the patent MX2011008552, a composite with the solid wood residue and PET polymer resin. The processing of this composite contemplates the following steps: verification of the materials with respect to their consistency and humidity; weighing polymer and sawdust; the components are mixed at high temperature (240-265 ° C) in the double screw extrusion process to ensure homogeneous form; and the compound is transferred to a mold at high temperature, to be compressed in a press, resulting in the final product.

The São Paulo Research Foundation (2009), in patent PI 0802175-9 A2, describes the use of wood residue flour, a by-product of non-impregnated wood, Pinus Caribe type, Pinus Elliott, eucalyptus, together with the MDF (Medium Density Fiberboard) residue, and its composite with virgin thermoplastics, post-consumer and industrial trimmings. The present invention demonstrates the development of manufacturing processes of polymer concentrates of wood-MDF composites used for the achievement of injected components and the extrusion of profiles from these composites in geometries pre-defined by the matrices.

As explained in the cited works, it is evident the viability of the reintegration of the residues produced by the furniture industry. However, the cases presented are incompatible with the reality of micro and small companies. Some factors contribute to this conclusion, highlighting the economic unfeasibility caused by the use of expensive production processes such as injection or extrusion and the difficulties of commercial exploitation of objects produced on a small scale. In addition to economic factors, it should be considered that the reintegration of waste into production processes or recycling should also minimize environmental and social impacts in relation to the final disposal of waste.

3 LIGNO PROJECT: MATERIAL COMPOSITE OF WOOD AND MDF RESIDUE APPLIED TO DESIGN

The results of the diagnosis carried out at the woodworking in the city of Belo Horizonte showed that the companies surveyed generated large volumes of MDF and solid wood (Corrêa et al. 2004). Therefore, these materials were used as a raw material for the development of the composite material. As already seen, most of the Brazilian furniture industries (74%) are microenterprises (Abimóvel 2006). In this way, the use of transformation technologies compatible with the reality of these furniture companies, both from a productive and economic point of view, was prioritized with the objective of making possible the reuse of wood and MDF residues by the woodworking companies themselves.

Considering the above, from the Ligno Project, a composite material was developed as an alternative for the reuse of solid wood residues and MDF. The Ligno also promoted the development of the design of two products for experimentation of the composite material. The manufacturing process of the products occurs by thermoforming in a closed matrix and the morphology of the products is exactly the same as the counterform of the matrix used. In this way, it is possible to obtain complex shapes for the application in products of high added value, without the need for further machining or treatment. The adhesive used is polyvinyl acetate (PVA), because it is economically more accessible, environmentally suitable and not widely exploited in this type of application.

Experiments of composite application in the design of products evidenced that the new material is an alternative for the reuse of the furniture residues; to manufacture new products; to postpone the early disposal of waste; and to generate an increase in income for both micro and small furniture industries and the local community.

3.1 *Materials and methods*

The tests performed for the composite development and application of the material in the design of products fulfilled the following sequence of steps:

a) Residue selection: MDF residues were grouped/sieved dry using a suspended vibrating sieve in order to match the particle sizes. b) Development of the composite material: the manufacturing of the composite slabs for the removal of the specimens utilized two groups of the MDF residue (particle size> 0.106mm and <0.106mm), and a group of the Peroba residue (particle size> 2.36 mm), which represented more quantity in the grading and two types of adhesive PVA and Vegetable Polyurethane. Firstly, as an experimental plan, the tests were performed in an open-plate heated plate press, then made the necessary investment in the manufacture of the molds to produce the products and the plate for the removal of the specimens. The initial pressing tests used 12 treatments as presented in Table 1.

The procedures for the manufacture of the plates included: weighing the residue and the adhesive; coating, ie spraying the adhesive on the residue with a paint spray gun and mixing the dough to homogenize the material; weighing the quantity, to reach the density of 650 kg/m3; col-soil formation in the press

Table 1. Specification of each treatment, demonstrating the type of residue, particle size, adhesive used and percentage.

Treat-ment	residue	size	adhesive	%
1	MDF	>0,106	PVA	20
2	MDF	>0,106	Vegetal polyurethane	20
3	MDF	<0,106	PVA	20
4	MDF	<0,106	Vegetal polyurethane	20
5	Peroba	>2,36	PVA	20
6	Peroba	>2,36	Vegetal polyurethane	20
7	MDF	>0,106	PVA	40
8	MDF	>0,106	Vegetal polyurethane	40
9	MDF	<0,106	PVA	40
10	MDF	<0,106	Vegetal polyurethane	40
11	Peroba	>2,36	PVA	40
12	Peroba	>2,36	Vegetal polyurethane	40

(Figure 1); Pressing: 6 minutes with 12 tons of pressure at 190 ° C. All plates were pressed with 6mm thickness and around 220mm width and length. The heated plate press was used at this time to avoid an initial investment in a specific mold.

c) Physical-mechanical tests for characterization of the material: after the composite development, static flexural strength tests were performed in accordance with ABNT NBR 14810 (Brazilian Association of Technical Standards 2013). For each treatment, five specimens measuring 6mm x 50mm x 170mm were used. d) Development of product design for composite application: alternative products were developed for the application of the composite as a raw material (monomaterial). In this stage, drawings were used, three-dimensional models made by hand, virtual modeling of some pieces and also, for some design alternatives, 3D printing was done. After the selection of alternatives, the projects for the manufacture of three steel molds were elaborated. One cup-making mold, another bowl-making mold and a third plate-making mold, 170 x 170 x 6mm, on which specimens (SAMPLE) would be removed. e) Production of products: for each type of product, various tests were carried out varying the quantity and type of residue and adhesive, the time of pressing and the temperature applied. The composite pressing process was performed by means of closed matrices with temperature control of the mold parts (female and male) independently using a heating.

3.2 *Results*

The results of the static bending strength tests of the MDF and Peroba wood composite boards with open matrix use can be seen in Figure 2. It was verified that the plates with greater amount of adhesive (40%) had higher values for the resistance module (MOR); and there was great variability of the results due to the use of open matrix. Although it has been known since the beginning of the research that the use of PVA adhesive would be more economical and more environmentally friendly (as it is water based) when compared to vegetable polyurethane adhesive, there was a lack of

Figure 1. Formation of the press mattress.

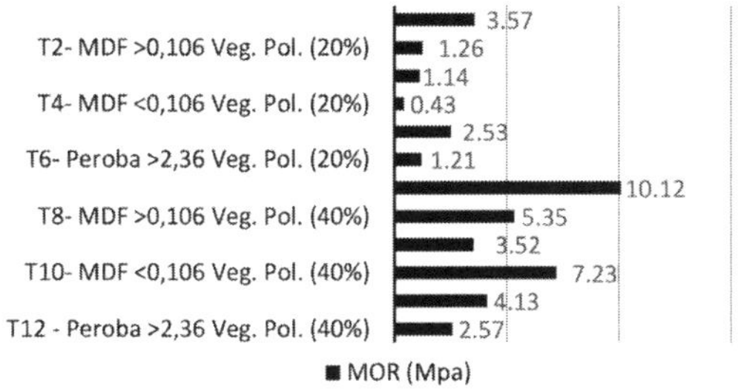

Figure 2. Result of the static flexural strength test of each of the twelve treatments using an open matrix.

knowledge about the technical feasibility of combining furniture waste with PVA and vegetable polyurethane adhesives. The tests performed with both resins showed that, in addition to the economic and environmental advantages, the use of PVA adhesive also presented better results of static flexural strength in all types and compositions of waste compared to the use of vegetable polyurethane adhesive. Given the above, it was decided to continue the tests using the PVA adhesive. It was observed that there was great variability of results, which may be related to the procedures used in the manufacture of plates without the use of mold, open matrix, although it was tried to establish the same procedure for the production of all plates.

New tests of resistance to static bending were performed in a flat and closed matrix with 170mm x 170mm x 6mm using MDF residues from group 1 (<0,106mm) and two groups of residues from another wood, the Jequitibá, group 1 (> 0,106mm) and group 2 (>2,36mm). For the purpose of comparison, the same tests were performed with a test piece of a MDF plate of national manufacturer of the same dimensions. The results can be observed in Figure 3.

The results of the static flexural strength tests performed on composite plates with closed matrix use were better and less discrepant than those performed with open matrix. The difference presented in the MOR results between the two tests was influenced exclusively by the change in the type of matrix used (open or closed) and not by the type of softwood (Peroba or Jequitibá). The composite developed, when pressed in a closed matrix, has approximate resistance of that presented by commercially available MDF boards.

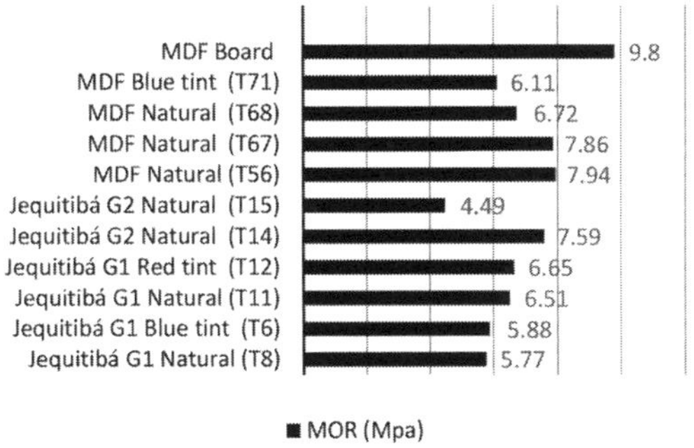

Figure 3. Result of the static flexural strength test of the composite MDF and Jequitibá wood with closed matrix.

Several products were developed during the generation of alternatives for composite application. The criteria for selection of products have prioritized the testing of the composite material and were limited by the budget available for the manufacture of molds. In view of the above, two products were selected for production: cup holder (Figure 4) and Bowl (Figure 5).

The experiments for the molding of the developed products allowed us to conclude that the pressing time, the amount of adhesive, the type of release agent and the temperature used should be different for each type of waste. For example, for the manufacture of products with MDF residue, the best

Figure 4. Cup holder produced with MDF waste composite from closed mold.

Figure 5. Bowl produced with composite of different residues: A) MDF residues; B) solid wood waste from Peroba; C) bowl assembled demonstrating the embedded parts; D) pieces with residue of MDF dyed of different colors.

results were obtained with 40% PVA adhesive in relation to the weight of the MDF residue, pressing time of 15 minutes in closed mold subjected to a temperature of 90 ° C and use of water based release agent. For the manufacture of the products with the solid wood residue (Peroba and Jequitibá), the best results were obtained with 30% PVA adhesive, 10 minutes pressing time in closed mold submitted to temperature of 120°C and use of an non-stick as a release agent.

4 CONCLUSIONS

The Ligno composite and its application for the development of new products is an alternative for the reuse of solid wood and MDF remains by the furniture micro-companies. The Ligno Project can be disseminated and transferred to the industries, which will be able to take advantage of its waste from the investment in arrays, some equipment and with the incorporation of the design. The application of furniture waste in new products allows the recycling of waste to be carried out by the carpenters themselves, since the technology used and the investment needed to transform waste into new products are compatible with the reality of furniture industries. The ease of technology transfer highlights the potential of the Ligno implementation and can be considered as the most significant contribution of this work in relation to the analogous solutions to the addressed problem.

The new products can generate economic gains for carpenters, and employment and income for the local community, since product design confers a high added value and allows the use of waste that would be discarded. The recycling process is always viewed with caution, since it demands the separation and transportation of materials and the process of transformation also consumes energy and new materials. However, the manufacturing process adopted for the composite makes it possible to use 100% of the raw material. This is because the production process does not demand use of machining and therefore does not generate new waste (zero waste). In addition, the conversion of furniture waste into new products carried out by the carpenters themselves avoids the transport of waste to other processing sites, extends the useful life of the material and puts it back on the market.

The results showed that there is significant contribution of this research to the area of Design due to the development of formal alternatives from the composite, generating new products with added value; to the area of Materials Engineering due to the potential of application for this new technology, and especially for the furniture sector, for the possibility of the use of MDF and solid wood residues.

The Ligno Project still has some limitations. Partial results showed that the composite material is an alternative for the recycling of furniture waste. However, it is still necessary to carry out the analysis of the life cycle of the compost and the products in order to verify if the recycling of furniture waste presents significant levels of environmental, social and economic improvement when compared to the destination of waste.

Due to the composite's ability to conform to complex shapes, future research should examine the economic and environmental feasibility for its use in substitution of virgin raw materials. In addition, other research should: develop new product design alternatives for its application; analyze suitable finishes for each type of product developed; perform moisture and degradability tests of the composite; and, mainly, to carry out the analysis of the life cycle of the company and the Ligno products.

The patent application for the Ligno composite was requested as well as the industrial protection for the developed products (configuration applied in bowls, configuration applied to the cup holder) and the application for registration of the Ligno brand.

ACKNOWLEDGMENT

This work was carried out with the support of the National Council for Scientific and Technological Development (CNPq), Brazil; the Research Centre for Architecture, Urbanism and Design (CIAUD), Lisbon School of Architecture, University of Lisbon and FCT, Foundation for the Science and Technology, Portugal.

REFERENCES

Abimóvel. 2006. Panorama do Setor Moveleiro no Brasil. www.abimóvel.org.br.

Araújo, G. M. G. de. 2012. *Desafios para aplicação da metodologia do berço-ao-berço ao ciclo de vida de móveis de MDF e MDP* (Unpublished master's thesis). Programa de Pós- graduação em engenharia urbana e ambiental, Rio de Janeiro: Brazil.

Associação Brasileira de Normas Técnicas. 2004. NBR 10004: Solid waste: classification. Rio de Janeiro, p. 71.

Associação Brasileira de Normas Técnicas. 2013. NBR 14810-2: Medium density particleboards Part 2: Requirements and test methods. Rio de Janeiro, p. 69.

Braga, J. C. & Moreira da Silva, F. & Ferrão, L. & Paschoarelli, L. C. 2017. Insertion of Sustainable Strategic Design in the Furnishings of Micro and Small Enterprises. *The International Journal of Designed Objects* (11) 3: 1–16.

Ciatec. 2013. *Proceso y producto obtenido a partir de una mezcla aserrin y polimero reciclado utilizando desperdicio del procesado de madera.* MX2011008552(A).

Corrêa, G. R. & Duarte, A. L. & Abreu, L. G. 2016. Resíduos da Indústria Moveleira: Diagnóstico das Empresas Associadas ao Sindmov-MG. In *Anais do 12º Congresso Brasileiro de Pesquisa e Desenvolvimento em Design; Belo Horizonte*, 4-7 October 2016. São Paulo: Blucher.

Fiksel, J. R. 2009. Design for environment a guide to sustainable product development. New York: McGraw-Hill.

Fundação de Amparo à Pesquisa do Estado de São Paulo. *Processo de obtenção de compósitos termoplásticos com resinas virgens ou recicladas reforçados com farinha de resíduo de madeira e/ou mdf e suas aplicações em produtos perfilados e componentes moldados por injeção para indústria moveleira, construção civil, decoração e paletes*. BR PI 0802175-9 A2.

Galinari, R. et al. 2013. A competitividade da indústria de móveis do Brasil: situação atual e perspectivas. *BNDES Setorial* (37)1: 227–272.

Ibá. 2017. Relatório anual 2017. Retrieved from http://iba.org/images/shared/Biblioteca/IBA_RelatorioAnual2017.pdf.

Iemi. 2016. Brazilian Furniture. Retrieved from http://www.brazilianfurniture.org.br/sobresetor.

Jeswiet, J., & Hauschild, M. 2005. EcoDesign and future environmental impacts. Materials & Design, 26(7), 629–634.

Leão, M. de S. 2009. Fatores de competitividade da industria de móveis de madeira no Brasil. Revista da madeira (1) 119: 4–11.

Leonard, A. 2011. The story of stuff: the impact of overconsumption on the planet, our communities, and our health-and a vision for change. New York: Free Press.

Lovins, L. H. & Lovins, A., & Hawken, P. 2008. Natural capitalism: creating the next industrial revolution. New York: Back Bay Books.

Ministério do Desenvolvimento, Indústria e Comércio Exterior. 2002. *Diagnóstico da cadeia produtiva madeira e móveis*. Brasília: IPT/MDIC.

Moreira da Silva, F. & Braga, J. C. 2018. Strategic Approach to Implement Sustainability in the Joineries of the City of Uberlândia, Brazil. In Broega et al (eds), *Reverse Design: A Current Scientific Vision from the International Fashion and Design Congress*: 365–372. London: Taylor & Francis.

Nahuz, M. A. R. Resíduos na indústria moveleira. In *3º Seminário de produtos sólidos de madeira de eucalipto e tecnologias emergentes para a indústria moveleira; Vitória, setember* 2005. São Paulo: IPT.

Hexion. 2010. *Wood composites, methods of production, and Eumethods of manufacture thereof.* US7803855B2.

Design education

Industry 4.0 – Shaping The Future of The Digital World – da Silva Bartolo et al. (eds)
© 2021 Taylor & Francis Group, London, ISBN 978-0-367-42272-1

A multi-platform virtual practice for education in chemical engineering

F. Yang & W. Li
University of Surrey, Guildford, UK

M. Heshmat
University of Surrey, Guildford, UK
Sohag University, Sohag, Egypt

E. Alpay, T. Chen & S. Gu
University of Surrey, Guildford, UK

ABSTRACT: Virtual reality (VR) and immersive technologies have become the frontier of research and development. Amongst the popular applications of these new technologies, training and education have become beneficial to many areas. Through the use of sophisticated design and methods, despite complex hands-on operations and laboratory environment, latest technologies can be applied to chemical engineering. In this paper, we describe our framework which develops a virtual chemical plant using VR, digital twin and robotic controls, as well as extend such framework to multiple platforms such as web browsers and mobile devices to boost its accessibility.

1 INTRODUCTION

During the last decade, the use and development of virtual reality (VR) and other immersive technologies has drastically increased (Rubio-Tamayo, Barrio, & Garcia, 2017) (Schofield, 2012) (Akbulut, Catal, & Yildiz, 2018). Initially led by the gaming industry, high-technology companies such as Facebook, Google and Microsoft have recognised the potential of VR in various aspects and have been investing in VR. Facebook's acquisition of Oculus VR for 2 billion dollars in 2014 provides a good example of this (Williams, 2015).

As technology develops and becomes more affordable, the popularity of VR will continue to rise and expand to more areas. Moreover, arguably one of its most recognisable strengths is in the area of training and education (Rubio-Tamayo, Barrio, & Garcia, 2017) (Schofield, 2012) (Akbulut, Catal, & Yildiz, 2018) (Jong, Linn, & Zacharia, 2013) (Zacharia & Constantinou, 2008) (Jaakkola, Nurmi, & Veermans, 2011) (Zhang & Linn, 2011) (Deslauriers & Wieman, 2011). There have been several reported cases where VR has improved traditional teaching. By offering immersive environments with full 3-D models, data visualisation has been made more intuitive and straightforward. We see these examples in teaching astronomy and celestial analysis (Chou, Hsu, & Yao, 1997), product design (Berg & Vance, 2016) and building-structural analysis (Chou, Hsu, & Yao, 1997).

Nevertheless, training and teaching in many aspects, particularly in chemical engineering, have been restricted. This is because hands-on experience, laboratory practice and instrument operation are some fundamental-learning blocks that have been challenging to digitalise. However, as technology develops, we recognise more opportunities for applying VR to education in chemical engineering. It is practically difficult to provide students full access to laboratories all the time, and some exercises such as operating chemical plants require full support and tutoring from teachers due to health and safety concerns. Because of the excessive requirements of human resources and costs, the number of hands-on exercises provided to students is limited.

VR technologies nowadays can simulate laboratory procedures and materials to offer similar physical hands-on experiments. They can be serious virtual games (Elliman, Loizou, & Loizides, 2016) which has some specific objectives, such as surgical skills training (Baby, Srivastav, Singh, Suri, & Banerjee, 2016), or enhancements to traditional education to neurosurgery (Pelargos, et al., 2017), dental courses (Mirghani, et al., 2018) and mechanical engineering (Hashemipour, Manesh, & Bal, 2011). Well-developed VR framework can even provide network connections suitable for multiplayer scenes to develop teamwork abilities. Therefore, VR technologies have become a viable alternative to the traditional teaching procedures of chemical engineering education (Ouyang, et al., 2018) (Probst & Reymond, 2018) (Perry & Bulatov, 2010) (Shin, Yoon, Lee, & Lee, 2002) (Cibulka, Mirtaheri, Nazir, Manca, & Komulainen, 2016) (Plesu, et al., 2004) (Ley, 2018) (Bell & Fogler, 2004). Besides, many trials and well-controlled comparisons state there are no differences between physical and virtual laboratories (Minner, Levy, & Century, 2010) (Linn, Lee, Tinker, Husic, & Chiu, 2006) (Wiesner & Lan, 2004) (Klahr, Triona, & Williams, 2007). At the same time, VR technologies provide additional virtues. For example, it mitigates safety concerns of many experiments (Buttussi & Chittaro, 2018), demonstrates unobservable processes (Zacharia & Constantinou, 2008) (Jaakkola, Nurmi, & Veermans, 2011), removes perplex details by altering reality and highlighting essential information (Trundle & Bell, 2010). As education providers, we may better record students' interactions via digital devices hence better diagnose and feedback to students (Linn, Lee, Tinker, Husic, & Chiu, 2006).

In this paper, we describe VR development for chemical engineering education application that exploits the latest technology in this area. In Section 2, we outline our framework in general and explain our aims. In Section 3, we describe in detail our implementation of each major stage within the framework. A conclusion with future work is presented in Section 4.

2 OVERVIEW OF OUR FRAMEWORK

There is an actual training chemical plant, namely the Fluor Pilot Plant, within the Department of Chemical and Process Engineering at the University of Surrey. It utilises the whole process to produce a very pure saline solution from a contaminated salt feed including chemical reaction, solids handling, filtration, heat exchange and CO2 capture by gas absorption. The basic aim of our work is to provide a virtual platform which replicates the Pilot Plant and to offer all of its usages and operations. This virtual plant will contain all instruments and operational as similar as possible to their physical counterparts.

The architecture of the virtual platform is displayed in Figure 1. The software development consists of four essential parts: 3-D model of the Pilot Plant, VR-enabled graphics rendering functionality, machine-human interactions, animation and conceptual design of the digital contents. These four essential parts form a basic VR environment allowing the user to participate in the training of using this chemical plant. Furthermore, we adjust reality to offer additional information which is a key benefit of VR mentioned in the previous section. For example, we make pipelines transparent, so flows can be observed. The final goal is to enable the user to operate the chemical plant from start to finish fluently within the immersive environment. At the current stage, we develop individual scenarios to reflect some components of the whole usage of this chemical plant. For example, the introduction to its different process loops and training in the operation procedures of certain alarms.

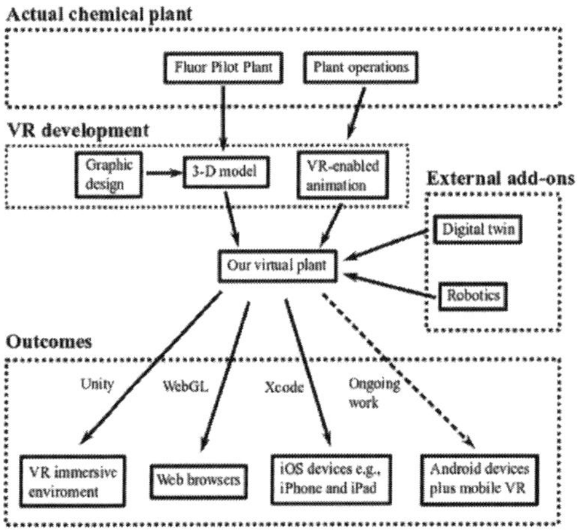

Figure 1. Architecture of the virtual platform.

There are two additional elements that we consider as novel. Firstly, we use a simulator to compute the chemical process in real time alongside the VR model. This simulator is sometimes termed a digital twin. The motivation for this feature is because only using computer animations for demonstrating purposes can be inaccurate and misleading. It is also very difficult to design such set of animations for dynamic control of a chemical plant where the number of situation and combinations of parameters can be enormous. On the other hand, software for process simulations of the chemical plant is well established, with very efficient and sophisticated commercial software available. Having a digital twin for the VR chemical plant is the solution forward. This digital twin will run and simulate the chemical process live and parameters will be connected to the VR system. Hence a change to the VR chemical plant, e.g., the user's manipulations of valves or buttons, will affect the chemical process simulations within the digital twin. These simulations will be computed in very short timeframes, typically within seconds, and we use the results of simulations to lead the changes in VR animations that should reflect such as changes in fluid flow, temperature and pressures.

The second novel addition to our framework is the use of robotics for monitoring and inspection. For future chemical plants, robots of various mechanisms will be inevitable, and we would like to investigate a framework forward which integrates teaching of how to control robots.

Robotics applications have much potential in chemical process engineering, for example, inspection tasks (Fukuda, 1990), manipulation (Lisca, Nyga, Balint-Benczedi, Langer, & Beetz, 2015), handling hazardous materials (Trevelyan, Kang, & Hamel, 2008), and environment monitoring (Bogue, 2011). With advanced technologies such as artificial intelligence, robotics, the present generation of students needs to be more prepared for technological changes than any before. The humans will share the workplace with robots. As a result, it will be very useful and interesting for the students to share their practical working time in the pilot plant with robots. The students will be much more enthusiastic about a robot in their studying environment. We should teach students how to interact with the robots at the same time we teach them how to operate on the chemical plant. A lot of attention has been given to human-robot interaction (Goodrich & Schultz, 2007), especially in a working environment (Green & Eklundh, 2003). Lewis, Sycara, & Walker (2018) identify that extensive training and education is required for those who work with robots. Which lead to a better and more comfortable human-robot interaction. In other words, students who experience robots (and robotics) in their studies, will find it much easier to interact with robots in the work environment. Currently, we use mobile robots with RGB-D camera for inspection. The development involves connecting our VR chemical plant to the robotic software using our internet servers.

Our main platform consists of a VR head-mounted display from HTC Vive, a high-end computer and latest Nvidia graphics processing units (GPUs). It is located at the Department of Chemical and Process Engineering and offers training in VR with its digital twin and robotic control. At the same time, having specific hardware associated with our development, we realise the issue of accessibility to its trainees. Therefore, we developed a web-browser and an iOS version of our framework, so it can be remotely accessed from anywhere and anytime. This flexibility is key to future education. We are also exploring the possibility to run our software on Android devices and offer VR experience via techniques such as Google Daydream on mobile devices.

3 IMPLEMENTATION DETAILS

In this section, we describe the details of our implementation of the five major components of our VR framework. Section 3.1 focuses on the development and design of 3-D models which recreates the Surrey Pilot Plant in the VR environment. Section 3.2 describes the design of human-machine interaction. Section 3.3 illustrates the techniques involved extending our high-end-desktop-based VR framework to different platforms. Section 3.4 has the details on the digital twin which dynamically runs simulations of the chemical process. Section 3.5 demonstrates controls over robotics.

3.1 *Our three-dimension model of the Surrey Pilot Plant*

The design and implementation of our 3-D model are done by using open-source software, namely Blender (Blender, 2019). The development involves drawing 3D shapes using points, planes, defining normal, creating texture, UV mapping etc. Apart from these typical design steps, the important issue is to understand the 3-D model will be rendered (i.e. draw by computer) at least 90 times per second to avoid VR-related motion sickness. This is a heavy computation burden even for the latest GPUs on the market. The situation intensities if one wish to extend this 3-D environment to platforms with less computational power, such as a web-browser or a mobile device. Hence, although there are plenty of detailed 3-D models, they will not be suitable for our purpose, simply because of the computational limit. Due to this reason, we developed our model from scratch within Blender. We pay particular attention to the size of each model and try to reduce as much as we can to save the computational requirement in the end when it comes to rendering and at the same time, trying to ensure its graphic quality. In Figure 2, we show the comparison of a photo of the real Surrey Pilot Plant (on the left) and the screenshot of our 3-D model (on the right).

Our model consists of 107,376 vertices and forms 219,621 faces in total. There are 214 separate objects in the scene. It occupies at least 99.51 megabytes in memory depends on its further relations with external graphical and physical engines.

3.2 *Unity-based interactive elements and virtual reality development*

Unity is commercial software and can be freely used for education and research purposes (Unity, 2019). It offers a fully integrated, cross-platform virtual reality engine and the necessary rendering system for stereoscopic equipment. The rendering system is based on Direct3D, OpenGL and OpenGL for Embedded Systems. Unity is open to customisations and we adopt to its C# scripting style to fully implement our interaction design. Typical controls such as rotating valves, pressing buttons can be done easily. Furthermore, employing Unity's particle system, we develop flows of various materials and use animations to recreate the internal sceneries inside a chemical plant and its piping network. The user's control commands such as opening a valve will have direct effects on the related animations. We demonstrate our web-version of the framework at http://industry-sixth-sense.eps.surrey.ac.uk/WebGLBuild/.

Figure 2. Our 3-D model of the Fluor Pilot Plant. A photo of the real plant is on the left, and a screenshot of our 3-D model is dis-played on the right-hand side.

3.3 *Extensions to other digital platforms*

As illustrated in Section 2, we aim to develop an educational framework that can be accessed without geographical and temporal restrictions. Therefore, we adopt our framework for multiple platforms. Web-browsers have become a fundamental tool for people to access information, we employ a technique called WebGL (WebGL, 2019) to extend our framework to web browsers. WebGL is a JavaScript API for rendering interactive 3D and 2D graphics within compatible web browsers without additional plug-ins. The best supported brand is Firefox, but it has claimed others like Google Chrome Safari and IE are also suitable. Due to the nature of web browsers, the graphical output will be limited to computer monitors and controls will be via keyboard and mouse. However, this is compensated by its wide accessibility and popularity. Hence, we believe it is a great tool to establish remote teaching and learning procedures. We include a screenshot of our web-browser version of Surrey Pilot Plant in Firefox in Figure 3.

Due to the rising popularity of mobile devices and smartphones in recent years, we feel opportunistic to extending our framework. Recent development shows the possibility to run our framework on iOS devices such as iPhones and iPads via Unity and Xcode (Xcode, 2019). We include a screenshot of our work on an iPad in Figure 4. We are actively looking at extending it to Android devices and making use of its full functionalities such as VR via Google Daydream, for example.

Figure 3. Our Surrey Pilot Plant framework in Firefox.

Figure 4. Our Surrey Pilot Plant framework on an iPad.

3.4 *Chemical simulation from Aspen Plus as a digital twin*

We described the motivation of having a digital twin in our framework in Section 2. Here we illustrate that our choice, namely Aspen Plus and Aspen Plus Dynamics (Aspen Plus, 2019), are software that can perform mass and energy balance calculations for dynamic process simulation created by Aspen Technology. It provides a large thermodynamic database, multiple thermodynamic methods, flexible and advanced unit operation modules and a wide range of applications including electrolyte system simulation (Lam, Klemes, Kravanja, & Varbanov, 2011). Rigorous and shortcut models are available for simulating key equipment such as packed absorption columns and heat exchanger, depending on your requirement on the computational cost which the tutorial effectiveness depends largely on. It makes an ideal tool software for dynamic simulation of the Surrey Pilot Plant.

3.5 *Robotics for inspection*

The robot investigated in this study is a Pioneer 3-AT robot for visual inspection of a chemical plant. This robot is a four-wheel-drive robot which is equipped with a SICK LMS200 laser range finder and a RealSense D435i camera. The laser range finder used for obstacles avoidance and navigation tasks. The D435i camera implements a stereovision depth imaging and 6 degrees of freedom inertial measurement unit. Besides executing the commands from the VR system, the robot should patrol around the chemical plant, performing the routine inspection tasks.

The software used to control this robot is the Robot Operating System, ROS (Quigley, et al., 2009). ROS is an open-source, meta-operating system for generic robots. It provides hardware abstraction, low-level device control, implementation of commonly-used functionality, message-passing between processes, and package management. It also provides tools and libraries for obtaining, building, writing and running code across multiple computers. The maneuverer stages of our robot include using the navigation stack available in ROS to build a map for the environment, then the robot will be able to dynamically and autonomously navigate around obstacles.

4 CONCLUSION

VR and immersive technologies have been a research focus as well as a growing product on the market. We have reviewed their influences on many areas and discussed particularly their usage and potentials in chemical engineering. AVR framework has then been presented to describe training on a chemical Pilot Plant located at the University of Surrey. This is combined with a digital twin which provides live simulation results for the dynamic chemical process. As a forward-planning scheme, we also include robotics in our framework that are able, for example, to carry out inspection duties via a camera feed.

As part of future work, we will expand the functionality of our framework to cover all aspects of the operations of our chemical plant. Currently, we are developing individual scenarios to reflect parts of the process operation. Extensive evaluation of this innovative learning technology will then be undertaken to explore both the student and teacher experience relative to traditional plant operation teaching methods. As mentioned above, the framework will also be expanded to other platforms such as mobile Android devices.

ACKNOWLEDGEMENT

Yang, Heshmat and Gu are supported by the Engineering and Physical Sciences Research Council Grant (EP/R001588/1).

REFERENCES

Akbulut, A., Catal, C., & Yildiz, B. (2018). On the effectiveness of virtual reality in the education of software engineering. Computer Applications in Engineering Education, 26(4), 918–927.

Aspen Plus. (2019, 1 18). Retrieved from Aspen Tech: https://www.aspentech.com/en/products/engineering/aspen-plus

Baby, B., Srivastav, V., Singh, R., Suri, A., & Banerjee, S. (2016). Serious games: An overview of the game designing factors and their application in surgical skills training. In M. N. Hoda (Ed.), 3rd International Conference on Computing for Sustainable Global Development INDIACom (pp. 2564–2569). New Delhi: BVICAM.

Bell, J. T., & Fogler, H. S. (2004). The application of virtual reality to chemical engineering education. IEEE Virtual Reality (pp. 217–218). Chicago, USA: IEEE.

Berg, L. P., & Vance, J. M. (2016). Industry use of virtual reality in product design and manufacturing: a survey. Virtual Reality, 21(1), 1–17.

Blender. (2019, 1 18). Retrieved from Blender Homepage: https://www.blender.org/.

Bogue, R. (2011). Robots for monitoring the environment. Industrial Robot: An International Journal, 38(6), 560–566.

Buttussi, F., & Chittaro, L. (2018). Effects of different types of virtual reality display on presence and learning in a safety training scenario. IEEE Transactions on Visualization and Computer Graphics, 24(2), 1063–1076.

Chou, C., Hsu, H. L., & Yao, Y. S. (1997). Construction of a virtual reality learning environment for teaching structural analysis. Computer Applications in Engineering Education, 5(4), 223–230.

Cibulka, J., Mirtaheri, P., Nazir, S., Manca, D., & Komulainen, T. M. (2016). Virtual reality simulators in the process industry: a review of existing systems and the way towards ETS. The 9th EUROSIM and the 57th SIMS (pp. 495–502). Oulu, Finland: EUROSIM & SIMS.

Deslauriers, L., & Wieman, C. (2011). Learning and retention of quantum concepts with different teaching methods. Physical Review Physics Education Research, 7, 010101.

Elliman, J., Loizou, M., & Loizides, F. (2016). Virtual reality simulation training for student nurse education. 8th International Conference on Games and Virtual Worlds for Serious Applications VS-Games (pp. 1–2). Barcelona, Spain: IEEE.

Fukuda, T. (1990). Plant inspection and maintenance robots. Advanced Robotics, 4(2), 169–178.

Goodrich, M. A., & Schultz, A. C. (2007). Human-robot interaction: a survey. Foundations and Trends in Human-Computer Interaction, 1(3), 203–275.

Green, A., & Eklundh, K. S. (2003). Designing for learnability in human-robot communication. IEEE Transcations on Industrial Electronics, 50(4), 644–650.

Hashemipour, M., Manesh, H. F., & Bal, M. (2011). A modular virtual reality system for engineering laboratory education. Computer Applications in Engineering Education, 19(2), 305–314.

Jaakkola, T., Nurmi, S., & Veermans, K. (2011). A comparison of students' conceptual understanding of electric circuits in simulation only and simulation-laboratory contexts. Journal of Research in Science Teaching, 48(1), 71–93.

Jong, T., Linn, M. C., & Zacharia, Z. C. (2013). Physical and Virtual Laboratories in Science and Engineering Education. Science, 340(6130), 305–308.

Klahr, D., Triona, L. M., & Williams, C. (2007). Hands on what? the relative effectiveness of physical versus virtual materials in an engineering design project by middle school children. Journal of Research in Science Teaching, 44(1), 183–203.

Lam, H. L., Klemes, J. J., Kravanja, Z., & Varbanov, P. S. (2011). Software tools overview: process integration, modelling and optimisation for energy saving and pollution reduction. Asia-Pacific Journal of Chemical Engineering, 6(5), 696–712.

Lewis, M., Sycara, K., & Walker, P. (2018). The role of trust in human-robot interaction. In H. Abbass, J. Scholz, & D. Reid (Ed.), Foundations of Trusted Autonomy. Studies in Systems, Decision and Control. Springer, Cham.

Ley, S. V. (2018). The engineering of chemical synthesis: humans and machines working in harmony. Angewandte Chemie, 57(19), 5182–5183.

Linn, M. C., Lee, H.-S., Tinker, R., Husic, F., & Chiu, J. L. (2006). Teaching and assessing knowledge integration in science. Science, 313(5790), 1049–1050.

Lisca, G., Nyga, D., Balint-Benczedi, F., Langer, H., & Beetz, M. (2015). Towards robots conducting chemical experiments. 2015 International Conference on Intelligent Robots and Systems (pp. 5202–5208). Hamburg: IEEE/RSJ.

Minner, D. D., Levy, A. J., & Century, J. (2010). Inquiry-based science instruction-what is it and does it matter? results from a research synthesis years 1984 to 2002. Journal of Research in Science Teaching, 47(4), 474–496.

Mirghani, I., Mushtag, F., Allsop, M. J., Al-Saud, L. M., Tickhill, N., Potter, C., Manogue, M. (2018). Capturing differences in dental training using a virtual reality simulator. European Journal of Dental Education, 22(1), 67–71.

Ouyang, S.-G., Wang, G., Yao, J.-Y., Zhu, G.-H.-W., Liu, Z.-Y., & Feng, C. (2018). A Unity3D-based interactive three-dimensional virtual practice platform for chemical engineering. Computer Applications in Engineering Education, 26(1), 91–100.

Pelargos, P. E., Nagasawa, D. T., Lagman, C., Tenn, S., Demos, J. V., Lee, S. J., Yang, I. (2017). Utilizing virtual and augmented reality for educational and clinical enhancements in neurosurgery. Journal of Clinical Neuroscience, 35, 1–4.

Perry, S. J., & Bulatov, I. (2010). The influence of new tools in virtual learning environments on the teaching and learning process in chemical engineering. Chemical Engineering Transcations, 21, 1051–1056.

Plesu, V., Isopescu, R., Onofrei, R., Josceanu, A. M., Dima, R., & Morcov, S. (2004). Development of an e-learning system for chemical engineering education. International Conference on Engineering Education and Research (pp. 1251–1256). Ostrava: TSB-TUO.

Probst, D., & Reymond, J.-L. (2018). Exploring drugbank in virtual reality chemical space. Journal of Chemical Information and Modeling, 58(9), 1731–1735.

Quigley, M., Conley, K., Gerkey, B. P., Faust, J., Foote, T., Leibs, J., Ng, A. Y. (2009). ROS: an open-source robot operating system. ICRA Workshop on Open Source Software.

Rubio-Tamayo, J. L., Barrio, M. G., & Garcia, F. G. (2017). Immersive environments and virtual reality: systematic review and advances in communication, interaction and simulation. Multimodal Technologies, 1(21), 1–20.

Schofield, D. (2012). Experiences with virtual learning. Journal on Interactive Systems, 3(1), 18–26.

Shin, D., Yoon, E. S., Lee, K. Y., & Lee, E. S. (2002). A web-based, interactive virtual laboratory system for unit operations and process systems engineering education: issues design and implementation. Computers and Chemical Engineering, 26, 319–330.

Trevelyan, J. P., Kang, S. C., & Hamel, W. R. (2008). Robotics in hazardous applications. Springer Handbook of Robotics.

Trundle, K. C., & Bell, R. L. (2010). The use of a computer simulation to promote conceptual change: a quasi-experimental study. 54(4), 1078–1088.

Unity. (2019, 1 18). Retrieved from Unity3D: https://unity3d.com/.

WebGL. (2019, 1 18). Retrieved from Mozilla: https://developer.mozilla.org/en-US/docs/Web/API/WebGL_API.

Wiesner, T. F., & Lan, W. (2004). Comparison of student learning in physical and simulated unit operations experiments. Journal of Engineering Education, 93(3), 195–204.

Williams, A. (2015). Reality check [virtual reality technology]. Engineering & Technology, 10(2), 52–55.

Xcode. (2019, 1 18). Retrieved from Apple: https://developer.apple.com/xcode/.

Zacharia, Z. C., & Constantinou, C. P. (2008). Comparing the influence of physical and virtual manipulatives in the context of the physics by inquiry curriculum: the case of undergraduate students' conceptual understanding of heat and temperature. American Journal of Physics, 76(4-5), 425–430.

Zhang, Z. H., & Linn, M. C. (2011). Can generating representations enhance learning with dynamic virualizations? Journal of Research in Science Teaching, 48(10), 1177–1198.

Industry 4.0 – Shaping The Future of The Digital World – da Silva Bartolo et al. (eds)
© 2021 Taylor & Francis Group, London, ISBN 978-0-367-42272-1

Can Digital Fabrication education affect the South African Science, Technology and innovation objectives through curriculum infiltration?

S.P Havenga, P.J.M van Tonder & R.I. Campbell
Department of Technology Transfer and Innovation, Vaal University of Technology, Vanderbijlpark, South Africa

D.J de Beer
DSI Chair: Innovation and Commercialisation of Additive Manufacturing, Centre for Rapid Prototyping and Manufacturing, Central University of Technology, Free State, Bloemfontein, South Africa

ABSTRACT: South African additive manufacturing is increasingly observed as an indispensable strategic component of INDUSTRIE 4.0 development. The South African government plans to implement this through its proposed white paper for Science, Technology & Innovation. Challenges identified highlights skills training, formal educational programs and implementation as developmental needs. Existing resources, syllabi and infrastructure for educational technology subjects are outdated. Substantial funding has been invested to provide infrastructure, but very limited focus was placed on formal advanced manufacturing education. This is aggravated by overburdened curricula, unskilled educators in technology and understaffed schools, creating time constraints. As part of an ongoing study, this paper investigates pedagogical solutions to address these limitations through the Idea2Product laboratories, which supply educators with support material and extensive training. Positive solutions are indicated to infiltrate present curricula without extensive disruption. Obstacles are also identified that may disrupt the assimilation of these science, technology & innovation objectives.

1 INTRODUCTION

1.1 *The state of AM education in SA (CAD)*

Additive Manufacturing (AM), also known as Digital Fabrication (DF), is identified as a key element in South Africa (SA) Industry Policy Action Plan (2017-2020), which identifies AM as a disruptive technology that can bring about fundamentally needed change (De Beer & Booysen 2018). This should include changes in current education system.

AM continues to grow according to the Wohlers report, with an estimated 6000 new three-dimensional printers (3DP) entering SA in 2017 alone. 90% of these were desktop 3DPs, which commonly, is used in the education sector (De Beer & Booysen 2018).

The SA Department of Science and Innovation (DSI), organized an international review on the newly launched Collaborative Programme on Additive Manufacturing (CPAM). Positive results were identified to the extent where further funding has been granted for the next three years.

Even though this is the case, extremely little documented formal pedagogical development and training interventions nor policy changes are taking place towards SA technology education. This can be seen in the DSI 10-year development program/plan (2008 to 2019), where it is noted that very little success was achieved to align with international implementations of the same nature (Department of Science and Technology 2017).

The SA National System of Innovation (NSI) however, report success between 1996 to 2016. According to the minister of Science and Technology: Recent reviews show that the SA-NSI has made significant progress between 1996 and 2016. She based this information on the notion that the STI landscape has grown and that technology orientated academic publications have expanded three-fold (Department of Science and Technology 2018). However, these academic publications mostly focus on tertiary education outputs and the maker spaces application of digital fabrication.

Where collaborations between educational entities and working groups/industry partners in SA take place, the focus still just remains on technology transfer in the form of information sharing, which is very similar to the USA Education Outreach Roadmap (Wohlers Associates 2018).

1.2 *Background to the study*

In the proposed DSI white paper for Science Technology and Innovation (STI), the minister of Science and Technology identifies challenges, stating that the NSI is not yet reaching its full potential through inclusivity. She mentions that the Fourth Industrial revolution is emerging in SA and that the risks and opportunities it holds should be addressed (Department of Science and Technology 2018). To remedy the above-mentioned lack of intervention a new

Idea2Product (I2P) training model was proposed to not only create technology transfer but to serve as a catalyst for formal pedagogical development.

These I2P laboratories were originally developed by Professor Deon de Beer at the Vaal University of Technology in 2011. The main driving force was to uplift the poor communities of the Southern Gauteng region, but also introduce AM principles through open-sourced (OS) technology transfer models. This was made available to anybody interested in product development (Havenga et al. 2018). One of the I2Ps main objectives includes the development of hybridized educational and infrastructure for the SA platform (De Beer & Campbell 2017).

To date, more than thirty I2P labs are reported to function globally, in countries such as SA, Singapore, New Zealand, Germany, Sweden, the UK and the US (Wohlers Associates 2018). An estimated 2500 students and public participants are serviced annually, through the original I2P laboratory at VUT. However, the training provided is unaccredited on the OS platform and it is marginalized through geographical limitations. Exposure is therefore limited to a broader audience and less technology learning spaces get involved. It is with all the above in mind that the I2P advanced model was created.

To address this technology-transfer obstacle, an accredited training model was developed but before an implementation pilot program could be executed the I2P had to identify and address the limitations in the SA education sector.

1.3 *Limitations*

Limited acceptance towards the adoption of AM has been observed in SA due to the perceived inflated costs, as well as a lack of knowledge and skills for emerging technology education, which is concurrent with recent research (Pacey 2013, Serlenga 2015 & Campbell et al. 2011). Similarly, to this idea, some research show that a lack of educational skills can present limitations to the adoption of technology in curriculums, which is exactly the case in SA (Ford & Minshall 2016).

A very important need to address is to overcome the lack of consistent access to technology-based education for school learners. Campbell and de Beer recommend that an AM roadmap be extended to promote the availability of digital technologies in pre-university education. Such an initiative will provide an early foundation for CAD and technology awareness, to build expertise in undergraduate and postgraduate students (De Beer & Campbell 2017). Through initiatives such as the I2P laboratories, this is done very successfully, however, no pedagogy development or infiltration has taken place in the FET education band. No formal 3D printing curriculum apart for the international Design for Additive Manufacturing course has taken place, which is not an accredited training program.

When the SA CAPS curriculum document was investigated, it was found that CAD is only presented as a choice module in the FET band and nothing appears in the GET (middle school) band (EGD CAPS 2011). To exaggerate this finding, educators are unskilled in the field of emerging technologies and schools are understaffed, which leads to even further time constraints. In such a system how can the envisioned implementation, as proposed by the DSI, be put in motion in an overburdened educational environment? A possible solution can be found, if the I2Ps advanced system infiltrates the existing curriculum and not just focus on extracurricular OS systems, as proposed by other research in the field (Schelly et al. 2015).

e aim of this research is therefore to develop a training model which will enhance educator skills, supply infrastructure to the education system, make use of the existing curriculum to develop digital fabrication infiltration (Department of Science and Technology 2018), and cautiously steer away from the keychain syndrome effect as seen in present technology education research (Blikstein 2013). Lastly, the paper will pose to align with the currently expected objectives set by the STI white paper (Department of Science and Technology 2018).

2 METHODOLOGY AND TRAINING MODELS

This paper is an introduction to a much larger study undertaken in SA. This study takes place through the I2P laboratory and is a collaboration between the Vaal University of Technology (VUT) and the Gauteng Department of Education (GDE). Due to the sensitive nature of the pilot study, the complete training model will not be disclosed in this paper.

The research is written from an interventional approach within a qualitative paradigm. The idea to infiltrate the current technology education system in SA had to be thoroughly investigated against the current trends seen around the globe (Freire 1974, Schelly et al. 2015, Blikstein 2013, Ford & Minshall 2016, Hedberg & Freebody 2007, Despeisse et al. 2016 and Verner & Merksamer 2015).

More importantly, a focus was placed on investigating emerging pedagogical models for technology. These models include Bliksteins' three-pillar model consisting of experiential learning, constructionism and critical thinking; Verners' CDIO (Conceiving, Design, Implement & Operate) system and Schellys' constructionist approach. The research relies heavily on emerging ideas of disruptive innovation, but without being a disruption in an overburdened and very fragile SA education ecosystem.

2.1 *PHASE 1: Creating a pilot program*

The appropriate educators were sought to negotiate the notion of intervention. To ensure compliance with the local department of education, collaboration and partnership were established with the GDE and a Memorandum of Understanding (MoU) was signed with the I2P at VUT. This partnership facilitated the sourcing of educators. The subject advisor of Engineering, Graphics & Design (EGD), Mr. Willem Goodchild identified seventy-two schools with appropriate educators. Of those seventy-two educators, thirty educators decided to partake in the pilot program. Of the thirty educators, eighteen completed all the workshop sets. Of the eighteen educators that completed the pilot workshops, another three left before phase one was completed due to personal undisclosed reasons.

Pre-workshops commencement and during the pilot study, Learner-Teacher-Support-Material (LTSM) was developed and incorporated into the EGD CAPS document. This led to the further development of an accredited 3DP training model program.

This pilot phase included the training of both theoretical and practical skills. Assessments were completed for both formative and summative testing. A five-day pilot workshop was hosted and followed by twelve one day workshops over six months.

2.2 *PHASE 2: Reflect and adapt*

During the workshops, the educators were informally interviewed, and an organic interventionist approach was used to adapt the workshops as needed. Data collected during the workshops and current literature on 3DP workshops were used to devise a reflective survey (Schelly et al 2015, Blikstein 2013 and Verner & Merksamer 2015). The respondents completed this survey anonymously online. The responses are reflected below and were used to see if such a model will be able to achieve the aim and objectives set out by the STI white paper.

3 RESULTS

Only fifty percent of the participants completed the pilot project, due to various obstacles. They include time constraints, work pressure, obligations, lack of skill development fear, logistical infrastructure to attend workshops, stigmatization of the project value and educators leaving schools for undisclosed reasons. Fifty percent of respondents completed the survey, based on the informal interviews conducted during the workshops and consisted of several open-ended questions given to all the educators and educational specialists after the workshops. The results are:

3.1 *What was learned?*

Place the most prominent feedback received was that educators learned practical 3D printing operations (Figure 1).

3.2 *Can educators use new knowledge acquired?*

Over ninety percent of educators specified that they will be able to apply what they learned in their classrooms post-workshop. Seven percent were cautious due to fear of time constraints and pressure to meet deadlines (Figure 2).

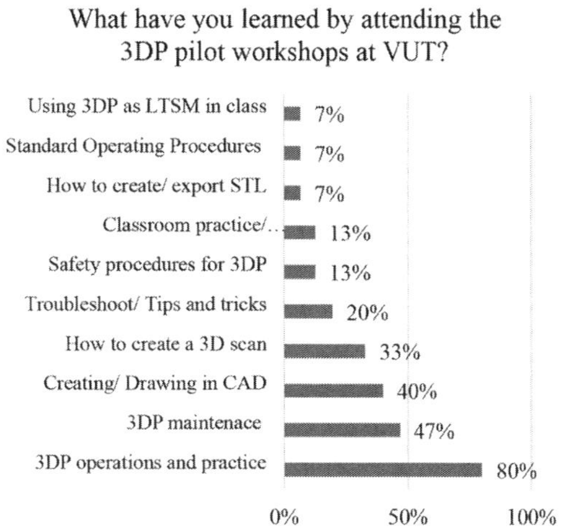

Figure 1. Skills acquired during the workshops in an ascending order of importance.

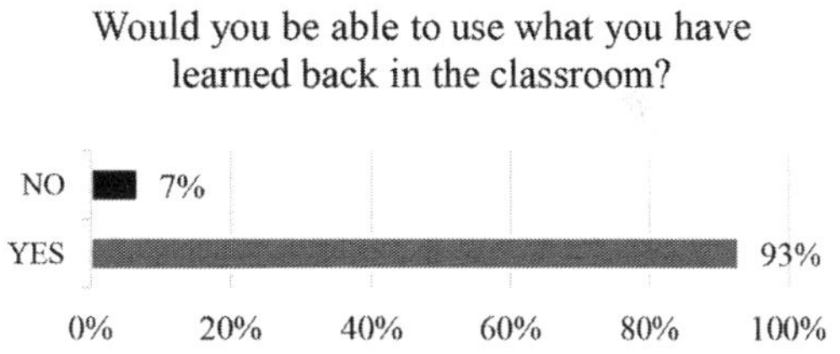

Figure 2. Most educators would be able to transfer 3DP skills.

3.3 *How can digital fabrication be used?*

Forty percent of the respondents decided that practical 3D printing is the best way to use Digital Fabrication in the classroom. The remaining sixty percent felt that 3D printing is best utilized in the classroom as a kinesthetic Learner-Teacher-Support-Material (Figure 3).

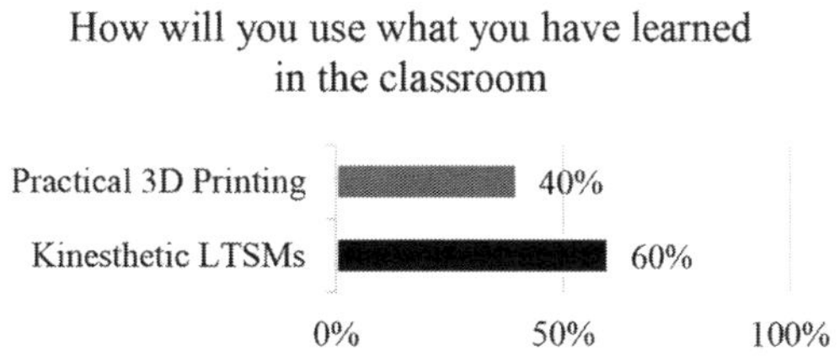

Figure 3. LTSMs are the preferred method of using 3DP in the classroom.

3.4 *Can digital fabrication benefit STEM?*

Ninety-three percent of the respondents agreed that Digital Fabrication will indefinitely benefit STEM education in SA on some or other level. Only seven percent was again reluctant due to fears of time constraints during the classroom time or that the technology might fail (Figure 4).

3.5 *Should the curriculum be updated?*

Ninety percent of the respondents believed the current Technology based subjects contain all the core knowledge needed, but that there is a lack of any emerging technologies and techniques seen in these subjects. Stigmatization surrounding the inflated costs of 3D printing and fear of not getting through the curriculum timeously was the reason for seven percent of the respondents not wanting the curriculum to be updated (Figure 5).

3.6 *Is the curriculum overburdened?*

Sixty percent of the respondents believed the curriculum is overburdened with outdated skills knowledge. This idea comes from technical tools and helping aids for the classroom seen as outdated. The remaining forty percent argued that limitations particular to 3rd world countries make the update of equipment difficult. Both sides, argue that the bulk of knowledge to be disseminated in the limited time available makes the system overburdened (Figure 6).

3.7 *Is time a factor?*

Respondents were apprehensive to link whether there is enough time during lessons to introduce new skills. Sixty-seven percent decided to not answer, while twenty percent agreed there is too little time. Only thirteen percent of the respondents argued there is enough time during the lesson time. The main argument for the latter came from subject area specialists that are CAD proficient or have already introduced Digital Fabrication to their learners (Figure 7).

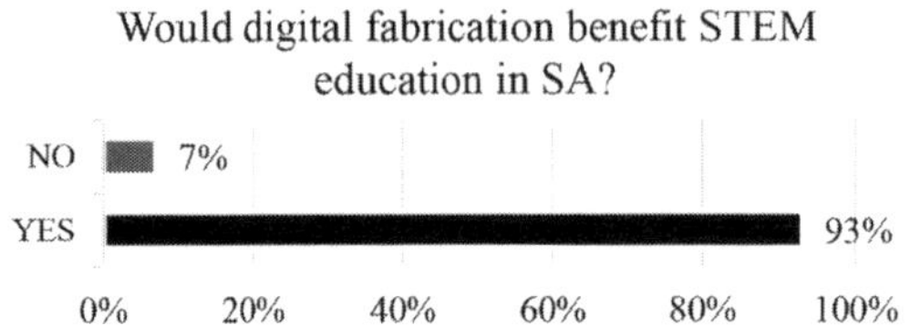

Figure 4. Educators feel Digital Fabrication will benefit STEM development in SA.

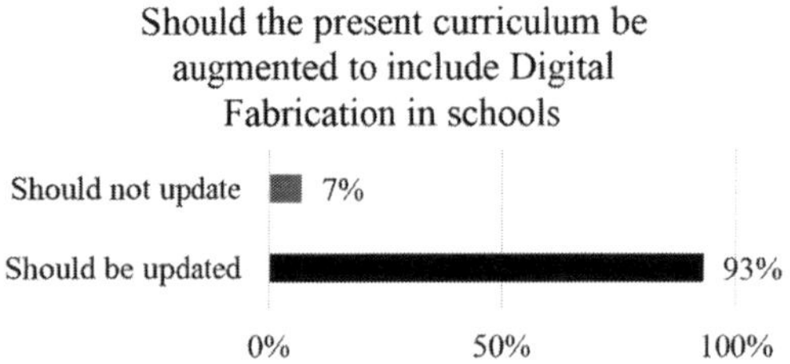

Figure 5. Curriculum update are indicated as priority.

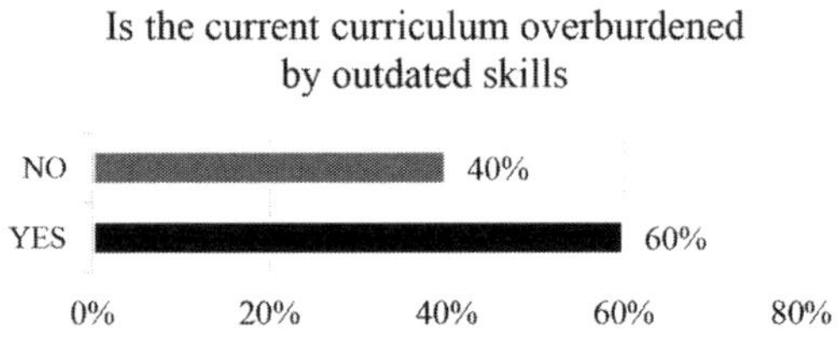

Figure 6. Educators are aware that the education system is overburdened.

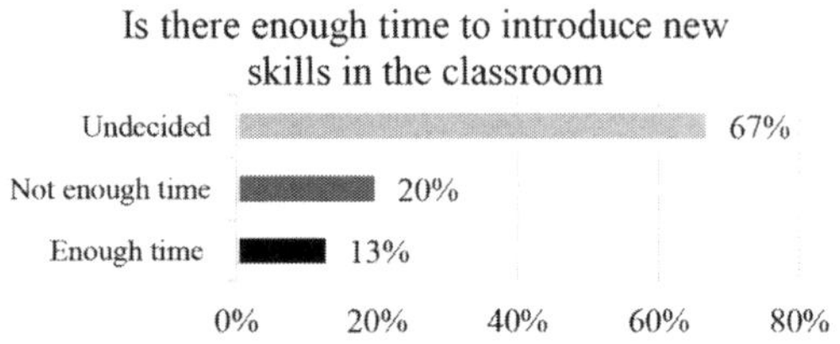

Figure 7. The appearance of time constraints visible when applying skills transfer.

3.8 *Who should use digital fabrication?*

Debates took place around the topic of participation and it was clear that respondents had strong opinions due to personal experience. Fifty-six percent of the respondents favored educators using Digital Fabrication in the classroom. The remaining forty-four percent wanted learners to experience 3D printing from a constructionist point of view, with objectives aligned with experiential learning (Figure 8).

3.9 *Can this influence learners?*

Many respondents had different ideas as to how learners will be affected by using LTSMs. Of

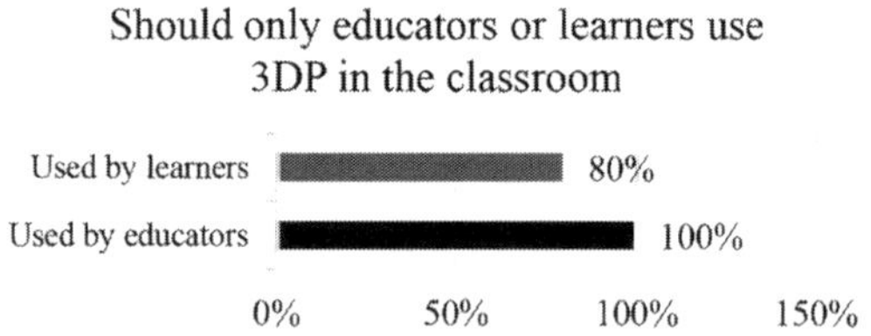

Figure 8. Respondents favored educators use 3DP in the classroom.

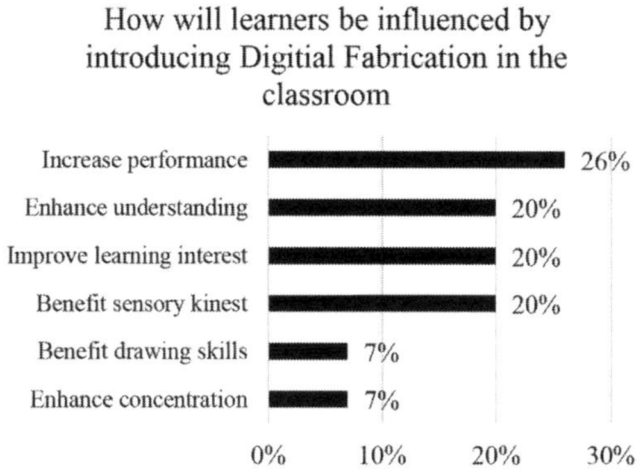

Figure 9. Educators feel learners will benefit most by introducing Digital Fabrication.

importance is that all respondents voted the influence to be positive. Twenty-six percent agreed that it will enhance learners' performance. Twenty percent felt it would improve learners' understanding, other twenty percent thought it would enhance learners' subject interest and a further twenty present discussed the benefits from a kinesthetic constructionist viewpoint. Drawing skills and enhanced concentration voted for only seven percent of the opinions (Figure 9).

3.10 *Open vs proprietary CAD?*

A surprising sixty percent of the respondents favored proprietary CAD software above the OS platforms. Due to the high cost of proprietary software packages, it was expected that respondents would rather make use of free CAD software, however, only twenty-seven percent chose this option. Thirteen percent of the respondents decided to not decide as they had very little knowledge of CAD systems (Figure 10).

3.11 *What next?*

Of all the respondent's eighty percent suggested more practical workshops to gain experience in the future. Seven percent of the respondents respectively wanted more schools to get involved, other seven percent pointed out more advanced CAD software be used and lastly six percent wanted other Digital fabrication technology to be included in the study (Figure 11).

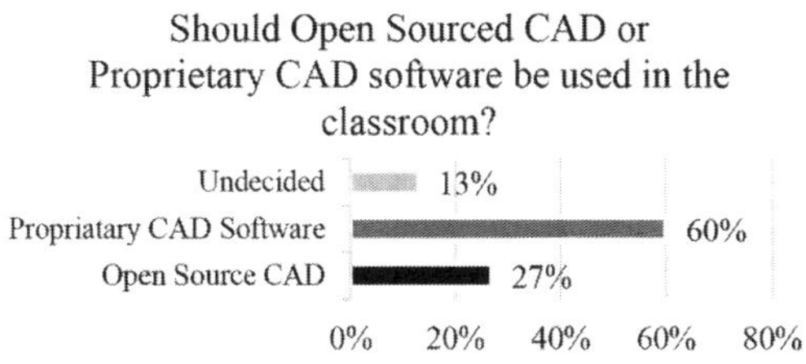

Figure 10. Educators favor proprietary CAD software, regardless of the high costs involved.

3.12 *What constraints and limitations can be expected?*

The major limitation according to the respondents are time constraints and the fear that 3D printing will take too much time in a classroom. This notion accounted for forty-seven percent of the respondents, while seven percent were cautious to introduce Digital Fabrication out of fear of new technology failure. The remaining forty-six percent were uncertain about what to expect (Figure 12).

3.13 *Conclusion:*

It is apparent that an advanced LTSM program will be able to infiltrate the current technology curriculum. Educators are favoring the system even though there is a margin of reluctance and uncertainty, as expected with a change in system.The full extent and impact of such an interventionist program can only become evident once the system has been piloted in FET schools. Educators will have to provide evidence of such a successful integration of Digital Fabrication and fed back to the GDE/DSI. More importantly is whether this curriculum infiltration system can address the current envisioned objectives of the DSI. The I2P labs are successfully implementing ten out of the fourteen objectives set by the DSI. They are in no order: The I2P focuses on inclusivity and transformation

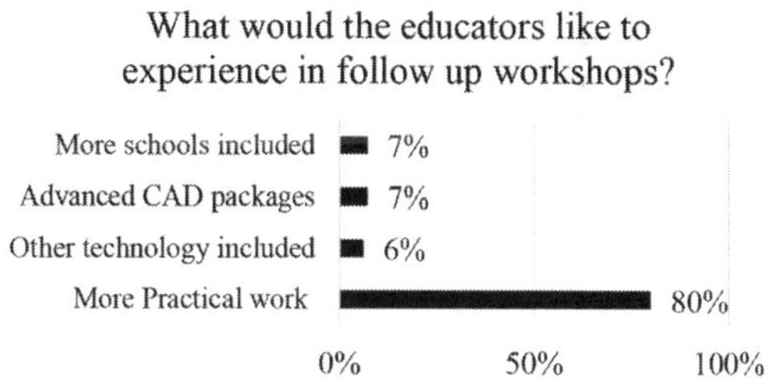

Figure 11. Educators indicate more practical experience exercise time during workshops.

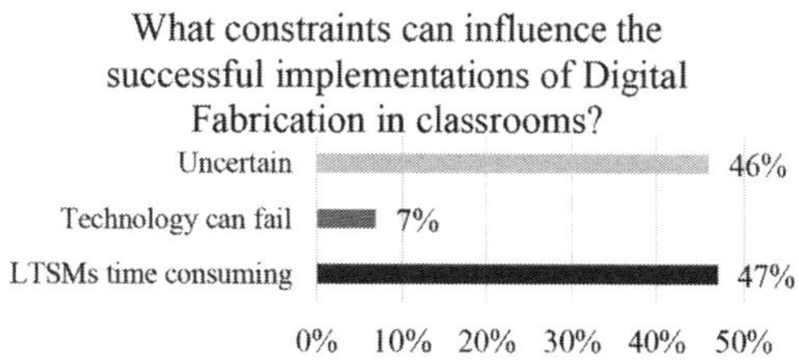

Figure 12. Almost 50% of the respondents' fear time as the prominent constraint.

of technology education. It enhances the innovative culture in society and government by performing technology transfer. It provides formalized monitoring and evaluation systems through its accredited training program by means of quality management and learning system. It continuously develops an environment of innovation and encourage start-up of SME's and entrepreneurs. Innovation ecosystems are nurtured through partnerships created with other higher education institutions on grassroot level. The I2P provide training that delivers human resource development, making use of innovation approaches. And, it encourages inter- and multidisciplinary knowledge collaborations through the training workshop systems. Funding, although not a focus of this paper, emerged as an ongoing obstacle for technology transfer. The DSI and other government institutions are making more funding available, certain impoverished schools still do not benefit. This is due to geographical, logistical and other environmental factors that need to be addressed in future research.

ACKNOWLEDGEMENTS

The authors would like to thank and acknowledge the continued support of the Southern Gauteng Science and Technology Park, the Gauteng Department of Education, The Vaal University of Technology and the Central University of Technology.

NOTE: The Department of Science and Technology (DST) has changed to the Department of Science and Innovation (DSI).

REFERENCES

Blikstein, P. 2013. Digital Fabrication and 'Making' in Education: The Democratization of Invention. In J. Walter-Herrmann & C. Büching (Eds.), FabLabs: Of Machines, Makers and Inventors. Bielefeld: Transcript Publishers.

Campbell, R.I., De Beer, D.J. & Pei, E. 2011. Additive manufacturing in South Africa: building on the foundations. Rapid Prototyping Journal, 17 (2), pp. 156–162.

De Beer, D.J. & Campbell, R.I. 2017. Using Idea-2-Product Labs ® as a strategy for accelerating technology transfer. International Journal of Technology Transfer and Commercialisation 14(3-4), pp.249–261.

De Beer, D.J. & Booysen, G. 2018. White Paper on Science, Technology & Innovation. Available from: www.dst.gov.za and www.gcis.gov.za. Accessed on 28 October 2018Wohlers Report, Wohlers Associates, INC. Accessed on 28 October 2018.

Department of Science and Technology. 2017. The Ten-Year Plan for Science and Technology. Available from: http://www.dst.gov.za/index.php/resource-center/strategies-and-reports/143-the-ten-year-plan-for-science-and-technology Accessed on 10 June 2017.

Department of Science and Technology. 2018. White Paper on Science, Technology & Innovation. Available from: www.dst.gov.za and www.gcis.gov.za. Accessed on 28 October 2018.

Despeisse, M., Baumers, M., Brown, P., Charnley, F., Ford, S.J., Garmulewicz, A., Knowles, S., Minshall, T., Mortara, L., Reed-Tsochas, F. & Rowley, J. 2016. Unlocking value for a circular economy through 3D printing: a research agenda. Technological Forecasting & Social Change, Forthcoming.

EGD CAPS. 2011. National Curriculum Statement policy (Final draft). Accessed on 21 April 2018. Available from: https://www.education.gov.za/LinkClick.aspx?fileticket= pqcxCwUTO98%3D&tabid=570&portalid=0&mid=1558 ISBN: 978-1-4315-0583-8.

Ford, S. & Minshall, T. 2016. 3D printing in education: a literature review. Institute for Manufacturing. University of Cambridge, United Kingdom. Centre for Technology Management. Accessed on 5 May 2018. Available from: https://www.researchgate.net/publication/308204294_3D_ printing_in_education_a_literature_ review DOI: 10.13140/RG.2.2.16230. 83526.

Freire, P. (1974). Pedagogy of the oppressed. New York, Seabury Press.

Havenga, S.P., De Beer D.J., Van Tonder, P.J.M., Campbell, R. I. & Van Schalkwyk, F. 2018. INTERNATIONAL Conference on Additive Technologies. (7; 2018; Maribor) Proceedings [Elektronski vir]/7th International Conference on Additive Technologies iCAT 2018, Maribor, Slovenia, 10. - 11. October. ISBN 978-961-288-789-6.

Hedberg, J. G. & Freebody, K. 2007. Towards a disruptive pedagogy: Classroom practices that combine interactive whiteboards with TLF digital content. Retrieved 15. 08.07,from. Melbourne: Le@rning Federation. Available from: http://www.thelearningfederation.edu.au/verve/_resources/towards_a_disruptive_pedagogy.pdf.

Pacey, E. 2013. Additive manufacturing: Opportunities and constraints. Published by Royal Academy of Engineering, Prince Philip House. Available from: www.raeng.org.uk/AM ISBN: 978-1-909327-05-4.

Schelly, C., Anzalone, G., Wijnen, B. & Pearce, J. M. 2015. Open-source 3-D printing Technologies for education: Bringing Additive Manufacturing to the Classroom. Journal of Visual Languages & Computing. 28 (2015) 226–237. DOI: 10.1016/j.jvlc.2015.01.004.

Serlenga, P. & Montaville, F. 2015. Five questions to shape a winning 3-D printing strategy. Bain & Company, Inc.

Verner, I. & Merksamer, A. 2015. Digital Design and 3D Printing in Technology Teacher Education.

Wohlers Associates. 2018. Wohlers Report 2018: 3D Printing and Additive Manufacturing State of the Industry Annual Worldwide Progress Report. Wohlers Associates, Inc. Colorado. Available from wohlersassociates.com. Accessed on 10 October 2018.

Industry 4.0 – Shaping The Future of The Digital World – da Silva Bartolo et al. (eds)
© 2021 Taylor & Francis Group, London, ISBN 978-0-367-42272-1

TERA SABI: Meta-design between design education and craft industry

L. Soares & E. Aparo
Polytechnic Institute of Viana do Castelo, Portugal & CIAUD - The Research Centre for Architecture, Urbanism and Design, Lisbon, Portugal

J. Teixeira & R. Venâncio
Polytechnic Institute of Viana do Castelo, Portugal

ABSTRACT: This paper is focus oriented on cultural universe of Guinea-Bissau to contribute to the valorization and the dissemination of traditional knowledge. This study is based on design-methods, handicraft culture and design for periphery. The authors intend to prove that it is possible to innovate and promote the craft of Guinea-Bissau with a didactic project, developing projects with a productive network between Portugal and Guinea Bissau. This notion may reinvent a model established on craft-design collaboration that aims to redesign all craft industry. The project is developed with undergraduate students and artisans. This study contributes to local craft expansion and design practice. This research also contributes to S2M 2019 as supply to reuse, remanufacturing, disassembly and recycling techniques and think permanently on design education issues. With this idea in mind, the authors aim to prove that the connections between the academy and the local context can stimulate sustainability, innovation and internationalization.

1 INTRODUCTION

1.1 *The natural scenario of Guinea Bissau*

The meaning of material culture can be interpreted as a bridge between the being and his/her environment, that includes the spirit of the place. In this sense it seems pertinent to grasp the etymological meaning of the word Guinea. As Belchior (1962) states, the word Guinea may have derived from the name of the Ghana Empire. On the other hand, the birth of the word Guinea was an occasion for the Portuguese to know these lands as the domain of the Guineans. These lands were also nominated by the inhabitants. In 1446 the Portuguese landed in the region and named it *Guiné Portuguesa*. In Amorim & Paladino (2012) in 1697, due to the threat of occupation of the region, especially by the French and English, the Portuguese Crown founded in that region a village, Bissau, which grew and became an important supplier of slaves, especially for the American continent in the following centuries.

Today, the country has good natural resources but it is considered a poor country, see AA.VV. (2016).

According to the strategic plan for the development of tourism in Guinea Bissau, see Cabral (2015), the principal supporters of Guinea-Bissau are the European Union, the World Bank and UNDP that have already developed studies on the potential of tourism and have concluded that the private operators are the main actor in their exploration. The political instability of the last twelve years has prevented the drafts and the funds of projects in this area. The Guinean economy refers to the exploitation of agriculture.

The plantations cover an area of 175,000 hectares and have an annual growth rate of 4%. In Cabral (2015) therefore, this crises situation may be understood as an opportunity to define new signifiers of a referent strongly related to natural incomes. On the one hand, the Guinean economy refers to the exploitation of minerals. On the other hand, the country has enormous and growing potential for traditional resources such as fishing or agriculture. In fact, on daily Diário de Notícias (2019) it is possible to read that according to the report 'Dynamics of Development in Africa-Growth, Employment and Inequalities of 2018′, among West African countries, (…), Guinea Bissau is the country with the highest percentage of employees in agriculture (90%). Another reference activity is fishing. As reported by Juvanio P. Cabral (2015) in the market of Bissau, it is estimated that a discharge of 15 tons per day, of which 60% is supplied by artisanal fishermen. The main species that arrive are of little value, so the market of species of higher value is much reduced. In addition, the landscape of the country is still virgin with singular and unique fauna and flora.

With this idea in mind the Guinean tourism minister told the Portuguese radio TSF (2017) that the tourism strategy of the country is focused on the 88 islands that can triple the number of tourists from 2016 to 2017, from 40 to 120 thousand. He added that only 20 of these islands are inhabited, there are unspoiled beaches, luxurious forests and unique animals such as the saltwater hippos, which are unique in the world.

1.2 *Craft culture as a cultural sign of a place*

Guinea Bissau presents different forms of craftsmanship ranging from wood carvings to wicker works, through pottery and musical instruments, to the most interesting weaving artifacts. The cloths, for instance, are much more than a garment, assuming a strong cultural and religious dimension. Guinean handicrafts are also characterized by the use of wood as raw material to produce sculpture of a ritualistic or religious dimension.

A case that supports this truth is the fork (furkidja). As M. Parente Augel (2007) argues forks are wooden figures, coarse or finely carved, anthropomorphic - also female figures - in a stylized or very simplified form, as simple stake. These characteristics are excellent and exceptional occasions for all actors - artisans, designers, entrepreneurs and their reality - work together, transmuted into each other. It can be an opportunity for companies to establish craft-design alliance, associating their image with the values of the place. However, the absence of local development policies forced people to take their own livelihood measures, gradually giving up the practice and the ability to make the wonderful cloth Pano Manjako. Brito et al. (2010) claims that the economic value has fallen considerably in favor of other activities, including the cultivation of cashew trees, fishing, carpentry or small-scale trade.

Therefore, this article addresses the concept of meta-design, see Mendini (1969) and the design for the periphery approach methodology by Bonsiepe (1983). Firstly, because it addresses the idea of making something reproducible, so that in a second stage it would become form. Secondly, because it contributes to the development of the craft industry of the country.

2 THE GROWTH OF SOME CREATIVE AREAS

2.1 *The competence of semiotics in design*

In Latour (2008), the role of semiotics competence may be a logical consequence to redesign the identity of a place. Today, it is remarkable to look at the past as basis that happens to guide us to vision the future. The cross-culture venture between past, culture and time maybe a key to decipher and transform reality. In fact, in Guinea Bissau there is a new generation in the diaspora who insists on writing their pride to have Sabi, their pride of being Guinean, which needs to be read and studied in Guinea-Bissau, see Leite (2014).

2.2 *The expansion of creative areas*

In the arts, the Guinean painter Sidney Cerqueira stands in the spontaneous realism with which he deals with the social and cultural themes of Guinea. In music, for instance, the musician and artist Lilison Cordeiro uses as raw material materials found in the territory of Guinea.

In the literature, the Guinean culture presents interesting cases in Kriol and in Portuguese. In this context it is possible to name different authors like Vasco Cabral, Agnelo Regalla or, more recently, Domingas Sami.

In the Cinema one of the movies that stands out is 'Nha Fala' from director Flora Gomes (2002). His work relates the Guinean myth with the current reality.

According to Darame (2018) at Deutsche Welle, a Germany's international broadcaster in music, apart from those who define the country's history, such as José Carlos Schwartz, there are interesting young musicians. One of them is Mû Mbana, who makes music using the traditional instruments of the Bijagós islands.

Another area that qualifies the country is the Tina dance. According to the journal Intercultural Dialogues Tina is one of the oldest Guinean dances, much practiced by the Creole citizens of Cacheu, for the communication, sensitization and entertainment of the northern community. Schwarz & Levy (2011) declare that today it is sung and danced in almost the whole country. At the moment, according to what the Director-General of Guinea-Bissau Cornelio da Silva told the Portuguese newspaper Observador (2019), the Government of Guinea-Bissau intends to present to UNESCO an application for (...) classification of the country's Carnival and traditional Dance of Tina as Intangible Heritage of Humanity.

In the fashion sector there are some young designers that stand out, see Maveau (2017). In Camará (2016), Djanaina Vaz or the designers Alfa Canté and Braima Sori Ba use the traditional cloths of Guinea as reference to create contemporary pieces.

3 TERA SABI PROJECT

3.1 *The theoretical context*

This project is centered on the cultural and social universe of Guinea-Bissau and aims to contribute to the valorization and dissemination of traditional products, techniques and knowledge of Guinea-Bissau. The project 'Tera Sabi' intends to interpret the essence of the message, contributing to the development of the periphery countries by making society aware of the center's problems. It would be anachronistic, ambiguous and deceitful to think a project from the center to the periphery. This practice could deprive the designer of the understanding of the real conditions that influence the methodology and the design processes.

Figure 1. Analysis of Guinea Bissau context. Source: Ermanno Aparo & Liliana Soares.

3.2 *The productive context*

The idea of approaching a reality from the periphery (Guinea Bissau) to a reality of the center (the North of Portugal) begins with the identification of common points between the two realities.

In particular, in center reality it seems important to benefit from ethnographic production on its artisanal productive dimension, analyzed and reported in the 20th century by Portuguese ethnographers such as Fernando Galhano or Benjamim Pereira as Campos et al. (2014) explain. The past exists to guide us in the future we want to draw, in the sense of connoting material culture with references and identity. Craft culture can be integrated in design process. It means, using craft thinking and know-how to work with the materials. Thus, the designer accepts the role of a reflective practitioner, see Schön (1983), contributing to identify the values of sustainability. However, according to the Gaia Giuliani (2007) most designers guide their own efforts only to those 10% of people who belong to rich countries, thus ignoring the rest of the planet. This reflection may be an opportunity to always think about the Planet and the people who inhabit (and in all living beings as well), during all the year and not only in specific and seasonal moments, such as international conferences or solidarity actions sponsored by companies of the center.

According to Aparo & Soares (2012), with this philosophy in mind, it means that design can create products that guide the user to a new ethical thinking, able to develop a function and at the same time to value the use, offering an ethically consistent and valid semantics when related to the context of reference. In this sense, the design assumes the responsibility of, through the product design or design service, to contribute to the education of people, as well. The Tera Sabi project could create important occasions of useful reflections capable of stimulating the dialogue between diverse communities that accompany the world panorama proposing new lifestyles and, mainly, new ways to consume and to exist, see Aparo & Soares (2012).

4 THE DESIGN FOR DEVELOPMENT

4.1 *Cultural transfer*

Previous studies of Margolin (2009), Manzini (2005), Buono et al. (2018) sustain that the designer is the professional capable of transfer knowledge from one area to another, allowing to find solutions in design scenarios. The basic idea is to begin with an analytical reading of artisanal activities, framing this productive universe in the logic of Maldonado (1991) called intermediate technologies. It means technologies which consider the various realities of underdevelopment and which seeks to provide flexible answers for each situation particular.

With this idea and acting as a reflective practitioner, the designer can contribute to create sustainable solutions. Therefore, 'TERA SABI: meta-design between design education and craft industry' is developed between the Academic domain and business environment, in order to create products with the genius Loci of Guinea-Bissau.

Figure 2. Analysis of the productive context of Northern Portugal. Source: Ermanno Aparo & Liliana Soares.

4.2 *The academic project*

The project was carried out done with almost 40 students, 4 professors and several artisans. Students work in team work focus on different sectors of design world, such as: travel accessories, fashion accessories, home accessories, toys, objects for free time.

It starts from a common analytical basis, but the subtends answers were aimed at generating different solutions.

All participants were cautioned that this is a project for development, and not a project from the center to the periphery. In a relevant interview, see Fathers (2003), the designer Gui Bonsiepe stressed that design for poor countries should be read as design in the poor countries or design of the poorest countries. It means that the center does not have the universal magic formula of industrial design that should be distributed to the inhabitants of the periphery and that the development agencies recognize as being underdeveloped.

With this idea in mind, the students used intermediate technologies, for instance, the technology developed locally by the local community with native inputs, see Bonsiepe (1983).

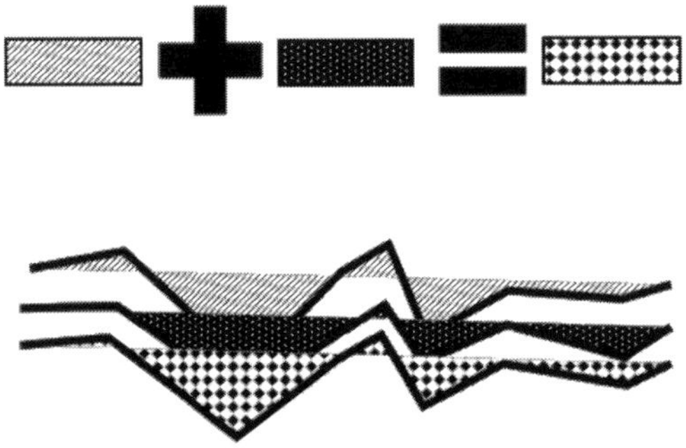

Figure 3. Design synthesis. Source: Ermanno Aparo & Liliana Soares.

4.3 *Results*

In this study, the design action was focused on generating new markets. In Celaschi et al. (2005) reinventing craft by design may produce new luxury ethical.

This design strategy is determined by the sociocultural context of its origin. Therefore, on the one hand, this study intends to produce development, to define new markets opportunities, focusing on internationalization. As Vitor Margolin (2009) states design for development needs to extend its synthesis of reasons, from an emphasis on poverty relief to the inclusion of strategic product creation for the global sphere. It means to create alternatives solutions that can rise up both cultures and stakeholders.

Semiotics as a competence of design may create an opportunity for places under development. As Francesco Morace (2008) declares the qualitative standard of the surrounding world is grounded in the new concept of sustainable emotion. Aesthetic production intercepts deeply with ethical themes, becoming an expression of sustainability. The clarity of the processes and the character of the products contribute fully to the success of the project. Ecology guides a way of living and not just a way of thinking, defining the new project creativity. The fact, is that the large levels of aesthetic qualities of care productive of well done, singularity and innovation are connected to the spirit of a place.

Today, product development refers to the sign-function as Eco (1990) states and that becomes the main function. "Today, values such as solidarity, creativity and participation can unite people to achieve common goals, creating miraculous developments." (Yunus cit in Aparo & Soares 2012). In fact, this sign-function can materialize the action, transforming it into something useful both for those who design and for those who use and enjoy it.

5 CONCLUSIONS

This paper presents a meta-project as create the foundations of a pilot project, proposing design for development. The abilities of designers have been improved and design skills are significantly applied to diverse markets.

In Soares et al. (2017), the topic of stimulating local manufacturing requires design knowledge, thus design for development may reinvent a local craft production. The inputs created between the periphery and the center may reinvent themselves and create opportunities for both, producing alliances and to start new markets territory for those activities. For design school this become an occasion to transform students in citizens with social responsibility for the real world. The production of goods using intermediate technologies may be an opportunity to rise up productive culture of Guinea Bissau. The products be present abroad.

With this pilot project the authors intend to challenge all participants to act in a profitable way. This is also a way to explore new market opportunities, applying knowledge and experiences achieved from this study as an incentive for additional research.

ACKNOWLEDGEMENTS

This work was carried out with the support of Polytechnic Institute of Viana do Castelo (Portugal), non-governmental organization Tiniguena (Guinea-Bissau), Integrated Rural Development Association of the Vale do Lima - ADRIL (Portugal) and Research Centre for Architecture, Urbanism and Design - CIAUD, Lisbon School of Architecture, University of Lisbon (Portugal).

WEBGRAPHY

https://www.dn.pt/lusa/interior/guine-equatorial-e-o-unico-lusofono-em-africa-com-previsao-de-recessao-ate-2020—bad-10455058.html (18/01/2019)

https://www.tsf.pt/sociedade/interior/o-povo-a-biodiversidade-e-os-bijagos-os-trunfos-do-turismo-da-guine-bissau-8535367.html, (15/01/19).

https://observador.pt/2019/01/04/guine-bissau-vai-propor-carnaval-e-danca-de-tina-para-patrimonio-imaterial-da-humanidade/, (21/01/19).

REFERENCES

AA.VV. (2016). *Perspectivas Económicas em África: cidades sustentáveis e transformação estrutural*. Lisboa, Portugal: OECD publishing.

Amorim, C. & M. Paladino (2012). *Cultura e Literatura Africana e Indígena*. Curitiba, Brasil: IESDE.

Aparo, E.& L. Soares (2012). *Sei progetti in cerca d'autore*. Firenze, Italia: Alinea.

Bonsiepe, G. (1983). *A Tecnologia da Tecnologia*. São Paulo, Brasil: Edgard Blücher.

Brito, B.R., J.R. Pinto & N. Alarcão (2010) *Abrindo trilhos, tecendo redes: reflexões e experiências de desenvolvimento local em contexto lusófono*. Lisboa: Portugal, Gerpress.

Buono, M., S. Capece, & E. Laudante (2018). Design e Artigianato 4.0 Identità culturale territoriale ed innovazione. *MD Journal Design e Territori*. *5*, 28–39;

Cabral, J. P. (2015). *El sector pesquero artesanal de Guinea Bissau*. Alicante, España: Universidad de Alicante.

Camará, S. (2016). Guineense alfa canté é sucesso da moda afro-europeia. *J. O Democrata*. 27 Nov.

Campos, A.M., C. Saraiva, J.Y. Durand & J.A. Botelho (2014). *Caminhos e diálogos da antropologia portuguesa: homenagem a Benjamim Pereira*. Viana do Castelo, Portugal: Câmara Municipal de Viana do Castelo.

Celaschi, F., A. Cappellieri & A. Vasile (2005). *Lusso versus. Italian design, beni culturali e luxury system: alto di gamma & cultura di progetto*. Milano, Italia: Franco Angeli.

Darame, B. (2018) Mû Mbana pretende divulgar instrumentos tradicionais dos povos bijagós. *Deutsche Welle*. Sep.

Dias Belchior, M. (1962). Sobre a origem do termo Guiné. *Boletim Cultural da Guiné portuguesa*. *65*, 41–56.

Eco, U. (1990). *O signo*. Lisboa, Portugal: Editorial Presença.

Fathers, J. (2003). Peripheral Vision: An Interview with Gui Bonsiepe Charting a Lifetime of Commitment to Design Empowerment. *Design Issues*. *19*. *Issue 4*, 44–56.

Giuliani, G. (2007). Arriva il design sostenibile per il Terzo Mondo cambierà la vita a cinque miliardi di persone. *La Repubblica, Tecnologia & Scienze*. Giugno.

Gomes, F. (2002) Nha fala. Fado Filmes (Portugal), Les Films de Mai (France), Samsa Films (Luxemburg).

Latour, B. (2008). A Cautious Prometheus? A Few Steps Toward a Philosophy of Design (with special attention to Peter Sloterdijk). *In Proc. Annual International Conference of the Design History Society*, Falmouth, UK. pp. 2–10.

Leite, J.E.B.C. (2014). *A literatura Guineense: contribuição para a identidade da Nação*. Coimbra, Portugal: Universidade de Coimbra.

Maldonado, T. (1991). *Disegno Industriale: un Riesame*. Milano, Itália: Campi del Sapere Feltrinelli.

Manzini, E. (2005). Un Localismo Cosmopolita. *SDI Magazine 2*, 1–6.

Margolin, V. (2009). Design para o desenvolvimento: para uma história. *Arcos Design*. *4*, 1–6.

Maveau, R. (2017). Lisbonne, porte d'entrée d'une mode africaine éthique. *Le Point. 7 March*.

Mendini, A. (1969). Metaprogetto, si e non. *Casabella*. *333*, 4–15.

Morace, F. (2008). Il futuro nasce in Brasile. *7th Floor. 11*, v.2. 62–67.

Parente Augel, M. (2007). *O desafio do escombro: nação, identidades e pós-colonialismo na literatura da Guiné-Bissau*. Rio de Janeiro, Brasil: Garamond.

Schön, D. (1983). *The reflective practitioner: How professionals think in action*. New York, The United States of America: Basic Books.

Schwarz, C. & I. Levy (2011). Cacheu caminho de escravos. *Diálogos Interculturais. Nov.* 53–88.

Soares, L., E. Aparo & F. Moreira da Silva (2017) Either/or: reflecting design thesis orientation. *In. Proc. 19th International Conference on E&PDE*. Oslo: Norway, pp. 674–679.

Brito, B.R., J.R. Pinto & N. Alarcão (2010) *Abrindo trilhos, tecendo redes: reflexões e experiências de desenvolvimento local em contexto lusófono*. Lisboa: Portugal, Gerpress.

The materials platforms and the teaching of product design

A. Miranda Luís & P. Dinis
CIAUD, Lisbon School of Architecture, Universidade de Lisboa, Lisbon, Portugal

ABSTRACT: With new materials-technological systems, an increasing attention is given to Product Design, making the evaluation of the programmatic contents offered by design Faculties necessary, forcing a better preparation of the contents that are taught. Current dissemination of the products and services of the industries in web platforms provide a wider dissemination and communication about materials, processes, and their applications, and can contribute to better support to teaching in design. The close connection between Faculties and Research Centers and industry is an opportunity to the improvement of the teaching and learning of design. In this research, we selected some studies cases of material platforms. In this paper we present the study of nine platforms previously identified and studied, and discuss about their objective and the target audience of each one. This work will question the need to create a "support for design students" that enhances their knowledge.

1 INTRODUTION

The main purpose of this article is the study of one of the great themes of the research work of the first author. Revisiting the history of teaching and the current methods applied in curricular units, we aim to compare not only how Portuguese design schools teach their students, but also the skills that are acquired by the students in a 3-year undergraduate degree.

"The great reform undergone in recent years in Portuguese (and European) higher education introduced by the Bologna Process, on the one hand imposes temporal constraints on higher education studies, on the other, opens space for the adoption of actions audacious and innovative methodologies that allow an effective development of competences in the students that promote constant and permanent learning and abilities throughout life as a vertiginous technological and social transformation imposes on us. It is up to all of us educational agents to participate in this." (Freely translated from Oliveira, 2012)

This research tries to contribute to the implementation of an applied methodology that aims to contribute to the enrichment of skills acquisition at the technical material level in the training of a product designer, promoting creativity and innovation in teaching design materials.

In the 1990s, the European Union (EU) initiated a program of global modernization of education, so that there could be a degree and diploma compatibility within the EU space, allowing the "achievement of the title of Europe of Knowledge" (freely translated from Dias, 2017).

With the entry of the Boot-Off Process, "which emerged by signing the Bologna Declaration" (Caeiro, 2010 *apud* Dias, 2017), the way of teaching will have to be adapted to the new system.

However, according to Abílio Silva, a professor at the University of Beira Interior, in Portugal the changes in the way of teaching will have to take into account the "class context" since, according to it, "we have to adapt the way we convey the knowledge to the students who are ahead of us in a class context. I feel the need to adapt my language to what I interact with the students. The first class and the 17th are not the same, the expression is not the same" (Silva, 2018).

The Bologna Process is thus an important governmental measure for educational institutions since it aims at promoting higher quality education (Dias, 2017). However, according to Silva (2018), "the truth is that, with the Bologna process and with the new technologies, the alteration that was made was the teacher began to use PowerPoint or to show a film. I also did this. But what is gained by this?" Thus, after 20 years of signing the Bologna Statement (May 1999) and after 10 years of implementation in the vast majority of higher education institutions, we can question the changes felt in Design studies, ascertaining what new tools can be added to teaching in product design, namely through the use of material platforms that the market and international institutions offer.

In this paper we report a study that looked at how design schools and research centers have developed teaching in the area of design and technologies. We analyzed material platforms currently available at European level, and present and discuss the results.

2 TEACHING IN PRODUCT DESIGN

Higher education in general, launched by the European Union, must be modified in its genesis in order to foster innovation, promoting competitiveness for not only economic growth, of each individual country, but also of the European Union (Raulik-Murphy, 2013 *apud* Dias et al. 2013). We need and want more and better design. For that we need to promote a competitiveness and relation between technologies and materials (Figure 1).

Dias (2013) in her article about design higher education, mentions how Claudia Monteiro states that the industry has been unreceptive to hiring young people trained according to the Bologna Process (3 years of education), because they consider their academic training as being insufficient and incomplete to start a professional career. However, Monteiro mentioned, it is important to promote in the academic field more and better design thinking (Figure1): "This situation presents itself as a cut that was previously filled by stages that could be professional or curricular, a circumstance that today only happens at the Master's level (2nd cycle)" (Dias, 2013).

"Design's strength lies in its breadth, its interaction with so many sources of inspiration and expanded areas of practice. Through technology, designers are in constant contact with new materials and processes that stimulate ideas for original products." (Ramalhete 2012, *apud* Raizman, 2010)

Thus, and as Doordan (2003) stated, with the partnerships created between industry and Laboratories, promotes a much more extensive set of available knowledge and resources than was previously known to designers. Ramalhete (2012) assumes that most authors (for example, Howard, Culley, & Dekoninck, 2008) consider that there is no one practice that can be considered the best one.

It is understood that "Investing in knowledge is investing in the future" (Gago, 1990). Material libraries are investigative, "living" structures that allow access to a high amount of information on materials that can be an important support to teachers and design students. Education in general needs to follow the evolution of industry, thus making academies multi-faceted teaching structures bi-polarized, i.e., science must open its horizons, and "cross-knowledge, role of extreme importance in this change of mentality" (Freely Translated from Moreira da Silva, 2017).

The knowledge and development strategies of teaching in the area of materials and its presentation to students with the proper practical and evolutive accompaniment in the industry may not be a viable task without the support of a tool that will minute update, such as a digital tool, accessible anywhere in the world.

According Veiga, Silveira, & Antunes (2013), material libraries or material plartforms arise from the need of students and teachers to gather useful and adequate information, such as technical specifications, creation of selection and testing methods for materials, which are interconnected with Course Unit (CU) curricula, to privileged in training.

The connection between industries and universities (Figure 2) should increasingly be strengthened, and "based on knowledge transfer processes." However, it should be noted that, in the age of technology and industry 4.0, "this interaction is precarious and there are problems of communication and poor evaluation of the competencies proposed by educational institutions, compared with the expectations of companies and those achieved, in reality, by the designers (freely translated from Dias, 2013). In addition, considering that in 2009 the technologies were beginning to be developed, "it becomes necessary to streamline the methodological response, in order to integrate this new knowledge into the curriculums (of curricular units of design) or to relativize them in the matrix of materials" (freely translated from Pereira, 2009).

3 MATERIAL LIBRARY

According to Veiga, Silveira, & Antunes (2013), material libraries, by having information organized with cataloging, generally acquire the form of "Materioteca, through the cataloging of materials in the market and seeks the interaction interdisciplinary approach that allied with the creativity of the designer help in the development of new alternatives. The more comprehensive the information about the materials, the better the choice of the raw material."

In a similar line of thought as Veiga et al., Doorman (2003) also reported that "Design is the process by which abstract ideas assume concrete form and thus become active agents in human affairs. One of the critical parameters in any discussion of designed artifacts is material: what something is made of and

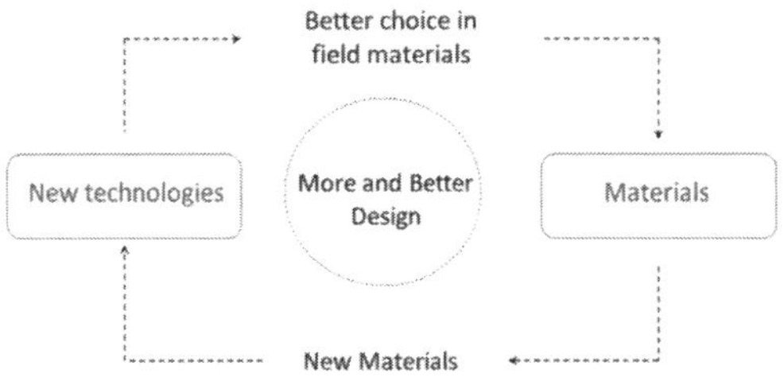

Figure 1. Design evolution scheme with the interconnection between materials and technologies/platforms. [source: author].

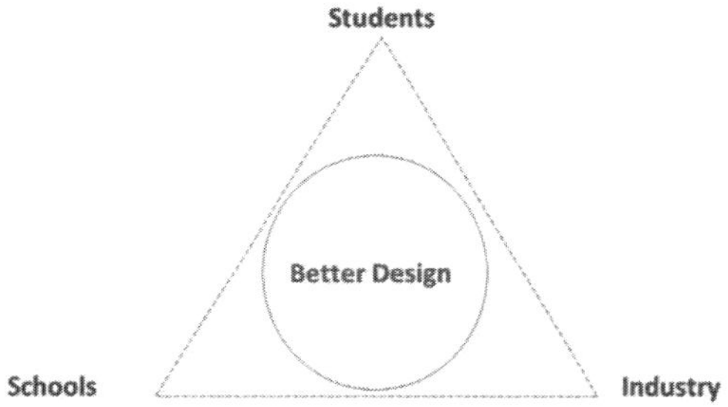

Figure 2. Connection between schools, students and industry, promoting better design. [source: author].

how the material employed affects the form, function, and perception of the final design. In a broad sense, the story of materials, their discovery and subsequent manipulation constitutes a significant thread in the history of civilizations and often provides a common point of reference for cultural discourse in general."

In short, the importance of knowledge of the material in the development of critical thinking and the act of creation is important in the construction and manipulation of materials that will lead to a better result.

One of the methodologies applied was the survey methodology, which aims to obtain data, by various means such as mail, telephone, email, face-to-face questionnaires individually or in groups. According to Moreira da Silva, "a good questionnaire must be capable of validating and measuring factors of interest, induce respondents to cooperate in the study, and obtain information accurately" (Moreira da Silva, 1999).

We realized the need for students to understand how they research and look for materials for their work, what material platforms they knew, and used. That way, we collected data from a pool of 107 students, wherein 67 were considered valid according to what was stipulated by the investigation team for the study. 79% of the respondents were female and 21% male and attended the first cycle of studies, i.e., an undergraduate degree in Design. 97% of respondents were between 17 and 24 years of age, and 39% were students of 1st year, 28% of 2nd year, and 33% were in the 3rd year.

The students were questioned about the importance of materials, and 53% indicated that materials classes are very important, and in terms of technologies the level of importance increased to 58%. Over 50% responded that materials and technologies are important for their construction as a design professional.

One of the concerns of this research is the fact that universities do not have material libraries that is useful to students, that this should not be a watertight library of materials. It will have to be updated and we consider this to be a possible obstacle to its creation.

We aimed at researching how design schools and research centers have developed teaching in the area of design and technologies. We analyzed material platforms and,identify their qualities and weaknesses. According to Oliveira (2012), we can see that in several countries, projects have been developed that serve as a bridge between industries and academies (teachers and students).

Based on the responses to the survey mentioned above, we selected nine platforms/sites as case studies. These were important to measure the relationship that platforms have with Design Schools and industries.

With the research carried out, we realized that in the European Union there are several material platforms, but most are not related to Educational Institutions; however, more than 31% of the students have never heard of any of the material platforms presented in the study.

It should be noted that Aalto University has done a remarkable job in the interconnection of the university with the industry, having developed a material platform for this purpose – which was launched in 2017 and is currently still in the experimental version. Its aim is to "develop, and support collaborative and multidisciplinary practices, internal collaboration, and industry collaboration. The platform is a think tank that receives input from stakeholders inside and outside Aalto University in relation to materials research, education, and outreach (Aalto University, 2019, online).

Analyzing the nine material platforms available in the market, (see Table 1 – Riba, Produtspec, Material-NL, Archiproduts, Materialon, Ces-edupack, efunda, Institute of Making, Aalto University Material Platform – considered by the team, the most important), we realized that all have information on their own websites, and somehow have a system of classification. We also note that only one of the nine was developed in a university (Aalto University material platform), which raises questions related to the lack of interest of the academies in these matters that should be investigated. There are, however, platforms used by engineering students, such as the Ces-edupack platform, which demonstrates that the materials engineering areas are more organized and focused on the theme. The platform itself provides a free space for students; however, its information is very focused on teaching engineering materials, even though it is called "CES-EduPack is a unique set of teaching resources that support Materials Education across Engineering, Design, Science and Sustainable Development" (Ces-edupack, 2019).

Table 1. Synthesis of the analysis of material platforms we considered most important. [source: author].

	site	Social media	Classification	University link
Riba	o	x	o	x
Produtspec	o	o	o	x
Materia-NL	o	o	o	x
Archiproduts	o	o	o	x
Material on	o	x	o	x
Efunda	o	x	o	x
Institute - making	o	x	o	x
Aalto University	o	o	for students	o
CEs – Edupack	o	o	o	x

In Portuguese universities, when we asked 1st cycle students about the existence of libraries of materials, in their libraries, 57% said no, 39% said yes, and 4% of respondents did not respond.

Regarding a library of virtual materials (material platforms or virtual materials), 85% say that their faculties do not have a database of materials and virtual technologies, which is a very expressive number.

4 CONCLUSIONS

We realize that previous research has identified the need to organize and provide information on materials and their properties' descriptive technical data and characteristics so that the platforms can be used by their students.

We conclude that Portuguese students have some difficulty in associating the virtual material libraries as material platforms, which may have conditioned some responses.

We found that the students of the design degrees interviewed still have a vision that they should only study what is taught in the classes by the professor, also demonstrating naivety in the search for more innovative solutions for their projects.

Thus, we believe that the use of platforms or material libraries can bring benefits for the educational system in product design in Portugal.

ACKNOWLEDGEMENTS

This work was carried out with the support from the Research Center of Architecture, Urbanism and Design (CIAUD), from the Lisbon School of Architecture of the University of Lisbon (FA.ULisboa) and the FCT, Foundation of Science and Technology.

REFERENCES

Aalto University, A., 2019. Aalto University Materials Platform. [Online] consult: march-2019 Available at: https://www.aalto.fi/aalto-materials-platform/materials-platform-mission-and-vision.

Almendra, R., 2016. Documento de apoio às aulas de Crítica do Objeto. Em: Lisboa: s.n. Lisbon School of Architeture.

Ces-edupack, 2019. granta design. [online] [consult: june-2018] Available at: https://grantadesign.com/education/ces-edupack/.

Dias, A. C., 2017. Transferência de conhecimento em design: Estratégias de aproximação das instituições de ensino superior aos mercados e sociedades Portuguesa. Lisboa: Lisbon School of Architeture at Universidade de Lisboa. [online] [consult: June 2017] Available at: https://www.repository.utl.pt/handle/10400.5/15271.

Dias, C. e. A. M. C., 2013. Ensino superior do design e indústria em portugal: estudo do ensino e práticas do design industrial. Porto, s.n. [online] [consult 18 03 2019]. Available: https://ud13.files.wordpress.com/2014/11/ud13_livro_de_actas.pdf.

Doorman, D. P., 2003. On Materials, [Online] [consult: june-2017] Available at: http://www3.nd.edu/~ddoordan/onMaterials.pdf: Design Issues.

Gago, J. M., 1990. Manifesto para a Ciência em Portugal. Lisboa: Gradiva.

Moreira da Silva, F., 2017. Investigar em Arquitetura Urbanismo e Design - uma polinização cruzada. in: A Lingua que Habitamos, April, pp. 83–96.

Moreira da Silva, F., 1999. *Colour/space: its quality management in architecture; the colour/space unity as an unity of visual communication. Tese de Doutoramento.* Manchester: Universidade de salford.

Oliveira, P., 2012. A identidade difusa dos materiais metálicos: Contributos para um ensino holistico sobre materiais em cursos superiores de design. FA-ULisboa ed. Lisboa: Tese de Doutoramento. Lisbon School of Architeture at Universidade de Lisboas.

Pereira, J. A., 2009. Considerações sobre Metodologias do Processo Tecnológico e Produtivo, Lisboa: Lisbon School of Architeture at Universidade de Lisboas.

Raizman, 2010, p. 408 apud Ramalhete, P. 2012, Metodologia de seleção de materiais em design: base de dados nacional. p. 348. Aveiro: Universidade de Aveiro [Consult: March 2019], [available: file:///C:/Users/HP/Downloads/7478.pdf].

Ramalhete, P., 2012. Metodologia de seleção de materiais em design: base de dados nacional.. Aveiro: Universidade de Aveiro.

Silva, A., 2018. Entrevista a Abilio Silva [Entrevista] (04 04 2018). Universidade da Beira Interior, Covilhã, Portugal.

Veiga, E. R., Silveira, Y. & Antunes, D., 2013. Laboratório de materiais: um acervo para a sustentabilidade.in: ANAIS do 9º Coloquio de Moda. Fortaleza, Brasil: 9th fashion Conference. [Online] [Consult: April 2019] [available: http://www.coloquiomoda.com.br/anais/Coloquio%20de%20Moda%20-%202013/COMUNICACAO-ORAL/EIXO-8-SUSTENTABILIDADE_COMUNICACAO-ORAL/Laboratorio-de-Materiais-um-acervo-para-a-sustentabilidade.pdf].

Industry 4.0 – Shaping The Future of The Digital World – da Silva Bartolo et al. (eds)
© 2021 Taylor & Francis Group, London, ISBN 978-0-367-42272-1

Action Research: A strategical methodology for applied research

F. Moreira da Silva
CIAUD, Universidade de Lisboa, Lisbon School of Architecture, Lisbon, Portugal

ABSTRACT: In research, when crossing the concepts of smart systems, sustainability and manufacturing, we are immediately directed to applied research that has substantially changed the research paradigm. In order to obtain a real and immediate knowledge transference, design education at research level is increasingly focused on practical, interventionist and experimental methodologies. Action Research, as a support to different project areas, has been consubstantiating as one of the main methodologies in applied research. The societal drivers for some Industry 4.0 technologies are aligned with Action Research types. This paper underlines the advantages of using this methodology when it requires insight, reflection and personal involvement with the topic to explore. The researcher introduces changes in the project until the effectiveness of the taken actions has been explored and the adjustments implemented. It's a research strategy which deals with real-world problems in a participatory and cyclical way, to produce knowledge and action.

1 INTRODUCTION

The Action Research methodology is based on results that aim to improve methods used in educational contexts, social sciences, community and others. Also known as participatory research, it requires insight, reflection, and personal involvement with the topic being explored. This methodology is conducted in real world contexts by the people directly involved with the problem or situation being investigated. Action research is not a soft data analysis, but a deliberate assessment of the impact that specific actions have on a given performance, being used at the level of applied research that has substantially changed the research paradigm, mainly at post-graduation level.

The present paper underlines the advantages of using this methodology which is a research strategy which deals with real-world problems in a participatory and cyclical way, in order to produce knowledge and action when it requires insight, reflection and personal involvement with the topic to explore, especially when crossing the concepts of smart, sustainability and manufacturing at design level.

2 WHAT IS ACTION RESEARCH?

Action research is known by many other names, including participatory research, collaborative research, emancipatory research, action learning, and contextual action research, but all are variations on a theme. Simply put, action research is "learning by doing" - a problem is identified, we do something to solve it, see how our efforts were successful, and if we are not satisfied, try again. "Active research … aims to contribute both to the practical concerns of people in an immediate problematic situation and to promoting the goals of social science simultaneously. Thus, there is a double commitment in action research to study a system and collaborate system members to change it into what is considered a desirable direction." (Gilmore et al. 1986)

What separates this type of research from general professional practices, consulting or solving daily problems is the emphasis on scientific study, that is, the researcher studies the problem systematically and ensures that the intervention is informed by theoretical considerations. Much of the researcher's time is spent in refining the methodological tools to meet the demands of the situation and in collecting, analyzing and presenting data on a continuous cyclical basis.

Action Research is defined by O'Leary (2015) as: "A research strategy that addresses real-world problems in a participatory, collaborative and cyclical way, to produce knowledge and action". He refers to a research methodology that works towards a type of change. As their objectives are geared towards change, not just for the collection of knowledge, active research studies are often based on everyday issues and are concerned with practical solutions to these problems.

Elements of action research studies include: problem identification; problem research and its probable causes; development of an answer to the problem; implementation of the proposed solution; observation of the implementation of the solution; reflection on results (and starting over if necessary).

After a problem or subject to be examined are recognized four repeated steps: Plan, Act, Observe, and Reflect (O'Brien 1998). The changes must be made in the plan, and the steps repeated until the researcher is satisfied that he/she has fully explored the effectiveness of the actions taken and the adjustments implemented. The methods of observation are left to the discretion of the researcher or researchers. Interviews, quantitative data collection, annotations and discussions are all possible observation and reflection techniques.

In the literature, the discussion of action research tends to fall into two distinct fields. The British tradition - especially that related to education - tends to see action research as a research aimed at improving direct practice. For example, Carr & Kemmis (1986) provide a classic definition: "Action Research is simply a form of self-reflective inquiry undertaken by participants in social situations in order to improve the rationality and justice of their own practices, their understanding of these practices, and the situations in which practices are carried out". Many people are drawn to this understanding of action research because it is firmly located in the realm of the practitioner - it is linked to self-reflection. As a way of working, it is very close to the notion of reflexive practice coined by Donald Schön (1983).

The second tradition, perhaps most widely discussed in the field of social welfare - and certainly the broader US understanding is action research as "the systematic collection of information destined to bring about social change" (Bogdan & Biklen 1992). They also say that their practitioners gather evidence or data to expose unfair practices or environmental hazards and recommend actions for change. In many respects, for them, it is bound up with traditions of citizen action and community organization. The practitioner is actively involved in the cause for which the research is conducted. For others, this commitment is a necessary part of being a practitioner or member of a community of practice.

3 ACTION RESEARCH OBJECTIVES

We can consider Action Research's main objectives: to improve practice through continuous learning and progressive problem solving. It involves action, evaluation, and critical reflection, and - based on the gathered evidence - changes in practice are then implemented; a deep understanding of practice and the development of a well-specified theory of action; being participatory and collaborative, it is performed by individuals with a common purpose and is an improvement for the community in which the practice is incorporated through participatory research; to be based on specific situations and context; develop reflections based on interpretations made by participants; to create knowledge through action and at the point of application; to involve problem solving.

Action research involves a systematic process of examining the evidence. The results of this type of research are practical, relevant and can inform the theory. Action research is different from other forms of research because there is less concern about the universality of results, and more value is attached to the relevance of the results to the researcher. Critical reflection is at the heart of action research, and when this reflection is based on careful examination of evidence from multiple perspectives, it can provide an effective strategy for improving the organization's work forms and the entire organizational climate. It can be the process by which an organization learns.

We conceptualize action research as having three outcomes - at the personal, organizational, and academic levels. At the personal level, it is a systematic set of methods for interpreting and evaluating a person's actions in order to improve practice. The action-research process involves the progressive solution of problems, balancing efficiency with innovation, thus developing what has been called an "adaptive" form of specialization. At the organizational level, action research involves the understanding of the system of interactions that defines a social context. Action research goes beyond self-study because actions, outcomes, goals and assumptions are located in complex social systems.

The action researcher begins with a theory of action focused on the intentional introduction of change into a system with assumptions about outcomes. This theoretical test requires careful attention to data and skill in interpretation and analysis. Activity theory, social network theory, system theories, and evaluation tools such as surveys, interviews, and focus groups can help the researcher gain a deeper understanding of change in social contexts within organizations. At the academic level, the action researcher produces validated findings and assumes the responsibility of sharing those findings with those in their environment and with the larger research community. Many people gain experience in their workplace, but researchers value the process of building knowledge through ongoing dialogue about the nature of their findings.

4 ACTION RESEARCH EVOLUTION

Kurt Lewin is generally credited as the person who coined the term Action Research: "The research required for social practice may be better characterized as research for social management or social engineering. It is a type of action research, a comparative research on the conditions and effects of various forms of social action and research that lead to social action. A research that produces nothing but books will not suffice" (Lewin 1946). His approach involves a spiral of steps, each of which is composed of a circle of planning, action and discovery of facts about the outcome of action. Lewin was a German social and experimental psychologist and one of the founders of the Gestalt School. He was concerned about social problems and focused on participatory group processes to deal with conflicts, crises and changes, usually within organizations.

Eric Trist, another major contributor to the field since that post-war era, was a social psychiatrist whose group at the Tavistock Institute for Human Relations in London was engaged in applied social research, initially for the civil repatriation of German prisoners of war. He and his colleagues tended to focus more on large-scale multi-organizational problems. Both Lewin and Trist applied their research to systemic change within and between organizations. They emphasized direct professional-client collaboration and affirmed the role of group relations as the basis for problem solving. Both were avid proponents of the principle that decisions are best implemented by those who help them do so.

5 ACTION RESEARCH PRINCIPLES

According to Winter (1989), in implementing Action Research, we must take into account its six fundamental principles: 1) Reflexive criticism - an account of a situation, such as annotations, transcriptions, or official documents, will make implicit assertions to be authoritative, that is, imply that it is factual and true. The principle of reflexive criticism ensures that people reflect on issues and processes and explain the interpretations, prejudices, assumptions, and concerns about which judgments are made. In this way, practical reports can give rise to theoretical considerations; 2) Dialectical criticism - reality, particularly social reality, is consensually validated, that is, shared through language. Phenomena are conceptualized in dialogue, so a dialectical critique is necessary to understand the set of relations between the phenomenon and its context, and between the constituent elements of the phenomenon; 3) Collaborative resource - participants in an action research project are co-researchers. The principle of collaborative resource presupposes that the ideas of each person are equally significant as potential resources for the creation of interpretative categories of analysis negotiated among participants. This makes perceptions particularly harnessed by observing the contradictions between many points of view and within a single point of view; 4) Risk - the process of change potentially threatens all previously established ways of doing things, thus creating psychic fears among practitioners. One of the most prominent fears comes from ego-risk arising from the open discussion of their interpretations, ideas, and judgments. The initiators of action research will use this principle to alleviate others' fears and invite participation, indicating that they will also be subject to the same process and that, whatever the outcome, learning will occur; 5) Plural Structure - the nature of research incorporates a multiplicity of views, comments and criticisms, leading to multiple actions and possible interpretations. This plural structure of inquiry requires a plural text for the report. This means that there will be many explicit explanations, with commentaries on their contradictions and a range of action options presented; 6) Theory, Practice, Transformation - for action researchers, theory informs practice, practice refines theory, in a continuous transformation. In any scenario, people's actions are based on assumptions, theories, and hypotheses implicitly maintained, and with each outcome observed, theoretical knowledge is improved. Both are interconnected aspects of a single process of change. It is up to the researchers to explain the theoretical justifications for the actions and to question the bases of these justifications.

6 ACTION RESEARCH: USE AND RESEARCH PARADIGM

Action Research is used in real-world situations rather than in experimental studies, since its main focus is to solve real problems. It can, however, be used by social scientists for preliminary or pilot research, especially when the situation is too ambiguous to formulate a precise research question. Mainly, however, according to its principles, it is chosen when circumstances require flexibility, the involvement of people in research or change must occur quickly or holistically.

The Action Research approach involves an intervention planned by a researcher and evaluated in order to discern whether or not this action produced the expected consequences. In other words, the researcher acts according to his or her beliefs or theories, so it is appropriate for designers. From their intervention and subsequent evaluation, researchers aim not only to contribute to existing knowledge but also to help address some of the practical concerns of people or clients who are trying to cope with a problematic situation. In this approach, therefore, the intervention of the researcher is an intrinsic part of the research design. Generally, the intervention can be perceived as analogous to the independent variables, and its consequences are treated as dependent variables whose change is monitored and evaluated.

Action Research often involves the use of control groups to allow cause and effect elucidation through the control of extraneous variables, as it implies the analysis of the direct intervention of the researcher. The incidence of the independent variable is constituted by the action of the researcher. Thus, it implies pre and post-treatment, measurement and comparison of the dependent variable in the experimental group and availability of an equivalent control group.

The essential difference between the Action Research methodology and the others lies in the close collaborative relationship of the former, where there is mutual agreement at each stage of the Action Research sequence. The diagnostic stage is ideally carried out jointly by the action researcher and the "client" if the research results have the client's commitment to the job and if the results have a fair chance of successful implementation; the researcher performs expert diagnostics and the client provides data.

The action-stage of the action research sequence is primarily client-driven, with the action-researcher's support, but the action plan is agreed upon together and the work is usually published. Finally, the evaluation phase where the work is reviewed, when new problems can arise, and the problem-solving sequence is recycled.

Action research is clearly an important approach to design research, particularly because of its stated goal of meeting both the practical concerns of designers and, at the same time, generalizing and adding to the theory. Most researchers using this methodology want to do an immediately useful job and at the same time move away from the specific one so that their research can be more widely used. Those who apply this approach are practitioners who want to improve their understanding of practice, social change activists trying to put together an action campaign or, more likely, academics.

We can locate Active Research in three types of Research Paradigm: Positivist, Interpretative and Praxiological.

6.1 *Positivist paradigm*

The main research paradigm for the last few centuries has been that of logical positivism. This paradigm is based on several principles, including: a belief in an objective reality, whose knowledge is only obtained from data of the senses that can be experienced directly and verified among independent observers. The phenomena are subject to natural laws that humans discover in a logical way through empirical tests, using inductive and deductive hypotheses derived from a body of scientific theory. Their methods depend heavily on quantitative measures, with relationships between variables commonly shown by mathematical means.

6.2 *Interpretive paradigm*

During the last half century, a new paradigm of research has emerged in the social sciences to break the constraints imposed by positivism. With its emphasis on the relationship between concept formation and socially engendered language, it can be called an interpretative paradigm. Containing qualitative methodological approaches such as phenomenology, ethnography and hermeneutics, it is characterized by a belief in a socially constructed reality, based on subjectivity, influenced by culture and history. However, it still maintains the objectivity ideals of the researcher and researcher as a passive collector and specialized data interpreter.

6.3 *Praxiological paradigm*

While sharing a number of perspectives with the interpretive paradigm, and making considerable use of its related qualitative methodologies, there are some researchers who feel that neither positivist paradigms are epistemological structures sufficient to frame Active Research. Instead, a Praxiological paradigm is seen as where the main affinities lie. Praxis, a term used by Aristotle, is the art of acting upon the conditions one faces in order to change them, addressing disciplines and activities predominant in the ethical, social and political life of the people. Aristotle contrasted it with Theory - which incorporates the sciences and activities that are concerned with knowing themselves. He considered that both, Praxis and Theory, are equally necessary. Knowledge is derived from practice, and knowledge-informed practice, in an ongoing process, is a cornerstone of Action Research (Achari 2014).

In Action Research, as an Interventionist Methodology that is, there is no neutrality on the part of the researcher, since it is a fundamental element in solving a problematic situation, in finding solutions. Hence the fact that this methodology is so used when the investigation implies a situation of practical project development, as is the case of disciplinary areas that use the project, such as architecture or design.

7 TYPES OF ACTION RESEARCH

There are four types of active research, which should be considered, depending on its application: Traditional, Contextual, Radical and Educational.

7.1 *Traditional Active Research*

Traditional Action Research has resulted from Lewin's work within organizations and encompasses the concepts and practices of Field Theory, Group Dynamics, T-Groups and the Clinical Model. The increasing importance of labor relations led to the application of action research in the areas of Organizational Development, Quality of Life at Work (QVT), Sociotechnical Systems (e.g. Information Systems) and Organizational Democracy. This traditional approach tends towards the conservative, generally maintaining the status quo with respect to organizational power structures.

7.2 *Contextual Action Research*

Contextual Action Research, also sometimes referred to as Action Learning, is an approach derived from Trist's work on relationships between organizations. It is contextual insofar as it implies reconstituting the structural relations between the actors in a social environment; based on domains as it attempts to involve all affected parties and stakeholders; holographic, since each participant understands the functioning of the whole; and emphasizes that the participants act as designers and co-researchers. The concept of organizational ecology comes from contextual action research, which is more a liberal philosophy, with social transformation occurring by normative consensus and incrementalism.

7.3 *Radical Action Research*

The radical current, which has its roots in Marxist "dialectical materialism" and Antonio Gramsci's praxis guidelines, has a strong focus on emancipation and overcoming power imbalances. The Participatory Action Research, often found in liberationist movements and international development circles, and Feminist Action Research struggle for social transformation through advocacy to strengthen peripheral groups in society.

7.4 *Educational Action Research*

Educational Action Research is based on the writings of John Dewey, the great American educational philosopher of the 1920s and 1930s, who believed that professional educators should community. Its practitioners operate mainly outside educational institutions, and focus on curriculum development, professional development and the application of learning in an essentially social context. Often, researchers using University-based Action Research work with primary and secondary school teachers and students on community projects.

8 ACTION RESEARCH TOOLS

Action Research is more a holistic approach to problem solving, rather than a single method for collecting and analyzing data. This allows several different search tools to be used as the project is conducted. These various methods, which are generally common to the qualitative research paradigm, include: keeping a research diary, collecting and analyzing documents, participant observation records, surveys, structured and unstructured interviews, and case studies.

To use this methodology, it has to be a changing system, where there is a single control and a single authority to introduce changes, which is the researcher himself. There must be a good hypothetical model to test the use of active research and there must be access to results.

9 ACTION RESEARCH AND INFORMATION TECHNOLOGY

In the past ten years, there has been a marked increase in the number of organizations that are making use of information technology and computer mediated communications. This has led to a number of convergences between information systems and action research. In some cases, it has been a matter of managers of corporate networks employing action research techniques to facilitate large-scale changes to their information systems. In others, it has been a question of community-based action research projects making use of computer communications to broaden participation.

Much of the action research carried out over the past 40 years has been conducted in local settings with the participants meeting face-to-face with "real-time" dialogue. The emergence of the Internet has led to an explosion of group communication in the form of e-mail, computer conferences, v-mail and video conferencing. While there have been numerous attempts to use this new technology in assisting group learning, both within organizations and among groups in the community, there is a dearth of published studies on the use of action research methodology.

Action Research is a form of collective self-reflective research conducted by participants in social situations in order to improve the rationality and justice of their own social or educational practices, as well as the understanding of those practices and situations in which practices are carried out.

The approach is only action research when it is collaborative, although it is important to realize that group action research is achieved through critically examined action of individual group members.

The reason it should be collective is open to some question and debate, but there is an important point here in relation to the commitments and orientations of those involved in action research.

10 CONCLUSIONS

The strength of action research lies in its focus on providing solutions to practical problems and its ability to empower practitioners by engaging them in research and subsequent development or implementation. Practitioners may choose to research their own practice or an outside researcher can be involved to help identify any problems, seek and implement solutions, and systematically monitor and reflect on the change process and results. Action research involves a spiral of self-reflective cycles of planning a change; acting and observing the process and consequences of the change; reflecting on these processes and consequences and then replanning; acting and observing; reflecting; and so on …

The main contributions of Action Research are related with the fact of being a practical research methodology to be used in applied research, such as all the research procedures that involve project development. It is possible to underline its main characteristics: a set of practices that respond to people's desire to act creatively in the face of practical and often pressing issues in their lives in organizations and communities; calls for an engagement with people in collaborative relationships, opening new 'communicative spaces' in which dialogue and development can flourish;

draws on many ways of knowing, both in the evidence that is generated in inquiry and its expression in diverse forms of presentation as we share our learning with wider audiences; value oriented, seeking to address issues of significance concerning the flourishing of human persons, their communities, and the wider ecology in which we participate; living, emergent process that cannot be predetermined but changes and develops as those engaged deepen their understanding of the issues to be addressed and develop their capacity as co-inquirers both individually and collectively; it has participatory character, a democratic impulse, and a simultaneous contribution to social science (knowledge) and social change (practice).

REFERENCES

Achari. P.D. (2014). *Research Methodology: a guide to ongoing research scholar in management*. India: Horizon Books Asia.

Bogdan, R.C., & Biklen, S.K. (1992). *Qualitative Research for Education: An Introduction to Theory and Methods*. Boston: Allyn and Bacon.

Carr, W. & Kemmis, S. (1986). *Becoming Critical: Education, Knowledge and Action Research*. London: Falmer.

Gilmore, T., Krantz, J. & Ramirez, R. (1986). Action Based Modes of Inquiry and the Host-Researcher Relationship, *Consultation 5.3* (Fall 1986): 161.

Lewin, K. (1946). Action Research and Minority Problems. *Journal of Social Issues* 2: 34–46.

O'Leary, N. et al. (2015). Closing the theory–practice gap physical education students' use of jigsaw learning in a secondary school. *European Physical Education Review* 21(2): 176–194.

O'Brien, R. (1998). An Overview of the Methodological Approach of Action Research. *Overview of Action Research Methodology*. Retrieved February 15, 2019, from http://www.web.ca/robrien/papers/arfinal.html

Schön, D.A. (1983). *The reflective practitioner: how professionals think in action*. Michigan University: Basic Books.

Winter, R. (1989). *Learning from Experience: Principles and Practice in Action-Research*. Philadelphia: The Falmer Press.

Industry 4.0 – Shaping The Future of The Digital World – da Silva Bartolo et al. (eds)
© 2021 Taylor & Francis Group, London, ISBN 978-0-367-42272-1

Fashion design course, a collaborative online learning experience

M.G. Guedes & A. Buest
2C2T Centro de Ciência e Tecnologia Têxtil da University of Minho, Guimarães, Portugal

ABSTRACT: Recognised as a complex system, fashion is affected by other complex phenomenon such as globalization, overconsumption, human labour exploitation, sustainability, etc. Fashion designers are constantly challenged to find the appropriate resources, and to work collaboratively to solve unpredictable problems. Fashion design education was expected to embrace fashion's non-linear dynamic, its changeability and adaptability. Nevertheless, traditional models of education still value a fixed, controlled and linear acquisition of knowledge and skills. It is this paper view that there is an opportunity to explore new models of education for the fashion sector. The collaborative, online learning model for fashion design (COL4Fashion) was elaborated in previous papers, part of a broader research about fashion design higher education. In this paper, the COL4Fashion will guide the development of a fashion design online course, discussing not the content, but the learning components, its limitations and potentialities in promoting a more meaningful and prospective learning experience.

1 INTRODUCTION

Fashion sector, as many others, is influenced by disruptive technological changes that affects consumer behaviour and business models. Broader wireless access allows and will continue to push connectivity, instigating the adoption on mobile, augmented and virtual reality experiences (WGSN Vision Team, 2018). This will change the way consumers relate to products, companies and brands. Influential fashion brands have social media presence, with millions of followers, that create content around them and demand more ways to participate in their creative process, personalising the final product (Global Fashion Industry Statistics, 2018). Meanwhile fashion designers work connected worldwide, influenced or partnered with social media 'influencers', resort to specialised information platforms, 3D software, and data-driven e-commerce (Statista, 2018a; Statista 2018b). In this context, it is valid to assume that, fashion designers, fashion consumers and fashion learners are part of a community that takes advantage of the online and mobile connectivity to exchange information, to work with multidisciplinary and multicultural teams, in a faster, efficient and cheaper way. It is also valid to assume that the changing behaviour towards the production and consumption of fashion, will require a new kind of 'fashion designer', with capabilities like agency, innovation, creativity, and problem-solving. (Guedes, & Buest, 2018 c, d).

The problem is that fashion education is still preparing (Edelkoort, 2017) 'the atelier diva', designers that work alone and disregard the collaborative, interactive, exchange economy. Fashion design traditional education models, supported by teacher-centred pedagogies and fixed curriculum, can only educate for what is already known. They cannot educate the next generation of concept workers, "creators and empathizers, pattern recognizers, and meaning makers" (Pink, 2005, location nº 82). The traditional perspective for fashion education also presents a barrier to explore learning in online environments, without the materiality of the atelier. While we do not argue against the atelier, we also acknowledge that the structural elements of the fashion product (textile, texture, cut, shape, colour, etc.) can be manipulated digitally, just like in other areas of design. The hypothesis is that, in the light of new technologies, learning fashion design can take advantage of collaborative online environments. Online learning environments have the potential to help fashion design learners develop autonomy and agency capabilities, establish connections that leads to new understandings not only about the fashion sector, but about themselves as designers. These environments can foster learning processes emerged from spontaneous connections and multiple social interactions within the learning community, or from real-life challenges and experiential opportunities.

This paper is part of a broader research about fashion design education in the European Higher Education Area (EHEA), especially offered in online environments. It intends to discuss the applicability of the *collaborative, online learning model for fashion design* (COL4Fashion), devised in a previous study (Guedes & Buest 2018a), exposing its limitations and potentialities, in an online course, so this could be further tested. By presenting an online course in fashion design, this paper intends to discuss not its content, but how the COL4fashion model would be implemented, and represent an alternative way of learning fashion, so that professionals can meet and foresee the rapid demands of the sector.

The following section presents the methodological choices, grounded in previous studies and complemented by information collected from the current offer in fashion design, within EHEA. Next, the results present the curricular structure present in seven fashion design courses, to understand their educational model. Then it discusses the components of the COL4fashion model applied in a fashion design online course, to explore how this could create a more flexible and negotiated curriculum to foster a collaborative learner-defined, personalised learning experience in the field. The final section presents the conclusions and limitations.

2 METHODOLOGICAL FRAMEWORK

As previously stated, this paper is part of a broader study about fashion design education that is grounded in Heutagogy or the study of self-determined learning (Blaschke, 2012; Blaschke & Hase, 2016; Dick, 2013; Hase & Kenyon, 2007; Hase, 2009), but also in constructivism (Schön, 1983) and connectivism (Siemens, 2005). These theoretical approaches place learners in the centre of the learning process, supported by the learning community, connected through technology, that allows learning by discover, share knowledge and information, and co-creating content. In a previous study (Guedes & Buest 2018-b), Heutagogy helped outline a set of learning principles that would be adequate for the specificities and complexities of the fashion system, while fostering competence and mainly capability acquisition. They can be summarised as:

- Reflective Thinking - critically understand a 'problem'; defy learner's own beliefs.
- Research and Interpretation - explore social reality with an interpretive, inquiry attitude.
- Creativity and Imagination - (re)interpret an existing reality and imagine new solutions.
- Collaboration and Communication - construct knowledge collectively and collaboratively.
- Complexity and Uncertainty - thrive under non-linear and unpredictable circumstances.

These principles were essential to elaborate the *collaborative online learning model for fashion design*, COL4fashion (Figure 1). The COL4fashion model presupposes an open learning environment, in which outcomes are broader and negotiated, as well as the assessment. The assessment is then, part of a contextualized, reality-based, flexible learning experience.

Learning starts at the 'Learning Discovery', aimed to identify learners' backgrounds and needs, as well as to provoke questioning (Learner-centred), while forming a learning community. Learning Assessment (LA) and Learning Outcomes (LO) are negotiated from the Learning Contract (LC3), considering the Personalised

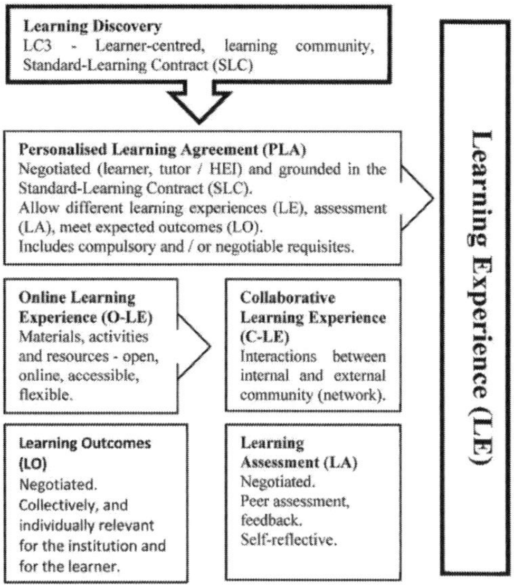

Figure 1. Collaborative online model for fashion design (COL4Fashion), adapted from Guedes & Buest, 2018-a.

Learning Agreement (PLA), all that will underpin the learning experience (LE). In this context, learners are not expected to correctly reproduce acquired knowledge, but to address and solve new problems critically and creatively and propose innovative solutions.

The five learning principles and the COL4fashion model were essential to acknowledge that, when organising an online course, it wouldn't make sense to establish a stable set of contents, a linear sequence of topics, neither fixed outcomes nor expected behaviours. Therefore, it was necessary to identify and understand the curricular structure of the current offer in fashion design courses (within EHEA).

From the initial 371 fashion programmes selected within the EHEA, 67 presented fully curriculum and syllabus that could provide information for this study. The focus was in the first cycle of the Qualifications Framework of the European Higher Education Area (QF-EHEA). So, the courses selected for this paper, are BA Hons, or degrees that give 180 and 240 ECTS. Another criterion was to select higher education institutions included in the following professional and academic rankings for the year 2018, since they can reflect institutional or programs notoriety in the sector:

- Academic Ranking of World Universities, ARWU
- Business of Fashion
- CEOWORLD Magazine Rankings
- CWTS Leiden Ranking
- Fashionista Fashion School Ranking
- QS World University Rankings
- Times Higher Education World University Rankings, THE
- US News Rankings

Finally, this study selected programmes offered in English, so the information provided about their curriculum and syllabus could be fully comprehended. From the fashion courses selected, 'fashion project' or similar course units were selected and studied, since they require acquisition of technical, social, communication skills, but also presupposes a reflective ability to use these in every new project.

The final step was to generate a 'sketch' of a fashion design online course, in which the learning components were organised to foster learner's agency, autonomy and ability to use the acquired knowledge and skills in different, contextualised circumstances. The COL4fashion model provided the necessary specifications to create an online course that could be further tested (as a prototype), so that the concepts and solutions presented in this paper can be verified.

3 RESULTS

3.1 *Fashion design education current offer*

The final sample below is composed by seven fashion design courses.

They provided an insight of the current curricular structure, so that we could understand if this was focused on knowledge and skills acquisition and/or this was organised to be aligned with the current and future requirements of the fashion sectors.

- Istituto Europeo di Design IED, Barcelona, Spain. Bachelor of Arts (Hons) in Fashion
- Istituto Europeo di Design, Milan, Italy. Fashion Design DAPL (*Diploma Accademico di I Livello equipollente ai diplomi di laurea universitari*).
- Institute Français de la Mode (IFM), Paris, France. Bachelor of Arts in Fashion Design.
- Ravensbourne University London, London, United Kingdom. BA (Hons) Fashion.
- University for the Creative Arts (UCA), Epsom, United Kingdom. BA(Hons)Fashion.
- Università Telematica San Raffaele Roma, Italy, Online. Architecture and Industrial Design for Fashion, Laurea Triennale or 'Laurea di primo livello'.
- Università Telematica ECampus (UniE-Campus), Novedrate, Italy, Online. Design and Fashion Discipline (D.M. 270/04).

3.2 *Fashion design online course*

The 'sketch' of the fashion design online course (Figure 2) was elaborated to be delivered totally online, through a learning management system, such as Moodle or similar, for 8-12 weeks.

As previously stated, the analysis of the current offer in fashion design focused on 'fashion project', as a subject. So, the online course encompasses the phases of the project development, inspired by the Delft Basic Design Cycle (vanBoeijen & Daalhuizen, eds., 2010), which relates to the heutagogical premise of double-loop learning, that stimulates questioning and reflectiveness. This way it is expected that the course covers a wide range of subjects, and competences fashion designers must acquire in order to successfully develop their projects.

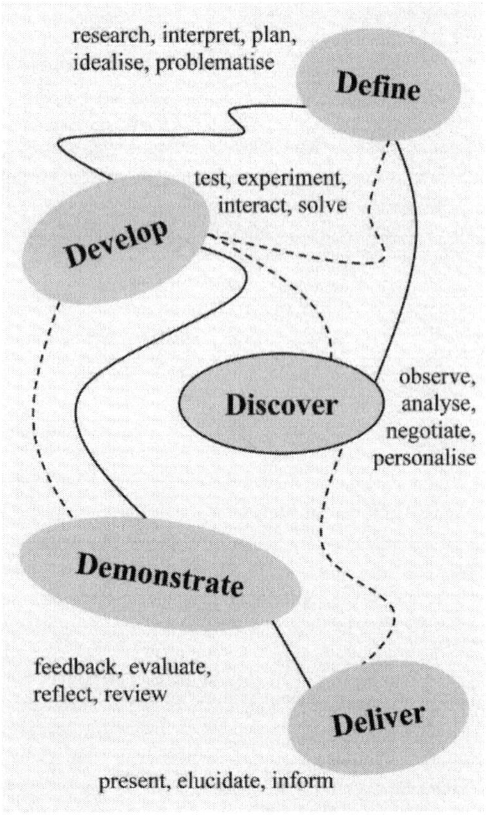

Figure 2. Fashion design online course.

The 'parts' of the course presented in Figure 2 were conceptualised to work in an interrelated manner, and to comprise both the collective and personalised learning experiences. Learning is expected to happen while learners are experimenting and critically involved in iterative cycles of analysis-synthesis, simulation-evaluation, from the initial ideas to solve a problem to the provisional and final design.

4 DISCUSSION

4.1 *Fashion design education current offer, analysed*

The courses selected and analysed in this paper provided an important sample since they are offered by well-respected and iconic institutions in the fashion sector.

An intriguing aspect during this research was to discover that, although fashion is an international sector, few programmes are taught in English, or are offered in foreign languages. This is even more critical when analysing the online programmes, which are both taught in Italian.

Another interesting aspect is that, in the programmes studied, fashion design adopts a completely different configuration to follow the national education and market guidelines requirements. This means that fashion design loses the opportunity to be fully internationalized and/or student-centred, meaning, fulfil the needs of foreign students.

Overall, the courses are aligned with Dublin Descriptors "expected achievements and abilities", comprising competences of the projectual, professional, technical, sociocultural, communicational and managerial levels.

Fashion project, as a subject, is present during the three years, except in the BA (Hons) Fashion, from Ravensbourne University, in which it is the main subject of the third year.

Subjects like fashion cutting, sewing, dressmaking, textile technologies, are very relevant in IED, during the three years, in IFM (in the first and second years) and in Ravensbourne University during the second and third years, similarly to the UCA.

In the online offer, San Raffaele programme presents 'Science of materials and Technology', which is a broad theme and can or cannot be considered relevant to fashion design learning. Both San Raffaele and Uni-ECampus emphasise history, anthropology and art, as well as management skills, due to their respective affiliation to architecture and literature faculties. Those subjects are more fashion contextualized in IED, UCA and Ravensbourne University programmes, and absent in IFM, which is mainly industry-focused.

Finally, professional experiences are very important in IFM (second year), and in the second and third years of UCA and Ravensbourne University programmes. Work placement or internship are absent in IED programme.

The analysis enlightened that fashion design curriculum is organized as a progressive accumulation of knowledge, meaning that separated learning units collaborate for the desired competence in that level (first, second or third year).

If 'fashion major project', included in the 3rd year, is presented as the culmination of all previously studied subjects, this means that student is supposed to acquire other competencies prior to be able to experience them in a relevant project. So, by the time students have the chance to face a fashion project, they may or may not use the range of skills they studied.

Furthermore, once the curriculum is fixed, with some courses offering optional subjects, students have to follow a teacher-centred pathway that can or cannot be aligned to the challenges they will face in the professional field.

Finally, the courses analysed are mainly content focused, with some exception in the case of Ravensbourne University London, that adopts a holistic perspective to the curriculum. Since students receive the content already prepared, what will motivate them to research further? To make connections with different areas? To question the reality?

4.2 *Fashion design online course*

This section analyses the different 'parts' of the fashion design online course relating them to each other and to the learning experience (LE) defined by the in the COL4fashion model.

First it is acknowledged that project development, as a subject, is already implemented in the fashion design curriculum, in both on-site and online models, as shown in the fashion design current offer. However, since this study is not concerned with the competence acquisition, which has been fairly accomplished by more traditional models of education, the main focus of this paper is to explore how a flexible, learner-centred, collaborative, personalised curriculum could help learners to develop their fashion projects and at the same time gain self-determined, autonomous and prospective capabilities that will help them navigate under uncertain professional scenarios.

Also, it is important to indicate that the online course is expected to create a flexible and dynamic curricular learning path, co-organised by the educational institution, the tutors and the learners, allowing the adaptation or incorporation of topics and resources brought by the learning community or by external sources, such as companies and experts. This means that course content, as well as learning outcomes and assessment opportunities will be defined together, but mainly in accordance with project requirements, with the function/use of the product to be developed, with what learners need to accomplish from the learning experience, etc. So, although all learners are required to move among all the parts in order to complete a fashion project (and the learning experience), in 'Discover' they are able to negotiate their own learning path, returning and advancing throughout it, mapping the content, resources, assessment and outcomes in a relevant manner for their personal and professional lives. So, 'Discover' relates with learners planning their own project development, but also to the Learning Discovery phase of the (COL4fashion), in which learners are introduced to the learning environment and the learning community. The tutor's role is to facilitate connections between learners, embracing each other's values, and defying their own, guide and provoke the discussion and questioning about the current and relevant issues of the fashion sector. Each learner is expected to uncover their own interest as well as help the colleagues on theirs. From the development of a fashion mood board, to a fashion trend forecast exercise or the exchange of relevant experiences as fashion professionals, it is not the theme of the activity that is the main focus, but how they offer opportunities for connection between specialists from different areas of the fashion sector. This is important because, by creating their own learning networks, learners became more independent from the course resources and from the tutor, relying on a much more extended community during the learning experience. By the end of this phase a personalised learning agreement (PLA) is negotiated among each learner, the tutor and the HEI. In the PLA document, learners expose their learning goals and needs, indicating how they intend to achieve that and what kind of outcomes would make sense for them.

Next, in 'Define' learners are expected to start the Learning Experience (LE, as indicated in the COL4-fashion model), interacting with the O-LE (materials and resources available online) as well as the C-LE (community and activities), that will help them understand and define the problem they wish to address. During the LE learners have access to relevant content provided by the HEI, by the tutor, by themselves, by the learning community or by their external network. So, the 'problem' can arise either from a personal or professional interest or from a company brief. While researching about the problem, the main concert of the tutors is to ensure that learners are activating *Reflective Thinking/Research and Interpretation principles*. The O-LE and C-LE continues in 'Develop', in which learners activate *Creativity and Imagination* and 'Demonstrate', which stimulates *Collaboration and Communication* principles. The online environment provides open databases, software (applications), digital tools and techniques, which learners can select and experiment to their projects. The learning community, external collaborators and the tutors are important to enhance learners' digital skills, to give feedback, and to maintain a diverse sociocultural context, triggering loops of learning and making the assessment (LA, as indicated in the COL4fashion model) part of the learning experience. Tutors assist learners to instigate *Collaboration and Communication/Complexity and Uncertainty* principles, as self-reflective capabilities used to test and analyse the results of their initial designs, deciding whether or not further testing and improvement are required, the problem need to be reviewed or even reformulated.

'Deliver' relates to the Learning Assessment (LA) and Learning Outcomes (LO) in the COL4fashion model, and it also triggers *Reflective Thinking/Collaboration and Communication/Complexity and Uncertainty* principles since learners are required to present the chosen solution, 'sell' their fashion design projects. It is important to emphasise that the Learning Outcomes from the project are not considered final, closed, definitive. They need to be verified against learners' and institutional expectations, as outlined in the Learning Contract and in the Personalised Learning Agreement, but it must be acknowledged their openness, as opportunities to generate new questions and challenges. In this sense the results of the project cannot be judged as 'right or wrong'. Firstly because, ongoing and personalised assessments were provided by the learning community and tutors throughout the LE, either using formative or summative feedback. Also, because assessments must be meaningful for the HEI, but mostly for learners, measurable their professional settings - e.g. understand how to apply upcycling in the company, resorting to textile leftovers or improve the product quality using simple augmented reality tools in distance meetings with the supplier. So, LA and LO support learners in reflecting upon their own process, their practice, whether they could achieve the competencies desired, while developing the project and most of all if they recognise themselves as capable to implement those competencies in different scenarios.

5 CONCLUSION

This paper intended to conceive a fashion design online course, based on the *collaborative, online learning model for fashion design* (COL4Fashion), so it can be properly tested. It is expected that this course represents an alternative possibility for fashion design education, not only because it is delivered online, but because it adopts a systemic dynamics that may enhance learners agency and self-determined capabilities, so that they can rapidly incorporate the outcomes in their academic and professional lives.

Still, it is important to acknowledge some limitations. Both HEI and learners might impose barriers to the model adopted by the proposed online course, to find it 'too open'. Another challenge is that, despite 'fashion people' (designers, consumers, learners, etc) are inclined towards technological affordances, especially for communication purposes, a certain level of comfort around technology will be desirable and may not be present. However, the big caveat will be to address and try to overcome the materiality of the atelier, resorting to accessible and affordable technological software.

Despite the challenges, the importance of this paper is that it envisioned an online course which will defy traditional higher education paradigms - to adopt a more adaptable curriculum, to propose flexible and negotiated assessments, yet maintaining the education validity and credibility. Furthermore, in order to grant learners more control over their own learning experience, tutors will need to adopt a design thinking posture, to help learners reflect while deconstructing problems, exploring solutions and making relevant choices.

The fashion design online course represents a provisional solution, an opportunity to rethink fashion education, especially in online environments and experiment with a more accessible, timely, updated learning experience, compatible with the technological changes in the sector.

ACKNOWLEDGEMENTS

This work is financed by Project UID/CTM/00264/2019 of 2C2T – Centro de Ciência e Tecnologia Têxtil, funded by National Funds through FCT/MCTES.

REFERENCES

Blaschke, L. M. 2012. Heutagogy and lifelong learning: a review of heutagogical practice and self-determined learning. *The International Review of Research in Open and Distributed Learning*, 13(1): 56–71.

Blaschke, L.M. & Hase, S. 2016. Heutagogy: a holistic framework for creating twenty-first-century self-determined learners. In B. Gros and M.M. Kinshuk (eds.), *The Future of Ubiquitous Learning. Lecture Notes in Educational Technology*: 25–40. Berlin: Springer Berlin Heidelberg.

Dick, B. 2013. Crafting learner-centred processes using action research and action learning. In S. Hase and C. Kenyon (eds), *Self-Determined Learning Heutagogy in Action*: 39–52. London: Bloomsbury Academic. https://media.8ch.net/file_store/74ebad3acfdd6bc81537b65bd93bb21974067e53d40c76353060bcc2581bb2e6.pdf/1497464358.pdf#page=50.

Edelkoort, L. 2017. *Anti-Fashion, a manifesto for the next decade* [web streaming video]. https://www.youtube.com/watch?v=LV3djdXfimI.

Global Fashion Industry Statistics. 2018. *Fashion United Business Intelligence*. https://fashionunited.com/global-fashion-industry-statistics.

Guedes, M. G. & Buest, A. 2018a. A collaborative, online learning model for fashion design. *11th annual International Conference of Education, Research and Innovation*, Seville, Spain, pp. 1216–1225.

Guedes, M. G. & Buest, A. 2018b. Heutagogical principles for fashion design higher education, *10th International Conference on Education and New Learning Technologies*, Palma, Spain, pp. 8603–8612.

Guedes M. G. & Buest, A. 2018c. Semiotic perspectives for a global fashion design higher education. *10th International Conference on Education and New Learning Technologies*, Palma, Spain, pp. 8613–23.

Guedes, M. G. & Buest, A. 2018d. The semiotics of fashion design: A proposal for higher education. *Critical Studies in Fashion & Beauty*, 9(2): 197–211.

Hase, S. & Kenyon, C. 2007. Heutagogy: A child of complexity theory, *Complicity: An International Journal of Complexity and Education*, 4 (1): 111–118. https://journals.li brary.ualberta.ca/complicity/index.php/complicity/article/view/8766/7086.

Hase, S. 2009. Heutagogy and e-learning in the workplace: Some challenges and opportunities. *Journal of Applied Research in Workplace E-learning*, 1 (1): 43–52. https://epubs.scu.edu.au/bus_pubs/72/.

Pink, D. H. 2005. *A whole new mind: Moving from the information age to the conceptual age*. New York: RiverheadBooks.

Schön, D.A. 1983. Design as a reflective conversation with the situation. In Schön, D.A. *The reflective practitioner: How professionals think in action*: 76–104. New York: Basic Books.

Siemens, G. 2005. Connectivism, a learning theory for the digital age. *International Journal of Instructional Technology and Distance Learning*, 2(1). http://www.itdl.org/Journal/Jan_05/article01.htm.

Statista 2018a. *Apparel Market in Europe*. https://www.statista.com/study/27600/apparel-market-in-europe-statista-dossier/.

Statista 2018b. *eCommerce Report 2018 – Fashion*. https://www.statista.com/study/38340/ecommerce-repor t-fashion/.

vanBoeijen, A. & J. Daalhuizen (eds.). 2010. *Delft Design Guide*. http://ocw.tudelft. Nl.

WGSN Vision Team. 2018. *The Message 2020, Part 1: Fix the Future*. https://www.wgsn.com/content/reports/#/Future+Trends/w/The_Vision/27534.

Industry 4.0 – Shaping The Future of The Digital World – da Silva Bartolo et al. (eds)
© 2021 Taylor & Francis Group, London, ISBN 978-0-367-42272-1

Design drawing/drawing design

A. Moreira da Silva
CIAUD, Lisbon School of Architecture, Universidade de Lisboa, Lisbon, Portugal

ABSTRACT: Nowadays some questions have been raised regarding the importance of hand drawing teaching within design education. Once, hand drawing and sketching ruled design curricula, because there simply was no alternative. The introduction of CAD, image editing, on-line search engines and other new technology advances have irrevocably changed the way we perceive and map the world around us. As a result, the status and value of hand-drawing has changed and it is important to reflect upon the functions, the teaching and the evaluation of the tools we use to express design. Drawing, within the design process, serves several functions. In this paper we intend to study this several functions and to understand the importance of the permanence of hand-drawing teaching classes in the design courses curricula despite the new technologies advances.

1 INTRODUCTION

This paper stems from a current post-doc research project motivated for the need of producing more knowledge and reflection on Drawing. Our main post-doc research question is about the nowadays importance of hand-drawing as a critical instrument and as an operative support within the design creative process.

We intend to understand how sketches can be powerful representations of new ideas, as part of a larger design process, as a key method for thinking, reasoning, and exploring solutions for design problems. Through the study of several statements from various authors we want to investigate the permanence of the sketching important role since mid-last century. From this theoretical approach we also aim to verify drawing teaching relevance for future designers despite the paradigm changes.

The main objectives of this current research are: to verify the relevance of the hand-drawing teaching in the formation of the future designers, and also, to verify the permanence and the importance of hand-drawing in designer's actual daily work, although the new assistive technologies constant progress.

We aim to make contribution on the study of creative design process, enhancing hand-drawing important value for design students and professionals.

2 DRAWING/SKETCHING/DESIGN

In this study we consider the close relationship between drawing and design and the relevance of drawing as stimulating instrument when sketching the first ideas and as critical verification of the several hypotheses.

As the designer must conceive and develop solutions for precise problems which may be of very different nature, sketching can work as operative support for inventive problem-solving during the creative process.

The complexity, characteristic of the transition from the idea to its materialization, is related to the actual extent of projects where a number of factors come into play, such as the ability to understand the context and imagine the solutions, the ability to know and take advantage of the processes and materials as means or vehicles through which the solution can be materialized, the ability to transform ideas into appropriate forms, having in mind the limits and material possibilities. Drawing holds, in itself, the making visible quality of the whole mental process underlying the conceiving stage, from the first sketches of a vague idea of we want to create until the final solution.

Drawing's transcendent importance lies in the capacity it gives to materialize abstract conceptualizations and to create the ideational basis for new forms and objects.

Within the complex process which takes place from how to imagine something new till its materialization, the designer can use free hand drawing as an indispensable mean to help him to develop the idea, as an operative support in the creative process. This attitude comes across the concept of drawing as mental activity, which uses the hand as an extension of the brain.

Sketching has been used long ago, especially since Italian Renaissance when drawing started to assume a more important role as the basis of creation. In this period drawing assumed its intellectual character and some authors of those times, like Alberti or Vasari, considered it like something that the hand achieves coming from the human brain. (Moreira da Silva, 2010)

Since then we can find sketches as the basis of creative production not only in the fine arts but also in design. Drawings have been used for generations of designers. The use of sketches not only helps a great deal in starting the ideas generating process, it also enables high quality concepts to begin with as initial ideas. These ideas can be accessed, combined, selected and developed.

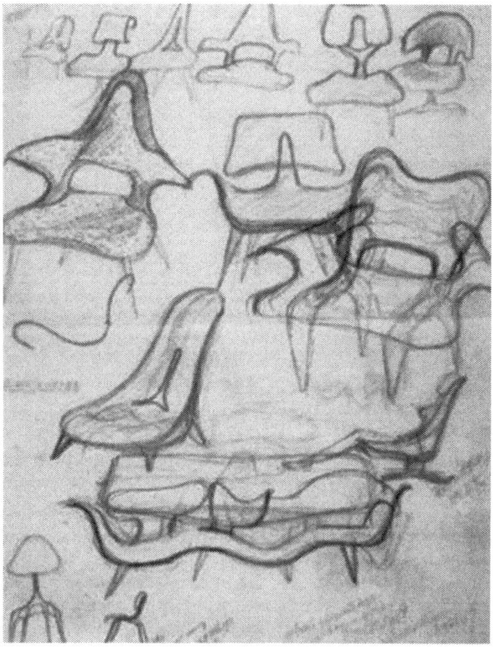

Figure 1. Charles Eames chairs sketches (1943).

Source: https://www.google.pt/search?q=Charles+eames+sketches&tbm. Accessed March 2019.

Several authors have written about the importance of drawing in design methodologies. Their general conclusion points out to drawing as an essential tool as it allows ideas to flow and provides a critical investigation of the several alternative solutions in creative processes.

Cross (2005) states that drawings are a key feature of the design process. At the early stages of the projects they are communications with oneself, and concludes that the conceptual thinking processes are based on the development of ideas through their external expression in sketches.

In *The Inexorable Rise of Drawing*, an article that analyses contemporary thinking about drawing, Annabel Tilley tries to identify the reasons why, nowadays, the drawing subject became so relevant. According to Tilley (2008), the interdisciplinary nature of drawing has been acknowledged nowadays and also recognized drawing's facility to allude to or to describe the intangible.

Sketching practice goes beyond the simple representation of ideas: it covers the development of intelligence from the perception and analysis of one's own drawing, a medium that is the reflection in action. (Schön, 2000)

Ortega and Weihermann (2017) state that when students use sketches, new ideas flow, qualities and relationships that were not previously imagined reveal themselves.

According to Mike Baskinger, drawing and sketching are an integral component to the development process for many designers. Sketching tends to be very engaging and invites others in for collaboration. Drawing by hand can enable to think differently about a subject or a design problem and can equip with greater persuasion and impact during collaboration. Hand-generated drawings can also provide a basis for transitioning into digital sketching in a variety of tools. The expediency and impromptu nature of picking up a pencil and letting ideas flow onto paper can be both powerful and compelling. Drawing ideas can serve to clarify, lead, and facilitate collaboration in meaningful ways. (Baskinger, 2008)

Sketching is fast and easy, allows to explore and refine ideas in a quick, iterative and visual manner. No need to learn any fancy design tools, just need to put a pencil or a pen to paper or marker to whiteboard and let the ideas flow. Sketching is a very simple and easy tool used by many creative professionals in any circumstance, because we can draw anywhere, anytime, with any medium.

The search for solutions, even for simple problems, implies that drawing studies in detail each phase of the process for obtaining the result we seek. The complexity of the required drawings emerges as the designer approaches the solutions which he considers ideal or that are possible. During the conception phase we should not disregard the importance of a trace, of a scribble, of a sketch. They can all contribute to the idea development.

Figure 2. Graphic design sketchbook layout.

Source: https://www.google.com/search?q=sketchbook+layout sketches+for+graphic+designers&tbm. Accessed March 2019

For a designer, the sketchbook is not just a place to draw but also a place to order thoughts, to gather graphically and visually information and to develop a design response. It emerges as a way of thinking through drawn lines as a creative process, revealing the central role that drawing can play within the development of any project.

"The real goal of sketching is about generating ideas, solving problems, and communicating ideas more effectively with others." (Rohde, 2011:n.p.)

Drawing, within design processes, serves several functions. We can narrow them down to four functions that are indissolubly interrelated: recording, exploration, communication and expression. These functions do not stand on their own, more often one drawing is able to combine several of them. (Lawson, 2004)

Figure 3. Sketching allows to explore a wide variety of ideas al at once.

Source: https://alistapart.com/article/sketching-the-visual-thinking-power-tool. Accessed March 2019.

The act of drawing allows that the reasoning and thoughts we have developed can be gradually translated and decoded throughout the drawn lines. Somehow, we struggle with our own ideas on the paper space. We scratch, we draw, we overwrite features, we configure, we represent, we visualize, giving physical form to our thinking. There is a direct link between the thought and the hand that performs the drawing: the hand as an extension of the brain, of the reasoning.

Rohde (2011) states that "sketching is the visual thinking power tool" and adding sketches to the process is a great way to amplify software and hardware tools. He concludes that the use of sketching during the creative process reveals the central role drawing plays within the development of a project and the importance of hand-drawing, particularly in an age dominated by digital media. For Rohde, sketches can work not only as a visual thinking tool but also as a primary language for capturing thoughts, exploring ideas, and sharing those ideas. Sketching provides a unique space that can help us think differently, generate a variety of ideas quickly, explore alternatives and enable constructive discussions with colleagues and clients.

For Wilkinson (2016) it can be a simple drawing to start, but something happens when sketching that encourages the creative thinking and ideas begin to flow, something begins to take form and "thinking through drawing" allows the evolution of several solutions and their critical analyses.

According to Norman (2010) the recent profound changes in project processes, due to the contemporary historical context where new technologies are so important, justify an extended discussion on the crucial role played by drawing in design courses, adapted to the new paradigm.

The fact of using new technologies does not invalidate the vital role played by hand-drawing, both at the initial stage of recording the first ideas and during their subsequent development and in the critical analysis of the different hypotheses.

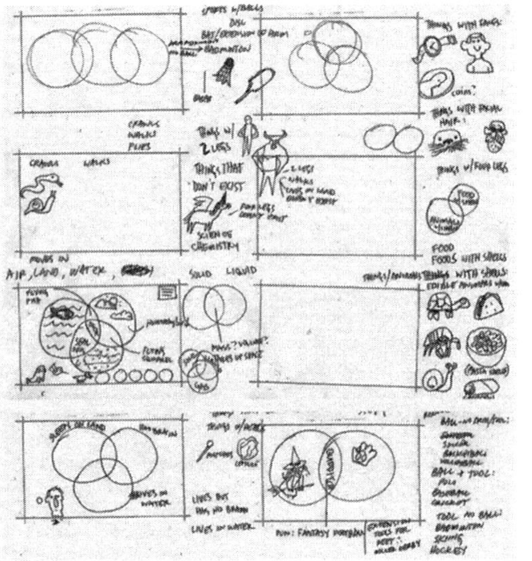

Figure 4. Mike Dutton's sketches for a Google doodle (2014).

Source: https://www.google.com/doodles/john-venns-180th-birthday. Accessed March 2019.

To Roome (2011) it is the possibility of cross-fertilization between traditional and digital platforms that is interesting and beneficial.

For Hamilton (2009) the value of hand-drawing is associated with the knowledge that drawing is the essence of creativity and its measure would lie in the individual ability to continuously express and create, using self-exploration and creative thinking through the process of drawing, linking together interesting ideas and technical know-how, enhanced by assistance and interaction with new technologies.

Domingos (2017) in his recent book, writes about the extraordinary advances of artificial intelligence and its promising future, but as he mentions throughout this book, artificial intelligence is developing in the direction of automatic learning, but not yet capable of creating something really new which is the very essence of human creativity.

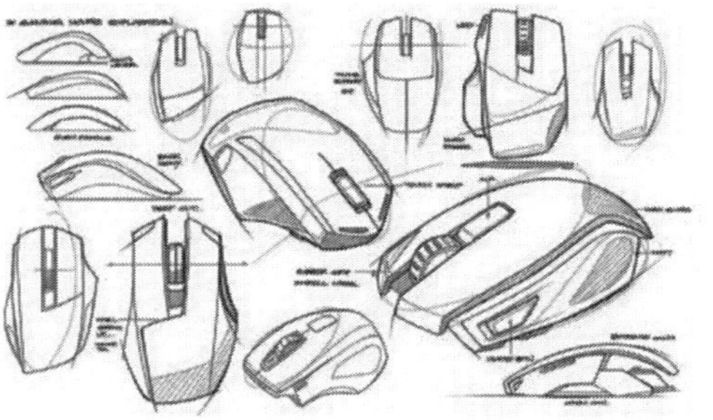

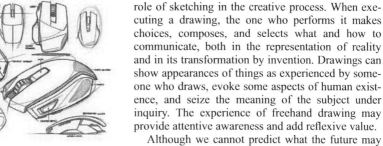

Figure 5. Gaming mouse sketches.

Source: https://medium.com/@haydmills/11-creative-professions-that-use-sketching-as-a-tool-in-their-design-process-and-why-995e40b2c6fb. Accessed March 2019.

Drawing holds, in itself, the making visible quality of the whole mental process underlying the conceiving stage, from the first sketches of a project vague idea to the creation of its final form. Drawing's transcendental importance lies in the capacity that it gives the designer to materialize abstract conceptualizations and to create the ideational basis for design forms and artifacts.

Drawing today serves as a primary medium for generating, testing and recording a designer own creative and conceptual musings about a project. If drawing is central for conceiving, it is also fundamental for defining how that conception is managed as it moves from its initial stages through its actual development and realization as a material form. Sketches illustrate various aspects of the creative design process. Therefore, the importance of drawing assumes a broad sense, conferring to the act of drawing the ability to become a means of multiple resources to practice the discipline of design.

3 CONCLUSIONS

From many authors' statements and according to the professional practice of many designers, we can conclude, from this initial research phases, that drawings are the key to the creative process in design. Drawing presents itself as an operative support for problem-solving and critical analyses in the design creative process.

We can also conclude that new technologies offer the designer new ways, allowing him to save time and facilitating many of the daily practice tasks, even becoming indispensable in many work phases. However, they do not replace sketching, which continues to assume a 'chameleon' shape during the various historical periods, in a constant time adaptation, and, especially by incorporating a critical dimension in design's creative process.

New technologies can be a stimulus for creative capacities, but it doesn't invalidate the important role of sketching in the creative process. When executing a drawing, the one who performs it makes choices, composes, and selects what and how to communicate, both in the representation of reality and in its transformation by invention. Drawings can show appearances of things as experienced by someone who draws, evoke some aspects of human existence, and seize the meaning of the subject under inquiry. The experience of freehand drawing may provide attentive awareness and add reflexive value.

Although we cannot predict what the future may bring in the Artificial Intelligence field, we believe that in the near future thinking machines will hardly replace the creative capacity that sketching can trigger. Drawings are able to trigger ideas while offering the practical tool to express those ideas. Sketches can be the key instrument of creation and control, also playing a critical role within the design creative process.

Despite the paradigm shift required by the changing times, we believe that hand-drawing will remain inseparable from the designers training and professional practice, assuming essential operating support in the project activity which continues being the design basis.

In the future, design education should pass through a systematic approach to hand-drawing in order to highlight and analyze the flexibility with which it adapts itself to various purposes, efficiently fulfilling a wide range of intentions, in a constant and vital adaptation to the continuous changes in the teaching and in the practice of the student and of the designer, given the new techniques and technologies.

To validate these assumptions, the current postdoc research can constitute a contribution for the understanding of the importance of hand-drawings permanence as the basis of the design creative process, although the constant progress of the new technologies.

REFERENCES

Baskinger, M. 2008. Pencils before Pixels, A Primer in Hand-generated Sketching. In *Interactions*, Vol. XV. https://mv122011.files.wordpress.com/2010/11/baskinger-pencils_pixels.pdf. Accessed February 2019.

Cross, N. 2005. *Engineering Design Methods, Strategies for Design*. Chichester, England: John Wiley & Sons, Lt.

Domingos, P. 2017. *A Revolução do Algoritmo Mestre*. Lisboa: Manuscrito Editora.

Hamilton, P. 2009. Drawing with technology in a digital age. https://www.lboro.ac.uk/microsites/sota/tracey/journal/dat/images/paul_hamilton.pdf. Accessed January 2019.

Lawson, B. 2004. *What Designers Know*. Oxford: Elsevier.

Moreira da Silva, A. 2010. *De Sansedoni a Vasari: um contributo para o estudo do Desenho como fundamento do processo conceptual*. Lisboa: Universidade Lusíada Editora.

Norman, D. 2010. Why Design Education must change. http://www.core77.com/blog/columns/why_design_edu

cation_must_change_17993.asp. Accessed January 2019.
Ortega, A.R., and Weihermann, S. 2017. Graphic dialogues: The progress of knowledge in architecture studio. In Kong, M. S. M. and Monteiro, M. R. (eds), *Progress(es) – Theories and Practices*. 111–114, London, UK: CRC Press, Taylor & Francis Group.
Rohde, M. 2011. Sketching: the Visual Thinking Power Tool. http://alistapart.com/article/sketching-the-visual-thinking-power-tool. Accessed February 2019.
Roome, J. 2011. Digital Drawing and the Creative Process. http://www./tracey@lboro.ac.uk. Accessed February 2019.
Schön, D. 2000. *Educando o profissional reflexivo: um novo design para o ensino e a aprendizagem*. Porto Alegre: Artes Médicas Sul.
Tilley, A. 2008. The Inexorable Rise of Drawing. http://www.transitiongallery.co.uk. Accessed March 2019.
Wilkinson, C. 2016. Thinking through drawing. http://vimeo.com/159085592. Accessed February 2019.

Urban spaces

Industry 4.0 – Shaping The Future of The Digital World – da Silva Bartolo et al. (eds)
 ISBN 978-0-367-42272-1

Modularization of integrated photovoltaic-fuel cell system for remote distributed power systems

C. Ogbonnaya & A. Turan
Department of Mechanical, Aerospace and Civil Engineering, The University of Manchester, UK

C. Abeykoon
Aerospace Research Institute, Department of Materials, The University of Manchester, UK

ABSTRACT: In the last decade, studies on how integrated photovoltaic-fuel cell (IPVFC) system can be used for grid-connected and distributed applications have intensified because the system has a huge potential application in the emerging hydrogen economy. Model-based and experimental works show that the system has potential applications for supplying reliable electricity based on solar hydrogen. However, no study has looked at the manufacturability of the system. In this study, modularization of the system is presented as a strategy to facilitate the ultimate commercialization of the system. Hydrogen management in the system is the most critical factor for utilizing the system for remote distributed power system. Discussions on the interconnections between modularity in design, manufacturability and supply chain integration were made as well.

1 INTRODUCTION

Solar energy is a sustainable form of energy because it is ubiquitous, clean and renewable (Ogbonnaya, Abeykoon, et al., 2019). However, technologies that depend on it suffer a major set-back due to occasional non-availability of the sun particularly at night or due to unfavorable weather conditions. To mitigate this deficiency, solar-based hybrid photovoltaic-fuel cell systems have been proffered as one of the possible solutions to the discontinuous nature of solar energy when acting as the prime mover. This implies that solar hydrogen can be produced and stored so that the hydrogen, as energy carrier, can be converted into electricity via electrochemical process in the fuel cell whenever power is demanded.

In the last decade, integrated photovoltaic-fuel cell (IPVFC) system has been studied from different perspectives ranging from its thermodynamic efficiencies, design, operations, and control. For instance, an attempt has been made to integrate photovoltaic, electrolyser and high temperature PEMFC operating at 1 bar and 60°C under the assumption of steady state, constant single home electricity demand of 1 kW, with equal hydrogen-oxygen stoichiometric ratio of 1.2 (Devrim & Pehlivanoglu, 2015). They used electrolyser and photovoltaic (PV) module with 88% voltage efficiency and 17.64% conversion efficiencies, respectively. The major finding is that the system needs power from the grid at winters for hydrogen production. This suggests that future stand-alone system needs to factor in the fluctuations of solar radiation at different seasons of the year since it directly affects hydrogen production if the system must be optimized for distributed generation completely independent on the grid and external supply of hydrogen. Establishing an appropriate solar radiation design point is also difficult as solar radiation is not constant in any location across the months of the year.

A design study (Lehman & Chamberlin, 1991) of photovoltaic-fuel cell energy system consisting of a PV with 96 cells in series with nominal 24 volts and 9.2 kW (Arco/Siemens M75 modules) integrated with alkaline electrolyser (Teledyne Energy System Altus 20) with 12 cells in series and proton exchange membrane fuel cell showed that the system can be integrated to work together as a reliable energy system. Although the study was at the initial stage at the time of reporting, the researchers inferred that the system could be autonomous if the monitoring and control systems are implemented seamlessly. Another IPVFC system that was successfully operated for six years as of 1999 in Julich Germany and was also used as a demonstration plant for renewable energy research is PHOEBUS (Meurer et al., 1999).

There are also model-based studies on different aspects of integrating photovoltaic modules and fuel cell for power generation. Mathematical modeling approach has been applied to model the integration of proton exchange membrane fuel cell (PEMFC), electrolyser, hydrogen tank, photovoltaic system based on solar radiation at Damascus (Saeed & Warkozek, 2015). They concluded that PV module of

20.441 kW was required to produce 2.104 kg of hydrogen required per day in order to produce a continuous power of 1 kW under 0.43 V across fuel cell voltage and at a temperature and pressure of 70°C and 1 atm, respectively. The challenge of predictability of solar hydrogen production was also acknowledged in the study.

Additionally, the reliability and power quality of IPVFC system could be improved if the fluctuations of radiation and load are hedged with power management control as suggested by these studies (Dash & Bajpai, 2015; El-Shatter et al., 2002; Ro & Rahman, 1998). The potential for grid integration of IPVFC using control strategy has been studied (Lajnef et al., 2013). It was established that the amount of hydrogen consumed by the fuel cell is proportional to the load per unit time and this also implies that the pressure in the hydrogen tank decreases as hydrogen is used from it. Different topologies of IPVFC systems were analyzed by Rekioua et al., (2014). Modeling and simulation approach (Özgirgin et al., 2015) was used to study the potential application of IPVFC in micro-cogeneration for domestic application. The study shows that although the system can serve as a distributed system, it under-produces energy during winter (October-March) and overproduces during summer (March to October); hence the recommendation that excess energy could be sold to the grid during summer. Results from a model-based study indicates that IPVFC system could be effective and sustainable as stand-alone integrated power systems (Ganguly et al., 2010).

Notwithstanding the general acceptance that the system is important in the imminent hydrogen economy, there is no current study that has considered the manufacturing perspective of the system. Manufacturability and cost effectiveness are essential for the commercialization of an engineering product such as IPVFC system. Consequently, in the current study, the system has been analyzed from a point of view of manufacturability which is ultimately necessary for the commercialization of the system. The contribution of this study is primarily to report a salient discovery that modularization philosophy in manufacturing could facilitate the commercialization of the system. This is based on the inherent modular nature of the system. The second contribution is to discuss the possible automations that can make the system suitable for a reliable and sustainable distributed power source in remote applications such as powering telecommunication equipment, mining equipment and researches in remote locations.

Lastly, from the recent global concerns on the impact of climate change, renewable energy has the greatest potential for reducing greenhouse emission such as carbon dioxide in the first place. Therefore, in order to keep the global warming average temperature below 2°C above pre-industrial levels in addition to limiting warming to 1.5°C as purposed by the Paris Agreement (Climate Focus, 2015) which was recently reviewed at Katowice in Poland in 2018, systems such as IPVFC systems which has zero emissions during operation should not only be innovated but commercialized as a matter of urgency.

The presentation of the contribution is divided into five sections. Section 1 introduces the background of the study. Section 2 describes the components of the system and how it operates under distributed mode. Section 3 discusses how the system can be modularized based on hydrogen generation, storage, and utilization. Section 4 discusses wider implications of the system's automation and manufacturability while section 5 concludes the study.

2 SYSTEM DESCRIPTION

The IPVFC system shown in Figure 1 is a potentially autonomous power system based on hydrogen energy and it is divided into three major sections: Hydrogen generation (HG), Hydrogen storage (HS) and Hydrogen utilization (HU) sections. The HG section consists of PV modules and electrolyser. The hydrogen tanks in the HS section stores the hydrogen produced from the electrolyser. The excess electricity from the PV is used to charge battery which can supply power directly to the load via the inverter or be utilized for regeneration of hydrogen from the electrolyser. The fuel cell utilizes the hydrogen to produce electricity depending on the demand. The system, when fully automated, functions as a Just-in-time system in which hydrogen is balanced in the three sections continuously to keep the system running. Therefore, appropriate sizing of the three sections must be such that hydrogen is continuously available without human intervention.

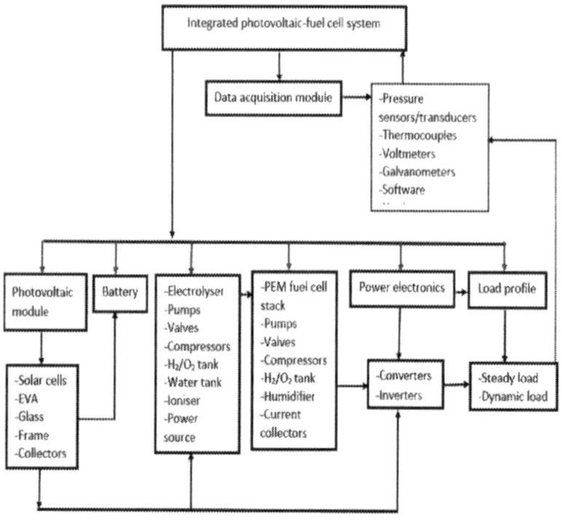

Figure 1. Components of integrated photovoltaic-fuel cell system.

3 MODULARITY IN SYSTEM DESIGN AND OPERATIONS

This section discusses the insights that can be drawn from our current study towards advancing the commercialization of integrated photovoltaic-fuel cell system. The overall aim of our current research is to produce a prototype of the system that can be utilized under distributed generation. The approach is bottom up as the system is made up of multi-components with multi-physics. The approach reveals the modularity of the design and its implication for manufacturing strategy and they are henceforth discussed.

3.1 *Modularity in manufacturing and supply chain integration*

Complex systems are often approached from the principles of modularity in design (Huang & Kusiak, 1998) and systems theory and system thinking (White, 2015). This allows the whole system to be broken down into smaller modules for design and optimization before they are integrated. The IPVFC system is a multicomponent system that requires diverse manufacturing processes and process technologies. Therefore, from a manufacturing perspective, it may not be competitive for a company to focus on the manufacturing of IPVFC system alone. Assuming that this is one of the products in the company's portfolio, it means that the system can be grouped into modules and outsourced to companies that are established in the components manufacturing. Figure 2 shows a simplified flowchart starting from when the customer places order.

The need to allow the IPVFC system, under potential commercial setting, to be co-created with the customers is to enable them to specify the load sharing ratio between the PV, Battery, and the Fuel cell. It is imperative to note that the inherent modularity of the PV, electrolyser and the fuel cell means that they are easily scalable. In the last one year, the Authors have modeled the photovoltaic module, proton exchange membrane electrolyser (PEME) and PEMFC in a manner that is biased towards its design for manufacture (DFA) and scalability (Ogbonnaya, Turan, et al., 2019c). Figure 3 shows model predicted power outputs of PV array connected in parallel with each model rated 40 W with 36 solar cells in series. PV modules can be added based of the overall power required from the PV component of the system for direct power supply to the load, electrolysis and charging of battery, depending on the design (Ogbonnaya et al., 2020; Ogbonnaya, Turan, et al., 2019a).

PEME uses electricity to produce hydrogen and oxygen through electrochemical process. PEMFC works in direct opposite of PEME as it uses hydrogen and oxygen to produce electricity. In both cases, there is also an inherent modularity in their design as the addition of cells in PEME increases the amount of

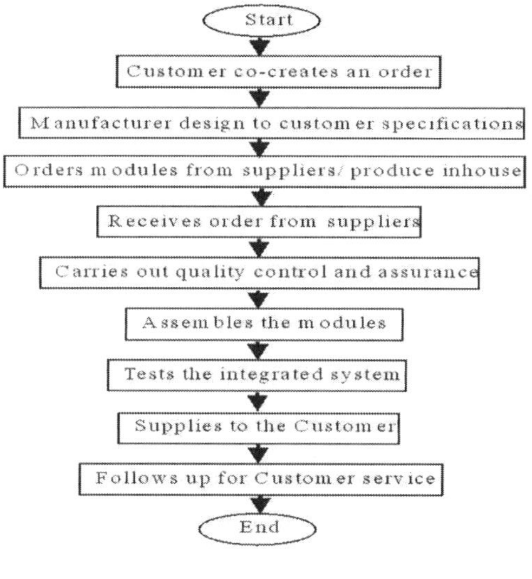

Figure 2. Simplified supply chain flowchart for integrating modularity in the manufacturing process.

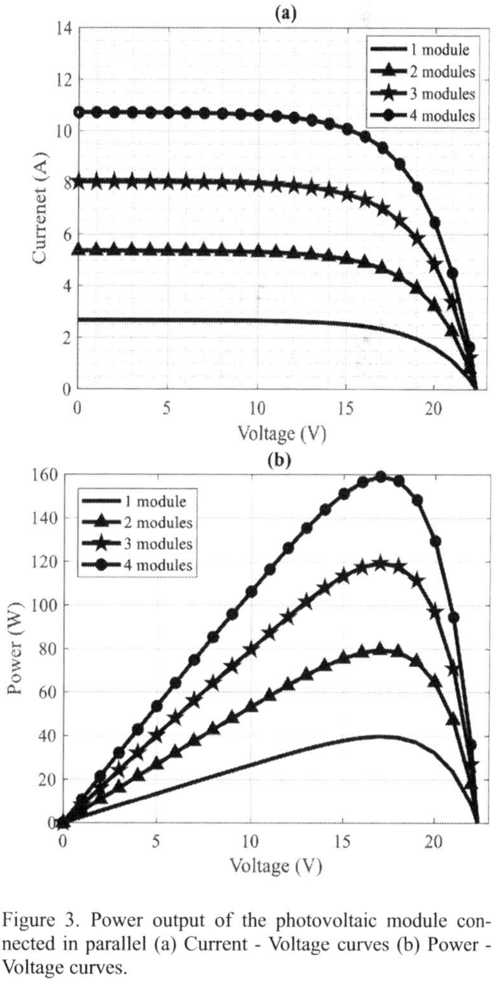

Figure 3. Power output of the photovoltaic module connected in parallel (a) Current - Voltage curves (b) Power - Voltage curves.

hydrogen produced whereas the addition of cells to the PEMFC increase the power density of the system. Equation (1) (Spiegel et al., 2007) shows that the power density ($W_{elect,stack}$) of the PEMFC depends on the number of fuel cells in series (N_{cell}), cell net potential (that is Nernst potential less activation, Ohmic and concentration losses (see Figure 4)), active area of the cell (A_{cell}) and current density (i_{cell}).

$$W_{elect,stack} = N_{cell} \times E_{cell} \times A_{cell} \times i_{cell} \quad (1)$$

Another benefit of viewing the system from modular design is the possibility of standardization of the modules. The efficiencies of the modules can be used to predict the performance of the modules thereby creating standard modules which will ultimately help reduce the choices customers can make as well as make the supply chain leaner and faster.

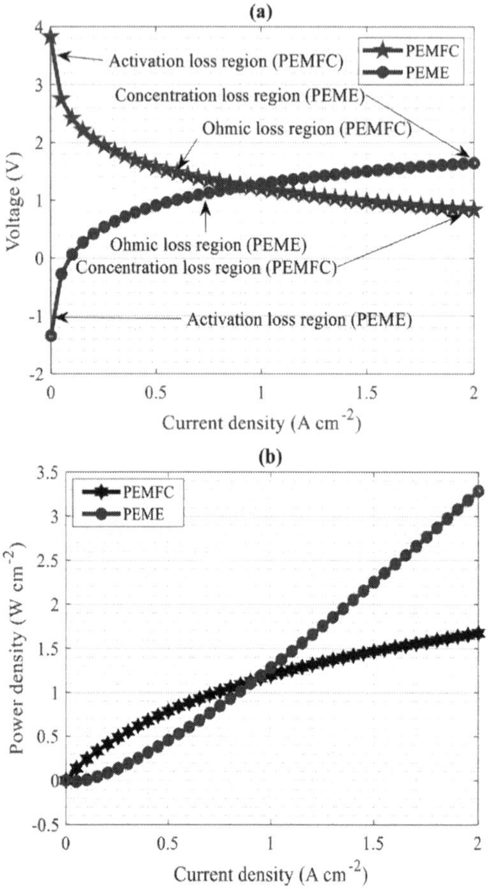

Figure 4. Polarization curves for proton exchange membrane electrolyser (PEME) and proton exchange fuel cell (PEMFC). (a) Current density - Voltage curves (b) Power - Current density curves.

Modularizing the system also allows for collaboration between companies that specializes in specific components. In addition, viewing the manufacturing strategy from the inherent modularity also means that global supply chain can support the manufacturing of the system as it does not matter which part of the world each of the modules are manufactured as long as the supply chain is efficient enough to support just-in-time processes that meet customers' needs.

Furthermore, components such as PEME and PEMFC can be manufactured in the same company since both use the same materials such as membrane, catalyst, anodes, cathodes, only that PEME is designed to split water into hydrogen and oxygen while PEMFC combines oxygen and hydrogen to produce electricity. The implication of this structural reverse is that the two modules can share manufacturing process technologies thereby driving tooling costs and space requirements down significantly.

3.2 *System automation*

Automating the IPVFC system is very necessary to make it fit for purpose in remote distributed power applications. This is because of the fluctuating nature of solar radiations and temperature which are key PV generation parameters as exemplified by solar radiation and temperature data of Gombe and Kano cities in Nigeria (see Figure 5).

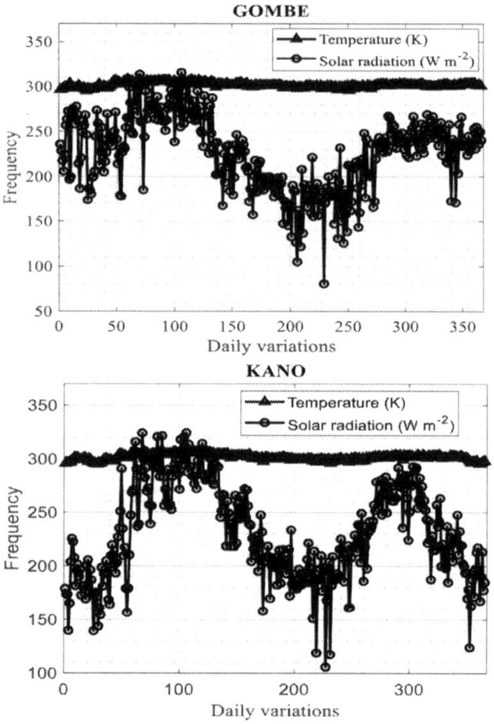

Figure 5. Daily solar radiation and temperature data for Gombe and Kano States in Nigeria.

Unlike internal combustion engines (ICEs) that can be controlled, solar energy availability cannot be controlled by man (Ogbonnaya, Turan, et al., 2019b, 2019c). Therefore, the potential control would be imposed on the human engineered system (i.e. the IPVFC system). This can be achieved through full system automation. The interface integration (see Figure 1) of the system requires that the power from the PV is inputted for electrolysis in the PEME and charging of battery but the processes should be terminated when the battery is fully charged and/or when the hydrogen tank is filled.

Furthermore, the rate at which hydrogen is consumed in the fuel cell needs to be integrated with the pressure sensor in the tank so that regeneration of hydrogen can be automatically initiated below the critical point. It is also required that power supply to the electrolyser should be met from the battery at night when there is no solar energy for the PV to produce the required electricity for electrolysis of water. The use of automation and control makes the system a potential hydrogen system with in-situ generation and utilization since hydrogen transportation is considered capital intensive and challenging (Ogden, 1999).

Data acquisition module is at the heart of the automation as the signals to each of the components are designed based on the algorithm of different foreseeable scenarios the system might encounter during operation. Data acquired can also be analyzed for the improvement of the future generations of the system.

3.3 *Modularity and operational efficiency*

The operational efficiency of the system depends so much on its level of automation. On one extreme, the system can be fully automated such that its operation can be without human intervention except during routine maintenance or during sudden breakdown. Still, diagnostic algorithm can be incorporated to allow the system self-repair in the first instance before human intervention can be involved. This is particularly necessary for the application of the system in harsh locations or distances that technicians cannot easily be deployed. Therefore, efficient remote operations require significantly reduced human interventions both for managing generation, storage, and utilization of the hydrogen. In practical terms, the system must be intelligent enough to resolve different scenarios that might impact on its continuous supply of reliable power for static as well as dynamic loads.

4 DISCUSSIONS ON THE CONSIDERATIONS FOR SYSTEM MODULARISATION AND AUTOMATION

From the foregoing, there are current issues and considerations that need to be contemplated before full commercialization of the system can be attained. Firstly, the system may not be competitive if the cost of automating it is too high. However, integrating the system with the grid could reduce the challenges associated with hydrogen generation in the absence of solar energy.

Another issue of concern is the current low conversion efficiency of the components. The energy conversion of a typical monocrystalline silicon solar cell is less than 10% (Sudhakar & Srivastava, 2014). The energy efficiency of PEMFC is still about 42.32% due to the losses during power generation (Hussain et al., 2005). Invariably, the overall system performance will increase as the components are improved in the future.

Again, any breakdown in the supply chain during manufacturing can be problematic as the system has to be integrated to produce power. Therefore, reliable supply chain must be utilized if the manufacturer cannot manufacture components in-house. Efficient supply chain would also reduce wastes associated with over-stocking of inventories since the customer is integrated into the supply chain.

Notwithstanding the enumerated challenges associated with producing the modularized IPVFC, the fact that solar energy is the most sustainable form of renewable energy source and that technologies that depend on it are clean, needs to be at the center of the consideration of this system (Ellabban et al., 2014). More so, hydrogen produced from other sources can also be applied in the system.

Maintainability of the system is high because of the modular design. A module can be replaced or detached and transported for repair without moving the entire system. Also, damaged module can be replaced without replacing the entire system.

5 CONCLUSIONS

In this study, a critical analysis and discussion has been made on how the modularization of integrated photovoltaic-fuel cell system can facilitate its commercialization. In particular, the system's potential in the imminent hydrogen economy has been highlighted. Potential manufacturing innovations and efficient supply chain integration based on the inherent modularity of the system has been discussed. It can be stated from the study that operational efficiency in distributed mode requires heavy automation which would ultimately increase the cost of the system. However, automation is necessary because the prime mover (solar energy) is unpredictable and the system needs to balance hydrogen in the generation, storage, and utilization sections for reliable power generation.

Modularization of the system would also help in standardizing the modules to make it easier for customers to choose a system based of their load requirements.

Future work by the Authors focuses on developing an integrated computational model for predicting the capacity of each component required for different power output scenarios.

ACKNOWLEDGEMENT

This work was supported the Petroleum Technology Development Fund (PTDF) Nigeria scholarship number PTDF/ED/PHD/OC/1078/17.

REFERENCES

Climate Focus. (2015). The Paris Agreement. Summary. *Briefing Note III, December*, 1–6.

Dash, V., & Bajpai, P. (2015). Power management control strategy for a stand-alone solar photovoltaic-fuel cell-battery hybrid system. *Sustainable Energy Technologies and Assessments*, *9*(1), 68–80.

Devrim, Y., & Pehlivano??lu, K. (2015). Design of a hybrid photovoltaic-electrolyzer-PEM fuel cell system for developing solar model. *Physica Status Solidi (C) Current Topics in Solid State Physics*, *12*(9–11), 1256–1261.

El-Shatter, T. F., Eskandar, M. N., & El-Hagry, M. T. (2002). Hybrid PV/fuel cell system design and simulation. *Renewable Energy*, *27*(3), 479–485.

Ellabban, O., Abu-Rub, H., & Blaabjerg, F. (2014). Renewable energy resources: Current status, future prospects and their enabling technology. In *Renewable and Sustainable Energy Reviews* (Vol. 39, pp. 748–764).

Ganguly, A., Misra, D., & Ghosh, S. (2010). Modeling and analysis of solar photovoltaic-electrolyzer-fuel cell hybrid power system integrated with a floriculture greenhouse. *Energy and Buildings*, *42*(11), 2036–2043.

Huang, C. C., & Kusiak, A. (1998). Modularity in design of products and systems. *IEEE Transactions on Systems, Man, and Cybernetics Part A: Systems and Humans.*, *28* (1), 66–77.

Hussain, M. M., Baschuk, J. J., Li, X., & Dincer, I. (2005). Thermodynamic analysis of a PEM fuel cell power system. *International Journal of Thermal Sciences*, *44* (9), 903–911.

Lajnef, T., Abid, S., & Ammous, A. (2013). Modeling, control, and simulation of a solar hydrogen/fuel cell hybrid energy system for grid-connected applications. In *Advances in Power Electronics* (Vol. 2013).

Lehman, P. A., & Chamberlin, C. E. (1991). Design of a photovoltaic-hydrogen-fuel cell energy system. *International Journal of Hydrogen Energy*, *16*(5), 349–352.

Meurer, C., Barthels, H., Brocke, W. A., Emonts, B., & Groehn, H. G. (1999). PHOEBUS—an autonomous supply system with renewable energy: six years of operational experience and advanced concepts. *Solar Energy*, *67*(1–3), 131–138.

Ogbonnaya, C., Abeykoon, C., Damo, U. M., & Turan, A. (2019). The current and emerging renewable energy technologies for power generation in Nigeria: A review. *Thermal Science and Engineering Progress*, *13*.

Ogbonnaya, C., Turan, A., & Abeykoon, C. (2019a). Numerical integration of solar, electrical and thermal exergies of photovoltaic module: A novel thermophotovoltaic model. *Solar Energy*, *185*.

Ogbonnaya, C., Turan, A., & Abeykoon, C. (2019b). Novel thermodynamic efficiency indices for choosing an optimal location for large-scale photovoltaic power generation. *Journal of Cleaner Production*, 119405.

Ogbonnaya, C., Turan, A., & Abeykoon, C. (2019c). Energy and exergy efficiencies enhancement analysis of integrated photovoltaic-based energy systems. *Journal of Energy Storage*, 26.

Ogbonnaya, C., Turan, A., & Abeykoon, C. (2020). Robust code-based modeling approach for advanced photovoltaics of the future. *Solar Energy*, *199*(February), 521–529.

Ogden, J. M. (1999). PROSPECTS FOR BUILDING A HYDROGEN ENERGY INFRASTRUCTURE. *Annual Review of Energy and the Environment*, *24*(1), 227–279.

Özgirgin, E., Devrim, Y., & Albostan, A. (2015). Modeling and simulation of a hybrid photovoltaic (PV) module-electrolyzer-PEM fuel cell system for micro-cogeneration applications. *International Journal of Hydrogen Energy*, *40*(44), 15336–15342.

Rekioua, D., Bensmail, S., & Bettar, N. (2014). Development of hybrid photovoltaic-fuel cell system for stand-alone application. *International Journal of Hydrogen Energy*, *39*(3), 1604–1611.

Ro, K., & Rahman, S. (1998). Two-loop controller for maximizing performance of a grid-connected photovoltaic-fuel cell hybrid power plant. *IEEE Transactions on Energy Conversion*, *13*(3), 276–281.

Saeed, E. W., & Warkozek, E. G. (2015). Modeling and Analysis of Renewable PEM Fuel Cell System. *Energy Procedia*, *74*, 87–101.

Spiegel, C., Hagar, L., Mckenzie, J., Hahn, L., Mcgee, M., & Ingle, S. (2007). Designing and Building Fuel Cells. In *Fuel Cells*.

Sudhakar, K., & Srivastava, T. (2014). Energy and exergy analysis of 36 W solar photovoltaic module. *International Journal of Ambient Energy*, *35*(1), 51–57.

White, S. M. (2015). Systems Theory, Systems Thinking. *Systems Conference (SysCon), 2015 9th Annual IEEE International*, 420–425.

Industry 4.0 – Shaping The Future of The Digital World – da Silva Bartolo et al. (eds)
© 2021 Taylor & Francis Group, London, ISBN 978-0-367-42272-1

Control systems of regional energy resources as a digital platform for smart cities

Yu. Koshlich, A. Belousov, P. Trubaev, A. Grebenik & D. Bukhanov
Belgorod State Technological University after V.G. Shukhov, Belgorod, Russia

ABSTRACT: The article presents an approach to design of regional segments of energy resources control systems using the example of public sector facilities. The basic techniques for determining the specific energy consumption rates and principles for monitoring their implementation at the organizational and technical level presented using the developed system for managing energy resources in the public sector facilities of the Belgorod region. Analysis of energy consumption by state institutions shows that the potential energy savings of about 25-35% of the cost of energy resources can be achieved through the introduction of intelligent mechanisms to predict energy consumption rates (energy consumption limits) and operational control of their performance. The article focuses on building a unified digital platform Smart City in the context of housing and public utilities based on the developed energy resources management system of the Belgorod Region.

1 INTRODUCTION

The transition to advanced digital, intelligent production technologies, the creation of systems for processing big amounts of data, machine learning and artificial intelligence caused, among other things, by the need to move to a new level of digital energetics, which ensures the interrelationship between the generation, distribution and consumption of energy resources and their effective management. The efficiency of the use of energy resources is one of the urgent problems in the energy sector. One of the promising directions in this area is the development and implementation of automated dispatch control systems (ADCS) that provide automated monitoring and centralized control of distributed power supply and life support facilities for buildings and energy management systems (EMS) that implement energy management at the organizational and technological level (CENEf 2008, GU IES 2003 & Ananyev 2017).

One of the approaches to increasing the efficiency of consumption of energy resources is the constant monitoring of the implementation of consumption standards. When considering the efficiency of energy consumption, there is usually a problem associated with the definition of baseline values or consumption rates. The article presents the developed method, which allows determining the efficiency of thermal energy consumption of a building taking into account its individual characteristics.

2 SMART CITY CONCEPTION

By order of the Ministry of Construction of Russia of October 31, 2018 No. 695/pr, the passport of the departmental project for the digitization of the city economy "Smart City" was approved (GU IES 2003). The project "Smart City" is implemented in the framework of the national project "Housing and urban environment" and the national program "Digital Economy". The "Smart City" project aimed at improving the competitiveness of Russian cities, creating an effective urban management system, and creating safe and comfortable living conditions for citizens. One of the important areas of the departmental program is the digital transformation of housing and public utilities in terms of energy conservation management in the public sector. The main tool for implementing the principles of Smart City is the widespread introduction of advanced digital and engineering solutions in urban and municipal infrastructure. The housing and utilities transformation involves building a unified digital platform that provides the functions of technological and organizational management of energy conservation. Since more than 50% of the cost of energy resources is the cost of heat supply to buildings, the important point is to determine the optimal level of energy efficiency and operational control of its provision at the organizational and technological level. To implement the control, it is necessary to solve two problems: to determine and predict the energy consumption of objects, taking into account the current environmental conditions and the parameters of buildings. To this end, a method has been developed for determining specific energy consumption parameters determining their energy efficiency on the basis of aggregated data about an object subject to comparable conditions.

3 METHOD

Energy efficiency of buildings, structures, structures is determined by the specific costs for their heating, which depend on two factors (CENEf 2008, GU IES 2003 & Berardi 2017):

- constructive, determined by the construction of the building and the materials from which it is made;
- regime, determined by the correspondence of the volume of heat actually supplied in the building, which is necessary for the quantity (Ananyev 2017).

The paper considers the relationship between the estimated (design) heat consumption for building heating with the specific heat consumption for heating and ventilation.

The thermal balance of the building can be represented as:

$$Q = Q_{wal} + Q_{win} + Q_{rf} + Q_{bsm} + Q_{vent} - Q_{ins} - Q_{int}, \tag{1}$$

where Q – heat consumption for heating; Q_{wal} – heat loss through the building walls; Q_{win} – loss of heat through windows and doors of the building; Q_{rf} – loss of heat through the attic floor or overlapping roof; Q_{bsm} – loss of heat through the basement or floor of the building; Q_{vent} – heat losses with ventilation; Q_{ins} – arrival of heat from solar radiation; Q_{int} – internal heat dissipation in the building from people, electrical appliances, lighting.

Losses of heat through the walls Q_{wal}, windows and doors Q_{win}, attic floor or overlapping roof Q_{rf}, through the basement or floor Q_{bsm} are determined by the equations of thermal conductivity in stationary conditions:

$$q = -\lambda \frac{\partial t}{\partial x}. \tag{2}$$

where q – heat flow, W/m^2; λ – thermophysical characteristic of a building material - thermal conductivity coefficient, W/(m·K), which is inversely proportional to the heat-shielding properties of the material; t is the temperature, K (°C); x – coordinate, m.

In building thermal physics, the concept of resistance to heat transfer (thermal resistance) R, m^2·K/W, is used which is directly proportional to the heat-shielding properties and it is determined through the coefficient of thermal conductivity λ and the layer thickness δ, m:

$$R = \delta/\lambda. \tag{3}$$

For a multi-layered enclosing structure, its resistance to heat transfer is:

$$R = 1/\alpha_H + \Sigma\delta_i/\lambda_i + 1/\alpha_B, \tag{4}$$

where α_H – coefficient of heat transfer from the external surface of the enclosing structure to the environment media, W/(m^2·K); α_B – heat transfer coefficient to the inner surface of the enclosing structure from the internal volume of the room, W/(m^2·K); δ_i, λ_i – thickness, m, and the coefficients of thermal conductivity of the layers, W/(m^2·K).

In the regulatory documents there are a number of correction factors and methods for calculating the resistance of heat transfer for:

- accounting for heterogeneity of enclosing structures;
- determination of the thermal resistance of attics, cellars, undergrounds, coverings and overlaps above the passages (arches);
- ventilated air interlayers.

For translucent structures (windows, stained-glass windows of doors, lanterns) in which a complex heat transfer process involving convection, thermal conductivity and radiation is carried out, the given resistance to heat transfer is taken from the reference data or test results in the laboratory given in the product certificates (Copiello 2017, Trubaev 2005).

For the building according to the normative literature, the given coefficient of heat transfer through the outer enclosing structures of the building (transmission coefficient of heat transfer), k_{tr}, W/(m^2·K) or k_{ht}, W/(m^2·K):

$$k_{tr} = \frac{1}{A}\sum n_i \frac{A_i}{R_i}; \; k_{ht} = K_k k_{tr}, \tag{5}$$

where A is the sum of the areas of all elements of the outer enclosures of the building envelope (walls, windows, attic and basements, etc.), m^2; A_i and R_i – the area of the elements and their reduced thermal resistance; n_i – coefficient that takes into account the difference in internal or external temperature of the structural element from the accepted internal or external temperature of the building is usually taken into account only for attics and basements; K_k – coefficient of compactness of the building (the ratio of the total area of the inner surface of the outer enclosing structures of building A to the enclosed heated volume V_{heat}).

The loss of heat with the air of the ventilation system is determined by the volume of air leaving the building and the temperature difference between the indoor air and the outside. The following methods can be used to calculate the volume of air:

- on the normative multiplicity of air exchange in the building (premises);
- according to the design capacity of the system of forced supply and exhaust ventilation;
- based on the results of measurements of the actual air flow in the ventilation system.

Based on the obtained air flow, the specific ventilation characteristic of the building k_{vent}, W/(m^2·K), having one dimension with the given (transmission) coefficient of heat transfer k_{ht}. Typically the value k_{vent} is 20-50% of the value k_{ht}.

Household heat input and heat input with solar radiation are also taken into account with the help of specific characteristics. Specific characteristic of internal heat of a building k_{int}, W/(m^2·K) is determined by the number of people and fuel-generating equipment inside the building (Kim 2017, Trubaev 2005). For residential buildings, the averaged heat dissipation values for 1 m^2 of the square q_{int}, constituent 10-17 W/m^2.

Calculation of specific internal heat release is performed according to the expression:

$$k_{\text{int}} = q_{\text{int}}A/(V_{heat}\Delta T) = q_{\text{int}}K_L/\Delta t, \qquad (6)$$

where Δt is the average temperature difference between the internal and external air for the heating period.

Specific characteristics of heat input into the building from solar radiation k_{ins} is determined according to the average solar radiation value in the normative literature during the actual conditions of cloud cover, taking into account the orientation in four directions. The receipt for the heating period determined by the expression:

$$Q_{rad} = \tau_1\tau_2\Sigma A_{tls\ i}I_i \qquad (7)$$

here Q_{rad} – heat input, MJ; τ_1, τ_2 – coefficients that take into account the penetration of radiation through translucent structures and the shading of openings by opaque elements; $A_{tls\ i}$ – area of translucent structures, oriented to the i-th side of the world, m^2; I_i – the flux of total solar radiation, the vertical surfaces coming in during the heating period under the actual cloud conditions, MJ/m^2.

Since for residential buildings, as a rule, glazing along oppositely located facades is symmetrical, the total intake from radiation can be determined as the average between the intake of radiation through each facade. Analyzing the climatic data of Belgorod, it can be concluded that this average value for opposite facades oriented to different directions of the world is practically the same. This is confirmed by the analysis of solar radiation data for 19 settlements, given in (Ilinsky, 1974). Therefore, when calculating heat losses for typical buildings, an average flux of total solar radiation and the glazing ratio of buildings can be used, and expression (7) can be written in the following form:

$$Q_{rad} = \tau_1\tau_2 k_{tls}AI_{avg}, \qquad (8)$$

The conversion to specific values made by the expression:

$$\begin{aligned} K_{ins} &= 11,6Q_{rad}/(V_{heat}\Delta tN_{hs}) = \\ &= 11,6\tau_1\tau_2K_kk_{tls}I_{avg}/(\Delta tN_{hs}), \end{aligned} \qquad (9)$$

where N_{hs} – duration of the heating season, days, I_{avg} – average solar radiation value, MJ/m^2.

The calculated specific characteristic of the heat energy consumption for heating and ventilation of the building, $q_{\text{от}}$, W/(m^3·K) according to the normative literature (SP 50.13330.2012) (Vatin 2012) should be determined by the formula:

$$q_{heat} = [k_{ht} + k_{vent} - (k_{int} + k_{ins})k_1]k_2, \qquad (10)$$

where k_1 – coefficients that take into account the thermal inertia of the enclosing structures and the efficiency of automatic regulation of heat supply in heating systems; k_2 – coefficients that take into account additional losses and reduced heat consumption of residential buildings in the presence of a quarterly heat energy for heating.

From expression (10) we can distinguish two parts-dependent and independent of the quantity Δt

$$\begin{aligned} Q_{heat} = &(k_{ht} + k_{vent})k_2 - K_k/\Delta t(q_{int} + \\ &+ 11,6\tau_1\tau_2k_{tls}I_{avg}/N_{hs})k_1k_2 \end{aligned} \qquad (11)$$

The second part of the expression for the given climatic conditions and operating conditions of the building will be constant. Therefore, for example, in order to reduce the specific consumption of heating in the building by two times in the conditions of the central Russia, where heat input is about 20% of losses through enclosure structures and with ventilation, it is enough to reduce heat loss by 40%, then in the northern regions where Heat input is 10%, losses must be reduced already by 45%.

4 IMPLEMENTATION

To address the issues of increasing efficiency through technological and organizational measures to improve energy efficiency, an approach to the synthesis of the energy resource management system has been developed (Belousov 2017). Structural model of the distributed energy resources management system, which presented in the Figure 1.

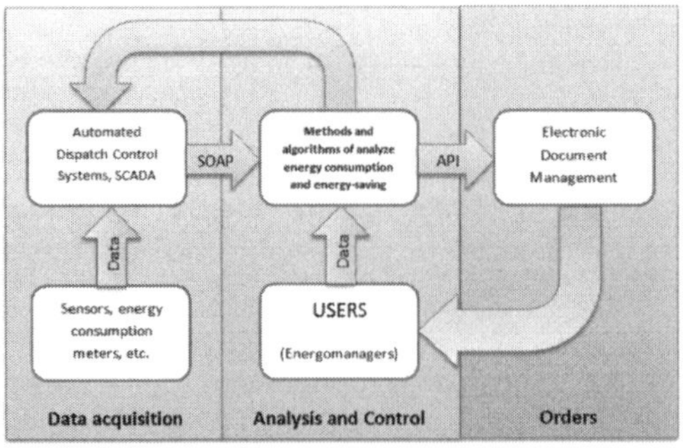

Figure 1. Structural model of energy resources control system.

The developed structure implemented in the software component of the distributed energy resources management system of the Belgorod region in the form of a web application.

The Figure 2 shows the screen form of the energy manager, which allows monitoring consumption efficiency in accordance with certain energy consumption norms and energy consumption limits. In the case of a significant deviation from the normative values, the system creates and sends an order to the electronic document management system, and also controls its execution.

The described approaches and techniques implemented in the software component of the EDMS (Figure 2).

The Figure 2 shows one of the screen forms that allow to evaluate the efficiency of consumption of energy resources in comparison with the established consumption standards (limits) and calculated standard values. Feedback is realized on the basis of interaction with electronic systems and electronic document circulation through the formation of appropriate orders for responsible persons.

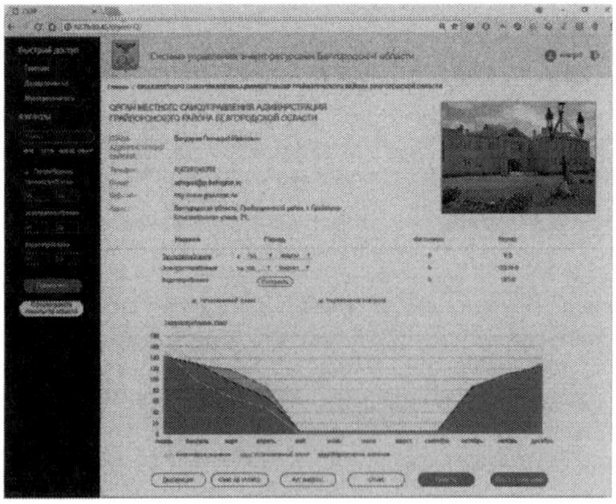

Figure 2. Screen of energomanager workstation.

5 CONCLUSION

The developed approach to the construction of energy efficient systems for control of distributed energy resources combining the advantages of both types of systems: automated dispatch control system and energy management system.

The presented technique is an attempt to determine the normative values of energy consumption for the institution. The developed software allows, at the organizational level, to effectively form control actions and create motivational aspects for maintaining the required level of energy consumption.

The main effect achieved by automating the calculation of energy consumption norms (limits) and monthly monitoring of their implementation. The implemented energy management system of the Belgorod region allows saving from 0.5 to 15% of the cost of energy resources of institutions, which in value terms is about 100 to 4 000 million Russian rubles annually.

Using the represented technique and realization methods of monitoring and control of energy efficiency of distributed facilities allows following to modern concept of energy saving and energy efficiency including ISO 500001.

REFERENCES

Ananyev, A.I. & Rymarov A.G. & Voitovich, E.V. & Latushkin, A.P. 2017. The influence of social factors on the required level of thermal protection of the exterior walls of residential buildings. *Messenger MGSU*, vol. 12 (7 (106)): 741–747. Moscow: MGSU publishing house.

Belousov, A.V. & Glagolev, S.N. & Koshlich, Yu. A. & Grebenik, A.G. 2017. Demonstrational zone of energy efficiency of BSTU after V.G. Shukhov. *17th International Multidisciplinary Scientific GeoConference: Energy and Clean Technologies*, vol. 17/issue 42: 167–174. Bulgaria: SGEM.

Berardi, U. 2017. A cross-country comparison of the building energy consumptions and their trends. *Resources, Conservation and Recycling*, vol. 123: 230–241. Nederland: Elsevier BV.

CENEf, 2008. *Energy efficiency in Russia: a hidden reserve. The report, prepared by experts from the World Bank, the International Finance Corporation and the Center for Energy Efficiency.* Moscow: CENEf.

Copiello, S. 2017. Building energy efficiency: A research branch made of paradoxes. *Renewable and Sustainable Energy Reviews*, vol. 69: 1064–1076. United Kingdom: Elsevier B.V.

GU IES of the Ministry of Energy of Russia, 2003. *Power engineering in Russia. Development Strategy 2000–2020 (Scientific basis of energy policy).* Moscow: GU IES of the Ministry of Energy of Russia.

Ilinsky, V.M. 1974. *Building Thermophysics. Fencing structures and microclimate of buildings.* Moscow: Higher education school.

Kim, Y.S. 2017. Building energy model calibration with schedules derived from electricity use data. *Applied energy*, vol. 190: 997–1007. United Kingdom: Elsevier B.V.

Trubaev, P.A. & Besedin, P.V. 2005. Criteria for the thermodynamic efficiency of cement clinker production from natural raw material. *Theoretical Foundations of Chemical Engineering*, vol. 39, vol. 6: 628–634. New York: Springer Publ.

Vatin, N.I. & Nemova, D.V. 2012. Influence of the level of thermal protection of enclosing structures on the magnitude of losses of thermal energy in the building. *Magazine of Civil Engineering*, vol. 8: 4–14. St. Petersburg: SPBSTU publishing house.

Industry 4.0 – Shaping The Future of The Digital World – da Silva Bartolo et al. (eds)
© 2021 Taylor & Francis Group, London, ISBN 978-0-367-42272-1

Climate Smart Cities in architecture knowledge nowadays

A.C. Figueiredo de Oliveira, J. Nicolau & C. Alho
Faculty of Architecture, University of Lisbon, Lisbon, Portugal

ABSTRACT: Climate change became a fascinating subject to urban environment. Our scope is to develop a research network, associating architecture schools, urban research laboratories, local authorities and NGOs on architectural solutions to deal with consequences of climate change, more focused on cities. Today, many entities are already working on issues related to the effects of extreme swings in climatic conditions. However, in recent years, many projects were interrupted when they faced strong opposition from public opinion, the media or lawsuits. Project proposals seem to address sensitive cultural values and consequently undergo years of discussion, only to be put aside eventually. Data related to historical and geographic records of specific lands are used to construct patterns of development, feed algorithms and integrate the support of artificial intelligence to design possible future scenarios. The final result aims to produce Sustainable Operative Solutions.

1 INTRODUCTION

Urban lands are facing new challenges when dealing with climate change. The problem affects the citizens, the environment and the local economy. Municipalities, stakeholders, Port Authorities and local communities often disagree upon their own needs. S.O.S. – Sustainable Open Solutions for European urban lands, a developed project for European Union aims to develop new solutions that emerge from the present necessities. Why is it urgent to share, systematize and upgrade knowledge on the subject?

The nest generations will face challenges that we don't imagine the results on them and as the effect of climate change is increasing at an odd rhythm which demands adaptation and transformation of vulnerable territories minute by minute, it is so urgent that no one wants to believe. In the last years, some interesting projects were interrupted because they face strong opposition coming from public opinion and either the media and lawsuits. Plans seem to address sensitive cultural values and consequently meet years of discussion, only to be put aside eventually. Such difficulties bring a loss of competitiveness on all sides, as well as decreasing quality of life for their citizens thus representing a problem within the European Community. To mitigate, climate change impacts, several solutions have been proposed to reduce greenhouse gas emissions, including modern, efficient energy alternatives and enhancing the use of sustainable energy sources (Hasan et al, 2018) and to complete the and increase the network, the young generations need to part of decisions, participation and knowledge. Communication is now and then part of all projects that need to have participation and need direct results and future application. Nowadays, a creation of data on the land uses historical and geographic records of specific locations to construct patterns of development, feed algorithms, and integrate the support of artificial intelligence to design possible future scenarios. S.O.S. Climate Land project gathers was based on partners from Portugal and The Netherlands, to share their best practices, including research on land that merge equipment's as well as public spaces and infrastructures, projects that succeed in conquering public support and integrating the signs of the collective identity. Climate change solutions are more successful when ensuring resilient urban environments and they achieve results are expected to engage citizens. The education about climate change challenges and solutions, and fostering possible changes in lifestyles can be part

Transnational examples and the exchange of best practices benefit from special conditions not available in a single city or a single institution. Local experts, municipal representatives, stakeholders, and international scholars are requested to work together, exchanging views, gaining new perspectives and discussing new interdisciplinary approaches, expanding and raising the level of the discussion, creating a think tank. The fundamental changes that are expected to take place in the climate system in the next decades are likely to have severe implications for the stability of the financial system (Dafermos et al, 2018).

Through an interdisciplinary methodology, a technique that are expected to be used often with future generations, where the S.O.S. Climate landfills will fill the gap in the understanding of how the different scales of urban and landscape planning, architectural design and technology will link in water-related strategies. They will impact each other in the definition of preventive action plans and in the enhancement of more conscious solutions to inform the community, human welfare, and socio-economic activities along those vulnerable territorial settings of the land.

2 EMERGY AT CLIMATE CHANGE

Nourishing a structured and necessary transfer of know-how and technology at several levels gives the research team excellency in research to the next move, enhancing a better position in the national and international research scenario. The relation between climate change and its potential effects on the stability of slopes remains an open issue (Palazzi et al, 2018).

The improvement in innovative nature of research towards a higher quality to reach excellence the research team focus on the urban land to prepare them better to face the challenges posed by climate change. To achieve this, a steady and interdisciplinary research agenda is part of the requirement, including environmental issues, smart technologies, strategies of resilience in urban design and culture. To reverse the vulnerability of urban land researchers, accomplish skills and know-how towards advancing knowledge and insights on how to produce strategies acting in a cross international formula and interdisciplinary solutions for the future.

European urban lands are a subject of growing impact upon the citizens and businesses concerning economic competitiveness and environmental quality. Sustainable Open solutions are in the center of the debate of land searching for resilient strategies in the context of difficulties stemming from environmental constraints brought by climate change. Land brings together some different topics since it relates two different worlds: territory and water. From the side of the water, there are maritime transportation requirements, natural ecosystems, the necessary reduction of carbon emission, control levels of pollution and sensitive ecological areas. Inland, the community, stakeholders and political leaders depend on solutions that require the collaboration of different fields of knowledge; physical, economic and social (Mello, 2002). The community's quality of life, environmental improvement and the increasing use of renewable energies are often in conflict with the necessary changes.

S.O.S. Climate land combines two strands of vibrant analyses—land research and disaster studies with base from the Japanese experts nowadays. To examine specific examples of modern urban lands reactions to the disaster and the disruption of urban structures and buildings. To envision how local, regional, national and international actors, both public and private collaboration to mitigate risks and enhance the resistance to face disasters.

The research project is growing from two approaches:

1) Develop research strategies aiming at designing policies and recommendations to meet the needs of urban lands regarding public spaces and selecting European best practices, to introduce researchers in general design and participatory processes on the transformation of nations. The parameters that influence the change of the land are shared with others to implement essential tools to deal with the production of future scenarios that integrate environmental, social, economic parameters to visualize and that encourage the community towards more participation in decision-making.

2) Develop new means of research based on historical records, geographic data, digital methods and co-creation with local scholars to engage them actively in research activities (co-research approach). Additionally, the collected information is used to increase the identification of patterns of previous urban development on the land. To visualize the future progression of the area depends on the development of the software produced by the team members. The new software identifies patterns of progress in the past to predict future scenarios and includes parameters directly related to climate change. The hypothesis produced for the coming transformation of lands is capable of engaging municipality representatives and especially technical staff to play a unique role and envision decision-making in the urban agenda.

The novelty of the approach is to overcome the problems that affect urban lands and join efforts of specialists from different disciplines that have recognized expertise in the subject and develop complementary research in the field of climatic transformation. The vulnerability of the land and the variety of implications require a method of synthesis and to process similar situations that the present institutions do separately, the project demands a working knowledge and know-how exchange among all partners solving specific problems that emerge from each land, via direct and broad discussion between experts:

Objective 1: Recognize gaps in current research, namely at geographical, historical, cultural, political and climatic among partners, to develop approaches for better understanding and enhance systematic means to relate people to the land.

Objective 2: Introduce experienced researchers to foster collaboration and provide a platform where communities, public spaces, and technology is addressed, based on new research fields, new forms of cooperation and new networks.

Objective 3: Upgrade the research group by enabling stronger networking between its research staff and top European researchers.

Objective 4: Progress the cross-fertilization and interaction between researchers from different institutions and with expertise in various fields.

Objective 5: Supply structured opportunities for developing scientific and personal relations, among and beyond the Project.

Objective 6: Provide the involvement of relevant stakeholders, and the dissemination of results to scientific, academic, practitioners and policy-makers at all levels (EU, national, regional and local), media and the general public.

Objective 7: Distribute know-how on aspects of humanities-led, collaborative sustainable urban development and new insights to improve downtown and social policies.

Objective 8: Promote the exploitation of advancements and Project results.

3 STATE OF THE ART

Sea level rise, high tides, storms, and floods enhance the vulnerability of urban land territories. The necessary transformations face years of discussions and lawsuits involving significant losses and costs, before cities adapting to change and effectively adapt lands to climate change. Resilience is transformative and in each transformation, tries to create a stronger, improved city (Yamagata, 2018). In recent years, most research projects have included regional, and municipality representatives and they have highlighted the importance of thinking beyond regulations to face the present challenges. The alternative thinking or so-called "out of the box," brings the risk to open the Pandora box with all the problems involved.

Cross-sectorial and interdisciplinary solutions are successful if dealing with conflict existing in every process of transformation in an environment free from current administrative procedures. However, for the future, lands will need to formulate lasting solutions efficient and imaginative strategies. The necessity to create more resilient cities, increase livable public spaces, promote urban natural environments and support a sustainable urban land is expected to improve healthier communities.

The necessary changes do not represent a problem. Transformation does not by definition compromise resilience. Quite the opposite, according to the Principal Researcher at the National Institute for Environmental Studies in Japan: Resilience is not a static state of a system. It is a process. A city is dynamic and is always changing. (...) Resilience is transformative, and in each transformation, tries to create a stronger, improved city's (Yamagata, 2018) that is, resistance itself can be understood as change. According to the United Nations Development program, the concept of human development is used in all research actions thus focusing on people, their skills and opportunities rather than depending only on resources or sufficient income. The primary goal is to expand the realm of possibilities so that urban lands can adapt, transform, develop skills and opportunities to be significant areas for the community. It goes therefore beyond economic features, and into reflect cultural, political, environmental and social, characteristics that influence the quality of human life in the context of climate transformation. Planning a land development, require the city officials or a developer to start by envisioning a network of well-connected multi-use public spaces that fit with the community's shared goals (Mostafa, 2017).

One of the central challenges is how to commit to citizens and land territories where opportunities for educating cities can be offered for everyone. It is directly linked with equality, sustainability inclusiveness cohesion and education for peace. Educative cities promote, policies and democracy, integrated and lifelong learning education based on knowledge on how modern cities recover from disasters (Campanella, 2005).

4 KNOWLEDGE SHARING TO ACHIEVE INNOVATIVE OBJECTIVES

The ground-breaking nature of the project comes from the methodology we have been testing since 2010 by bringing together municipality representatives, environmentalist, designers, geographers, cultural agents, and researchers to exchange knowledge and cooperate in finding solutions and define strategies. The methodology to meet those established goals includes, in particular for the transfer of knowledge, joint initiatives for capacity building, and trans-national collaborative activities. The selected parameters cover a wide range of data resulting from human activities and environmental conditions. The costs involved to mitigate and adapt significantly and encompasses a broad range of expertise. To process data from different fields of knowledge and interpret their consequences it is necessary to exchange data and to cross references with other experts. The various developing countries in Asia like China, Malaysia, South Korea, Singapore, and others are expanding their infrastructural facilities at an exponential rate (Purohit et al, 2015).

The problem of the land has been developed by each partner participating in the project, and it has been addressed from complementary perspectives, economic, cultural, data management, social and environmental. The diversity of themes covered by the partners is crucial for the success of the project. Expertise, exchange of interdisciplinary knowledge with others lead to a broader perception in the field of interest of a particular land. The trade is oriented to include preliminary research of the selected area, following three important guidelines, historical, cultural and geographical records. After the workshop edition, the dissemination of results can reach broader audiences, planning and designing adaptation programs and strategies, considering variables as mentioned earlier help to increase the efficiency of efforts (Jamshidi et al, 2018).

To have an imagination of the future is necessary to understand the past through the historical records available, that provide background on previous conditions. The geographic data is fundamental to relate urban territory with the natural region, as the water currents shape the landscape and the fluvial or/ and maritime activities. Historical urban and geographical research contributes to a rigorous evaluation of particular land potentialities: a way for products distribution, fishing activities or other cultural valences. With climate change, it is difficult to predict the impact on land areas as they are vulnerable and their resilience depends on competent but fragmented studies. S.O.S. Climate Land methodology allows to strengthen collaboration, to investigate new techniques, to cross-references and learn to access specific instruments and methods but adding the technology with the support from the nowadays techniques.

It will be directly with the workshops that the participation will brings together the interdisciplinary team that integrates permanent staff, scholars and researchers cooperating with the scientific committee to engage the present debate, the scientific committee invites external consultants and professionals from partner institutions to integrate the research group participating in the workshop with new technology systems. It will be into the workshop where it be debated the technologies to put the program in a level of excellency. Each workshop addresses one specific land location. A transnational multidisciplinary perspective contributes to the necessary integration of the local land urban environment. The cooperation with experts coming from international partner institutions provides the opportunity to develop a set of local proposals. During the event participants exchange views, gain new perspectives on the researched topic, discuss new approaches of resilience and have the opportunity to test their methods in an international environment.

The Conferences will gather local experts, public representatives, partners, and international scholars to share them researches on urban lands, and this is useful to exchange mutual visions and standard practices, which constitute a relevant tool for future research. Conferences are designed to disseminate the knowledge produced to the broader audience of policy-makers, stakeholders, environmental associations, local communities, media, and the general public and to improve the skills of its staff. The research community promotes the active participation in international science and technology related conferences by providing the framework to include host presentation of papers from external guests previously selected.

This is an important output for the research proposal since communication is oriented towards different audiences.

The overarching methodological framework is backed by the concepts of co-oriented and adaptive and comparative research. The scientific strategy encompasses measures for stepping up and stimulating scientific excellence and innovation capacity, that includes the development of research design and digital tools to playfully engaging companies and communities in research activities and create opportunities to gather with a specialist to collect the results of the researches.

- Dissemination and outreach activities - S.O.S. Climate land research community, promotes the scientific and public dissemination of the outcomes and findings. A website operating as a knowledge platform will be set at an early stage organized towards providing easy access to knowledge and know-how around the project's key areas and issues detected within the topic. It will also serve as central for social media activities. Further dissemination efforts include publications in relevant/indexed journals, policy briefs, studies reports, booklets and cooperation with other organizations, programs, networks, operating in the areas related to the project. The benefits from creative and innovative solutions contribute for most popular sustainable development of urban land in Europe. The creation of a body of knowledge on the subject of territory that contributes to the identification of long and short term solutions is sustained by the edition of the web-based material, production of documentaries, oriented towards the different audiences and publication of books. Everything is the central outcome of the research. The investment in communicating to reach new audiences through web tools, social networks and local media intends to disseminate and bring for public awareness, solutions that transform the land and the lives of their communities if local actors converge efforts and collaborate. It plays an essential role in this process because they influence the public opinion to think differently and attract the public interest to support the necessary transformations.

Our previous publications succeed to influence local administrations to take in consideration the results thus the challenges now facing climate change are required to receive feedback from local and international partners and the results of a comprehensive set of field observations, surveys, and controlled laboratory experiments demonstrated that the land suction, which remained thus far unexplored in soil mechanics (Sassa et al, 2013).

The activities will develop a strategy to maintain the cohesive and sustainable collaborative network beyond the Project S.O.S. Climate land clear strategy is focused on four main research lines:

1) environmental planning for climate change, the influence of geographic and historical factors on the construction of the new territories over the water. As an illustration the precarious condition and the complexity existing at the land, the analytical cartography will inform several stakeholders and citizens about the artificial landfill and the risks associated with extreme swings in climatic conditions.

2) economic impact. The unsuccessful implementation of the necessary transformation for land sites is simultaneously affected by discontinuous and contradictory decisions lead by politicians. Such situations conduct to buy an enormous waste of

effort. The economic costs are significant and represent a large percentage of the necessary investments thus highly permeable to political short term decisions. Consultants for this project have been involved technically but not politically at the land transformation process.

3) cultural influence and public space the top-down approach, based on desk-based macroscopic diagnostic and prospective analyses, plans and projects, has prevailed in urban development planning and decision making, thus neglecting the fact that the present city dynamics and future options depend on the changing presence, character, and activity of a significant number of individual and institutional actors as development stakeholders. It is especially evident in lands, one of the most valuable and indeed unique urban area.

4) data management. The team will collect data to produce visual animations of possible scenarios that envision the vulnerability of lands when exposed to climate change. To envision the implications of future projects, management of public spaces that merge equipment's, as well as squares, gardens and road structures, data, will cover previous plans and future scenarios. The development of possible land urban scenarios using technology spreads a broader understanding of land transformations. It is innovative to use material from different fields of knowledge to simulate new outputs. Data related to historical and geographic records will feed the information of specific land development. The information regarding the transformation of the built environment will allow the identification of patterns of development.

Such information will be used to feed algorithms and trace the storyline to speculate on climate change effects. Also, a pattern will be useful to integrate the support of artificial intelligence to visualize possible future scenarios. Though speculative this is useful for researchers to be in dialogue with citizens, practitioners, decision-makers and engage them in innovative solutions. It is not the cities that can be intelligent, but the societies that inhabit them that must be prepared (Oliveira, 2017).

5 CONCLUSIONS

The network action with researchers and specialists is the principal point and common tool to arrive on solutions that can be part of our future.

The decided working plan for the research was discussed and approved by all different team leaders and will be very important at the final results. The agreement about the number and dates of visits as well as about visiting persons is established according to their profile (MNG, ER, ESR) and specialization will create the direct quality of the project and it will be connected with the functions assigned to each participant are designed according to the work plan to meet their expertise and focus in the field of their scientific and scope of interests.

Furthermore, all the partners will welcome knowledge sharing in the structure of the S.O.S. Climate Land project that will directly provide an excellent opportunity to expand their field of research/expertise and enhance the scientific innovation of their institutions. The collaboration in the project will be profitable for all the members and to future researchers.

The strong consortium with a unique combination of expertise that is implementing the S.O.S. Climate Land (academic and non-academic) will be basilar to the network that is being created. All partners work at the forefront of their discipline areas and have a well-known scientific and leading reputation, thus ensuring research excellence. The consortium will train new researchers, expand research competencies in innovative design, partner with private sectors in research projects, thus providing substantial opportunities for real-world testing of the research and introduce creative/innovative practice research methodologies to a new generation. The project is prepared develop interdisciplinary and multi-sectoral competences that will have a significant impact on leading researchers that will be experts in the assessment of lands environmental adaptations and will provide successful water resilient strategies. The direct beneficiaries will also be staff, practitioners, policymakers. The S.O.S.'s extensive dissemination (website, media web, books, exhibitions, conferences, scientific articles,) will ensure that the results achieved will have open-access and will have widely shared worldwide.

REFERENCES

Campanella, [et al.] Vale - Introduction, The Cities Rise Again, in The Resilient City, How Modern Cities Recover from Disaster. New York: Oxford University Press, (2005).

Dafermos, Y. [et al.]- Climate Change, Financial Stability and Monetary Policy. Ecological Economics. ISSN 0921–8009. Vol. 152 (2018), p.219–219–234.

Hasan, Mohammad Mehedi & Wyseure, Guido - Impact of climate change on hydropower generation in Rio Jubones Basin, Ecuador. Water Science and Engineering. ISSN 1674–2370. (2018).

Jamshidi, Omid, [et al.] - Vulnerability to climate change of smallholder farmers in the Hamadan province, Iran. Climate Risk Management. ISSN 2212–0963. (2018).

Mostafa, Lobna A. - Urban and Social Impacts of Lands Development, Case Study: Jeddah Corniche. Procedia Environmental Sciences. ISSN 1878–0296. Vol. 37 (2017), p.205–205–221.

Oliveira, A. Claudia, Climate Smart Cities: Green Architecture and Sustainability applied on buildings. Evaluation systems in 'blank' smart cities and 'converted' smart cities. Lisbon University - Lisbon School of Architecture Lisbon University - Lisbon School of Architecture (2017).

Palazzi, Elisa, [et al.] - Implications of climate change on landslide hazard in Central Italy. Science of the Total Environment. ISSN 0048–9697. Vol. 630 (2018), p.1528–1528–1543.

Purohit, A. A., [et al.] - Effect of Mega Bridge on Hydro – morphodynamics of Land Facilities in Wide Estuarine Harbours- A Sustainable Development. Aquatic Procedia. ISSN 2214-241X. Vol. 4 (2015), p.341–341–348.

Sassa, S., [et al.] - Ecological geotechnics: Role of land geoenvironment as habitats in the activities of crabs, bivalves, and birds for biodiversity restoration. Soils and Foundations. ISSN 0038–0806. Vol. 53, n.º 2, (2013), p.246–258.

Yamagata, Yoshiki, Ayyoob Sharifi - Resilience-Oriented Urban Planning – Theoretical and Empirical Insight. Switzerland: Springer Cham, (2018).

Industry 4.0 – Shaping The Future of The Digital World – da Silva Bartolo et al. (eds)
© 2021 Taylor & Francis Group, London, ISBN 978-0-367-42272-1

Collaborative fabrication 1:1 scale prototyping and its space experiencing

F.C. Quaresma
Department of Architecture and Urbanism, Universidade Lusófona de Humanidades e Tecnologias (ULHT), Lisboa, Portugal

ABSTRACT: The development of sheltering to people with emergency needs, and the design of a small pavilion to support a Municipal running cycled path system in an academic context are showed in this paper. To this propose it was developed 3D models using visual programming and different techniques of digital fabrication. The natural scale prototypes (1:1 scale) built demonstrates the collaborative fabrication effort of different team members (students of architecture), the public administration and an industry partner. Different systems of communication and digital fabrication tools were implemented proving an efficient way of design and assembling the artifacts. As final result different prototypes were built simulating a real building experience.

1 INTRODUCTION

It is the goal of this paper to shows some of the results achieved through a design and digital fabrication processes and the development of three-dimensional models using visual programming and different technics of communication. The main goal of the experiment was to simulate a construction scenario of the design achieved. The natural scale (1:1 scale) prototypes produced had the objective to augment the collaborative group work related to the design, production and assembly of a prototype providing a simulation of a real building scale. The design process, as experiment, allowed students of architecture to have a pragmatic contact between public administration representatives (from a municipal urban planning department), as well with the industry (card board company) that provided the raw material to be used in the prototype. Also important was the programmatic themes that were approached, i.e. the sheltering to people with emergency needs as well the informal public sports equipment, exposing the students to such contemporary problematics.

2 METHODOLOGY

The experiments shown in this paper were performed during an academic course dedicated to 3D modeling through visual programming with Grass hopper (GH) and digital manufacturing at different scales (Kolarevic 2003) (Figure 1).

The first half of the course was dedicated to design processes. It was given special focus on parametric modeling fundaments, particularly on generative and parametric design that is a way to generate designs true a set of rules that may be explicit has an algorithm, among other subjects related to computational methods applied to different design systems (Coutinho 2013). Some programming knowledge was given, using GH, as how to use numeric data and Boolean operations, strings and constants (lists), mathematical operators and functions, conditions and repetitions, among others. With this set of programming concepts students were ready to develop there one designs i.e.: building composition elements; structural elements (pillars and beams); trusses; and mapping elements of a facade. They were also updated about techniques for the preparation of discrete elements for manufacturing using laser cutter and fused deposit modeling (FDM).

The second half of the course was based on the development of the design (architectonic project) using as the main tool the GH. Students at these earlier design stage developed hand sketches and small test conceptual models. A third phase of the course was dedicated to gain more knowledge on production processes focusing both on subtractive and additive technics. The first technic consisted in subtracting material from the raw material used to produce the prototype. In this case a laser cutter machine was the fabrication tool. The second, consisting in the addition of polymeric material (ABS) through a set of paths performing a scale prototype of the design development. (Pupo 2009). The final stage was centered on the assembly of the real scale prototype of one group elected proposal (Figure 2).

2.1 *Collaborative design process*

The design experiences performed by the different groups of students (composed by 3 members) started from an analysis of an urban problem (sociological, topological, planning, etc.). They had meetings with the technicians of the municipality that gave the

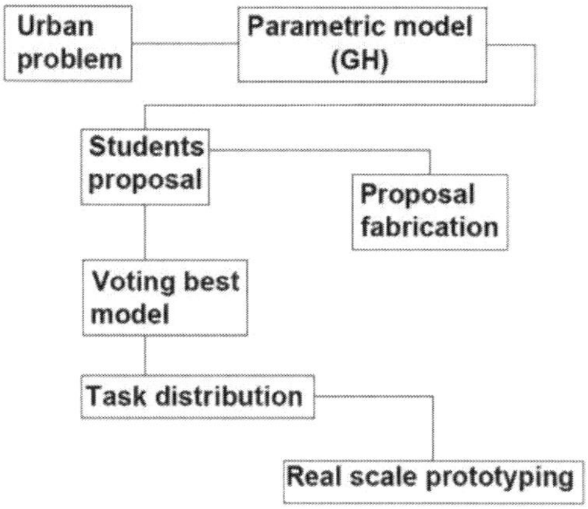

Figure 1. Development structure of the Digital Processes III discipline.

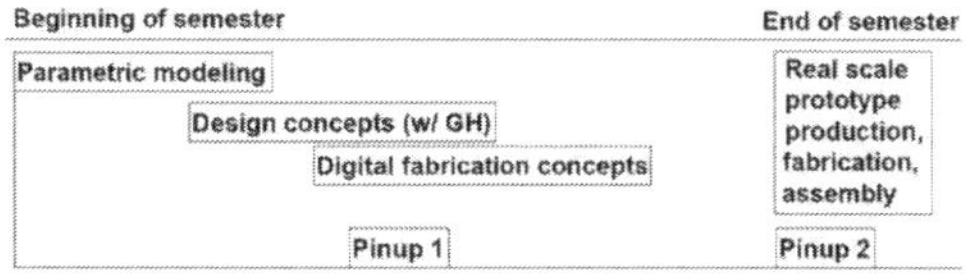

Figure 2. Structure of the main events of the Digital Processes III discipline.

students several inputs about building normative, construction standards and some insights about the strategic plans implemented by the municipality regarding the public needs.

The different prototypes showed in this paper had different design strategies performed among the members of each design groups. The shelter for people with emergency need had its generation principles related with the manipulation of a catenary curve that was rotated generating a surface with its configuration. Then, using the computation of this surface and a proto tile composed by a Boolean difference between a cone and a cube, all the parts were generated (Figure 3).

The sports support pavilion design used the origami logics. The main surface gave place to a set of folded parts that were assembled among other giving the shelter its structural support (Figure 4 and 5).

The designing of each group was then prototyped in a 1:100 scale using FDM and mainly laser cutter. The purpose was to simulate the way of construction in order to understand and to make a more straight analogy with the 1:1 scale construction process. With this method it was possible to them to show the feasibly of the assembly process in order to better understand it and, eventually, optimize it (Figure 6 and 7).

All models were presented to the students and the municipality technicians in order to be voted

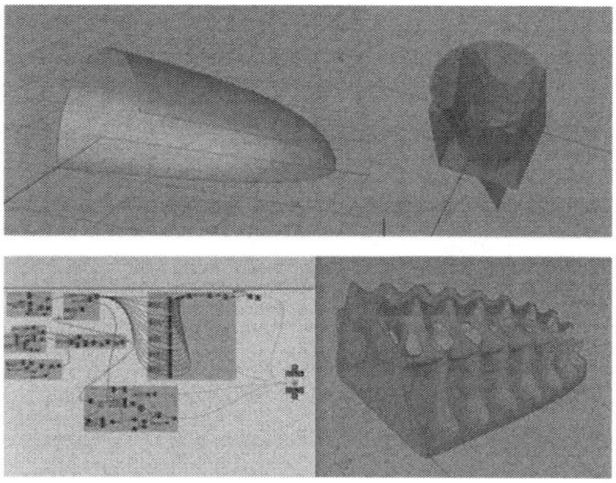

Figure 3. (a), (b), (c) and (d). Figure a is the main generic shape were the element showed in Figure b will be populate the previous shape generating the 3d model of the shelter for people with emergency need design generated with GH present in figue d. Figure c is related with the structure of the GH code.

Figure 4. Different views of the paper model testing the design with origami logic.

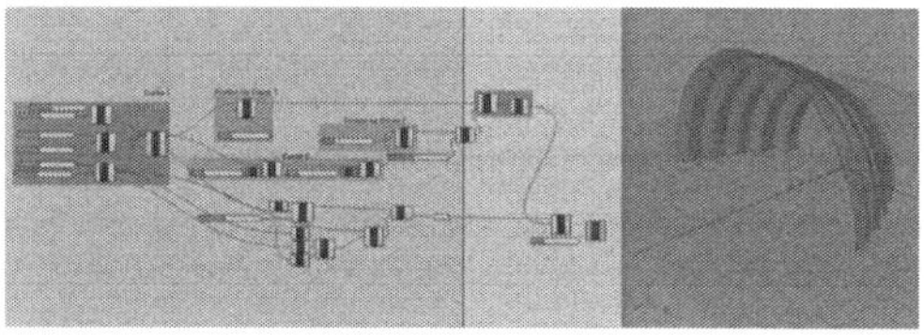

Figure 5. (a) and (b). Figure a is related with the structure of the GH code that generated a shape based in the origami logic aiming to produce the Sports support pavilion design showed in Figure b.

regarding the choice of the best solution to be fabricated. Following the collective decision the production of the real scale prototype and all related tasks were distributed among the different groups. Some

Figure 6. Shelter for people with emergency FDM model.

Figure 7. Sports support pavilion FDM model.

groups were responsible for all the logistics, namely the contacts with the cardboard factory administration and technicians. An informal protocol was set between the class and the company in order to have the type of card that better suited the prototype fabrication (10mm double brown cardboard). The raw material was then transported to the faculty to be cut, as the tool elected was an epilog 80x50cm laser cutter.

2.2 *The design problematic*

One of the issues that students were confronted to build was a shelter for migrants, serving as an emergency architectural element in an unspecified place. The students were asked to do a bibliographic surveying related to this specific topic having in mind the tragic events, killing thousands of migrants and refugees in the Mediterranean Sea as it was particularly important to reflect on this contemporary problematic.

Another program that students had to deal with was the sports equipment to be used in public spaces, by all the municipal citizens. This equipment was inserted in a wider running and cycling path system in the City of Odivelas, Portugal. For this purpose they made a field trip to the site, and they had meetings with the municipal urban planning technicians, were they received information about the strategic planning's. This plan was dealing with different issues concerning the local and the specific population needs to be carried out, namely, mobility and its impacts on new infrastructures. The goal of such municipal strategies was promoting better interurban passive traffic, leading to Paris Agreement regarding carbon 0 targets convergence.

3 RESULTS

The 1.1 scale prototypes were made using 8 mm double corrugated board. They were produced, cut and assembled at the Department of Architecture's to be presented at the end-of-course exhibition.

The Municipal support pavilion prototype was installed in the exhibition of the Municipal Prize of Architecture of the Municipality of Odivelas and transported to the university campus.

One of the objectives was provide students an experience of a full-scale prototype, having a thorough knowledge of the project that resulted in the three-dimensional model.

Another goal was to enable collaborative work regarding the development of something whose quality of the final result is the result of an ability to adapt to collaboration protocol. The main media used was the mobile phone to share drawings and descriptions, trough PDF files mainly as well augmented reality to visualize the proposal on site. These files contained the assembly drawings and notations of the assemblage procedure for each student teams. As result of extensive use of such media devices it was possible to avoid the use of paper printed documents to share design information (Figure 8 and 9).

3.1 *Producing, cutting and assembling the prototypes*

The prototypes developed consisted in a set of pieces generated in GH. Each one was marked with a label

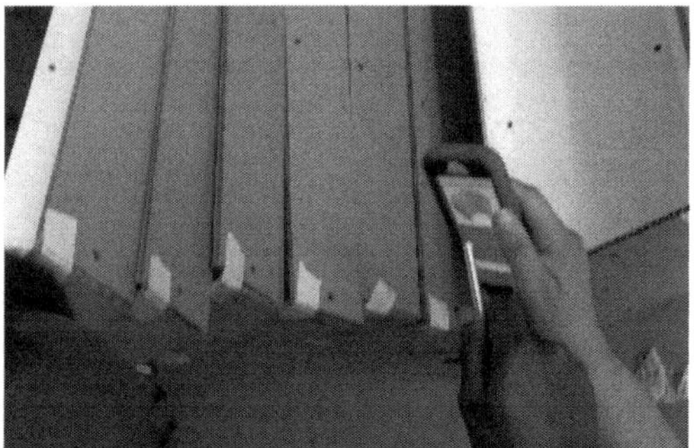

Figure 8. Using mobile phone to share drawings and descriptions for assembly.

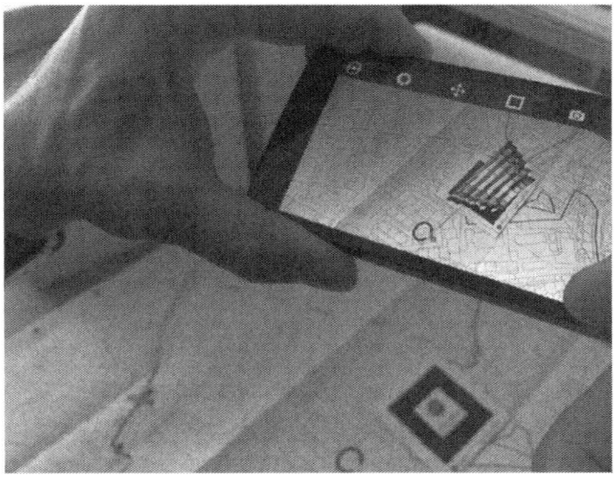

Figure 9. Using mobile phone to visualize the project with augmented reality.

(number). Those numbers were showing the sequences of assemblage of each of the components of the prototype. It was "randomly" mapped and generated a set of circumferences on the external face of the pieces that allowed cutting some openings and fenestrations similar to a previous experience carried out in FAUP at José Sousa class showed at CAADRIA 2013 (Sousa 2013).

Each of the pieces of the prototype was connected to each other by plastic clamps. The different pieces were produced in an Epilog Fusion 32 laser cutter with a cut area of 50cmx80cm.

One group of students was responsible for the cutting assistance of all 350 parts of all pieces. This work took approximately 12 hours divided for 3 different days (shifts).

The assembly process was divided in three main phases. First to connect the 5 different of each piece that was coming out of the laser cuter. Most part of the 25 students participated on this action that took circa 4 hours in one day. Secondly, all pieces were connected producing an arch. The team members of 4 students were responsible to assembly 3 or 4 arches. This task took circa 2 hours to perform. Because of the paraboloid shape of the prototype surfaces it was necessary to essay the final mockup assembly without the clams to see if the final shape was achieved.

Finally, all the arches were transported to the exhibition site, an assembled (Figure 10, 11 and 12).

All the prototypes produced might be removable and transportable in order to be possible to exhibit them in different places. In this way one of the models was part of the exhibition and further publication of the Municipal Prize of Architecture of the

Figure 10. Assembly the shelter for people with emergency prototype on site.

Figure 11. View of final assembly the shelter for people with emergency prototype on site.

Figure 12. Students experimenting the real scale prototype.

Figure 13. Students preparing the real scale prototype public exhibition.

Figure 14. Sports support pavilion prototype at Municipal Prize of Architecture exhibition.

Municipality of Odivelas 2017 (http://www.cm-odivelas.pt) (Figure 13, 14).

4 CONCLUSION

The academic experiences described in this paper had the objective of allowing the students 'contact with different manufacturing techniques, describing their project intentions through an algorithm using GH.

They understood the importance of collaborative work between teams of designers, and other actors as municipal technicians and industrial partners, that lead to the assembly, simulating the construction environment. They experienced the minimal use of paper in the communication and documentation of the project and other elements necessary for the execution of the 1:1 scale prototype to using Smartphones as main media.

Finally, we consider particularly useful and relevant that the curriculum of the architecture course may allow students to have an experience of constructing an architectural element in a 1:1 scale empowering their responsibility on the process of concepts materialization.

ACKNOWLEDGMENT

The author would like to thank all the students of the Department of Architecture of the Universidade Lusófona de Humanidades e Tecnologias (ULHT), in particular Madalena Cabral, Luís Campos, Simon Paukner, João Pedro Serafim and LABTEC monitors. Also thanks the SAICA Pack paper company and the Odivelas Municipality that over the last few years were key partners in this process. This paper was funded by the Experimental Laboratory of Architecture and Urbanism (LEAU) of ULHT.

REFERENCES

Coutinho, Filipe, Duarte, José and Krüger, Mário. 2013. "Digital fabrication and rapid prototyping as a generative process". Green Design, Materials and Manufacturing Process. A.A. Balkema Publishers – Taylor & Francis. The Netherlands. pp. 509–512.

Kolarevic, Branko. 2003. "Architecture in the Digital Age: Design and Manufacturing". Spon Press. pp. 65–71.

Pupo, Regiane. 2009. "Inserção da prototipagem e fabricação digitais no processo de projeto: um novo desafio para o ensino de arquitetura". Tese (Doutorado). Universidade Estadual de Campinas. Campinas, SP. pp. 191–197.

Sousa, José; Xavier, José. 2013. "Symmetry-Based Generative Design: A Teaching Experiment Open Systems: Proceedings of the 18th International Conference on Computer-Aided Architectural Design Research in Asia (CAADRIA 2013). Singapore 15–18 May 2013, pp. 303–312. http://www.cmodivelas.pt/anexos/areas_intervencao/urbanismo/premio_arquitetura_2017.pdf . Last access at 10/3/2020.

Industry 4.0 – Shaping The Future of The Digital World – da Silva Bartolo et al. (eds)
© 2021 Taylor & Francis Group, London, ISBN 978-0-367-42272-1

Prefabrication in construction: A preliminary study in India and Portugal

S. Vijayakumar
School of Technology and Management, Polytechnic of Leiria, Portugal

F. Craveiro
School of Technology and Management, Polytechnic of Leiria, Portugal
CIAUD, Lisbon School of Architecture, Universidade de Lisboa, Portugal

V. Lopes
School of Technology and Management, Polytechnic of Leiria, Portugal

H. Bártolo
School of Technology and Management, Polytechnic of Leiria, Portugal
CIAUD, Lisbon School of Architecture, Universidade de Lisboa, Portugal

ABSTRACT: Prefabrication/modular construction is critical for construction productivity, increasing its speed, efficiency and overall safety, while reducing environmental impacts. A review was conducted on relevant academic and industrial data concerning prefabrication/modular construction. An exploratory questionnaire survey of Indian and Portuguese construction practitioners was then conducted to get practitioners' views and opinions on the benefits and drawbacks of offsite construction, as well practitioners' forecasts on the future growth of the offsite construction sector in both countries. The large majority of practitioners are aware of the possibilities and potential of offsite construction, and most of them are familiar with its advantages and disadvantages. Findings suggest that most practitioners know the possibilities and potential of offsite construction, and most of them also recognize the advantages and disadvantages of its use. Recommendations for future highlight the need to further progress towards standardization and user-driven digital customization in offsite prefabrication factories.

1 INTRODUCTION

The construction industry plays an important role in the global economy, accounting for approximately 6% of the GDP worldwide (World Economic Forum 2016). It has a great impact on any country's economy, playing a fundamental role in major infrastructure development, generating huge employment and supporting other industrial sectors, critical for a sustainable economic growth (Best & De Valence 2002). The Engineering & Construction (E&C) sector plays a key role in creating value for society by designing innovative and sustainable solutions, though it is responsible for large global CO_2 emissions and a major consumer of raw materials. The most used materials in prefabrication are wood, concrete and steel, similarly to what happens to conventional construction.

The E&C industry has been reluctant in embracing digital technological opportunities, so productivity performance has consequently dropped (Mckinsey Global Institute 2017). Conversely, construction projects are getting increasingly complex and larger in scope, facing cost overruns and schedule delays, which combined with labor shortages have been pressing stakeholders worldwide to introduce new technologies and methods (Hoover & Snyder 2017).

Traditional construction work uses locally made technologies to perform a specific project, requiring a temporary organization with start and end (Dineshkumar & Kathirvel 2015), while offsite construction, comprising prefabrication, modularization, preassembly or offsite fabrication, brings a manufacturing approach to construction (Burgess et al. 2018), providing improved efficiencies through the use of repeatable components and processes, achieving a higher quality product at lower cost and in less time.

1.1 *Role of prefabrication in construction*

Dakwale et al. (2011) sustains that there is a growing demand for new construction methods, faster delivery times, lighter structural components, and lower life-cycle costs. Modular construction, based on the production of standardized structural components in a factory environment subsequently assembled onsite, include an innovative set of systems, which can be used separately or in hybrid systems depending on project complexity and scale. Prefabrication, including either completely modular buildings or individual prefabricated components, can increase productivity and efficiency with a high-quality level. However, these prefabrication techniques require a high level of knowledge and experience, coupled with a high quality of onsite application.

Faster construction times, efficiency of materials, and higher quality of construction are vital for countries like United States, Japan, and Western Europe, as middle class demands quality buildings, while developing countries like India use prefabrication systems essentially to produce fast and affordable houses (Kim 2009) in safer and controlled factory environments.

Prefabrication is seen as an important way to improve not only time, quality and safety, but also sustainability and value (Tam et al. 2007). The rise of construction products manufactured in factories, conveyed onsite for assembly and installation is now a common trend offering better quality control, more efficient site processes, greener manufacturing and reduced project costs. It has been used in advanced projects to create permanent buildings and facilities ranging from single-family homes to residential apartments and large commercial buildings (Chiang et al. 2006).

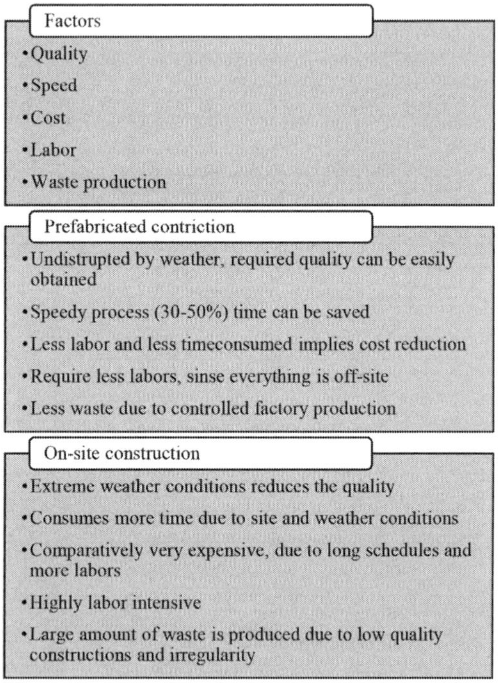

Figure 1. Prefabricated construction vs traditional one.

1.2 *Market demands for adoption of prefabrication*

The rise in global construction and infrastructure is one of the major drivers of the global prefabricated construction market. According to a market research report published by Markets and Markets (2018), the market size is projected to grow to USD 129.67 billion by 2023, mainly due to the rise in demand for complex structures at affordable rates, rapid urbanization, and industrialization. Rising construction activities in emerging economies and the need for sustainable methods are creating vast opportunities for modular construction. Several factors are considered towards adoption of prefabrication in construction, such as structural efficiency (Gustafsson & Hehir 2005, Saari & Heikkilä 2008), optimal use of materials (Prior et al. 2012, Asnaashari et al. 2009), adaptability (Van Acker et al. 2001) and protection of the environment (Forsberg & von Malmborg 2004).

1.3 *Prefabricated vs Traditional construction*

Modular construction can be made concurrently in a factory and onsite, allowing delivering projects much faster than traditional construction (Arashpour et al. 2017, KPMG 2016). Factory production, taking advantage of task repetition and requiring more rigorous and effective quality control, allows both the implementation of efficient and rational production processes and a faster return on investments, fundamental for commercial property owners (Charles & Sacks 2008).

Figure 1 illustrates major differences between prefabricated construction processes and traditional ones.

According to Shahzad et al. (2015), Ulubeyli et al. (2014) and Tam et al. (2007), prefabricating systems in a controlled factory environment, produced within enclosed working environments with reduced labor hours, have important benefits for building deconstruction, not demolition. Greener, sustainable building practices are a priority for a modern construction companies, so modular construction can be more appealing than conventional one.

1.4 *Limitations of prefabrication*

According to Goulding et al. (2014), offsite construction methods, though recognizing its suitability to improve quality and efficiency, can present significant perceived disadvantages among key stakeholders, while Jonsson & Rudberg (2013) sustain that major constraints and barriers of modular construction are mainly subjective, being perceived costs a major barrier to an increasing market share. On the other hand, Arif et al. (2012) affirm that one of the main limitations to prefabrication is higher capital costs, though its negative public perception is an important obstacle too.

Vijayakumar (2018) considers other barriers too, like additional transportation costs, inflexibility for late design changes and lack of adequate storage space for on-site assembly.

1.5 *Prefabrication in India*

Indian construction sector employs more than 35 million people, accounting for more than 8% of Indian GDP (Vernikos 2012). There are several performance issues, especially related to inefficiencies, mainly caused by lack of skilled labor, flaws in construction planning and Indian construction culture.

In 1948, prefabrication was launched in India with the Hindustan Housing Factory, specialized in the production of pre-stressed concrete railway sleepers, which replaced wood sleepers in Indian Railways. Shortly after, the company diversified its operations and production, and lately renamed Hindustan Prefab Limited (HPL), located in Delhi, where it prefabricates concrete for civil and architectural projects throughout India (Lal 1995). Today, it aims at moving towards large scale construction and general housing sectors to which prefabrication technology is appealing.

1.6 *Prefabrication in Portugal*

In Portugal, though the construction industry contribution to GDP is substantial, the sector comes after the tourism and the manufacturing sector, it fell to €1,629.70 million in the second quarter of 2018, having achieved €1.673 billion in the first quarter of 2018 (Trading Economics 2018). Sustainable construction and energy efficiency are important issues in Portugal. Major initiatives include clusters on sustainable construction, focused on eco-innovation and energy efficiency, promoting an efficient building renovation/rehabilitation, as well energy efficiency measures for infrastructures, co-financed by the European Commission.

After the 1755 great Lisbon earthquake, the "Baixa Lisbon district" was rebuilt using an anti-seismic design combined with innovative and prefabricated construction methods. Large multi-story buildings were made outside the city, transported in pieces and then assembled onsite. This novel process, which lasted until the mid-nineteenth century, enabling city inhabitants to live in new safety building structures.

Today, prefabrication is a major industrial process with an exponential increase in Europe, though Portugal is considered a moderate innovator regarding offsite construction as the construction sector has not yet recovered from the 2008 economic crisis (European Construction Sector Observatory 2018).

2 DATA COLLECTION AND ANALYSIS

In 2018, Vijayakumar (2018) conducted a survey to analyze the perceptions and progress of the adoption of offsite construction in India and Portugal. The data for this investigation was obtained from four main sources:

1. a literature review of existing academic and industrial surveys and publications;
2. a preliminary study conducted to define the final version of the questionnaire;
3. a main questionnaire survey of 50 Indian construction practitioners was conducted in English, including 31% architects, 23% engineers, 31% contractors and others (15%);
4. a main questionnaire survey of Portuguese construction practitioners was conducted in Portuguese, including 20% architects, 50% engineers, 15% contractors and others (15%).

Only part of the data obtained in this survey is presented in this work.

A non-probability sampling method (a mix of convenient sampling and judgment sampling) method was used. The overall intention was to analyze experts' perceptions of the use of offsite construction in India and Portugal. The main limitation of this survey method involved the availability of internet connection for accessing the web-based survey and the use of the English language, especially for Indian respondents, which may have contributed to a low response rate.

The questionnaire was emailed to professionals from Indian and Portuguese construction companies and other construction stakeholders (engineers, architects, contractors and others construction practitioners). Out of all valid responses from companies/organizations working in the field of prefabrication, 61% belonged to Portuguese construction experts and 39% to Indian ones.

Respondents were asked to rate on a scale of 1 to 5, from the highest to the lowest level, each factor determining the perceived importance of the factors included in each question. The data analysis and final reporting was conducted using descriptive statistics and visualization analytics through MS Excel 2007 and IBM SPSS statistics 23. A sample distribution of the mean difference was calculated to test the significance of the differences among the results obtained for both countries.

2.1 *Results' summary*

Table 1 summarizes the most/least important advantages and limitations for Portuguese and Indian construction practitioners. Most experts outline 'Fast return on the investments' as an unimportant advantage, selecting 'improved quality control' and 'health and safety' as the most important advantages of prefabrication. Indian Engineers, contractors and other construction experts highlighted 'lack of suppliers' as the most important disadvantage, whereas for Portuguese stakeholders 'Higher initial cost' and 'Resistance to change' are the most important barriers. Portuguese architects, Indian contractors and other Indian stakeholders stressed 'Poor connections with other

Table 1. Most/least important advantages and disadvantages of offsite construction.

	Advantages		Disadvantages	
Respondents	Most Important	Least important	Most Important	Least important
Indian architects	Improved quality control	Fast return on investments	Repetitive design leading to monotony	Not flexible enough.
Portuguese architects	Improved quality control	Fast return on investments	Higher initial cost	Poor connection with other elements
Indian engineers	Reduction of design time	Fast return on investments	Lack of suppliers	Lack of industry expertise
Portuguese engineers	Reduction of construction time	Fast return on investments	Resistance to change	Lack of suppliers
Indian contractors	Improved quality control	Fast return on investments	Lack of suppliers	Poor connection with other elements
Portuguese contractor	Improved health and safety	Fast return on investments	Higher initial cost	Lack of industry expertise
Other Indian stakeholders	Improved quality control	Fast return on investments	Lack of suppliers	Poor connection with other elements
Other Portuguese stakeholders	Reduction of construction time	Reduction of design time	Resistance to change	Higher initial cost

elements' as a limitation, whereas Indian engineers and Portuguese contractors underlined 'Lack of industry expertise' as a limitation to the introduction of modern prefabricated techniques.

Construction stakeholders were also asked to comment on both "the barriers contributing to a slow adoption of prefabrication in construction" and "the use of offsite manufacturing and BIM to overcome challenges facing the construction sector".

More than 50% of the respondents from India identified transportation/logistics as the biggest barrier for the adoption of prefabrication. As a result of construction modernization, existing construction supply chains only dedicated to the delivery of raw materials to sites, need to adapt to variations in demand on construction sites designing new optimized modular construction supply chains (Hsu et al. 2018). Apparently, the cost of transportation and shipping regulations is perceived as a barrier to the adoption of prefabrication by Indian construction practitioners. Another limitation regards highly densely populated urban areas in India, which make it more difficult to ship large components through narrow roads or gateways. The other logistics barrier, according to Indian construction professionals, is the Payload for heavy vehicles in India, which is highly taxed, adding cost to the transportation of large and bulky components, requiring extra caution when carried out across roads.

"Repetitive design leading to monotony" is cited by Indian architects as one of the reasons why clients/customers do not consider prefabrication as a good alternative, hesitating in adopting new modern modular technologies instead of traditional construction ones, mainly due to lack of information. Contractors from India also referred the difficulty in finding trained and qualified labor workers. Training new skilled workers is time consuming and expensive, which can lead to an increase in project costs, being very difficult to convince clients to adopt new costly methods.

Portuguese respondents highlighted "higher initial cost" and "lack of funding" as the major barrier for the adoption of prefabrication. In the past, Portuguese banks apparently restricted the funding to projects incorporating new construction methods. Another obstacle also referred by some Portuguese stakeholders was the high wages required by skilled workers. Similarly, to Indian clients, Portuguese companies hesitate in applying new construction techniques mainly due to a deficient knowledge of the technology. The lack of knowledge on prefabrication and standardization technologies, as well the lack of regulatory codes is also referred by Portuguese professionals as a major limitation to its adoption. Transportation and shipping costs also plays a vital role, as well higher capacity cranes required onsite, which are expensive increasing the budget. Another important limitation for Portuguese Engineers is the behavior of the structure, reporting that prefabricated structures sometimes does not perform as intended, because the links between the various structural elements, often not well executed, so the structure do not behave like a monolithic structure. Design issues are also pointed out by Portuguese architects, referring that the lack of design flexibility limits creativity.

Perceptions regarding the benefits of BIM implementation for the application of offsite construction will be the object of another work.

2.2 *Discussion of results*

According to most of the respondents, transport costs have been neglected regarding prefabrication optimization. In general, the transport of components can be carried out in two ways, either by the transport of components onsite by the crane in the prefabrication factory or the construction site, or by transporting components between different locations by truck. Most respondents highlighted higher initial costs by renting heavy capacity cranes and trucks add on to the project budget, though costs can be reduced by fabricating the components on-site. In this case, there is no controlled factory environment. Indian stakeholders pointed out another solution, an increased number of prefabricating factories can reduce the distance between factory and the assembly site. According to them, less tax regulations should be applied on vehicles handling heavy prefabricated components.

Setting up more prefabrication factories and material outlets, could be a good solution not only to solve the issue of transportation but also the problem of 'lack of suppliers' and 'Lack of industry expertise'. When more factories setup, it can increase the competition which might result in reduction of the final product cost, which has direct impact on initial costs. With larger numbers of prefabrication factories there will be more opportunity for new designs and components, which can eliminate the 'Repetition leading to monotony'.

Portuguese architects think that it is possible to conceive ingenious solutions with a good esthetic value, that are economical to build and efficient from the structural point of view, combining all the advantages of the prefabrication industry, with a good articulation between designers, prefab companies and contractors. Some Portuguese stakeholders stressed adequate planning for modular service construction as an important step to further progress on prefabrication acceptance.

3 CONCLUSIONS

Rapid urbanization, climate change and resource scarcity can impact the health and wellbeing of society, so the innovation has been a priority to the European economy. Growing sustainability concerns are encouraging new strategies for climate mitigation and resilience, promoting sustainable design of new buildings to enhance future re-use and resources efficiency, cutting this way carbon emissions.

Most of E&C organizations recognize the need to drive digital transformational change in the construction sector. The global adoption of new construction techniques and technologies, and the emergence of new business models, allow responding to the urgent need for new housing and infrastructure, as well increasing its productivity and efficiency.

Globally, the demand for new techniques is rising towards faster delivery times, lighter structural components and lower life-cycle costs, so offsite construction has been emerging as a key technology for faster, safer and cheaper construction. Its use has been intensified by the potential to improve productivity and efficiency without reducing quality. On the other hand, there is a growing awareness of the importance to reduce the use of scarce natural resources and construction waste. Construction operations have great impacts on the environment, with the deterioration of natural resources and generation of huge quantities of waste. Manufacturing processes produce less waste in comparison with conventional building practices, so design for manufacture can be a good solution, as it is a much more efficient and controlled process with materials ordered and cut to size in the factory.

In this work, an exploratory questionnaire survey of Indian and Portuguese construction stakeholders was conducted to identify the main views and opinions on the benefits and drawbacks of prefabrication, as well practitioners' forecasts on its future in both countries. Differing opinions of Indian and Portuguese stakeholders seems to depend on both culture and technology.

Shorter onsite construction times and improved quality were considered the major advantages, and perceived additional cost of offsite was the main barrier to its use. The lack of skilled workers in the construction industry is an opportunity for the increased use of offsite. For the offsite market to develop however, the negative perceptions of offsite need to be overcome. Transportation/logistic issues were also considered a major barrier for the adoption of prefabrication. Most respondents were aware of the possibilities and potential of prefabrication in the future.

In the future, BIM will potentiate its usage, making planning and designing much easier and efficient, by allowing to compare virtual and real design, improving overall project quality and enabling the prefabrication of larger and more complex parts.

REFERENCES

Arashpour, M., Too, E & Le, T. 2017. Improving productivity, workflow management, and resource utilization in precast construction. 9th International Structural Engineering and Construction Conference: Resilient Structures and Sustainable Construction. ISEC Press.

Arif, M., Bendi, D. Sawhney, A. & Iyer, K.C. 2012. State of offsite construction in India-Drivers and barriers. Journal of Physics: Conference Series 364(1): 012109.

Asnaashari, E., Knight, A., Hurst, A., & Farahani, S.S. 2009. Causes of construction delays in Iran: project management, logistics, technology and environment. Proceedings of the 25th Annual ARCOM Conference.

Best, R. & De Valence, G. 2002. Design and construction: Building in value. Oxford: Butterworth-Heinemann.

Burgess, G., Jones, M. and Muir, K., 2018. What is the role of offsite housing manufacture in a digital Built Britain. Cambridge Centre for Housing & Planning Research.
Charles, E. & Sacks, R. 2008. Relative productivity in the AEC industries in the United States for on-site and off-site activities. Journal of construction engineering and management 134(7): 517–526.
Chiang, Y.H., Chan, H.W., Lok, K.L. 2006. Prefabrication and barriers to entry- a case study of public housing and institutional buildings in Hong Kong. Habitat International 30 (3): 482–499.
Dakwale, V. A., Ralegaonkar, R. V. & Mandavgane, S. 2011. Improving environmental performance of building through increased energy efficiency: A review. Sustainable Cities and Society 1(4): 211–218.
Dineshkumar, N. & Kathirvel, P. 2015. Comparative Study on Prefabrication Construction with Cast In-Situ Construction of Residential Buildings. International Journal of Innovative Science, Engineering & Technology 2(4).
European Construction Sector Observatory 2018. Country profile Portugal. European Commission.
Forsberg, A. & von Malmborg, F. 2004. Tools for environmental assessment of the built environment. Building and Environment 30: 223.228.
Goulding, J.S., Pour Rahimian, F., Arif, M. & Sharp, M.D. 2014. New offsite production and business models in construction: priorities for the future research agenda. Architectural Engineering and Design Management 11(3): 163–184.
Gustafsson, D. & Hehir, J. 2005. Stability of tall buildings. Master's Thesis, Chalmers University of Technology, Göteborg, Sweden.
Hoover, S. & Snyder, J. 2017. A New Era for Modular Design and Construction. FMI.
Hsu, P., Angeloudis, P. & Aurisicchio, M. 2018. Optimal logistics planning for modular construction using two-stage stochastic programming. Automation in Construction 94: 47–61.
Jonsson, H. & Rudberg, M. 2013. Classification of production systems for industrialized building: a production strategy perspective. Construction Management and Economics 32: 1–17.
Kim, T. 2009. Comparison of prefab homes and a site-built home: Quantitative evaluation of four different types of prefab homes and a site-built home. Southern California University, USA.
KPMG 2016. Smart Construction - How offsite manufacturing can transform our industry. KPMG.
Kumar, B. and Paul, S.R. 2016. The quality consciousness in engineering education 'an alarming attention'. International Journal of Science & Technology 6(3).
Lal, A.K. 1995. Handbook of Low Cost Housing. New Age International Publishers.
Markets and Markets 2018. Modular Construction Market by Type, Material, End-use sector, and Region - Global Forecast to 2023. Markets and Markets.
Mckinsey Global Institute 2017. Reinventing construction: a route to higher productivity. Mckinsey Global Institute.
Prior, T., Giurco, D., Mudd, D., Mason, L. & Behrich, J. 2012. Resource depletion, peak minerals and the implications for sustainable resource management. Global Environmental Change 22: 577–587.
Saari, A. & Heikkilä, P. 2008. Building Flexibility Management. The Open Construction and Building Technology Journal 2: 239–242.
Shahzad, W., Mbachu, J. & Domingo, N. 2015. Marginal Productivity Gained Through Prefabrication: Case Studies of Building Projects in Auckland. Buildings 5: 196–208.
Tam, V., Tam, C., Zeng, S. & Ng, W. 2007. Towards adoption of prefabrication in construction. Building and Environment 42(10): 3642–3654.
Tam, V.W., Tam, C., Zeng, S. & Ng, W.C. 2007. Towards adoption of prefabrication in construction. Building and Environment 42: 3642–3654.
Trading Economics 2018. Portugal GDP From Construction. Trading Economics.
Ulubeyli, S., Kazaz, A. & Er, B. 2014. Planning Engineers' estimates on labour productivity: theory & Practice. Procedia Social &Behavioral Sciences 119: 12–19.
Van Acker, A., Valaitis, K. & Ambrasas, A. 2001. Modern prefabrication techniques for building structures. The 7th International Conference. Vilnius Gediminas Technical University.
Vernikos, V.K. 2012. Optimising Building Information Modelling and off-site construction for civil engineering, Civil Engineering 165(4): 147.
Vijayakumar, S.B., 2018. Prefabrication in construction: an exploratory study of advantages and challenges in India and Portugal. Master's thesis, Polytechnic Institute of Leiria, Leiria, Portugal, retrieved from IC-Online (http://hdl.handle.net/10400.8/3888)
World Economic Forum 2016. Shaping the Future of Construction: A Breakthrough in Mindset and Technology, Geneva, WEF Report.

Industry 4.0 – Shaping The Future of The Digital World – da Silva Bartolo et al. (eds)
© 2021 Taylor & Francis Group, London, ISBN 978-0-367-42272-1

From conflicts to synergies between mitigation and adaptation strategies to climate change: Lisbon Sponge- City 2010-2030

N. Pereira & C. Alho
University of Lisbon, Lisbon, Portugal

ABSTRACT: During the past thirty years, European cities have been addressing global climate change and local impacts through testing and implementation of mitigation and adaptation strategies. Lisbon Downtown is no exception, with ten masterplans under execution since 2010 with completion scheduled until 2030, valued at one billion euros of public investment. However, the gap between mitigation and adaptation strategies is not yet sufficiently studied in parallel to *nuances* vulnerability and risk mitigation, resilience and adaptation. In Lisbon Downtown, these plans are being implemented separately, consequently compromising the effectiveness of public investment. This paper analyses the current urban development actions in Lisbon Downtown in order to identify potential conflicts and synergies. The results suggest that the largest public investment in Lisbon on flooding mitigation towards a Sponge City will conflict with the new cruise ship terminal and old Downton building stock, therefore increasing risk and vulnerability factors.

1 INTRODUCTION

The phenomenon of extreme anthropogenic climate change arising from carbon civilization (National monitoring, reporting and evaluation of climate change adaptation in Europe 2015, Lomborg 2001) is a relevant research area of contemporary Urbanism in line with stabilized paradigms of Sustainable Development (Bentivegna and Lombardi 1997, U. Nations 2015).

It is increasingly evident that the origin of the disruption of global climate patterns stems from the ecological footprint of man (Berry et al. 2008, Conway 2012, Weart 2012).

As more than 80% of the old continent's population lives in urban areas, and European cities are composed of a fabric that has been developed around a historic center (Biesbroek 2009), it should be relevant to investigate how the problem of ACC (Anthropogenic Climate Change) affects this unique territory (Bulkeleya 2013). New strategies for SUD (Sustainable Urban Development) are being implemented with the environmental issue being affirmed within the general public (Burdett 2008, Hoornweg 2011).

Since the end of the last century, the concern of decision makers has focused on mitigation, on the causes of CC (Climate Change) (Dulal 2012). However, a different approach seems to have arisen in the last decade to the problem focused on adaptation to CC that is, on the effects or impacts (Costa 2013). Lisbon is no exception and it´s Downtown area, The *Baixa Pombalina* (Coelho 2010) is currently the lab territory where several plans have been implemented simultaneously since 2010 with the completion of construction stage expected in 2030 (Enova 2017).

The MACC (Mitigation and Adaptation to Climate Change) strategies have definitely entered the agenda of European cities (Landauer 2015), but some disarticulation has occurred in their design, construction and monitoring stages (Larsen 2012). The priority intervention criteria are not yet stabilized and adaptation strategies seem to gain traction across Europe at the expense of mitigation measures (UKCIP 2011).

Both strategies have a common denominator, that of extreme anthropogenic climate change (Flannery 2005). Mitigation is part of a global agenda of civilizational decarbonization and focused on the causes of the problem and adaptation is part of local agendas tailored to specific contexts (Klein 2005), focused on effects, impacts, risks and resilience (Prasad 2009, Rosenzweig 2014). Mitigation conceived in a long-term implementation perspective (high budget) and adaptation in a short-term implementation perspective (low budget).

In this way, different, sometimes contradictory strategies are implemented simultaneously, where conflicts or synergies can emerge unexpectedly, especially in a vulnerable area of the city as well as the historical center (current regulatory framework is insufficient to address these new problems) (CML 2017). In all cases, there seems to be a consensus about the relationship of conflict and synergy that the simultaneous implementation of MACC strategies encompasses, thus adding the problem of unpredictability (and trade-off) to the equation to be solved (Viguié 2012).

On a global scale, studies in this area seem to suggest that there are currently two blocks with sometimes contradictory approaches: the West (Europe-led) adaptation-oriented and the Far East (China-led) focused on mitigation (Klein 2007, Liu 2013, Oliveira 2013, Warren 2017, Zimmerman 2011).

2 PROBLEM

On the theoretical framework, the main matrix indicators of synergy or conflict often arise with a low degree of evidence, as well as the indicators of consensus among the scientific community (IPCC 2007, IPCC 2014). As a result, this commonly adopted reference model does not refer to synergies or trade-offs, making a clear distinction between mitigation and adaptation. This research argues that in the matrix of the latest IPCC (Intergovernmental Panel on Climate Change) report there is research to be done.

At the empirical level, related to the MACC plans of Lisbon Downtown on the blue, green and gray infrastructures, some problems appear evident: From the Blue infrastructure (Leboeuf 2015): the Lisbon General Drainage Plan was announced at a public hearing as an innovative and element of synergy between adaptation and mitigation. This premise contradicts several models that typify it as Adaptation Plan in conflict with Mitigation and Biodiversity.

A careful analysis of the PGDL- Lisbon General Drainage Plan (CML 2017) allows to conclude that the main storm water sewage (5m diameter) will divert the storm water from the *Baixa Pombalina* Basin, into *Santa Apolónia* area, site of new built Cruise terminal with capacity for five large size cruise ships (Figure 1), conflict omitted in PGDL (Figure 2).

From the Green infrastructure (Enova 2017): images from the end of the 19th century reveal the existence of trees in *Terreiro do Paço* (The Commerce Square), which are aligned according to a well-defined rule, which has since been cut off for the purpose of car parking, however, recent urban regeneration work on the site has not foreseen its replacement (CML 2017).

On the theoretical level, the properties of trees and green roofs on buildings are characterized as synergistic elements that function as CO_2 sinks

Figure 1. Lisbon´s new cruise ship Terminal.

Figure 2. Observed conflict: Terminal (mitigation) vs. Pipeline (adaptation). The projected 5m Ø storm water pipeline ending on the worst possible location with 30m width.

(Energy Cities 2014, Newman et al. 2009), absorb or retain water and regulate the temperature in addition to producing shade. In this area, the MACC plans for the territory are also not sufficiently highlighted (Berry et al. 2008).

From the Grey infrastructure (CML 2016): several studies' results classify the insulation of buildings as a measure that benefits mitigation but conflicts with adaptation. In a climate like Lisbon, well-insulated buildings do not require mechanical means of heating or cooling which reduces CO_2 emissions (Kabisch 2016, Ramaa 2012, Romero 2011).

However, the same measure also helps users to adapt to temperature variations. This seems to be an example of synergy between the two strategies. In the case of *Baixa Pombalina*. According to future building regeneration plans, lime and cork have been considered on the external painting, materials with insulation properties with low ecological footprint (Enova 2014).

In the field of energy transition, there are doubts regarding the effectiveness of ZER 1 (Reduced Emissions Zone) and installation of solar panels in the coverage of buildings at a World Heritage zone. It should be noted, however, that the solar panel has synergistic properties as it reduces the carbon footprint by the micro-generation of energy and isolates the roof of the building from direct sun exposure, decreasing its interior temperature.

In the area of mitigation, another measure with a high degree of uncertainty will be the management of IoT (internet of things) Mega data for raising privacy issues (Rifkin 2014).

Thus, the theoretical and empirical problematic that this article explores in the area of Urban Engineering is framed. These preliminary considerations reveal the importance to deepen the study of relevant synergy models between adaptation and mitigation to CC, with the hypothesis to qualitatively improve specific and significant MACC Plans currently underway in *Baixa Pombalina*.

The paper outlines the specific theme of the common space between mitigation strategies and adaptation to CC in historical centers with water fronts and investigates the specific problem around the resulting conflict/synergy relationship (Laukkonen 2009).

3 OBJECTIVES

At the theoretical level, this paper contributes to extend existent knowledge on synergistic impact matrices between mitigation and adaptation strategies to CC (IPCC 2014), suggesting new indicators on the green, blue and grey infrastructures (Khun 2012).

At the empirical level, this paper targets a qualitative improvement of the main CC mitigation and adaptation plans with direct impact in the Downtown area, which has been implemented since 2010 with the completion of work scheduled for 2030. The selected Plans are the Cruise Terminal (Compact City theory/CC Mitigation) and the Lisbon General Drainage Plan (CC Adaptation/extreme weather events).

The article analyzes the MAAC reference model as well as the reference framework of interactions adopted by the IPCC in the 2014 report (IPCC 2014).

4 METHODOLOGY

This research work in the area of Urbanism covered several areas of Engineering, such as Environmental, Naval, Civil, Hydro and Energy, thus giving it a transdisciplinary profile according to a holistic approach. A panel of experts has followed the development of this study since its inception.

On this paper, the theoretical approach was carried out through reviewing published scientific literature and reference models that are resistant to the test of refutation (Popper 2002).

From the literature review, this study extracted arguments that seek to explain the nature of the problem (conflict vs. synergy) in its theoretical constructs, thus directing critical reflection to stabilized synthesis models (IPCC 2007).

This method was followed by the collection of information relevant to the objectives proposed in published scientific books and articles, theses, reports and project memos, cartography, statistical data, etc.

Subsequent empirical research work was carried out according to quantitative and qualitative on-site analysis. Contacts with local experts were established in face to face meetings or using digital technology. After the data collection and interpretation, it was possible to cross-reference the information with the theoretical reference model (Table 1) and to test the research hypothesis (Synergy between Mitigation and Adaptation).

The theoretical model was tested in Lisbon Downtown, (ongoing World Heritage Site evaluation process by UNESCO) (CML, 2017) according to a positivist and empirical research methodology.

Table 1. MACIS model (excerpt): Minimization of and Adaptation to Climate Change Impacts on Biodiversity (Berry et al. 2008).

Positive	Effect on biodiversity	Negative	Mitigation
Win-lose-win	Improved Transport Efficiency	Win-lose-lose	
Sweet Spot Mitigation& Adaptation Sinergy	Urban Intensification		
Lose-win-win	Flood Control Infrastructure	Lose-win-lose	Adaptation

On this territory, a number of MACC plans have been phased in since 2010. The mitigation plans are: the Sharing Cities project (partnership with Milan and Bordeaux) (McLaren 2015), the Solar Map of *Baixa Pombalina* (CML 2012), the Lisbon Energy Matrix (Enova& IST 2014), the new Cruise Terminal and the New Terminal of the Lisbon International Airport (direct connection with the analyzed territory via future jetfoil lane). The Adaptation measures are described in the EMAAC manual of Lisbon - Municipal Strategy for Adaptation to Climate Change (Enova 2017). CML also presents a synergy plan between adaptation and mitigation in the context of the blue infrastructure: The PGDL - *Plano Geral de Drenagem de Lisboa* (Leboeuf 2015).

The essential regulatory instruments are drawn from the new PDM (Municipal Zoning Master Plan), the PPSBP (Downtown Heritage Site Protection Plan) (CML 2017) as well as the provisions of the ZER 1- Reduced Emission Zone 1 (Enova 2005). Access to other relevant information was made possible by key actors from CML and Port of Lisbon.

5 RESULTS AND DISCUSSION

From the theory on the cause-effect relationship between mitigation and adaptation it is possible to make logical deductions on the interaction between the two projects where conflicts will emerge with higher impact. The compact city theory with mixed-use buildings as a means to mitigate CO2 emissions has been the subject of widespread consensus among the scientific community since the 1990s (Curwell 2008, Seto 2015). However, with the assertion of strategies to adapt to climate change, models emerged suggesting a lower density of soil occupation. The island heat effect is reduced, the area of permeable soil is increased, and the risk of flooding is reduced (Hamin 2009).

The problem of circulating distance gradually fades with the introduction of digital technologies, green energy and 3D printing (Townsend 2014).

European historical cities are adopting scientific and technological advances, and the Lisbon Downtown is an example (there are also concerns in these areas

related to the preservation of the identity of the heritage safeguarded by the legislation in force).

The reference matrix (IPCC) used by researchers reveals some fragilities and needs further study on result demonstration and indicators. The matrix that typifies benefits/synergies (blue) or side effects/conflicts (red) between mitigation and adaptation should include black trade-offs in cases of lesser evidence and consensus. According to the IPCC report (IPCC 2014), compact development and infrastructure preserves open spaces outside the city, which creates synergies between mitigation and adaptation. However the city needs space for the three infrastructures: green, blue and grey as this study reveals. In this case a tradeoff solution is a feasible scenario (Ministero dell′ Ambiente 2015).

In the present case study it is self-evident that the current solution generates negative outcomes. This is a case of elementary conflict between mitigation (terminal) and adaptation (PGDL) due to overlapping issues.

The terminal was not designed to serve as the main drainage sewage fall of Lisbon. This solution entails an increase in risks and vulnerabilities that can lead to fatalities, material damages and ecological accidents (with inefficient allocation of funds from the public purse) (Müller 2010).

Accidents occurring with vessels are frequent in the Tagus estuary and well known. And there are known cases of public works that stopped several years due to project deficiency (metro blue line of Commerce Square).

The cruise terminal project dates back to 2010 and the construction has been completed in 2018 (in which the author participated having coordinated a design team of architecture and engineering). The Lisbon General Drainage Plan dates from 2016 was revised in July 2017 and presented in a public session on August 24 (summer vacation).

However, the current solution (building a concretewall at the end of the 30m wide sewage slow down the storm water) is not yet satisfactory. In fact, it seems to bring more problems by not solving any (the water can raise the pavement at the limit). On top of this, experts alert that the unloading structure will always be subject to the impact of mooring in a turbulent area of 300m quay and 4,000 passengers cruise ships, 24/7. Other risk factor to consider are the impact of construction works on the normal operation of the quay/international area, not to mention related vulnerabilities related to terrorist attacks. Nevertheless, evidence that mitigation and adaptation can benefit each other is becoming increasingly accepted among researchers suggesting that there are alternatives to the existing conflict scenario (Larsen 2009). Synergy scenario requires institutional cooperation among the key players, with the relocation of the storm water

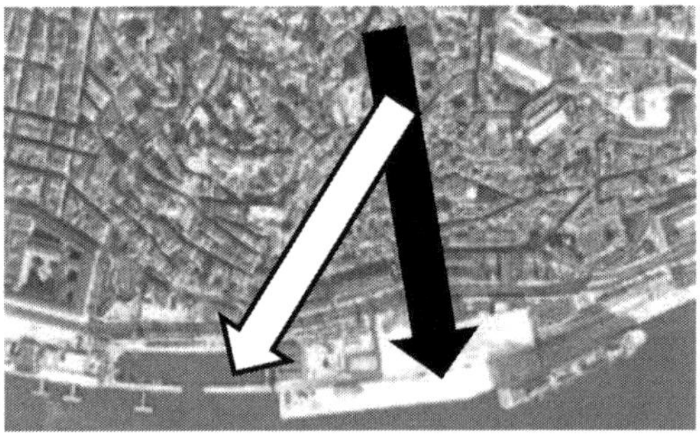

Figure 3. Storm water pipeline (black) diverted from cruise ship terminal to adjacent dock (white): possibility of integration in the river promenade green corridor by microgeneration of hydropower (mitigation& adaptation synergy).

drainage sewer on its last section to the adjacent small dock, a navy facility (Figure 3).

The storm water drainage sewer will be able to produce and store hydroelectric power at dedicated points to and feed some equipment, preferably from the cruise terminal. The passenger terminal is a mitigation project (compact city, low energy consumption and reduction of CO2 emissions), and it complements other synergistic projects, such as the green corridor of the river (CO2 sink, reduction of the heat effect, permeable soil, retention basin and increase of biodiversity) through its landscape green design (Williams 2012).

With the two suggested solutions supported on a scientific basis, the hypothesis that it is possible to mitigate the evident conflict and to suggest the synergy between the two is valid with a high degree of probability (Solecki 2011). The aim of the two projects is to tackle the problem of climate change in a Sustainable Urban Development framework (institutional, social, economic and environmental dimensions) (Deakin& Curwell 2004). Results validate the hypothesis that it is possible to create synergies between mitigation and adaptation to climate change in Lisbon Downtown (Wamsler 2015).

6 CONCLUSION

On the theoretical level this article contributes to the scientific community by demonstrating that on IPCC impact matrix, the "sectoral mitigation measure" creates a "Land-use competition" effect, which in this case is negative.

This paper suggests the inclusion of a trade-off indicator, between conflict and the synergy indicators in the IPCC impact matrix in cases of low evidence and consensus (IPCC 2014).

New areas of research are explored with the integration of sustainability, mitigation and adaptation principles and criteria (Figure 3).

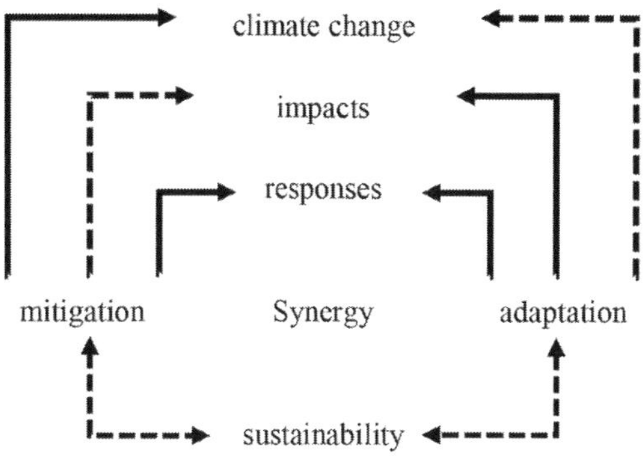

Figure 4. Emerging areas of future research (dashed).

The empirical research shows that the Lisbon Municipality has made a notable effort in recent years in the field of addressing climate change. However, the solution adopted for the drainage tunnel between the city's green hill and the Tagus River creates a conflict with the existing Cruise Terminal that does not respect international safety standards because it endangers human lives or fosters ecological disasters (Vale& Campanella). In this research we propose an alternative route that will solve the problem if there is cooperation among the key players.

This work offers a positive contribution to the international and multidisciplinary team working in Lisbon Downtown plans that is behind the plans underway in Lisbon Downtown framed by a Sponge City model, namely the PGDL. There is evidence that it is possible to make a case study of synergy between mitigation and adaptation of the two contingent Plans.

Research relating these two projects with other projects underway, valued at over EUR one billion and on late design stages are yet to be done.

Academia and other European cities may adopt or be subject to similar research methodology, thus increasing overall sustainability indicators on storm water risk mitigation urban development projects (Figure 4).

REFERENCES

Bentivegna, V.& Lombardi, P. 1997. Evaluation of the Built Environment for Sustainability. London: E& FN Spon.

Berry, P. et al. 2008. Minimisation of and Adaptation to Climate change: Impacts on biodiversity (deliverables 2.2 and 2.3). University of Oxford.

Biesbroek, R. 2009. The mitigation- adaptation dichotomy and the role of spatial planning. Habitat International, UK: Elsevier, 33: 230–237.

Bulkeleya, H. 2013. Revisiting the urban politics of climate change Environmental Politics. Environmental Politics, UK: Frank Cass Publications, 22(1): 136–154.

R. Burdett. 2008. The endless city. London: Phaidon

Costa, J. 2013. Urbanismo e adaptação às alterações climáticas. As frentes de agua. Lisboa: livros Horizonte.

CML. 2016. ClimAdaPT, Manual Avaliação da Vulnerabilidade Climática do Parque Residencial Edificado, Lisboa: CML.

CML. 2017. Plano de Pormenor de Salvaguarda da Baixa Pombalina. Lisboa.

CML. 2017. Lisboa Histórica, Cidade Global (candidatura UNESCO Património Mundial). Lisboa.

CML. 2017. Elementos adicionais. Estudo de Impacte Ambiental dos Tuneis do Plano Geral de Lisboa. Monsanto- Santa Apolónia e Chelas- Beato. Lisboa: CML.

Coelho, C. 2010. O Tecido da Baixa, Excepção ou Regra na Cidade de Lisboa. Rio de Janeiro: Gazzaneo, 1: 312–327.

Conway, O. 2012. Merchants of doubt. How a handful of Scientists obscured the truth on issues from tobacco smoke to global warming, London, NY, Berlin, and Sydney: Bloomsbury Publishing.

Curwell, S. 2008. Sustainable Urban Development, London& NY: Routledge, 1–4.

Deakin, M.& Curwell, S. 2004. The BEQUEST Framework: the Vision and Methodology of a Collaborative Platform for Sustainable Urban Development. Geneva: Editions du Tricorne, 91–104.

Dulal, H. 2012. Greenhouse gas emission reduction options for cities: Finding the Coincidence of Agendas between local priorities and climate change mitigation objectives. Habitat International, UK: Elsevier Ltd., 38: 100–105.

Enova. 2005. Matriz Energética do Concelho de Lisboa. Lisboa: CML.

Enova, L.& IST. 2014. Matriz Energética de Lisboa 2014, edição revista e corrigida. Lisboa: Lisboa E- Nova, Agencia de Energia e Ambiente de Lisboa.

Enova, L. 2017. Estratégia Municipal de Adaptação às Alterações Climaticas de Lisboa. Lisboa: CML.

Flannery, T. 2005. The Weather Makers. How man is changing the climate and what it means for life on earth. New York: Atlantic Monthly Press.

Hamin, E. 2009. Urban form and climate change: Balancing adaptation and mitigation in the U.S. and Australia. Habitat International, UK: Elsevier, 33: 238–245.

Hoornweg, D. 2011. Cities and Climate Change: Responding to an Urgent Agenda. Washington DC: World Bank Publications.

IPCC. 2007. Summary for Policymakers in climate change: The Physical Science Basis. Cambridge& NY: Cambridge University Press.

IPCC. 2014. Climate Change 2014: Impacts, Adaptation and Vulnerability. NY: Cambridge University Press.

IPCC. 2014. Climate Change 2014: Impacts, Adaptation, and Vulnerability. Summaries, Frequently Asked Questions, and Cross-Chapter Boxes. A Contribution of Working Group II to the Fifth Assessment Report of the Intergovernmental Panel on Climate Change. Geneva: World Meteorological Organization.

IPCC. 2014. Climate Change 2014: Impacts, Adaptation and Vulnerability. Part B: Regional Aspects. NY: Cambridge University Press.

IPCC. 2014. Climate Change 2014: Mitigation of Climate Change. Geneva, Switzerland: World Meteorological Organization.

IPCC. 2014. Climate Change 2014: Mitigation of Climate Change, contribution of working Group III to the Fifht Assessment. NY: Cambridge University Press.

Kabisch, N. 2016. Nature-based solutions to climate change mitigation and adaptation in urban areas: perspectives on indicators, knowledge gaps, barriers, and

opportunities for action. Ecology& Society, CA: Resilience Alliance, 21(2): 39.
Khun, T. 2012. The Structure of Scientific Revolutions. Chicago: The University of Chicago Press, Ed. 4.
Klein, R. 2005. Integrating mitigation and adaptation into climate and development policy: three research questions. Environmental Science & Policy, NL: Elsevier BV, 8: 579–588.
Klein, R. 2007. Climate Change Risk: an adaptation and mitigation agenda for Indian cities. Environment& Urbanization, USA: Sage Publications, 20: 207–229.
Landauer, M. 2015. Inter-relationships between adaptation and mitigation: a systematic literature review. Climatic Change, NL: Reidel Pub. Co, pp. DOI 10.1007/s10584-015-1395-1.
Larsen, K. 2009. Climate change scenarios and citizen participation: Mitigation and adaptation perspectives in constructing sustainable futures. Habitat International, UK: Elsevier, 33: 260–266.
Larsen, S. 2012. Mind the gap in SEA: An institutional perspective on why assessment of synergies amongst climate change mitigation, adaptation and other policy areas are missing. Environmental Impact Assessment Review, NL: Elsevier, 33: 32–40.
Laukkonen, J. 2009. Combining climate change adaptation and mitigation measures at the local level. Habitat International, UK: Elsevier, 33: 287–292.
Leboeuf, Y. 2015. Plano Geral de Derenagem de Lisboa 2016-2030. Lisboa: CML.
Liu, F. 2013. Integrating mitigation of air pollutants and greenhouse gases in Chinese cities: development of GAINS-City model for Beijing. Journal of cleaner production, NL: Elsevier BV, 58: 25–33.
Lomborg, B. 2001. The Skeptical Environmentalist: Measuring the Real State of the World. NY: Cambridge University Press.
Cities, E. 2014. C. Mayors, 30 Energy Cities' proposals for the energy transition of cities and towns. Energy Cities.
McLaren, D. 2015. Sharing Cities. MIT Press eBooks. Massachusetts: MIT Press eBooks.
Müller, B. 2010. Urban Regional Resilience: How Do Cities and Regions Deal with Change?, Dresden: Springer Science & Business Media.
Ministero dell´Ambiente. 2015. Strategia Nazionale di Adattamento ai Cambiamenti Climatici. Rome.
National monitoring, reporting and evaluation of climate change adaptation in Europe. 2015. Luxembourg: Publications Office of the European Union.
Nations, U. 2015. COP 21- Conference of the Parties. Paris: FCCC.
Newman, P. et al. 2009. Resilient Cities: Responding to Peak Oil and Climate Change. Washington, DC: Island Press.
Popper, K. 2002. The Logic of Scientific Discovery. NY: Routledge (1st Ed. Vienna: Springer, 1935).
Oliveira, J. 2013. Promoting Win-Win Situations in Climate Change Mitigation, Local Environmental Quality and Development in Asian Cities through Co-benefits. Journal of Cleaner Production, NL: Elsevier BV, JCLP 3614: 1–17.
Prasad, N. 2009. Climate Resilient Cities: A Primer on Reducing Vulnerabilities to Disasters. Washington, DC: World Bank Publications.
Ramaa, V. 2012. A review of climate change, mitigation and adaptation. Renewable and Sustainable Energy Reviews, NL: Elsevier BV, 16: 878–897.
Rifkin, J. 2014. Zero Marginal Coast Society. New York: Palgrave Macmillan.
Romero, P. 2011. Cities in transition: transforming urban centers from hotbeds of GHG emissions and vulnerability to seedbeds of sustainability and resilience. Current opinion in environmental sustainability, NL: Elsevier BV, 3: 113–120.
Rosenzweig, C. 2014. Hurricane Sandy and adaptation pathways in New York: Lessons from a first-responder city. Global Environmental Change, NL: Elsevier BV, 28: 395–408.
Seto, K. 2015. The Routledge Handbook of Urbanization and Global Environmental Change. London & NY: Routledge.
Solecki, L. 2011. Climate change adaptation strategies and disaster risk reduction in cities: connections, contentions, and synergies. Current opinion in environmental sustainability, NL: Elsevier BV, 3: 135–141.
Townsend, A. 2014. Smart Cities. New York: W. W. Norton& Company.
Vale, L.& Campanella, J. 2005. The Resilient City: How Modern Cities Recover from Disaster. New York: Oxford University Press.
UKCIP. 2011. AdaptME: Adaptation monitoring and evaluation. Oxford: Pringle, 2011.
Viguié, V. 2012. Trade-offs and synergies in urban climate policies. Nature Climate Change, UK: Nature Publishing Group, 2: 334–337.
Wamsler, C. 2015. Mainstreaming ecosystem-based adaptation: transformation toward sustainability in urban governance and planning. Ecology& Society, CA: Resilience Alliance, 20(2): 30.
Warren, R. 2017. The role of interactions in a world implementing adaptation and mitigation solutions to climate change. Philosophical Transactions, UK: Royal Society Publishing, 369: 217–241.
Weart, R. 2009. The Discovery of Global Warming. London: Harvard University Press.
Williams, J. 2012. Zero Carbon Homes: A Road Map. New York: Earthscan.
Zimmerman, R. 2011. Climate change mitigation and adaptation in North American cities. Current opinion in Environmental Sustainability, NL: Elsevier BV, 3: 181–187.

Industry 4.0 – Shaping The Future of The Digital World – da Silva Bartolo et al. (eds)
© 2021 Taylor & Francis Group, London, ISBN 978-0-367-42272-1

Designing inclusive hospital wayfinding

M. de Aboim Borges
CIAUD– Research Center of Architecture, Urbanism and Design,
Lisbon School of Architecture, Universidade de Lisboa, Lisboa, Portugal

ABSTRACT: This paper relates to a research project developed in an Ophthalmology hospital, aimed for elderly patients with low vision condition. A pilot project was implemented in order to measure how this group interacted with a wayfinding system, that combined boards with written information displayed on walls and haptic textures applied on pavements. In order to build this orientation system, different visual and haptic tests were produced, so that a final solution could be achieved to promote an independent, safe and autonomous way on the usage of this space by this group of patients.

1 INTRODUCTION

1.1 *Context*

Design valences can play a major role in the deconstruction of the built environment so subjects and particularly elderly with visual impairments, may act and use it in an independent and safe way, namely hospitals. Elderly people are those who more often access medical care, due to the different pathologies that are related to ageing. We are observing worldwide an increasing population of very old people, explained by the advances in medical research.

But although all the medical progressions, people age due to a physiological decline. Kong *et al.* (2014) defines ageing as "a complex process of progressive decline in overall physiological functions, resulting in a diminished capacity to withstand internal and external damage and an increased susceptibility to diseases and risk of death". Ageing brings changes in the elderly's daily living due to a combination of the different aspects related to autonomy and mobility.

Table 1. Perceptive ageing in different sense modalities (Fontaine, 2000, p. 64).

STIMULI	AGE EFFECT
Taste	Very strong
Smell	Very strong
Kinesthesia	Very strong
Touch	Strong
Temperature	Strong
Pain	Strong
Equilibrium	Very strong
Vision	Very strong
Hearing	Very strong

Those changes are reflected in how they perceive the world through vision and hearing, but also how they manage balance and dexterity in order to pursue their lives. A bended posture to the front represents a form of walking preventing falling and a better observation of the ground.

As in Table 1, some senses are strongly affected with ageing, in response to stimulus perception.

Also, some activities are compromised due to loss of muscle mass in result of lack of exercise.

After some gerontologists, at 65 years of age, is denoted the ageing threshold or advanced age beginning, because at this age some psychological and physical changes become visible (Stephanidis 2010), whereas other academics divide older people in four groups - the *young old* with ages between 60 and 69 years, the *middle age old* between 70 and 79 years, the *old old* from 80 to 89 years and *very old* with more than 90 years old (Stephanidis 2010). One thing that is clear, is that the individual variability of sensorial, physics and cognitive functions raise with age and each of these functions have a decline at different speeds, depending on several factors, being the most evident a socioeconomic status and life conditions (Stephanidis, 2010).

In the ophthalmology hospital, where this study was held, the patients observed were affected by a process of congenital, professional or age-related vision loss, that limited them in their independent orientation and in their daily living, where some patients were dependent on other to live their life.

2 RESEARCH WORK

2.1 *Ageing factors*

As after WHO (World Health Organization) in 2012, the world is facing a unprecedent situation, that briefly we will have more aged people than children and those elderly with extremely advanced age, something never seen before. This factor is associated with life expectancy and lower birthrate. WHO estimates that between 2000 and 2050, the proportion of people worldwide with 60 years or more will duplicate from the actual 11% to 22%, passing from 605 million to 2 billion people. After data in 2012, there were 285 million people with visual impairment, 39 million were in a blind situation and 246 million people had low vision.

This longevity in result of medical progressions and socioeconomic conditions brought an apparent well-being. But reality demonstrates that the majority of the elderly demonstrate a dependency related to problems of health. Age related changes are applied in a general way to all, but may differ from individual to individual over varied life conditions and genetical constitutions (Robert and Fulop, 2014).

Ageing means a complex process of a progressive decline of all physiologic conditions and results in a diminishing capacity to support internal and external damages and in a growing possibility of getting sick and the risk of dying. It is a broader concept that includes physical body changes, mind and mental physiological capacities changes, socio-psychological changes in the way we think and act and changes in how we are seen and what is expected from us (Tracy *et al.*, 2010).

Understanding ageing, the one people feel, is understanding senescence, a process where normal cells begin an irreversible multiplication limitation after a limited number of cellular divisions (Robert and Fulop, 2014).

Another critical aspect is the lack of strength and muscle mass, responsible for walking difficulty. In an adult, on average, the strength reduction is of 10% per decade from 30 years of age, growing into 15% per decade from 60 years of age. Therefor doctor's recommendation for doing physical exercises, in order to avoid incapacities and mobility difficulties. But other physiologic systems present similar decline, such as the cardiac and breathing systems, renal and liver systems, cognitive and sensorial systems, as well as vision and hearing (Guccione, *et al.*, 2000).

A study developed by INE (National Statistics Institute) in Portugal in 2001 (INE, n.d.) which included visual impairment data, showed that 1,6% of the Portuguese population had visual impairment, and that within this percentage 3% had very severe problems of vision. Low Vision Specialized Center (CEBV, n.d.) in Portugal, in their website disclosed that between 400 and 500 thousand people could not read a paper even with correct glasses, 200 thousand had slight low vision and that 50 thousand suffering of severe low vision pathologies were not identified.

But other factors affect the elderly, such as decline in mobility, dexterity and managing equilibrium difficulties; odors perception and identification; high frequencies perception, palm hand touch sensibility and thermal sensibility.

One decline that affects the elderly and that is vital for a life quality is vision. Vision is the sense that allow us to apprehend reality, shorten distances and enjoy everything around us.

In an article, Elias and Vasconcelos (2008) describe the senses used when drawing and this can be extrapolated on how we use our senses in everyday life "the sort of 'seeing' I mean is an observation that utilizes as many of the five senses as can reach through the eye at one time. Although you use your eyes, you do not close up the other senses – rather, the reverse, because all the senses have a part in the sort of observation you are to make (...)". This is a synesthetic and haptic way of drawing, a sensory way of seeing that activates all the senses to represent what is observed in a certain moment and that all the senses are activated to co-operate in the construction of what is observed.

But as vision begins to fail, also other senses must collaborate in how we walk, orient and perceive the environment around us. This is what Gibson (1986) refers when he says "looking with the eyes alone is merely looking at, not looking around" because we see the environment with the "eyes-in-the-head-on-the-body-resting-on-the-ground". In his words "vision is a whole perceptual system, not a channel of sense".

Warren (Barraga, 1985) affirms that vision reaffirms motor activities allowing the integration of motor with visual functioning, on promoting movement, in spatial exploration and recognition, demonstrating a clear interpretation of senses integration, combined and preserved in the brain. This language acquisitions facilitates sensory impressions integration, where this language will allow an exchange with the other senses, clarifying and verifying sensory impressions. The visual act is an easy one, because by processing a stimulus from different sensory channels, we relay on the one that is more precise and exact for doing that specific task that for what we are preparing to do (Cattaneo and Vecchi, 2011). Vision allow us to calibrate, regulate and coordinate movements in the environment, such as locomotion and hand gestures. Elderly' vision capabilities are altered with changes with retina photoreceptors loss, being this loss more expressive in peripheral vision with the number of fovea rods decline, but in contrast, central vision cones maintain relatively intact (Birren and Shaie, 2006). Visual acuity between 60 and 87 years, even with a healthy ocular and an optimal refractive situations, shows reduced levels (Schieber, 1994).

2.2 *Color perception*

Colour perception is affected by the degradation of pupil size and focusing ability related to crystalline's natural elasticity loss and also other factors such as ocular media opacity and the amount of light received in the aged retina (Pinheiro and Moreira da Silva, 2009). This aged vision causes a progressive opacity, loss of transparency and yellowing, provoked by eye's fluid decolorating that result in the loss of some chromatic contrasts, making us feel like looking through a yellow filter (Stephanidis, 2010).

The elderly need more time to recover from different light intensity transitions in spaces, because the ageing eye is more sensitive to clarity effects, they need higher levels of contrasts, around three and a half more than a 20 ton 30 years individual (Pinheiro and Moreira da Silva, 2009).

In low vision pathologies these aspects assume a greater dimension, for they are related with central or peripheric vision, respectively related to red/green and blue/yellow colors perception. Low vision pathologies affect chromatic perception, color contrast and the visual field are highly limited, beyond that in some there is a substantial loss in color interpretation needing a higher intensity of light to distinguish them, and opposition other pathologies the excess of light creates difficulties in the apprehension of color and also space interpretation. Each pathology has its characteristics in terms of color perception. Cataract is a crystalline opacification, resulting in a blurry image. Glaucoma is related with peripheral vision loss and at a late stage it is known as tunnel vision, which means that only a small part ("window") of the eye is able to see. Diabetes is responsible for some eye pathologies like diabetic retinopathy that is characterized by a weakness on the eye veins that pour small quantities of fluid into de retina turning it dim. ARMD Age related macula degeneration is clearly a ageing pathology, resulting in a distorted vision, affecting central vision making difficult for patients to read/see or write. These are many more all affecting vision and color.

2.3 *Touch/haptic perception*

In the treaty "The sense of touch" edited in 1834, Weber referred the interdependency of de senses in order to obtain convergent sources of information (Ittyerah, 2013, p. 104). Senses are physiologic tools destined to perceive information from the environment (El Saddik *et al.*, 2011).

Touch is the first sense that a child experience and react, by reacting to heat, cold, roughness and softness (El Saddik *et al.*, 2011). Haptic perception is related to the study of human sensitivity and manipulation through tactile and kinesthetic sensations for object recognition (El Saddik *et al.*, 2011). The sense of touch is different from all other senses acting like a closed circuit process (loop) and with one bidirectional channel of sensing and acting, dependent of a physical contact and the receptors distributed along the body (El Saddik *et al.*, 2011). It is trough touch that we understand what is constant, the approximate, the invariant, the general, which means that the sensory content is noticeable from form geometry (Paterson, 2005). Also Pallasmaa (2005) reinforces this importance of the sense of touch on the experiencing and understanding of the world seen by our eyes. Vision reveals what touch already knows, and that we could think of touch as the inconscient of vision (Pallasmaa, 2005)

The sense of touch is the one on which we support and trust for developing a vast number of daily activities, such as using a mobile phone, play an instrument, prepare meals, grab and browse a book, beside an infinity of other tasks that we do by synchronizing with vision to raise perception quality, with hearing contributing to an environmental better perception in result of sound produced by events in space (El Saddik *et al.*, 2011).

Our haptic system is composed by 4 components: a mechanic being the most significant the system hand/arm; a sensory that includes varied classes of receptors and skin nerve endings in our skin, joints, tendons and muscles; a motor, where the brain analysis and activates the adequate commands to the muscles for movement initiation; and cognitive that responds through the efferent nerves or impulse conducers, emanated from the nervous system. This system perceives two types of information: tactile or cutaneous and kinesthetic or proprioceptive. These information sources are however perceived in combination of both. Cascio and Sathian research, showed that fine textures haptic perception is assigned to vibration in result of skin movements and stimuli, while coarse textures are basically perceived by their object' geometrical properties (El Saddik *et al.*, 2011).

Haptic system is generally interpreted as a perceptual system mediated by two afferent subsystems, cutaneous and kinesthetic, that involves active manual exploration and where the haptic system is particularly effective on processing surface material characteristics and objects, controlled by the own individual (Lederman and Klatzky, 2009).

Plantar touch is physiologically similar to palm hand touch regarding the existence of the same type of receptors, the same that exist in glabrous skin (Visell *et al.*, 2009). The larger differences, include higher thresholds on foot activation and a larger receptive field than the in the hand touch (Kennedy and Inglis, 2002). This functional differences are related with prehensile dexterity on the hand, for a broader and sustainable force feet are subjected in locomotion and in systematic differences on activity type executed by the hand (grabbing, manipulation, precise exploration) in contrast with feet (posture, balance, sensorimotor mobility) (Visell *et al.*, 2009). Foot sole receptors are sensitive to contact pressure and to potential changes in pressure distribution and the integration of these somatic sensors, seem to contribute with important information of body positioning related to the surface of support (Kennedy and Inglis, 2002). Foot sole mechanoreceptors play a fundamental role in equilibrium control for they are the frontier between body and pavement (Kavounoudias *et al.*, 1998).

2.4 *Design response*

Collecting bibliographic references in the different areas were of main importance for developing this research project. In order to purpose a wayfinding/ orientation system for the elderly with low vision in hospital environment, the participation of the administration, patients, ophthalmology doctors, auxiliary teams and nurses was the most important step, because they could refer all accessibility problems, pre-exams locations and ways of hospital functioning. These testimonies were registered to begin the design process, in the definition of the written information and mobility limitations for the definition of haptic pavement textures and flows within the building.

For developing a hospital orientation system for these patients, it was necessary to evaluate their space perception, namely legibility and readability vision distances, color and contrast color perception, dexterity, mobility and haptic foot touch perception. Each of these issues had to be tested and measured, so the orientation system would be efficient.

At a first stage of the research work, an analysis and diagnostic of users flows within the building - Instrumental Value evaluation - was carried out identifying which circuits were common to the different pathology's treatment areas and medical staff. In the same evaluation, was evaluated the type of information proposed to users for promoting access to the different areas of the building, such as doctor's offices for pathologies treatment and medical examination rooms locations.

In order to evaluate patients' limitations to sight, color perception, haptic perception and mobility, among others, a method of analysis was developed – Waysensing – composed by a battery of tests and inquiries to the patients involved in the study.

The whole research project named Percept Walk, can be defined as the visual and haptic sensorial capacity that allows the human being the explicit collection of information from the physical involvement, in particular the spatial organization, in order to act in conformity.

Inclusive and user-centered design methodologies supported the research project, to identify and know how this group of people, with severe vision loss, interacted with the hospital environment and how they were able to shift, and some by their own using public transportation, for reaching the hospital. Some patients observed lived alone, since their partner had already gone, or were apart from their families.

As Huppert (2003) stated "the need for design change is not limited to consumer products. It should also be a priority for designers involved in the public services. From design of printed matter and communications and information technology, to the design of transport, housing and public buildings, a better understanding of users' needs can dramatically improve the independence and quality of life of the vast number of older users. But for these endeavors to be most effective, we need to go beyond the numbers, to understand the lifestyles of today's older adults, as well as their physical and mental capabilities".

3 EMPIRICAL WORK

3.1 *Coping with reality*

The empirical project was built in phases, beginning with a questionnaire to evaluate habitational aggregate, health, pathologies and visual acuity, plantar haptic perception to textures and limitations on daily living.

After responding the questionnaire tests of legibility and readability were made, for identifying which words in different fonts, sans serif, and letter size were best viewed.

The pathologies observed were retina dystrophy, amblyopia, bilateral optic atrophy, bilateral optic retinopathy, diabetic retinopathy, ARMD – age-related macular degeneration and glaucoma, all presenting a very low visual acuity.

As pre-results fonts with classification of Neo Grotesque (Helvetica LT Std Roman) and Humanist (Myriad Pro Regular, were the most easily read and also the size best read at a distance between 20cm and 100cm was with 40mm of height.

A second phase tested chromatic vision, interpretation of pictograms that were related with the hospital environment information and a first approach to plantar haptic perception through different shapes. For color and contrasts perception tests were presented 16 plates, each with a pair of colors in contrast using Pantone color code, and following literature evidences from the color theories of Young-Helmholtz, Ewald Hering and Itten. A total of 20 plates of pictograms, related to hospital services, were presented having each plate 3 versions of representation - positive, negative and with no frame - but all printed in color blue against white background.

Plantar haptic tests were made for exploring plantar haptic shape perception and texture height by patients using their ordinary shoes. The majority of patients, 61%, had glaucoma pathology, 35% had retinopathy pathologies, 4% maculopathy and 4% high myopia, and 30% had between 68 and 74 years and 22% had 75 years or more. As pre-results the colors in contrast best viewed were (background color/text color) anthracite grey/white, yellow/green, yellow/ black, anthracite grey/beige, being all the other pairs of contrasted colors been also seen, but some were commented that "some colors flicker against the background and is difficult to read the word".

For the pictograms the representation in negative with frame, represented 79% as the best read.

For the four haptic shapes proposed (round, rectangular, square and triangle) the round and the rectangular shapes had 52% and 61% respectively, with just 23% for the square shape and 18% for the triangular.

The third phase aimed for legibility and readability of signs applied on the colors most voted by the patients and tests for evaluation of new haptic textures based on the shapes most sensed. For the graphic material, plates were produced and located in the decision points defined in the Instrumental Value evaluation and were available for real test. Haptic tests were prepared for real test also, but due to feet dragging of the elderly, the height was reduced in order to reduce the risk of imbalance or even falls.

The last phase was devoted only to haptic tests, for the information proposed on phase 3 was well accepted by users and prove to be efficient in the deconstruction of the building. Haptic tests were refined and color was introduced in the three textures defined. Height was reduced to 3 mm, and proved to be enough to be perceived and avoided imbalance by all patients.

This final phase of tests has permitted the installation of a pilot project, for testing in real life the effectiveness of the sensorimotor wayfinding system. As for the graphic, the locations were defined for the haptic plates and the objective of the haptic information was to promote visualization of the graphic information system on walls, at a lower height to be seen by elderly that walked bent to the front, to orient patients to the waiting room and from the waiting room to pre-exams room. Everyone has to an eye exam before medical consultation and most of the patients lose themselves-

3.2 *Conclusions*

One of the points this research project aimed was that with an effective wayfinding system, human resources such as auxiliary staff, nurses and doctors, would have more time to do their real work and not being obliged to orient people within the building. Another research question was that with a sensorimotor orientation system, combining graphic and text plates with haptic pavement information textures, patients would more easily find they way. Both of these questions were answered and the system, who also combined color stripes on the pavement, proved to be effective in orienting everyone. The user-centered design method proved that some answers given by patients were important to design the system.

The fact that most of them began following the orientation system, has contributed to an autonomous mobility and independency, something that all wanted, and this fact made patients self-confidente on their capabilities of using this space and self-esteem and well-being was observed among them.

The final solution confirmed that the system brought clear benefits in the orientation and distribution of users within the building.

REFERENCES

Barraga, N. (1985) 'Disminuidos visuales y aprendizaje', Madrid. ONCE. Available at: http://sid.usal.es/idocs/F8/FDO23237/disminuidos_visuales_y_aprendizaje.pdf (Accessed: 21 January 2013).

Birren, J. E. and Shaie, K. W. (eds) (2006) *Handbook of the Psychology of Aging*. 6th edn. London: Elsevier Academic Press (1).

Cattaneo, Z. and Vecchi, T. (2011) *Blind vision: the neuroscience of visual impairment*. London: MIT Press.

CEBV (no date) *CEBV - Centro Especializado em Baixa Visão*. Available at: http://www.cebv.pt/cebv_catalogo.html (Accessed: 4 October 2016).

El Saddik, A. *et al.* (2011) *Haptics technologies: bringing touch to multimedia*. Ottawa, Canada: Springer Science & Business Media.

Elias, H. and Vasconcelos, M. C. (2008) 'Approaching Urban Space Through Drawing', in Lucas, R. and Mair, G. (eds) *Sensory Urbanism Proceedings 2008, January 8th & 9Th. Sensory Urbanism*, University of Strathclyde, Glasgow: The Flâneur Press., pp. 55–64. Available at: http://strathprints.strath.ac.uk/id/eprint/16493.

Fontaine, R. (2000) *Psicologia do Envelhecimento*. Lisboa: Climepsi Editores (Psicologia).

Gibson, J. J. (1986) *The Ecological Approach to Visual Perception*. New Jersey: Lawrence Earlbaum Associates, Inc. (LEA).

Guccione, A. A., Avers, D. and Wong, R. (2000) *Geriatric physical therapy*. 2ª ed. Missouri: Elsevier Health Sciences Mosby.

Huppert, F. (2003) *Designing for older users*. Edited by J. Clarkson et al. London: Springer Verlag London Ltd.

INE (s.d.) *INE*. Available at: http://www.ine.pt/xportal/xmain?xpid=INE&xpgid=ine_main.

Ittyerah, M. (2013) *Hand preference and hand ability: evidence from studies in haptic cognition*. John Benjamins Publishing.

Kavounoudias, A., Roll, R. and Roll, J.-P. (1998) 'The plantar sole is a 'dynamometric map'for human balance control', *Neuroreport*, 9(14), pp. 3247–3252. Available at: http://journals.lww.com/neuroreport/Abstract/1998/10050/The_plantar_sole_is_a__dynamometric_map__for_human.21.aspx (Accessed: 15 March 2016).

Kennedy, P. M. and Inglis, J. T. (2002) 'Distribution and behaviour of glabrous cutaneous receptors in the human foot sole', *The Journal of Physiology*, 538(3), pp. 995–1002.

Lederman, S. J. and Klatzky, R. L. (2009) 'Haptic perception: A tutorial', *Attention, Perception, & Psychophysics*, 71(7),pp. 1439–1459. Available at: http://link.springer.com/article/10.3758/APP.71.7.1439 (Accessed: 15 April 2013).

Pallasmaa, J. (2005) *The eyes of the skin: architecture and the senses*. London: Wiley.

Paterson, M. (2005) 'The Forgetting of touch', *The Forgetting of touch, Angelaki: Journal of the Theoretical Humanities*, 10(3), pp. 115–132. doi: http://dx.doi.org/10.1080/09697250500424387.

Pinheiro, M. C. and Moreira da Silva, F. (2009) 'Comunicacão Visual e Design Inclusivo–Cor, Legibilidade e Visão Envelhecida Visual Communication and Inclusive Design–Colour, Legibility and Aging Eye'. Available at: http://www.faac.unesp.br/ciped2009/anais/Design%20Social%20e%20Inclusivo/Comunicacao%20Visual%20e%20Design%20Inclusivo.pdf (Accessed:11 January 2013).

Robert, L. and Fulop, T. (2014) *Aging: Facts and Theories*. Ettlingen: Karger, 45–93.

Schieber, F. (1994) *Recent Developments in Vision, Aging, and Driving: 1988-1994*. Michigan: Citeseer. Available at: http://citeseerx.ist.psu.edu/viewdoc/download?doi=10.1.1.133.4554&rep=rep1&type=pdf (Accessed: 27 January 2013).

Stephanidis, C. (2010) *The Universal Access Handbook*. Boca Raton: CRC Press.

Tracy, J. I. *et al*. (2010) *Handbook of Medical Neuropsychology - Applications of Cognitive Neuroscience*. Edited by C. L. Armstrong and L. Morrow. New York: Springer.

Visell, Y., Law, A. and Cooperstock, J. R. (2009) 'Touch is everywhere: Floor surfaces as ambient haptic interfaces', *Haptics, IEEE Transactions on*, 2(3),pp. 148–159. Available at: http://ieeexplore.ieee.org/xpls/abs_all.jsp?arnumber=5166445 (Accessed: 10 April 2013).

Industry 4.0 – Shaping The Future of The Digital World – da Silva Bartolo et al. (eds)

The building to let in Lisbon, a built resource for a sustainable city

V. Matos & C. Alho
Lisbon School of Architecture, University of Lisbon, Lisbon, Portugal

ABSTRACT: The building to let (16th century-1976) in Lisbon is a built resource for an efficient and sustainable city. Between the first half of 16th century and 1976 different models of building to let emerged in Lisbon. The urban rehabilitation of the city's heritage is fundamental to Lisbon environmental and urban sustainability, in which the building to let, an abundant legacy, plays an import role. The HBIM, Historic Building Information Modeling and the 3D printing play an important role in the urban rehabilitation of this heritage.

1 A BUILDING THAT SHAPED LISBON

Since king D. Manuel I, the architecture was defined by the Crown and the Senate of the City, resulting in a precise and theorized corpus of laws concerning architectonic polices that defined Lisbon's image and the image of the Portuguese territories. From the 16th to the 1st half of the 18th century, the division of the city's spare land in urban plots correlates to careful planning of the area and to the appearance of an organizational pattern of a multi-family building: the building to let. Between the first half of 16th century and 1976 different models of building to let emerged in Lisbon. The building to let would shape the image of the city.

The designation of *building to let* (in Portuguese, *edifício de rendimento*) was firstly used by Carita (1994), where he tried to trace the origins of this typology (Carita, 1994). The data set by Carita (1999) for the *Manuelino* urbanism in Lisbon helped us to draw, for the first time, the picture of the building to let from its origins until the 1st half of the 18th century (Matos 2003). In the 20th century the buildings to let presented to the Municipality of Lisbon were referred to as *prédio de rendimento* that is, *property for renting or letting*. Considering that *prédio*, means either an urban or rustic property (a land), we kept the designation employed by Carita (Matos 2009; 2018). In the 1950s, 1960s and 1970s, letting dwellings predominated in the city.

The Horizontal Property Act, Decree 40333, 14 October 1955, permitted the transition from a tenants' city to a city of owners. Embryonic credit and savings regimes were created for acquisition of dwellings, made by *Caixa Geral de Depósitos* and *Crédito e Previdência*, which targeted a very selective population. From 1976 onward a political inflection occurred: the credit for dwelling acquisition with government-subsidized loan was created. The investment focused on the building to let gave place to the investment centered on the promotion of the dwelling for sale. The city of landlords and tenants gave place to the city of the financial institutions and owners (Silva Nunes, 2005). However, the typology of a building vertical organized for different families remained and evolved until the present.

2 THE IMAGE OF THE BUILDING TO LET

The stone and lime buildings (16th to the 1st half of the 18th century). The legislation of the Old City's by-laws of Lisbon defined the conditions for letting dwellings and the minimum occupation per building. The document of 1785 of the City's by-laws of Bahia (Brazil), for the construction of houses shows us a type of building vertically organized in *loja*, *sobreloja*, first floor and second floor, similar to those in Lisbon for the above mentioned period. The regulations specified a pattern image for the building's façade, the interior of a dwelling and the building materials: chestnut wood floorboards, wooden roof deck, and a wooden floor above the bottom chord of the roof trusses. The legislation stipulated the existence of a chimney and masonry benches in each dwelling. It also defined the thickness of the walls (2.5 spans), the layout of the doorframes, the levelling of the urban plot and the stepping pattern of the quarter's façades whenever the ground slope presented more than 5 spans. The length and width of the building materials were standardized by the Weights and Measures Amendment of 1499, producing precise measures and rhythms in the façades and in the internal layout of the building. The urban legislation reinforced the sense of pattern by stating: *everything must be done according to a unique plan, as outlined by the Senate's building constructor, a building inspection which does not respect the ornate and beauty of the city cannot exist.*

The iconography and the best preserved buildings from the 16th to the 1st half of the 18th centuries give a precise image of the building to let regarding façade's tops, roofs, window shutters and balcony shutters, doors, garrets, openings, Lisbon's armorial bearings and lessee stone boards. Two groups of buildings appeared: one of the 16th and 17th centuries and another of the 1st half of the 18th century. The number of rooms found in the most typical buildings to let (3 or 4 rooms) matched the information presented in the demographic studies on Lisbon and Guimarães, the *post mortem* inventories, plus the information on the human geography of the Western Quarters of Lisbon. They revealed an average of 4,56 children per couple, a prolonged nursing period and a rare presence of slaves in the common family. The crises in the 17th century, the annual exodus of people to Brazil, the loss of many men during the war against Spain and the Lisbon's 1721 plague confirmed the internal layout for the majority of the dwellings (Matos, 2003).

The Pombalino building (1758-1860). The standard Pombalino building to let was five floors high: on the ground floor were located the shops and on the upper floors the dwellings; the higher floor had garrets in the alignment of the other openings. The buildings constructed on the main streets had bay-windows in the first floor and sill-windows in the second and third floors. The buildings built in the secondary streets had sill-windows in all floors. The façade´s plaster was painted. Stone pilasters were put in the corners of the quarter, separating each building to let. Ashlars were put around the openings. Within this normalization the buildings could vary regarding the gauges, the dimensions of the openings and the façade´s finishes, i.e., the design detail, the type and quality of the building materials employed (Teixeira & Valla, 1999). Though forbidden by the Marquis of Pombal, after his displacement in 1777, the balconies persisted as confirms the notarial register of 31st December 1791. The 9 of March 1835 edictal forbids them as well as dust-retainers. The Lisbon's 1930 legislation also forbids the balconies that go out from the doorposts, but an example of them still exists in Lisbon in Rua Serpa Pinto, 8 (Matos, 2018).

Figure 1. Balcony of Rua Serpa Pinto, 8, Lisbon.

The Pombalino cage structure, *gaiola*, was laid on the stone basement or on the foundations, with the help of ground-sills reinforced with stone-bolts. It was also fixed to the stone masonry corners by using bolts made of iron sheet. The gable walls were basically *frontal* walls and generally common to two buildings. Their thickness rarely changed as it progressed to the upper floors. The gable walls used the same pug masonry as the frontal walls (Lopes, 2005). Due to the slowness of the reconstruction of Lisbon after the 1755's earthquake, Baixa Pombalina presents a variety of situations regarding the internal layout of the buildings, the modules by urban plot and the location in the quarter (Barreiros, 2004).

A work contract of December 30th, 1791, for an urban plot in Rua dos Ourives do Ouro, shows us the internal image of a dwelling: a kitchen with its fixed stone slabs for pots, the main rooms, the dining room and the corridors (Matos 2009, 2018). The standard dwelling of Baixa had six rooms, inscribed in a rectangle, including the kitchen with direct access to the stairs. Some had an extra room acceding directly to the stairs, the independent room (Barreiros, 2004). Outside *Baixa Pombalina*, the buildings had to adapt to the dimensions of the existing urban plots, presenting earlier typologies or mixed solutions distant from the models of Eugénio dos Santos and Carlos Mardel (Carita, 1994).

Figure 2. Pombalino Building, Lisbon.

The gaioleiro building (1861-1930). The *gaioleiro* used an adulterated model of the Pombalino cage structure. The exterior walls were made of common uneven stone mortared masonry and the gable walls and inner yards of solid brick. The interior walls had brickwork confined within an orthogonal grid of uprights (in the contours of doors and tops) and transoms (at the level of the floor and intermediate levels), or wall paneling made of plastered planks. In the rear façades, balconies and verandas were built, supported by slender iron structures with pavements of beams and small domes of solid brick masonry joined by a mortar of lime and cement. The main walls became thinner as they progressed to the upper floors. In the exterior coatings, stone socles were replaced by mortar socles, and glazed tiles often cover the façade. The foundations are made of masonry and mortar of sand and lime (Appleton, 2005).

The notarial register of 5th August, 1861 sets the emergence of the *gaioleiro* building between 1861, when a new model of staircase emerged, and 1930, when the General Regulation of the Urban Construction for the city of Lisbon was approved. In 1858 the stairs were associated to a skylight, in 1861the new model shows a paved entrance with an arch preceding the staircase; in 1869 this model evolves: *the staircase would be in modern style of curved pieces of woods; the entrance would be paved in black and white stone in a chessboard pattern; the first step would be in square stone as the arch.* In 1867, the plan for the building of Rua Luz Soriano 75 materializes this model, a staircase with its well.

According to the notarial register of 5th August 1861, each floor will have a living-room, a sitting room, three bedrooms, a dining-room and a kitchen with a window-sill near the inner yard. The regulation of 1903 places the kitchen, the vestibules and ante-chambers near the inner yard. The 1902 regulation defines that the semi-basement floors must have their own latrine, a place for the oven and chimney. The dining-room was near the kitchen and from the corridor one could access to all rooms. The number of rooms per dwelling varied between six and eight. The independent room opened directly to the staircase. The regulation of 1930 stated that every room should have a door or window opening directly to the exterior, unless contiguous to compartments with openings or fan-lighted doors.

The notarial register of 1863 shows the free use of colours in the doors and door-posts. Apart from white, other colours, including orange, could be used. In 1866 the rooms were lined with wallpaper tile. From 1850 onwards the revetment of the façades with glazed tiles became current. They are a reflex of the romantic Lisbon. Embossed glazed tiles in their bas and fully embossed version made their appearance in the 2nd half of the 19th century. (Matos 2009; 2018). Examples of façades with glazed tiles in the turn of the 20th century are Rua do Poço dos Negros 147, Rua do Poço dos Negros 105-107, Rua da Imprensa Nacional 96; Rua da Imprensa Nacional 40, Rua da Imprensa Nacional 36-38, Rua Marcos Portugal 59-61, Rua Gustavo Matos Sequeira 18-36, Praça do Príncipe Real 1/Rua do Século 171, Avenida D Carlos I 82, Avenida D Carlos I 108, Rua Silva Carvalho 118-122, and Rua Silva Carvalho 124-126. The glazed tiles of Rua do Vale do Pereiro 2/Rua do Salitre 132 (1949) Rua Conde Rendondo 20-20B (1956) attest the evolution and renovation that glazed tiles would undergo.

The Estado Novo building (c.1930 – c.1940). The 1930's Regulation, active until 1951, established bland shade colours for the façades. Within this norm the colours could vary from different tones of pink, grey, blue, cream-coloured, yellow, green and were applied to modernist and Estado-Novo buildings. The rose tones dominated in the monumental Areeiro urban complex. In Rua do Salitre, Rua Sousa Viterbo (Bairro Lopes), which were non-monumental areas, a variety of colours was used. In Rua D. João V axis, an Estado-Novo urban complex with some monumentality, rose tones and a variety of bland shade colours were applied.

The descriptive memory of Carlos Mardel 106 (1939) helps to precise the colour palette of Estado-Novo interiors. In all the floors of the dwellings the floors would be in pine wood in British floor style. The floors of the kitchens and bathrooms would be made in hydraulic mosaics or in other building material suitable and approved by the municipal inspection. The walls would have glazed tiles with no decorations until the height of the doors and from that until the ceiling would be painted using an enamel paint. Each kitchen would have a sink made of withe limestone, a second sink for waste made of limestone, and a closet. The pantries would have shelves. The roof would be in Portuguese roof-tile, with eaves in Portuguese style. The structure of the roof would be made of pine wood. The Laura Santos book presented a picture of the modern kitchen showing building materials referred in the descriptive memories of Estado Novo buildings.

According to the 7th edition of the 1930's Regulation the garnishing of the exterior openings, when not in concrete or squared stone, could be in stone masonry or solid brick. This norm would constitute a path to modernity as can be seen in Rua de São Marçal 176 (1944-project, 1947-ampliation) and in Avenida da República 36-36F (1948), where the solid brick stands out as a strong red, full of modernity. This building material would be later used in the Costa Cabral Block (1953-1955), in Oporto, and also in the 1950s, in the big urban interventions in Lisbon, in Olivais, where it would play a major role in Rua Sargento Armando Monteiro Ferreira (1959-1968), Rua Cidade da Beira, Rua Cidade de João Belo (1969), thus becoming a building material associated to the Portuguese Modern Movement. The MG building (1959-1960, 1961-1968) in Rua Júlio Dinis, in Oporto, using red ceramic brick, whose expressivity is comparable to a Scharoun or Aalto work, became an iconic image of 1960s building to let. The Rua Rodrigo da Fonseca 206 (1960-1967), in Lisbon, completed that picture.

As for the construction materials, *the Estado Novo building* belongs to the historical constructive period of the first buildings with a partial structure in reinforced concrete, defined by Vieira & Lacerda, 2010. They have a mixed structure of masonry with some elements of reinforced concrete, the latter used in pavements of kitchens and bathrooms, stairs, balconies, building visors, consoles, pergolas and suspended bodies. The exterior walls are of ordinary stone masonry or stonework and the interior walls of brick. The pavements are of concrete and wood.

*The modern building (c.1951 – c.1960).*The Regulation of the General Urban Edifications, 1951, defined that the quality and nature of the building materials applied should assure the most adequate salubriousness and aesthetic conditions to the use of the building. The use of new building materials or construction techniques would imply an approval by the Laboratory of Civil Engineer.

The free use of colours and building materials legislated by the 1951's Regulation are illustrated by the modern buildings in Praça das Águas Livres 8-8I/Rua Gorgel de Amaral 1-1A, Calçada Engenheiro Miguel Pais 42, Rua da Imprensa Nacional 64-64D/Rua Marcos Portugal 91-91D, in the historical city. In Praça das Águas Livres 8-8I/ Rua Gorgel de Amaral 1-1A the revetment materials, as well as everything else in the building, were carefully chosen. In this building were used painted plaster, carpentries in varnished tola wood, vitrified pastille, hydraulic mosaic. The colours in the façades are bland shade colours: yellow, pink, grey, blue and a soft brown. Regarding Rua da Imprensa Nacional 64-64D/Rua Marcos Portugal 91-91D, the colour palette of its descriptive memory let us know that, in the interiors of the dwellings, in the living room and bedrooms, the floor would be in parquet, and in the service areas, the floors would be in hydraulic mosaics and the walls would have glazed tiles until 2.00m above the floor; the walls and ceilings would be plastered. The colours used in the façades were bland shade colours, black, soft brown, white and yellow. In the façades it was used vitrified pastille, mostly in the ground floor and in the corner of the building, combined with plastered surfaces in the majority of dwellings floors.

In Calçada Engenheiro Miguel Pais 42, the tridimensional glazed tiles in a strong and vibrant colour are fundamental in the construction of the façade of this small building, contrasting with the use of the white as main colour of the façade and the blue limestone in the ground floor, marking the entrance to the building. The colour palettes of these buildings helped to construct the image of the Lisbon of the epoch.

According to Agarez, the Regulation of 1951 produced the norm of the 45 degrees, responsible for the dissemination of the rectangular plan of the buildings, and the need to consider their exposure to the sun and dominant winds. This meant the end of the street-corridor, the acceptance of the principles of the Modern Movement (Agarez, 2009) and of a new model of building to let. The 1950s buildings are included in the historical-constructive period of the buildings with structure of reinforced concrete (c.1940 – c.1960), material ruled by the regulations of 1935 and 1943.

In Lisbon these buildings have an integral structure of reinforced concrete. The brick masonry is used on the exterior and on the interior resistant walls. The pavements are made of reinforced concrete. Our analysis showed a coexistence of innovative with traditional elements, specifically the requirement of a bathroom for the servants, which leads to an *Estado Novo* complexity in the private, social and service areas. The transitional modern type (1950-1970) prevails: the partial adoption of the modern matrix regarding the rationalization of the spaces, the use of the rectangular or quadrangular plan, the organization of private, social and services zones, the automatization of the living and the dining areas, though in a logic of spatial continuity. The pure modern matrix has a rectangular or quadrangular plan and the dwelling is organized in 3 zones, social, service and private areas; the social zone comprises the common room, the service zone includes the kitchen, and the private area has undifferentiated bedrooms sharing the same bathroom. There is a centrality of the common room, with the kitchen close to it and the private area kept apart. We based our analysis on plans of the 1950s' significant buildings. Pereira (2012) defined the types of dwelling according to the family type, based on middle class apartments advertised in the weekly newspaper *Expresso* from 1973 to 1999, which we applied to the building to let in Lisbon.

The variety and comfort of the rooms was planned accordingly with the richness of the zone. The Rua Braacamp 5/Rua Mouzinho da Silveira (1955) shows the contruction of the building to let in a wealthy zone of Lisbon. The project previewed a dwelling for the doorwoman and another for a tenant in the 1rst floor, shops in the ground floor, a storage room in the basement for one of the shops, and dwellings for the tenants in the remaining floors.

The Re-examination period buildings (c.1961 – c. 1976). In 1961 was published the Popular Architecture in Portugal. In CIAM was questioned the validation of the theories of the Athens Chart. It's a time of discovery of the value of the vernacular, its rationality, and the diversity of the existing solutions in Portugal. Fernando Távora, Álvaro Siza Vieira, Nuno Teotónio Pereira and Nuno Portas were among the architects whose modern works, previous to the Inquire to the Portuguese Regional Architecture, drew a new posture which marked the beginning of the 1960s (Fernandez in Becker et al; 1998). In the end of the 1950s the brutalist expression as formal model is felt in Portugal.

The period *c.* 1961- *c.* 1976 was the epoch of Tomás Taveira's Valentim de Carvalho Shop (1966-1969) breathing a very Pop language, where architecture took place on walls through forms, colours and poetry; of Gonçalo Byrne and Reis Cabrita's rationalist Pink Panther (1971-1975); of the Public Housing in Chelas Zone N2 plot 232 of Manuel Vicente (1973-1975 and 1976-80); of the Five Fingers (1973) of Vitor Figueiredo; and of Tomás Taveira's Condado Quarter, previously known as Zone-J (1975-1978 and 1975-1985). Supported by Curtis (Curtis,1996) for the international context, we consider that the term Regionalism in the Portuguese context, sometimes used to describe this period, does not encompass the variety of developments that took place and reduce part of them to something peripheral and provincial. Thus, we preferred the term *Re-examination period* to name this period.

The Mãe de Agua Block (1958-1963), in Rua Francisco Sales 17, is a small building, three floors high, preparing the entrance on a platform over which is arranged on pilotis the large seven-floors high curved block. The platform is the roof of the garage, the landscape for the benefit of the inhabitants, as in the modern Corbusian City. The dwellings have a left/right distribution with the following program: vestibule, living room, dining room, four main bedrooms, two main bathrooms, W.C., service vestibule, kitchen, cupboard, pantry, bedroom and bathroom of the housekeeper. The pavements used in the kitchens and bathrooms, pantries, cupboards and balconies are of ceramic mosaic of uniform colour. The vestibules, rooms and corridors are made of exotic wood parquet. The curtains and handrails of the stairs were made in concrete. As for the façades, the garnishing of the exterior openings were in marble. In the gable walls the structural grids reveal the constructive system and the division of rooms. The stone was used as revetment of the wall of the garage. Pre-fabricated vase flowers were used in the balconies. The roofs were made of concrete vaults. The remaining revetments were in plastered painted in bland shade colours. This building has a modern type plan, but its organization reveals a traditional-modern type plan, with a service area with a kitchen, a W.C. and a bedroom of the housekeeper.

The Rua Rodrigo da Fonseca 206 six floor high building (1960-1967) has two dwellings per floor and shops on the ground floor. It has a clear design that goes from the plan of the dwellings to the façade options. In the exterior, red brick and pressed stone were used in the front facade In back façades was used a painture made of a base of cement. The spatial organization the living-rooms and bedrooms overlook the streets and the other rooms face inwards. The building was built with shops and doorwoman dwelling in the ground floor, the upper floors had two dwellings per floor and finally it had a recessed floor for a single dwelling. The basement was used for storage and shops. As for the building materials, parquet was used in the bedrooms and remaining rooms, and hydraulic mosaics in the bathrooms, kitchens, entrances and shops. The stairs and landings were in stone. The walls were plastered and the bathrooms had glazed tiles until the high of the lintel of the doors. The walls of the staircase and elevators were in mosaic.

The architectural drawings from the approval process of corner building in Calçada da Estrela 76/Rua Doutor Teófilo Braga 2 (1962), show a different image of the building to let of the epoch, with its austere façades, building materials and concrete grids of its balanced upper body. The modified project presented in 1967 shows three types of dwellings: T1 (one bedroom per dwelling), T4 (four bedrooms per dwelling) and T5 (five bedrooms per dwelling). The first one has one bedroom, a living room and kitchen with a pantry. The other two have undifferentiated bedrooms sharing the same bathroom, a small social bathroom, a kitchen with pantry, from which we can access to a bedroom with a bathroom. This building has a traditional-modern type plan, with a service area with a kitchen, a W. C. and a bedroom, possibly for the housekeeper, as foreseen in the first project.

Figure 3. Re-examination period Building, Lisbon.

The building of Rua Tristão Vaz 31-31A (1970) was designed to be a six floor left/right dwelling built with a basement with the doorwoman dwelling and a shop. The pavements of the bathrooms, kitchens, balconies and shops used hydraulic mosaics, the bedrooms, and other rooms used parquet. The interior walls were plastered and painted, and in the kitchens and bathrooms were used glazed tiles until the minimum of 1.50m m high. The façades used lioz bush-hammered squared stone in the socle, and green Cinca mosaic in the balconies. In the revetments of the facades, mosaic of ivory colour was used. The type of plan of this building is a traditional-modern one, with the service area with a kitchen, a W.C. and a bedroom for the housekeeper. These buildings contrast with each other, showing the diversity of this period. They belong to the historical-constructive period of the buildings with a reinforced concrete structure of the 2nd phase *(c.1960 - c.1980)*. They present an integral structure of reinforced concrete (Vieira & Lacerda (coords.); 2010).

3 A RESOURCE FOR AN EFICIENT CITY

A fundamental key to the environmental and urban sustainability of Lisbon is the urban rehabilitation of the city´s heritage, a built resource that must be preserved to achieve an efficient and sustainable city. The building to let, an abundant built heritage, is fundamental in this process. The HBIM, *Historic Building Information Modeling*, plays an important role in the management and conservation of this heritage. The BIM provides information about the building, representing its physical and functional characteristics, simulating the development, appearance and performance of a built asset. Additionally BIM gives a complex analysis of proposed interventions in various scenarios and offers a framework of collaborative working processes, sharing datasets across a multidisciplinary team. The information provided by BIM can be given throughout the life cycle of the historic asset giving an unique management tool to the professionals in the urban rehabilitation. The HBIM can be used: to inform conservation, as a heritage management tool; as an archive and information resource, to help research.

The applications of HBIM include: forming an information repository for documentation and recording of activities; condition monitoring; preventive maintenance; asset management; heritage management; heritage interpretation; visitor management; intervention appraisal; work programming (conservation, repair, maintenance and reuse), construction simulation, project, security, fire safety, visitor safety and health and safety planning; disaster preparedness (Antonopoulou, & Bryan; 2017). The scenario is enhanced by the 3D printing. It is possible to reproduce a head of a putti, parts of a house, or a beam. Thus, it is possible to make replicas with significant saving of time and cost with an enormous quality of reproduction with less waste than the other reproduction methods. It is possible to create elements from precast concrete with any limitations to creativeness. Most of the architectural elements can be created in less than a day. The 3D models used to print them can be cataloged and archived for future use to make new replicas (Fenner & Esler 2018).

4 CONCLUSIONS

The building to let shaped the image of Lisbon from the 16thcentury to 1976. These buildings are a built and abundant resource, part of the Lisbon identity and memory, that must be preserved and rehabilitated for a sustainable and efficient city. If not protected, this heritage can be destroyed and part of the Portuguese, and consequently European, memory will be lost. The BIM technologies and fabrication technologies, namely the 3D printing can be a valuable tool to the urban rehabilitation of this heritage.

REFERENCES

Agarez, R. C. 2009. *O Moderno revisitado, habitação multifamiliar em Lisboa nos anos de 1950*. Lisboa; Câmara Municipal de Lisboa.

Alho, C.; 2005. *Lisbon Urban Planning and Sustainability*; 12 October 2005. Lisboa.

Appleton, J., 2005, *Reabilitação de edifícios "gaioleiros", um quarteirão em Lisboa*, Amadora: Edições Orion.

Barreiros, M. H., 2004. *Casas em cima de casas Apontamentos sobre o espaço doméstico da Baixa Pombalina* in Alçada, M. (dir.) Monumentos nº 21, pp. 88–97.

Carita, H., 1994, *Bairro Alto, Tipologias e modos arquitectónicos*, Lisboa: Câmara Municipal de Lisboa

Carita, H.; 1999, *Lisboa manuelina e a formação de modelos urbanísticos da época moderna (1495–1521)*, Lisboa: Livros Horizonte

Curtis, W. J. R..; 1996; *Modern architecture since 1900*; London; Phaidon.

Fernandez, S, 1998. Arquitectura Portuguesa 1961–1974. in Becker Annette, et al, 1998. *Arquitectura do século XX, Portugal*; Munchen: Prestil:. 55–57.

Lopes, V., 2005. Um plano de cores para o território da Baixa e as argamassas para a conservação de fachadas in Mateus, J. M. (coord.), *Baixa Pombalina: bases para uma intervenção de salvaguarda*, Direcção de Gestão Urbanística, Lisboa: Câmara Municipal de Lisboa.

Matos, V. P., 2003, *Subsídios para o estudo do edifício de rendimento em Lisboa, século XVI à 1ª metade do século XVIII*, Faculdade de Arquitectura da Universidade Técnica de Lisboa.

Matos, V. P., 2009, *O edifício de rendimento em Lisboa de 1758 ao dealbar do Estado Novo, da Cotovia ao Rato*; Universidade Lusíada de Lisboa, Faculdade de Arquitectura e Artes.

Matos, V. P., 2018. *Habitação coletiva de promoção coletiva de promoção cooperativa, critérios de autenticidade na sua conservação e reabilitação*, Faculdade de Arquitetura Universidade de Lisboa.

Pereira; S. M., 2012, *Casa e Mudança Social, uma leitura das transformações da sociedade portuguesa a partir da casa*; Casal de Cambra; Caleidoscópio

Silva Nunes, J. P., 2005. *Uma cidade de proprietários? Mudanças na distribuição dos estatutos de ocupação na área metropolitana de Lisboa entre 1950 e 2001*; Forum Sociológico, nº 13/14 (2ª série), 2005: 113–135.

Teixeira, M. C. & Valla, M., 1999, *Urbanismo Português, Séculos XIII - XVIII, Portugal-Brasil*, Lisboa: Livros Horizonte.

Vieira, J. & Lacerda, M. (coords.), 2010. *Kits- Património, Património Arquitectónico de Habitação Multifamiliar do século XX*; IHRU & IGESPAR.

Antonopoulou, S., Cons A., & Bryan P.; Historic England 2017 BIM for Heritage: Developing a Historic Information Model. Siwndon. Historic England.

Fenner & Esler; 01 may 2018; Innovations in Architecture: Restoring Historic Buildings Using 3D Printing.k.

Industry 4.0 – Shaping The Future of The Digital World – da Silva Bartolo et al. (eds)

BREEAM and LEED in architectural rehabilitation

A.P. Pinheiro
CIAUD, Lisbon School of Architecture, Lisbon, Portugal

ABSTRACT: There are several international systems on Energy Certification to evaluate the sustainability of buildings. Each system makes different approaches in order to achieve the desired score. However, the parameters that are considered in the assessment of their environmental impact vary, as there are also differences in the understanding of what sustainability is. Although the aim is to meet established benchmarks and to introduce important aspects of sustainable construction, this assessment is not always scientific and does not attain essential objectives. As a methodology we refer to a European (BREEAM) and American (LEED) system. By comparing the items in which the evaluation is objectified, we can better understand its strengths and weaknesses. Examples include three university buildings: one in Europe - Central Saint Martins in London, UK and the other in the USA - Sloan School of Management, MIT, Boston and The Fuqua School of Business at Duke University in Durham.

1 INTRODUCTION

1.1 *Context of research*

This paper is based on the comparative analysis done in the Author's PhD thesis and corresponds to an on going research.

Its importance is decisive nowadays, since the convergence criteria regarding sustainability and the Nearly Zero Energy Building (NZEB) have not been effectively objectified. It also alerts to scientific statements and market reasons that are sometimes irreconcilable and dangerous.

In addition, it focuses on the need for a reformulation of the principles of architectural design, which in the twentieth century dominated subsystem thinking. There is always a solution to solve what has not been solved at the outset. This process must be reversed during the life cycle of the building.

"We believe that sustainability needs to be at the heart of the design strategy in order to create exceptional projects." Stanton Williams.

It is essential to design energy efficient buildings, through effective constructive solutions (passive systems), avoiding the use of active systems as an energy complement. Thus, one should consider the principles of passive sustainability: water tightness, insulation, lack of thermal bridges, heat recovery, natural ventilation, shading and windows strategically and correctly insulated to use solar energy efficiently and allow ventilation (Pinheiro, 2017).

There are several Energy Certification Systems that have been developed and applied internationally with the objective of evaluating the sustainability of buildings, reducing their environmental impact. However, evaluation processes are generally optional, but may be required by the Owner to ensure the energy efficiency of the building and increase its market value.

Each Certification System normally has a standard evaluation sheet that must be adapted to suit different project types and different locations so that the process is credible and transparent.

Most systems essentially address environmental sustainability issues, although some also consider social and economic sustainability. The comparative analysis presented is focused on two systems: BREEAM and LEED.

The aim of these systems is to identify the buildings with the best energy performance, stimulating the search for greater sustainability and healthy comfort.

1.2 *BREEAM certification system*

The Building Research Establishment's Environmental Assessment Method (BREEAM) was developed by the BRE (Building Research Establishment) in 1990 in the United Kingdom. Its purpose is to function as a tool to measure the sustainability of new non-domestic buildings.

This system does not certify any particular product. It is essentially a holistic approach to measuring and improving all types of new and existing buildings and is updated every two years. There are several assumed a priori characteristics that buildings will have: low environmental impact; innovative solutions that reduce the environmental impact; proven environmental practices; better energy performance than required by the Regulations in force; progress towards corporate and organizational environmental objectives; cost reduction and maintenance; quality of life.

Currently the BREEAM evaluation is made according to 10 categories: Management; Health and wellness; Energy; Transport; Water; Materials; Waste; Land Use and Ecology; Pollution; Innovation.

1.3 *LEED certification system*

The other system we refer to is the Leadership in Energy and Environmental Design (LEED) Certification System created in 1993 by the USGBC (U.S. Green Building Council). The first Pilot Project of this Program was launched in August 1998.

LEED Certification is a voluntary, market-oriented Assessment System, and it is based on consensus. It promotes the approach between building and sustainability, defining and evaluating its environmental performance throughout the life cycle. It thus provides a classification standard for a green building in terms of design, construction and operation:

"Green buildings are designed to meet certain objectives such as protecting occupant health; improving employee productivity; using energy, water, and other resources more efficiently; and reducing the overall impact to the environment" (Kubba, 2012).

In 2006, a LEED project could reach a total of 100 points. In LEED 2009, two new bonus categories were introduced: Innovation in Design (or Building Operation) and Regional Priority.

This resulted in an increase of 6 points for Innovation in the Design (or Functioning of the building) and 4 points if it obtained classification as having Regional Priority. If the project reaches more than 4 credits in this Category, the Project team can choose the credits where they want to apply these points, and LEED can total 110 points.

This regional bonus demonstrates the importance that site conditions play in improving building practices and environmental design.

2 CASE STUDIES

2.1 *Central Saint Martins; Sloan School of Management; Fuqua school of business*

To exemplify these issues, three university buildings were selected for Case Studies: the Central Saint Martins building in London, United Kingdom, opened in 2011 to verify the BREEAM Certification in an European country; the Expansion of Sloan School of Management, MIT, Boston, USA, also inaugurated in 2011, to verify the application of the LEED Certification and to test sustainability practices in an American University building; and Fuqua School of Business at Duke University in Durham, completed in 2008, with several failures in Design, but even so, obtaining LEED Certification.

Central Saint Martins is an area of urban renewal and of barns' and traffic sheds' rehabilitation. These buildings, built in the mid-nineteenth century, had the function of storing grain coming from the wheat fields and then distributing it to London bakeries. They consisted of massive volumes of exposed brick with six-storey height and wooden floors. They had little daylight and low ceilings.

The Sloan School of Management building is an extension of two adjoining buildings from the early twentieth century, to which it relates through galleries. The construction of this building enabled the urban renewal of the area, which converted in outdoor garden spaces, 70,000 Sq Ft of asphalt car parks.

In Fuqua School of Business, the buildings are current, with reinforced concrete construction, having been built separately in different phases (1982, 1989, 1999, 2002, 2008), and later linked.

After comparing the approximations and Criteria of both Certifications, the conclusions are presented. The criteria is based on incidents of varying nature, covering different areas from the beginning of the project and during the construction process in order to guarantee the qualities of the buildings, including their integration on site. Thus, the following criteria should be mentioned.

2.2 *Integrated design process*

In the typical design process, the work is linear, so that each specialist works one after another. The integrated process includes all architects and engineers from the outset, so they can work more effectively as a team.

The Sloan School of Management project was the first to use this kind of process at MIT. It was verified that Central Saint Martins also complies with this principle. Fuqua did not have an Integrated Design Process.

2.3 *Passive sustainability*

Figure 1. Central Saint Martins. Inner Street. Photograph: Author, March 2014.

Central Saint Martins and Sloan School have natural lighting and natural ventilation. Fuqua has no passive sustainability.

2.4 *Certification*

Regarding Certification, Central Saint Martins has a Very Good Classification (59%) of BREEAM 2006; Sloan School has Gold Classification (earned 70 Points) from LEED 2009; Fuqua has Silver Classification (obtained 34 Points) from LEED 2005. This Certification is very focused on Construction and Engineering and less on Architecture.

2.5 *Site*

Central Saint Martins and Sloan School being Projects of urban renewal of brownfield sites had Points in this Criterion.

In Fuqua there was no choice of the site, since it is an extension of an existing building. Thus, they were penalized for this, not having Points in this Criterion.

2.6 *Design*

Central Saint Martins and Sloan School have sustainable design.

In Fuqua the Atriums were designed with natural light, but the Offices and Rooms have reduced natural light.

2.7 *Glass*

Central Saint Martins and Sloan School have high performance glass. In Fuqua we have no information.

2.8 *Shading*

Figure 2. Fuqua School of Business, Duke University, Durham. McClendon Dining Room with exterior shading system. Photograph: Author, January 2010.

All buildings have interior blinds.

In addition the American buildings still have exterior shading and screens to reduce solar heat gains, preventing a rise in temperature.

2.9 *External Thermal Insulation Composite System - ETICS*

Figure 3. Sloan School of Management, under construction, ETICS. Photograph: Author, January 2010.

Central Saint Martins used Apparent Concrete without External Thermal Insulation System - ETICS.

Sloan School used ETICS. In Fuqua we have no information.

2.10 *Materials*

Central Saint Martins used precast concrete and continuous floors, resistant to use, that involves a minimum of maintenance, allowing versatility in use.

The two American Case Studies use low-emitting materials, including adhesives, sealants, paints, and carpets. Fuqua also uses materials manufactured within a 500-mile radius, 20% of which are manufactured locally.

2.11 *Recycling*

All buildings have recycling bins and garbage collection areas.

In the American Case Studies the demolition remains and the debris generated during construction were recycled.

2.12 *Electrical installations*

Central Saint Martins and Sloan School used high-efficiency, low-energy lighting fixtures with motion and natural light sensors. In Fuqua, normal fluorescent lighting was used.

2.13 *Heating*

Central Saint Martins uses a system with a combined heat and power plan and gas boilers. It uses recovered thermal energy from the air treatment units that serve the University. The heat generated is harnessed for heating instead of releasing the energy into the atmosphere.

At Sloan School they use radiant floor heating, reducing energy consumption and noise, thus improving occupant comfort. Heat recovery methods incorporated into air conditioning systems were used cumulatively.

In Fuqua, were used HVAC (heating, ventilating, and air conditioning) and other equipment that does not produce CFC.

CFC, Chlorofluorocarbon is a synthetic, gaseous and non-toxic compound that is the main cause of the hole in the ozone layer.

2.14 *Indoor Air Quality - IAQ*

Central Saint Martins and Sloan School do not provide information.

In Fuqua, monitors were installed to monitor CO2. During construction, the ducts and applied materials were protected to avoid contamination. In the two weeks prior to the occupation, a test was carried out in the 100% fresh air building to confirm indoor air quality.

2.15 *Water efficiency*

Central Saint Martins and Sloan School used automatic faucets, urinals and low flow showers.

Sloan School uses a drip irrigation system connected to the central weather station to minimize the amount of irrigation water.

In Fuqua there was a 20% reduction in the water usage through the installation of self-flow devices in the pipes and through the use of low-flow equipment. The landscape design ensured water efficiency reducing around 50% of the water consumption through the use of native plants that require less maintenance and through the installation of a drip irrigation system.

2.16 *Rainwater management*

Central Saint Martins sanitary facilities and Granary Square fountains are fed by rainwater stored in cisterns.

Sloan School uses a drip irrigation system connected to the central weather station to minimize the amount of irrigation water.

Rainwater drains are filtered to improve the quality of the water that reaches the Charles River and prevent pollution; the green roofing includes an electronic leak detection system.

In Fuqua areas were provided to filter and reduce the amount of rainwater flow through vegetation or ponds built in landscaped areas. There was also a reduction of areas with waterproof pavements.

2.17 *Renewable energy*

Only the Central Saint Martins building included renewable energy from the beginning of the Project to the construction.

Sloan School only left the infrastructures for renewable energy at rooftop. In 2010, this building was built without having used renewable energy.

However, by 2016, renewable energy was already installed in Sloan School and Fuqua School, as one can see in the Plans accessed in Google Earth.

2.18 *Green roofs*

The Central Saint Martins building has no green roofs.

Sloan School has green roofs with extensive system (light, low maintenance and generally inaccessible green roofs). They are planted with species with little demand of humidity and substrate with 2 to 6 inches (5-15 cm).

Fuqua instead of green roofs used flower boxes on the terraces.

2.19 *Exterior green areas*

At Central Saint Martins plantations were reduced to a minimum to reflect the character of the original goods dock.

At Sloan School, the parking areas that were initially on asphalt were converted into a garden.

In Fuqua the exterior green areas were designed to reduce the Heat Islands, and trees were planted to create shade in the paved areas.

3 FAILURES IN PROJECTS

From the comparison presented on the effectiveness of the applied criteria, we can evaluate the strengths and weaknesses of the referenced buildings:

- At Central Saint Martins, the use of existing buildings compromised the achievement of a higher BREEAM score. This is in itself a contradiction because it recovers buildings that could have been demolished. Materials are also recovered and there is an economy in the process, in addition to preserving collective memory. These are factors that should be referenced and valued.
- The inner street of Central Saint Martins is naturally lit with translucent ETFE cover.
- The Sloan School of Management has not fulfilled several of the Credits related to the efficiency of water and materials.
- The Fuqua School of Business has several drawbacks: little natural lighting; low ceilings; materials of poor quality; Classrooms without outlets on the students' benches.

- The building does not meet the parameters of sustainable construction, and can't be considered a green building.
- Even so, it had LEED Certification, with Silver Classification.
- Much of the LEED score was acquired on items that are not specific to the building's architecture, but rather related to the various engineering and procedures used during construction.
- The Sloan School of Management and The Fuqua School of Business did not include renewable energy projects.
- None of the Case Studies is a Nearly Zero Energy Building - NZEB (Pinheiro, 2017).

4 ETFE

Ethylene tetrafluoroethylene (ETFE) is a fluorine-based plastic. It was designed to have high corrosion resistance and strength over a wide temperature range. ETFE is a polymer and its source-based name is poly (ethene-co-tetrafluoroethene).

ETFE has a relatively high melting temperature, excellent chemical, electrical and high-energy radiation resistance properties.

Among the many advantages it has, we can mention its price which is much cheaper than glass beyond its lightness and translucency.

However, there are dangerous drawbacks: when burned, ETFE releases hydrofluoric acid, highly toxic, posing a health risk. In case of fire it attacks the eyes and can cause blindness, penetrates the tissues and bones and can cause cardiovascular diseases, irritation and pulmonary edema.

Despite the danger to public health, the physical qualities of this material are always enhanced, without mentioning the dangers of its chemical characteristics.

5 CONCLUSIONS

The International Systems on Energy Certification imply that the buildings already have a set of sustainable characteristics, and are an inspiration to find innovative solutions that minimise the environmental impact.

BREEAM and LEED systems are important contributors to the sustainability of buildings.

Despite their scope and purpose, they do not always focus on essential aspects. Thus, they do not value the recovery of buildings as it happens in Central Saint Martins in London.

It is essential to privilege passive sustainability in the design of buildings.

Local materials should be used to favour sustainability.

The physical and chemical behaviour of the materials should be taken into account during the life cycle of the building.

None of the Case Studies is a Nearly Zero Energy Building - NZEB.

REFERENCES

Kubba, S. 2012. *Handbook of Green Building Design and Construction: LEED, BREEAM, and GREEN GLOBES.* Waltham: 27. Oxford: Elsevier.

Pinheiro, A.P. 2017. *Reabilitação Arquitectónica, Sustentabilidade e Design.* PhD. Lisbon School of Architecture, Lisbon: 39, 203-204, 209-2010, 273-344. http://www.apexgreenroofs.com/mit-building-e60/[Accessed 22 October 2015].

http://www.apexgreenroofs.com/mit-building-e60/ [Accessed 22 October 2015].

https://www.arqhys.com/construccion/plastico-etileno-tetrafluoro.html [Accessed 21 February 2019].

http://www.arts.ac.uk/csm/about-csm/[Accessed 14 December 2013].

https://www.atsdr.cdc.gov/MMG/MMG.asp?id=1142&tid=250 [Accessed 21 February 2019].

https://www.iasoglobal.com/pt-PT/etfe [Accessed 21 February 2019].

http://www.sloanvalve.com/Water_Efficiency/LEED_Qualification_Guide.aspx [Accessed 28 November 2014].

www.stantonwilliams.com/design/sustainability [Accessed 28 February 2014].

Author Index